Evolutionary Genetics

Evolutionary Genetics

Edited by Richard Arber

SYRAWOOD
PUBLISHING HOUSE
New York

Published by Syrawood Publishing House,
750 Third Avenue, 9th Floor,
New York, NY 10017, USA
www.syrawoodpublishinghouse.com

Evolutionary Genetics
Edited by Richard Arber

International Standard Book Number: 978-1-68286-406-7 (Hardback)

Cataloging-in-publication Data

Evolutionary genetics / edited by Richard Arber.
p. cm.
Includes bibliographical references and index.
ISBN 978-1-68286-406-7
1. Evolutionary genetics. 2. Genetics. 3. Molecular evolution. I. Arber, Richard.
QH390 .E96 2017
572.838--dc23

Printed in China.

TABLE OF CONTENTS

Preface VII

Chapter 1 Characterising the Phenotypic Diversity of *Papilio dardanus* Wing Patterns using an Extensive Museum Collection 1
Martin J. Thompson, Martijn J. T. N. Timmermans

Chapter 2 Genome-Wide Survey and Expression Analysis of Calcium-Dependent Protein Kinase in *Gossypium raimondii* 10
Wei Liu, Wei Li, Qiuling He, Muhammad Khan Daud, Jinhong Chen, Shuijin Zhu

Chapter 3 Insights into Alpha-Hemolysin (Hla) Evolution and Expression among *Staphylococcus aureus* Clones with Hospital and Community Origin 21
Ana Tavares, Jesper B. Nielsen, Kit Boye, Susanne Rohde, Ana C. Paulo, Henrik Westh, Kristian Schønning, Hermínia de Lencastre, Maria Miragaia

Chapter 4 Genome-Wide Analysis of the NADK Gene Family in Plants 34
Wen-Yan Li, Xiang Wang, Ri Li, Wen-Qiang Li, Kun-Ming Chen

Chapter 5 New Insights into 1-Aminocyclopropane-1-Carboxylate (ACC) Deaminase Phylogeny, Evolution and Ecological Significance 49
Francisco X. Nascimento, Márcio J. Rossi, Cláudio R. F. S. Soares, Brendan J. McConkey, Bernard R. Glick

Chapter 6 Adaptation of a Cyanobacterium to a Biochemically Rich Environment in Experimental Evolution as an Initial Step toward a Chloroplast-Like State 66
Kazufumi Hosoda, Masumi Habuchi, Shingo Suzuki, Mikako Miyazaki, Go Takikawa, Takahiro Sakurai, Akiko Kashiwagi, Makoto Sueyoshi, Yusuke Matsumoto, Ayako Kiuchi, Kotaro Mori, Tetsuya Yomo

Chapter 7 Adaptation to Fluctuating Temperatures in an RNA Virus is Driven by the Most Stringent Selective Pressure 72
María Arribas, Kirina Kubota, Laura Cabanillas, Ester Lázaro

Chapter 8 Evolutionary Origin of GnIH and NPFF in Chordates: Insights from Novel Amphioxus RFamide Peptides 83
Tomohiro Osugi, Tomoki Okamura, You Lee Son, Makoto Ohkubo, Takayoshi Ubuka, Yasuhisa Henmi, Kazuyoshi Tsutsui

Chapter 9 Inferring Gene Family Histories in Yeast Identifies Lineage Specific Expansions 94
Ryan M. Ames, Daniel Money, Simon C. Lovell

Chapter 10 Patterns of Variation at *Ustilago maydis* Virulence Clusters 2A and 19A Largely Reflect the Demographic History of its Populations 105
Ronny Kellner, Christian Hanschke, Dominik Begerow

Chapter 11 **Expression and Molecular Evolution of Two *DREB1* Genes in Black Poplar (*Populus nigra*)** 115
Yanguang Chu, Qinjun Huang, Bingyu Zhang, Changjun Ding, Xiaohua Su

Chapter 12 **Niche Construction on Environmental Gradients: The Formation of Fitness Valley and Stratified Genotypic Distributions** 128
Xiaozhuo Han, Cang Hui

Chapter 13 **Molecular Evolution of Candidate Genes for Crop-Related Traits in Sunflower (*Helianthus annuus* L.)** 137
Jennifer R. Mandel, Edward V. McAssey, Savithri Nambeesan, Elena Garcia-Navarro, John M. Burke

Chapter 14 **Assessment of the Extirpated Maritimes Walrus using Morphological and Ancient DNA Analysis** 147
Brenna A. McLeod, Timothy R. Frasier, Zoe Lucas

Chapter 15 **Impact of Gene Molecular Evolution on Phylogenetic Reconstruction: A Case Study in the Rosids (Superorder Rosanae, Angiosperms)** 161
Khidir W. Hilu, Chelsea M. Black, Dipan Oza

Chapter 16 **Evolutionary Divergence of Geographic Subspecies within the Scalloped Spiny Lobster *Panulirus homarus* (Linnaeus 1758)** 173
Shane D. Lavery, Ahmad Farhadi, Hamid Farahmand, Tin-Yam Chan, Ashkan Azhdehakoshpour, Vibhavari Thakur, Andrew G. Jeffs

Chapter 17 **Support Vector Machine Classification of Streptavidin-Binding Aptamers** 186
Xinliang Yu, Yixiong Yu, Qun Zeng

Chapter 18 **The Roles of Standing Genetic Variation and Evolutionary History in Determining the Evolvability of Anti-Predator Strategies** 191
Daniel R. O'Donnell, Abhijna Parigi, Jordan A. Fish, Ian Dworkin, Aaron P. Wagner

Chapter 19 **Rapid Birth-and-Death Evolution of Imprinted snoRNAs in the Prader-Willi Syndrome Locus: Implications for Neural Development in Euarchontoglires** 202
Yi-Jun Zhang, Jian-Hua Yang, Qiao-Su Shi, Ling-Ling Zheng, Jun Liu, Hui Zhou, Hui Zhang, Liang-Hu Qu

Chapter 20 **Substrate-Dependent Evolution of Cytochrome P450: Rapid Turnover of the Detoxification-Type and Conservation of the Biosynthesis-Type** 218
Ayaka Kawashima, Yoko Satta

Permissions

List of Contributors

Index

PREFACE

The study of transformation of genetic variation to evolutionary change is called evolutionary genetics. It is multi-disciplinary in nature consisting topics like evolution of genome structure, the genetic base of specialization and adaptation and genetic change amongst others. The various advancements in this field are glanced at and their applications as well as ramifications are looked at in detail in this text. It also provides an in-depth study of the various processes that are involved in evolutionary genetics like mutation, migration, selection, etc. The aim of this book is to present researches that have transformed this discipline and aided its advancement. This book, with its detailed analyses and data, will prove immensely beneficial to professionals and students involved in this area at various levels.

The purpose of the book is to provide a glimpse into the dynamics and to present opinions and studies of some of the scientists engaged in the development of new ideas in the field from very different standpoints. This book will prove useful to students and researchers owing to its high content quality.

At the end, I would like to appreciate all the efforts made by the authors in completing their chapters professionally. I express my deepest gratitude to all of them for contributing to this book by sharing their valuable works. A special thanks to my family and friends for their constant support in this journey.

Editor

1

Characterising the Phenotypic Diversity of *Papilio dardanus* Wing Patterns Using an Extensive Museum Collection

Martin J. Thompson[1,2]*, Martijn J. T. N. Timmermans[1,3]

1 Life Sciences Department, Natural History Museum London, London, United Kingdom, 2 Department of Zoology, University of Cambridge, Cambridge, United Kingdom, 3 Division of Biology, Imperial College London, London, United Kingdom

Abstract

The history of 20th Century evolutionary biology can be followed through the study of mimetic butterflies. From the initial findings of discontinuous polymorphism through the debates regarding the evolution of mimicry and the step-size of evolutionary change, to the studies on supergene evolution and molecular characterisation of butterfly genomes, mimetic butterflies have been at the heart of evolutionary thought for over 100 years. During this time, few species have received as much attention and in-depth study as *Papilio dardanus*. To assist all aspects of mimicry research, we present a complete data-derived overview of the extent of polymorphism within this species. Using historical samples permanently held by the NHM London, we document the extent of phenotypic variation and characterise the diversity present in each of the subspecies and how it varies across Africa. We also demonstrate an association between "imperfect" mimetic forms and the transitional race formed in the area where Eastern and Western African populations meet around Lake Victoria. We present a novel portal for access to this collection, www.mimeticbutterflies.org, allowing remote access to this unique repository. It is hoped that this online resource can act as a nucleus for the sharing and dissemination of other collections databases and imagery connected with mimetic butterflies.

Editor: William J. Etges, University of Arkansas, United States of America

Funding: This work was funded by NERC (www.nerc.ac.uk). MJT is funded by a NERC studentship and MJTNT by a NERC Postdoctoral Fellowship (NE/I021578/1). Part of this work was funded by the former Entomology department of the NHM London (www.nhm.ac.uk). The funders had no role in study design, data collection and analysis, decision to publish, or preparation of the manuscript.

Competing Interests: The authors have declared that no competing interests exist.

* E-mail: martt@nhm.ac.uk

Introduction

One of the core questions in evolution biology concerns the origin of morphological novelty – where do new phenotypes come from and what is the nature of the changes which give rise to them? Insect wing patterns are among the most amenable structures for investigating processes determining phenotypic evolution and adaptive colouration [1–5] and have provided many textbook examples on the genetics of adaptive change. In particular, wing pattern mimicry in butterflies has been used as a case-study in adaptation [6–8].

During the early 20th Century, evolutionary biologists such as Punnet, Fisher, Ford and Dobzhansky were integrating the theories of Mendelian genetics into studies of evolution and biological diversity and using this new perspective to synthesise novel mathematical and theoretical underpinnings of the science [9–12]. Throughout the Modern Synthesis, polymorphic mimicry in butterflies was cited repeatedly as offering a corollary of species formation and the origins of evolutionary novelty [13]. In particular, competing interpretations of mimetic butterflies were at the heart of disagreement about the expected step-size of evolutionary change: what is the magnitude of changes which are the basis for novel phenotypes or species [14–17]?

At the centre of this debate was a spectacular '*exemplar*' of polymorphic Batesian mimicry [18] the Mocker Swallowtail *Papilio dardanus* Yeats in Brown 1776. Females of this African butterfly mimic members of various Müllerian mimicry rings, largely Danaidae and Acraeini [19,20]. *P. dardanus* displays a very diverse array of wing patterns (Figure 1) and it is therefore not surprising that many forms were initially described as separate species. It was not until 1869 that it was recognised that these forms did in fact belong to one species [21–23].

The genetics of wing pattern determination in *P. dardanus* was studied intensively in the 1950s and 1960s by Clarke and Sheppard [24–27]. Their investigations built on previous breeding experiments (described in [19,20]) and the documented inheritance of the various phenotypes were vital for understanding the genetics of mimetic wing patterns in *P. dardanus*. Their controlled crosses, summarised in Figure 2, confirmed that within a single population inheritance of wing pattern generally behaved as a single Mendelian factor with large phenotypic effect and a characteristic dominance hierarchy, consistent with theoretical expectations [28–30]. However, when they mixed locally-adapted genetic backgrounds by crossing individuals from different subspecies or from populations with differing morph compositions, the dominance hierarchy between the forms broke down, often with largely unpredictable outcomes (Figure 2). This indicates that in addition to a large single effect locus, wing pattern in *P. dardanus* is affected by unlinked modifier genes that perfect mimetic resemblance to locally co-occurring models [31]. More recent

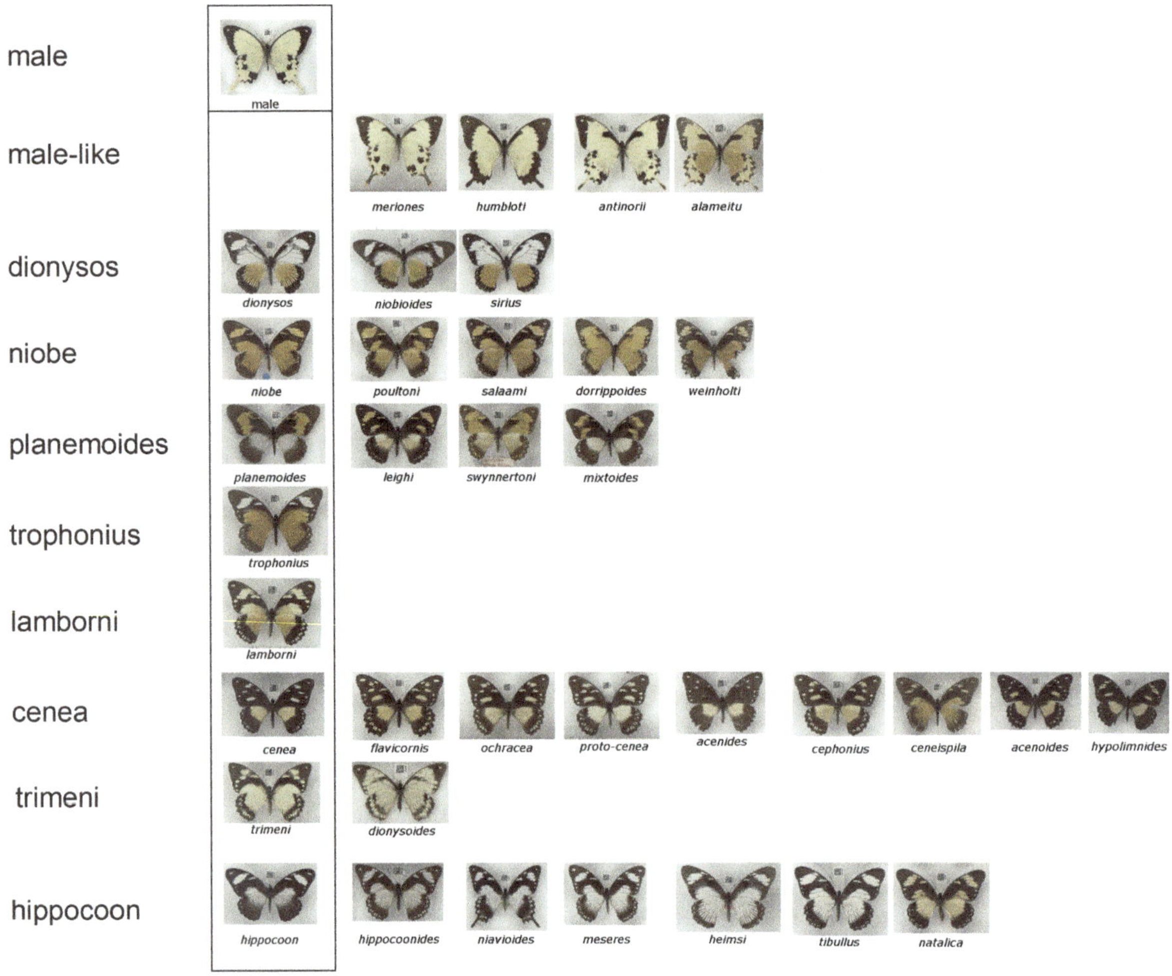

Figure 1. Female morphs of *Papilio dardanus* recognised in this study. Similar looking morphs are grouped into morph groups on each horizontal line, largely following the suggestions of Bernardi et al. [32]. Forms are present in order of the dominance hierarchy, apart from males and male-like forms (top two rows) and the f. *dionysos* group. In the case of the morph f. *dionysos*, no breeding work exists to support the conclusion of it being an allele at *H*, rather than being due to a second, unlinked locus. Not figured are forms which are not well represented in the NHM London collection, primarily some 'tailed mimic' forms of ssp. *P. dardanus antinorii* such as f. *vaccaroi* (tailed variant of f. *cenea*).

work has identified candidate genes for the mimicry switch locus – the transcription factor *invected* and its paralogue *engrailed* [32].

P. dardanus is noted for its variability across Africa [20]. The species is considered to be composed of between 11 and 18 subspecies (Table 1, [33,34]). Despite extensive interest for over 100 years, we have not yet answered many of the questions initially posed concerning its mimicry. It is unfortunate that much of the existing literature remains relatively inaccessible – in some cases due to its age (1920s–1950s), or its publication in non-English language journals (e.g. [33]), or in defunct or low-circulation journals [19,35]. Different authors report different numbers of female morphs, which makes obtaining a comprehensive overview of the system difficult (e.g. Grimaldi and Engel [36] mention 5 forms, whereas Clark et al. [32] suggest more than dozen forms exist). Such difficulties are undoubtedly compounded by the fact that some of the names used for *P. dardanus* subspecies are also utilised to describe female forms; this nomenclatural impediment is an unfortunate legacy of earlier works, a fact lamented as early as 1924 [19]. The fact that there are (relatively) few comprehensive image sets of the female forms is also problematic. Such figures as do exist (e.g. Figure 1 in Nijhout, 2003 [37]) typically demonstrate the major differences between the morph groups, but the existence of intermediates and the high levels of diversity within morph groups has not often been emphasised. Of particular note is the possibility for refinement of mimicry to become locally adapted to variations in model pattern [38]. This is well described for the forms f. *hippocoon* and f. *hippocoonides* [39] and it should be recognised that this may exist for other morphs of *P. dardanus*.

Much of the contemporary work on this species concerns the molecular characterisation of the mimicry switch locus: how a small genomic region can direct wing pattern development along divergent trajectories leading to widely divergent mimetic patterns, and what we can learn from such a switch locus regarding the evolutionary origin of novel morphology. If we are to understand the evolutionary processes giving rise to morphological diversity, a necessary preliminary step is to fully document and describe the extent of natural variation. As work continues to further characterise and understand the genetic basis and evolutionary biology of wing pattern evolution in *P. dardanus* [32], museum collections will play an important part in documenting the remarkable phenotypic diversity displayed by this species. To this end, we present a thorough analysis of all historical specimens held

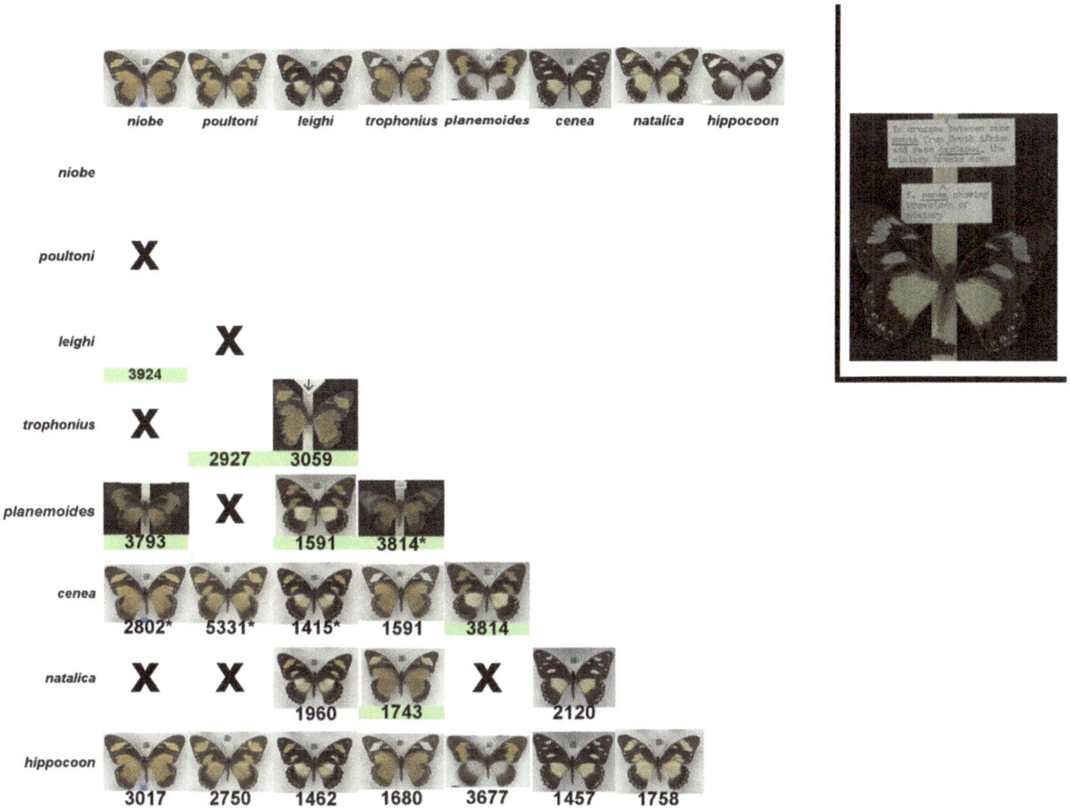

Figure 2. Inheritance of female wing patterns in *Papilio dardanus.* Overview of the crosses performed by Clarke and Sheppard in the 1950s and 1960s. Asterisks indicate crosses with variable outcomes that range from complete dominance to breakdown of mimicry. Green shading indicates incomplete dominance (production of intermediate phenotype). Where possible, intermediates are taken from the Clarke/Sheppard/Turner collection [40], otherwise insects were chosen from wild-caught specimens, which match the descriptions given in the published manuscripts. Most intra-subspecific crosses follow a strict dominance hierarchy, with only a few exceptions. For example, crosses between co-occurring forms f. *trophonius* and f. *natalica* produce butterflies where the dominant phenotype (f. *trophonius*) shows atypical tawny tinted forewing patches. In contrast, inter-subspecific crosses often produced aberrant butterflies. For example, on crossing a *P. dardanus dardanus* f. *cenea* with a male from a West African population of *P. dardanus dardanus*, which lacks this form, the female progeny no longer possessed accurate mimicry of *Amauris echeria*, but displayed an intermediate pattern (inset).

by the Natural History Museum (NHM) London. All specimens were digitised and databased. Metadata was used for the generation of concise maps detailing the distributions of most of the described subspecies. By making all this information freely available online, we hope to trigger collaborative scientific interest in the "*most interesting butterfly of the world*" [19].

Materials and Methods

All samples are permanently held by the NHM London under specimen numbers BMNH933783-BMNH937172.

The NHM collections are home to a large number of wild-caught specimens of *P. dardanus*, along with bred series from investigators such as E. G. L. van Someren and C.A. Clarke and P.M. Sheppard [40]. Specimen labels were copied verbatim for all *P. dardanus* and all individuals were given unique identifier labels (voucher numbers) in the form of both a numerical label and a 2D barcode, generated with Qdatamatrix [41]. Individuals were classified into morphs and subsequently organised into groups according to Bernardi et al. [33], with the exception of f. *leighi*, which was included in the f. *planemoides*-group based on its similarity to f. *planemoides* in the crosses of Clarke and Sheppard (summarised in Figure 2).

Drawers of specimens were digitised in the Sackler Biodiversity Imaging Lab at the NHM London using a SmartDrive SatScan machine (Huntingdon, Cambridgeshire, UK) and individuals were cut from the drawer-scan images manually in GIMP 2.8.2 (www.gimp.org). All data were stored in a MySQL database (http://www.mysql.com), accessible via http://www.mimeticbutterflies.org/.

Geocoding was done manually for all specimens, using internet resources (Global Gazetteer 2.2 http://www.fallingrain.com/world/, http://www.mapplanet.com and the Fuzzy Gazetteer http://isodp.hof-university.de/fuzzyg/). Only specimens where an unambiguous match to a location with a linear distance error of less than 40km were used. The geocoded dataset was imported as a point vector file from comma-separated values in ArcMap 10 [42] and projected into an Albers equal-area (Africa) space for validation, plotting and data management. Records were split into separate point vector layers corresponding to assigned morph, morph group or subspecies values. Sampling effort was estimated with a hot-spot analysis of a point density layer constructed in ArcMap 10 [42]. Plotting of Z-values from the Moran's I tool determined the most appropriate distance band for point density and hot-spot (Getis-Ord G_i*) analyses [43–45]. Briefly, varying distance bands were used to calculate Z-values; the distance band with the highest score indicates the scale at which there is the

Table 1. List of subspecies, morph groups and ranges, summarised from Bernardi et al. [32].

Subspecies	Clark & Vogler clade	Range	Morphs
dardanus	Western	Western and central Africa	hippocoon, dionysos, planemoides, trophonius, cenea, niobe, lamborni, niobioides, natalica, mixtoides
tibullus	Eastern	Eastern Africa from Kenya to Mozambique	cenea, trophonius, hippocoonides, salaami, natalica, trimeni, poultoni, lamborni, mixtoides
cenea	Eastern	South Africa	cenea, leighi, trophonius, hippcoonides, salaami, natalica
ochraceana	Western	Mount Marsabit and Mount Nyiru, Kenya	ochracea
flavicornis	Eastern	Mount Kulal, Kenya	flavicornis, atavica
antinorii	Eastern	Ethiopia and South Sudan	antinorii, niavoides, ruspinae, weinholti, vaccaroi
meriones	Indian Ocean	Madagascar	meriones
humbloti	Indian Ocean	Comoros	humbloti
meseres	Eastern	Lake Victoria area, Kenya, Uganda and Tanzania	hippocoonides, cenea, meseres, lamborni, planemoides, salaami, trimeni, mixtoides
polytrophus	Eastern	Central Kenya, East of the Rift Valley	hippocoonides, cenea, meseres, poultoni, lamborni, planemoides, trimeni, swynnertoni, carpenteri, mixtoides
byatti	Eastern	Somalia	
figinii		Eritrea	
alticola		Ethiopia and South Sudan	
hodsoni		Ethiopia and South Sudan	
sulphureus		Bioko, Sao Tome and Principe	
xanthocaudatus		Rift Valley, Kenya	
nairobianus		Kikuyu area, Kenya	

Subspecies in italics are not recognised by all authors, and have not been considered further in the present work. The f. *hippocoon* group is separated into forms related to f. *hippocoon*, f. *hippocoonides* and the intermediate form f. *meseres*. Membership of the Eastern, Western or Indian Ocean clade is given based on Clark & Vogler's analysis of cytochrome b [47].

highest clustering of sampled locations. The distance so obtained is therefore an indication of the 'clumpiness' of the data and provides an appropriate scale for the hot-spot and point density analyses.

Subspecies were assigned using existing designations where possible, supplemented by extending these designations to nearby areas and using characteristic wing pattern forms in certain subspecies (a combination of male patterning and the use of unique female forms: e.g. f. *tibullus* in subspecies *P. dardanus tibullus* or f. *flavicornis* in subspecies *P. dardanus flavicornis*). Polygons representing inferred distributions of taxa were estimated by constructing a convex hull from the point locality data using the minimum bounding geometry tool in ArcMap 10 [42]. The resulting polygon was clipped using ecosystem and isobioclimate data from USGS data ([46]; United States Geological Survey Africa Ecosystems mapping project, http://rmgsc.cr.usgs.gov/outgoing/ecosystems/AfricaData/). Briefly, ecosystem and isobioclimate data were downsampled to 20 km rectangles using block statistics from USGS raster map and ecosystem or isobioclimate classes were extracted for each specimen point locality locality using the 'extract values to points' tool in ArcMap 10 [42]. Ecosystem or bioclimate classes with zero *P. dardarnus* samples were clipped from the convex hull.

Results

General summary of the database

All wild caught *P. dardanus* that are preserved in the NHM pinned butterfly collection were digitised and databased. Voucher numbers, sex, subspecies designation and verbatim copied data labels were stored in a MySQL database that can be queried for morph and subspecies via a web-interface available at www.mimeticbutterflies.org. Individual images are also made available on this website. High resolution images are available upon request (e.g. for morphometric analyses).

We assessed the density of collection locations and the levels of coverage using a density and a hot-spot analysis (Figure 3). The results of both these analyses demonstrate that the area of East Africa surrounding Lake Victoria in Uganda and Kenya is much more densely sampled than other areas. This is also borne out in the specimen pie charts (Figure 4A) with Uganda having 528 specimens and Kenya 1035.

The database, which excludes the separate Clark/Sheppard/Turner collection [40] and other bred specimens, currently contains 3395 specimens from 25 countries (Figure 4A–C, note that islands of Pemba, Zanzibar and Bioko have been included separately), with 1576 females (Figure 4). Morphs used in the present study are given along with representative images in Figure 1 (grouped according to Bernardi et al. [33], with the exception of f. *leighi*). The collection contains a considerable number of intermediate, putatively non-mimetic female butterflies (Figure 4) which are believed to result from incomplete dominance or from breakdown of mimetic pattern (similar butterflies were produced in the crosses of Clarke and Sheppard; Figure 2).

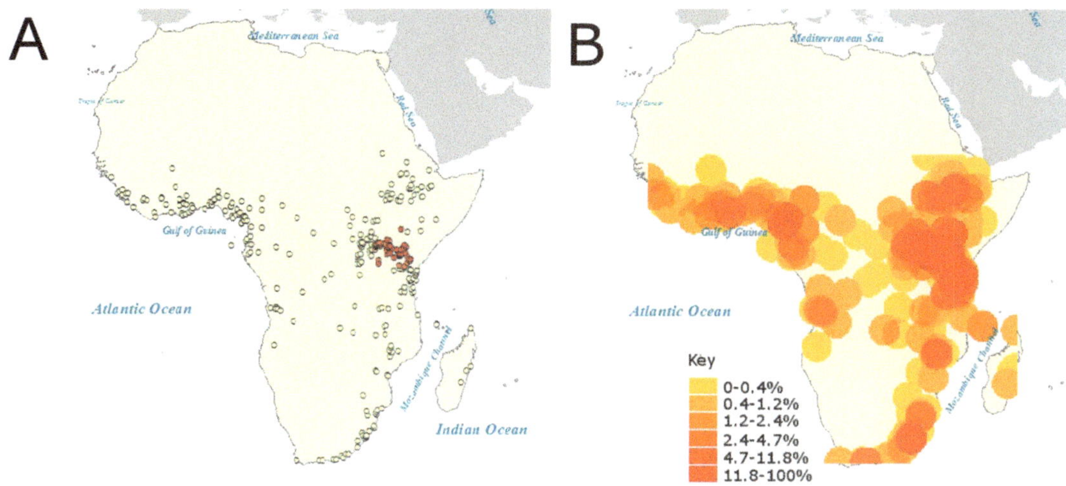

Figure 3. Summary of collection sample density. A) Hotspot analysis showing density outliers. Red dots indicate density more than 2.5 standard deviations above the average. No locations were labelled as outliers with low density. B) Heatmap of sampling density across Africa using circles of radius 280 km. Circles are coloured according to a scale of 7 quantile classes, which are labelled in the key with the proportion of density variation contained within each class.

Distributions of forms and subspecies of *Papilio dardanus*

Papilio dardanus is divided in several subspecies that show a distinct separation in male wing pattern between East and West Africa [34]. This division was later confirmed using genetic analyses [47] and is easily recognisable from the NHM collections material as shown by the males depicted in Figure 5. The museum collection also corroborates that the male patterns of the eastern "races" *P. dardanus tibullus* and *P. dardanus cenea* are very similar.

The dataset reveals that previous subspecies-centred treatments of female morph composition oversimplify the complex phenotypic diversity of *P. dardanus*. Not only are morphs distributed heterogeneously across Africa, but substantial variation exists within each subspecies in terms of the number of morphs co-occurring at sites (Figure 6A and 6B). This is the first time, to our knowledge, that distribution data for the various female forms has been analysed on a continental scale without reference to relatively subjective subspecies boundaries. Most phenotypes are not distributed over the full subspecies ranges, but are found in smaller areas only (Figure 6B): for example, f. *dionysos* is restricted to Western Africa and Cameroon and shows little overlap with other forms except for f. *hippocoon*, whilst the subspecies *P. dardanus dardanus* has its eastern limit in western Kenya and is highly polymorphic in this area.

In some cases, however, subspecies can be characterised by the presence of unique morphs. For example, f. *antinorii* (a male-like female phenotype) is endemic to subspecies *P. dardanus antinorii* from Ethiopia. This subspecies also contains unique tailed variants of the mimetic forms f. *trophonius* (f. *ruspinae*), f. *poultoni* (f. *weinholti*), f. *hippocoonides* (f. *niavioides*) and f. *cenea* (f. *vaccaroi*). The scarcity of many of these tailed forms prevents a systematic analysis of their distribution, but our data suggests that all are found within the range of *P. dardanus antinorii*. Unfortunately, it was not possible to georeference any specimens of *P. dardanus byatti*. This subspecies is likely of interest with respect to the evolution of polymorphic

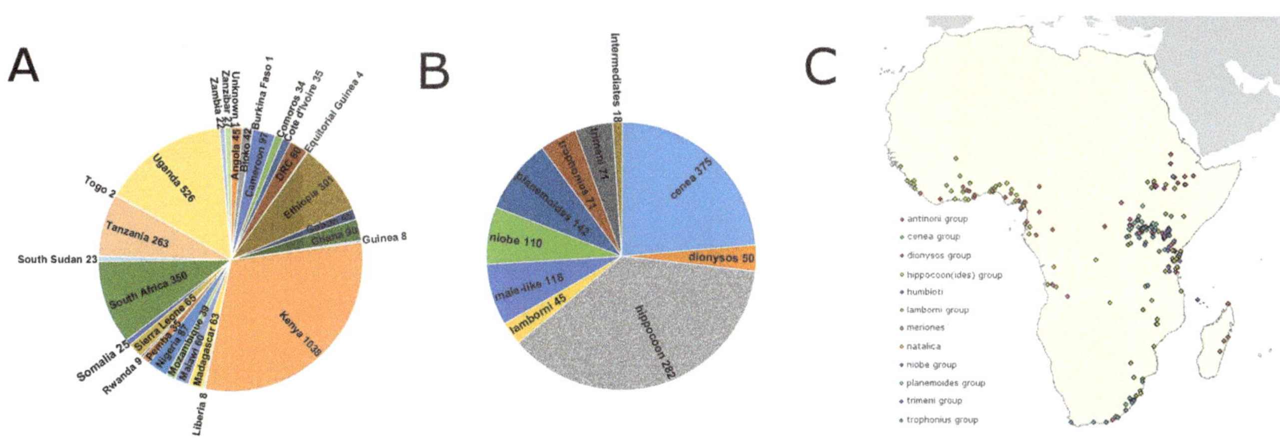

Figure 4. Summary of pinned *Papilio dardanus* preserved by the NHM London. A) Pie chart showing breakdown of NHM London *P. dardanus* collection by country. The islands of Pemba, Zanzibar and Bioko are considered separately from the African mainland. B) Pie chart showing contribution of each morph group among the 1575 female specimens in the collection. Not shown are the 1817 male specimens. The 'intermediate' class contains specimens with attributes of more than one morph group, making classification impossible. C) Distribution of point locality data across Africa, labelled by wing pattern.

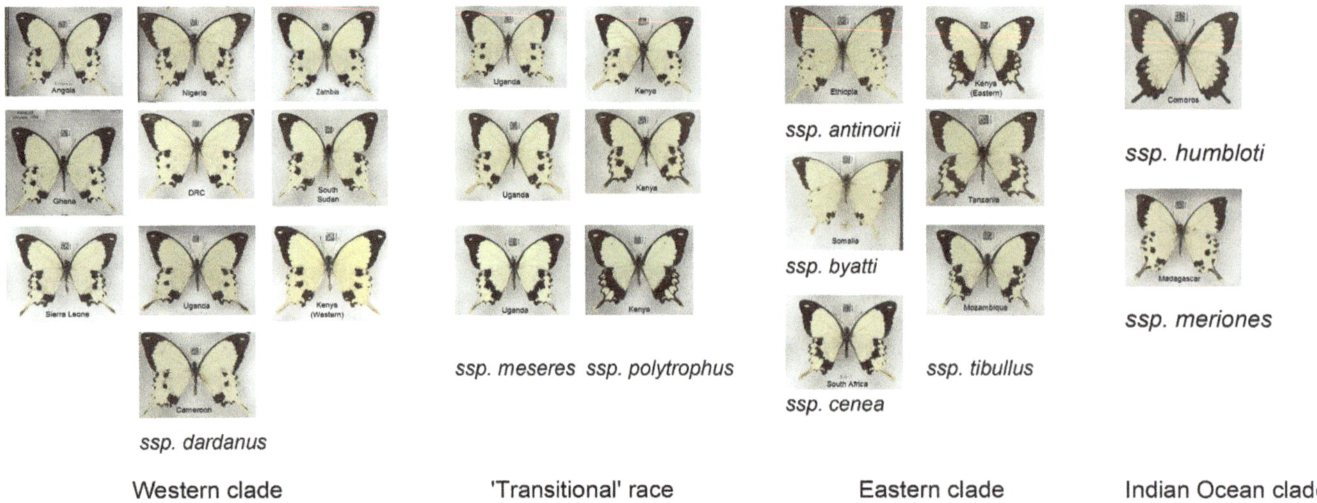

Figure 5. Typical male patterns for each of the subspecies of *P. dardanus.* Variation in male pattern was first used by Jordan [33] to divide *P. dardanus* into sub-species. The variability of pattern in the transitional subspecies *meseres* and subspecies *polytrophus* is noticeable.

mimicry in female *P. dardanus*, it being a mainland subspecies with only male-like females known (Vane-Wright, pers comm).

Imperfect mimics and the 'transitional race'

We used the NHM collection to investigate whether imperfect forms are over-represented in specific subspecies. We specifically focussed on the populations of around Lake Victoria and Mount Kenya as they are often described to contain a high proportion of 'imperfect' mimics that show a mix of patterns from several forms, or contain traces of the fluorescent yellow male pigment [19,20,25,48]. These subspecies, intermediate between Eastern and Western African lineages [34,49,50], have been suggested to be of hybrid nature occurring as they do at a contact zone between the West African subspecies [37]. We find that 'imperfect' forms are indeed associated with these two subspecies. Figure 7 shows a point-in-polygon overlay of point locality data of 'imperfect' individuals (the intermediate individuals in Figure 1 along with morph groups f. *lamborni* and f. *trimeni* and the individual imperfectly mimetic forms f. *dorrippoides*, f. *mixtoides* and f. *protocenea*) alongside the inferred polygon for the extent of the two 'transitional' subspecies *P. dardanus meseres* and *P. dardanus polytrophus*. Out of 145 imperfect mimetic specimens which could be geocoded, 131 were collected from within the extent of subspecies *P. dardanus meseres* and *P. dardanus polytrophus*. The remaining 14 were all from locations in Kenya (particularly towards the Kenyan coast) and Eastern Uganda (Figure 7).

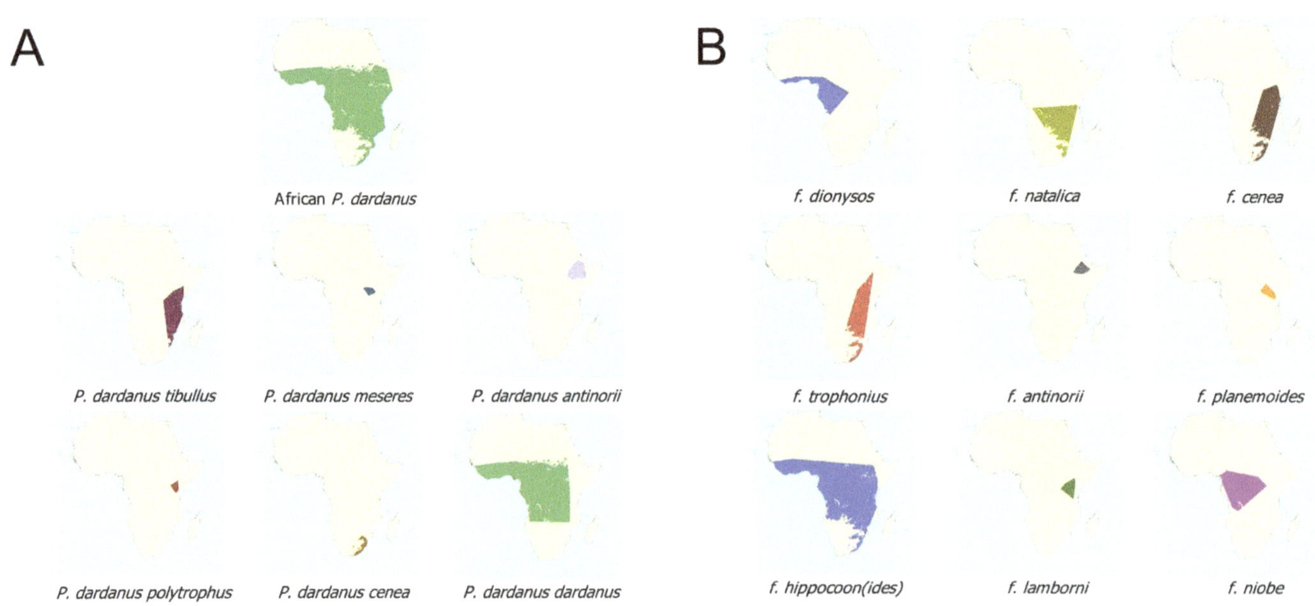

Figure 6. Distribution maps. A) Distribution maps showing the inferred range for the entirety of *P. dardanus* and the various described subspecies. B) Distribution maps showing inferred ranges of the morph groups of female wing patterns in *P. dardanus*. Substantial variation in the extents of subspecies ranges is present, and there is little correspondence between the subspecies ranges (A) and the morph ranges (B).

Figure 7. Localities of intermediate forms. Point locality for intermediate forms and imperfect mimics overlain over inferred polygon for subspecies *polytrophus* and *meseres*. Female forms plotted are: f. *trimeni*, f. *dorrippoides*, f. *dionysoides*, f. *proto*-cenea, f. *mixtoides*, f. *swynnertoni*, f. *carpenter* and f. *lamborni*, and intermediates between f. *lamborni* and other forms.

Discussion

Studies on mimetic butterflies have the potential to shed light on questions going to the core of contemporary evolution biology. These insects have been studied for over 100 years, informing our knowledge of the origin of novel traits and species, and the genetic architecture underlying adaptation. In particular, mimetic butterflies exemplify the multi-dimensional nature of adaptation to different environments. Polymorphic mimicry requires accurate matching of diverse wing patterns differing in several characters, likely determined by several genetic loci. However, the processes of recombination and the independent assortment of chromosomes mean that the loci involved may be broken up, leading to inaccurate mimicry, unless they are closely linked. The answer to this paradox is found in the concept of supergenes [51], tightly linked blocks of multiple co-adapted genetic elements. *Papilio dardanus* is renowned for the morphological diversity of its mimetic females and has been central to the development of the supergene theory. However, although supergenes have recently been shown to underlie polymorphism in other mimetic butterflies [6,8], their existence in *P. dardanus* remains unconfirmed. To promote *P. dardanus* as a study system, we present a systematic overview of variation displayed within the species accompanied by distribution maps. We expect this work to facilitate future research on the evolutionary genetics underlying mimicry by providing a concise overview of the variation displayed and by allowing identification of areas of interest within the range of *P. dardanus*. The presented distribution maps show that the degree of polymorphism present at a site varies greatly and the choice of which population to study will undoubtedly affect the ease with which we may come to understand the genetic control of polymorphic mimicry in this species.

Although the mimetic females of *P. dardanus* have received most scientific attention, subspecies have largely been defined on the basis of differences in male wing pattern. In 1905, Jordan [33] divided *P. dardanus* from the African continent (whilst excluding the populations in North East Africa) into five 'races', and described a clear separation between the western subspecies (*P. dardanus dardanus*) and the three eastern ones (*P. d. cenea*, *P. d. tibullus* and *P. d. polytrophus*). His fifth subspecies occupied the contact zone and showed intermediate traits (*P. d. meseres*, [50]). Jordan reports that the separation between the three eastern subspecies is less clear-cut and shows much morphological overlap – the male wing patterns of *P. d. tibullus* and *P. d. polytrophus* and the male genitalia of all three eastern subspecies are similar [49]). The collection material presented here is consistent with these subspecies descriptions. It shows the clear divergence between the East and West African populations (Figure 5) and corroborates that the eastern "races" share male wing pattern phenotypes (see also [33]). We do not believe that the existing subspecies designations should be discarded however. The subspecies as described reflect populations evolving somewhat independently of others, as reflected by the documented differentiation in male wing patterns and genitalia. The subspecies boundaries in *P. dardanus* are porous, allowing morph distributions to evolve independently in most cases, and precluding the accumulation of detectable genetic differentiation.

Female wing pattern variation in *P. dardanus* is often presented as following a clear-cut Mendelian hierarchy, a prerequisite of accurate polymorphic mimicry: intermediate phenotypes are non-mimetic and lack protection from predation [52]. The collection reveals this to be a simplification, and in fact the phenotypes arising from a mating depend on a multitude of factors including the particular alleles and the genetic background at loci other than the single, large effect mimicry switch (Figure 2). Even in single populations, intermediate heterozygote phenotypes exist, for example. f. *salaami* in *P. dardanus cenea* [24]. Such additional complexity is, we feel, often overlooked and yet it may be of great future interest as it allows insight into the role of modifier genes in the perfection of polymorphic mimicry and is an example of the evolution of dominance relationships.

Under the prevailing theory of secondary 'modifier' loci alternately perfecting the mimicry in different subspecies, contact zones where genetic backgrounds become admixed should display mimicry and dominance breakdown. This hypothesis is supported by our results, which show that the *P. dardanus meseres* and *P. dardanus polytrophus* populations, which are found in-between the diverged eastern and western mainland lineages [47], have very variable wing patterns (Figure 2, Figure 7). Almost all deviating phenotypes were collected from this region and the boundaries between subspecies are here indistinct, overlapping with *P. dardanus dardanus* (for *P. dardanus meseres*) or *P. dardanus tibullus* (for *P. dardanus polytrophus*).

Analyses of these 'transitional' populations may therefore shed light on the role of unlinked modifier loci in determining accurate mimetic wing patterns, similar to the insight gained from studying hybrid zones between *Heliconius* races [53]. Ongoing work to sequence and characterise genomes from several subspecies of *P. dardanus* is already underway. From the results presented here, we predict that the subspecies of *P. dardanus meseres* and *P. dardanus polytrophus* likely facilitate gene flow between eastern and western lineages. Fixed divergence between these lineages will be low, apart from at loci under disruptive selection (such as mimicry modifier loci) which would show a strong signal of differentiation [54,55]. Such a situation would be strongly reminiscent of the situation of the multiple, parapatric races of *Heliconius* species, which are differentiated from one another despite genome-wide gene flow [56].

An alternative hypothesis to explain the 'breakdown' of mimicry in this 'transitional' region is a reduction in effective selection for mimicry due to either a reduction in predation pressure or to local scarcity of distasteful model species [20,57]. The validity of the two hypotheses could be tested with the help of the distribution data, which will allow the identification of further contact zones between the two mainland lineages (e.g. around Lake Malawi, as suggested by Turner [49] on the basis of the genitalia of a single male butterfly) that may, or may not show breakdown in mimicry.

This study highlights the continuing importance of museum collections and demonstrates the utility of digitising collection information. The digitisation of the NHM London *P. dardanus* collection allows us to obtain detailed information on the biology and ecology of this species. However, we must recognise the incomplete nature of any collection: for example, as shown in Figure 3 large gaps exist in the distribution of sampled locations and demonstrates the variation in sampling density across Africa. By making the NHM London collection information openly available online, it may eventually be possible to fill in some of the spatial and/or temporal sampling gaps through the incorporation of specimens held by a number of different institutions. Yet, we believe it is already possible to begin to synthesise a more complete picture of the biology of *P. dardanus*. For example, precise information on the distribution of each of the mimetic phenotypes could be correlated with the distribution of their models to test the mimicry system at the ecological level. If shaped by natural selection, correlations between the ranges of mimetic forms and their models are expected to be strong. Discrepancies might indicate recent range shifts or other ecological factors being of importance (e.g. the absence of predators). The observed statistical patterns will assist molecular studies and potentially explain genetic changes, such as these indicating recent selective sweeps or population bottlenecks.

The staggering diversity of *Papilio dardanus* continues to captivate evolutionary biologists. The systematic overview presented here unlocks a single museum collection and reveals an unequal distribution of morph classes over sub-Saharan Africa. It also exposes the generally overlooked abundance of imperfect phenotypes. Poulton once noted on the origin of phenotypes in *P. dardanus* that "*the study is hindered at the outset by a complicated nomenclature*" [19] Here we followed the classification of Bernardi et al. [33], and grouped the female forms in 9 classes. This, however, remains an artificial classification, merely based on resemblance rather than the fundamental genetics of the system, ignoring the possibility that some phenotypes may have evolved convergently. Future molecular work on this intraspecific wing pattern radiation will reveal whether similar-looking colour patterns are indeed genetically related and share similar evolutionary histories or have arisen multiple times under strong mimicry selection.

Acknowledgments

We are grateful to Flavia Toloni and James Fage for assistance with curation, databasing and digitisation of specimens. We would like to thank Blanca Huertas, Dr David Lees and the NHM collections management staff for assistance with curation, nomenclature and taxonomy and Mrs Ramnik Arora, who arranged the collection in geographical sequence in the late 1970s. We thank Vladimir Blagoderov for help with scanning. We would additionally like to thank Prof Alfried Vogler and Dr Chris Jiggins for useful discussion regarding the interpretation of the results, and Steve Collins for assisting us in our work on mimicry in *P. dardanus*. Prof Dick Vane-Wright is thanked for his critical comments on an earlier version of manuscript and for sharing his profound knowledge on *Papilio dardanus* with us.

Author Contributions

Conceived and designed the experiments: MJT MJTNT. Performed the experiments: MJT MJTNT. Analyzed the data: MJT. Wrote the paper: MJT MJTNT.

References

1. Brakefield PM, Gates J, Keys D, Kesbeke F, Wijngaarden PJ, et al. (1996) Development, plasticity and evolution of butterfly eyespot patterns. Nature 384: 236–242.
2. Carroll SB (2011) Evolution. How great wings can look alike. Science 333: 1100–1101. doi:10.1126/science.1211025.
3. Brunetti CR, Selegue JE, Monteiro A, French V, Brakefield PM, et al. (2001) The generation and diversification of butterfly eyespot color patterns. Curr Biol 11: 1578–1585.
4. Cook LM, Saccheri IJ (2012) The peppered moth and industrial melanism: evolution of a natural selection case study. Heredity 110: 207–212. doi:10.1038/hdy.2012.92.
5. Beldade P, Brakefield PM (2002) The genetics and evo-devo of butterfly wing patterns. Nat Rev Genet 3: 442–452. doi:10.1038/nrg818.
6. Joron M, Frezal L, Jones RT, Chamberlain NL, Lee SF, et al. (2011) Chromosomal rearrangements maintain a polymorphic supergene controlling butterfly mimicry. Nature 477: 203–206. doi:10.1038/nature10341.
7. The *Heliconius* Genome Consortium (2012) Butterfly genome reveals promiscuous exchange of mimicry adaptations among species. Nature 487: 94–98. doi:10.1038/nature11041.
8. Kunte K, Zhang W, Tenger-Trolander A, Palmer DH, Martin A, et al. (2014) *doublesex* is a mimicry supergene. Nature 507: 229–232. doi:10.1038/nature13112.
9. Fisher RA (1930) The genetical theory of natural selection. London: Oxford University Press.
10. Punnett RC (1905) Mendelism. 1st ed. London: Macmillian and Company.
11. Ford EB (1931) Mendelism and evolution. London: Methuen & Co.
12. Dobzhansky T (1937) Genetics and the origin of species. New York: Columbia University Press.
13. Carpenter GDH, Ford EB (1933) Mimicry. London: Methuen & Co.
14. Fisher RA (1927) On some objections to mimicry theory; statistical and genetic. Trans R Entomol Soc London 75: 269–278. doi:10.1111/j.1365-2311.1927.tb00074.x.
15. Fisher RA (1930) The evolution of dominance in certain polymorphic species. Am Nat 64: 385–406.
16. Punnett RC (1915) Mimicry in butterflies. Cambridge: Cambridge University Press.
17. Goldschmidt RB (1945) Mimetic polymorphism, a controversial chapter of Darwinism. Q Rev Biol 20: 205. doi:10.1086/394883.
18. Davis FR (2009) *Papilio dardanus*: The natural animal from the experimentalist's point of view. In: Cain J, Ruse M, editors. Descended from Darwin: Insights into the history of evolutionary studies, 1900–1970. Philadelphia: American Philosophical Society. 221–242.
19. Poulton EB (1924) The most interesting butterfly in the world. J East African Nat Hist Soc 20: 4–22.
20. Ford EB (1936) The genetics of *Papilio dardanus*, Brown (Lep.). Trans R Entomol Soc London 85: 435–466.
21. Trimen R (1869) On some remarkable mimetic analogies among African butterflies. Trans Linn Soc London 26: 497–522.
22. Trimen R (1874) Observations on the case of *Papilio merope*, Auct. with an account of the various known forms of that butterfly. Trans R Entomol Soc London 22: 137–153.
23. Cohen A (2002) Roland Trimen and the 'merope' harem. Notes Rec R Soc 56: 205–218.
24. Clarke CA, Sheppard PM (1959) The genetics of *Papilio dardanus*, Brown. I. Race *cenea* from South Africa. Genetics 44: 1347–1358.
25. Clarke CA, Sheppard PM (1960) The genetics of *Papilio dardanus*, Brown. II. Races *dardanus*, *polytrophus*, *meseres*, and *tibullus*. Genetics 45: 439–457.
26. Clarke CA, Sheppard PM (1960) The genetics of *Papilio dardanus*, Brown. III. Race *antinorii* from Abyssinia and race *meriones* from Madagascar. Genetics 45: 683–698.
27. Clarke CA, Sheppard PM (1962) The genetics of *Papilio dardanus*, Brown. IV. Data on race *ochracea*, race *flavicornis*, and further information on races *polytrophus* and *dardanus*. Genetics 47: 909–920. doi:10.1111/j.1469-1809.1962.tb01305.x.
28. Charlesworth D, Charlesworth B (1975) Theoretical genetics of Batesian mimicry III. Evolution of dominance. J Theor Biol 55: 325–337.
29. Charlesworth D, Charlesworth B (1975) Theoretical genetics of Batesian mimicry II. Evolution of supergenes. J Theor Biol 55: 305–324.
30. Charlesworth D, Charlesworth B (1975) Theoretical genetics of Batesian mimicry I. Single locus models. J Theor Biol 55: 293–303.
31. Clarke CA, Sheppard PM (1963) Interactions between major genes and polygenes in the determination of the mimetic patterns of *Papilio dardanus*. Evolution 17: 404–413.
32. Clark R, Brown SM, Collins SC, Jiggins CD, Heckel DG, et al. (2008) Colour pattern specification in the mocker swallowtail *Papilio dardanus*: the transcription factor *invected* is a candidate for the mimicry locus *H*. Proc R Soc B Biol Sci 275: 1181–1188. doi:10.1098/rspb.2007.1762.
33. Bernardi G, Pierre J, Nguyen Thi H (1985) Le polymorphisme et le mimétisme de Papilio *dardanus* Brown (Lep. Papilionidae). Bull la Société Entomol Fr 90: 1106–1159.

34. Jordan K (1905) Der Gegensatz zwischen geographischer und nichtgeographischer Variation. Z Wiss Zool 83: 151–210.
35. Le Cerf F (1924) Catalogue annote des "types" et formes nouvelles des *Papilios* d'Afrique contenus dans la collection du "Hill Museum." Bull Hill Museum 1: 369–399.
36. Grimaldi D, Engel MS (2005) Evolution of the Insects. Cambridge: Cambridge University Press.
37. Nijhout HF (2003) Polymorphic mimicry in *Papilio dardanus*: mosaic dominance, big effects, and origins. Evol Dev 5: 579–592.
38. Larsen TB (1990) The butterflies of Kenya and their natural history. Oxford: Oxford University Press.
39. Ford EB (1964) Ecological Genetics. London: Methuen.
40. Clarke CA (1991) The Clarke/Sheppard/Turner genetic collection of butterflies at the Natural History Museum, London. J Lepid Soc 45: 222–225.
41. EDC GmbH QDataMatrix. http://qdatamatrix.sourceforge.net/.
42. ESRI (Environmental Systems Research Institute) (2011) ArcMap 10.
43. Getis A, Ord JK (1992) The analysis of spatial association. Geogr Anal 24: 189–206.
44. Moran PAP (1950) Notes on continuous stochastic phenomena. Biometrika 37: 17–23.
45. Ord JK, Getis A (1995) Local Spatial Autocorrelation Statistics: Distributional Issues and an Application. Geogr Anal 27: 286–306. doi:10.1111/j.1538-4632.1995.tb00912.x.
46. United States Geological Survey (2009) African Isobioclimates.
47. Clark R, Vogler AP (2009) A phylogenetic framework for wing pattern evolution in the mimetic mocker swallowtail *Papilio dardanus*. Mol Ecol 18: 3872–3884. doi:10.1111/j.1365-294X.2009.04259.x.
48. Clarke CA, Sheppard PM (1960) The evolution of dominance under disruptive selection. Heredity 14: 73–87.
49. Turner JRG (1963) Geographical variation and evolution in the males of the butterfly *Papilio dardanus* Brown (Lepidoptera: Papilionidae). Trans R Entomol Soc London 115: 239–259.
50. Carpenter GDH (1948) Notes on the males of *Papilio dardanus* Brown (Lep.) with the definition of a new transitional race and a redescription of *P. dardanus ochracea* Poulton. Proc R Entomol Soc London 17: 11–17.
51. Thompson MJ, Jiggins CD (2014) Supergenes and their role in evolution. Heredity. doi:10.1038/hdy.2014.20.
52. Turner JRG (1987) The evolutionary dynamics of Batesian and Muellerian mimicry: similarities and differences. Ecol Entomol 12: 81–95.
53. Nadeau NJ, Whibley A, Jones RT, Davey JW, Dasmahapatra KK, et al. (2012) Genomic islands of divergence in hybridizing *Heliconius* butterflies identified by large-scale targeted sequencing. Philos Trans R Soc B Biol Sci 367: 343–353. doi:10.1098/rstb.2011.0198.
54. Turner JRG, Johnson MS, Eanes WF (1979) Contrasted modes of evolution in the same genome: allozymes and adaptive change in *Heliconius*. Proc Natl Acad Sci USA 76: 1924–1928.
55. Nosil P, Funk DJ, Ortiz-Barrientos D (2009) Divergent selection and heterogeneous genomic divergence. Mol Ecol 18: 375–402. doi:10.1111/j.1365-294X.2008.03946.x.
56. Nadeau NJ, Martin SH, Kozak KM, Salazar C, Dasmahapatra KK, et al. (2013) Genome-wide patterns of divergence and gene flow across a butterfly radiation. Mol Ecol 22: 814–826. doi:10.1111/j.1365-294X.2012.05730.x.
57. Nijhout HF (1991) The development and evolution of butterfly wing patterns. Washington: Smithsonian Institution Scholarly Press.

2

Genome-Wide Survey and Expression Analysis of Calcium-Dependent Protein Kinase in *Gossypium raimondii*

Wei Liu[1]ⓢ, Wei Li[1]ⓢ, Qiuling He[1], Muhammad Khan Daud[2], Jinhong Chen[1]*, Shuijin Zhu[1]*

1 Department of Agronomy, Zhejiang University, Hangzhou, Zhejiang, China, **2** Department of Biotechnology and Genetic Engineering, Kohat University of Science and Technology, Kohat, Pakistan

Abstract

Calcium-dependent protein kinases (CDPKs) are one of the largest protein kinases in plants and participate in different physiological processes through regulating downstream components of calcium signaling pathways. In this study, 41 CDPK genes, from *GrCPK1* to *GrCPK41*, were identified in the genome of the diploid cotton, *Gossypium raimondii*. The phylogenetic analysis indicated that all these genes were divided into four subgroups and members within the same subgroup shared conserved exon-intron structures. The expansion of *GrCPKs* family in *G. raimondii* was due to the segmental duplication events, and the analysis of Ka/Ks ratios implied that the duplicated *GrCPKs* had mainly undergone strong purifying selection pressure with limited functional divergence. The cold-responsive elements in promoter regions were detected in the majority of *GrCPKs*. The expression analysis of 11 selected genes showed that *GrCPKs* exhibited tissue-specific expression patterns and the expression of *GrCPKs* were induced or repressed by cold treatment. These observations would lay an important foundation for functional and evolutionary analysis of CDPK gene family in *Gossypium* species.

Editor: Baohong Zhang, East Carolina University, United States of America

Funding: The works are support by the National Basic Research Program (973 program, No: 2010CB126006) and the National High Technology Research and Development Program of China (2011AA10A102, 2013AA102601). The funders had no role in study design, data collection and analysis, decision to publish, or preparation of the manuscript.

Competing Interests: The authors have declared that no competing interests exist.

* E-mail: shjzhu@zju.edu.cn (SZ); jinhongchen@zju.edu.cn (JC)

ⓢ These authors contributed equally to this work.

Introduction

The plant growth and crop production are adversely affected by common stress conditions, such as drought, low temperature and high salinity [1]. Adaptation of plants to these environmental stresses includes the perception of stress signals and subsequent signal transduction, leading to the activation of various physiological and metabolic responses [2,3,4]. As a ubiquitous second messenger in cells, calcium (Ca^{2+}) plays an important role in the signal transduction pathways. The transient changes of cytoplasmic Ca^{2+} concentration in response to various stresses were sensed and decoded by several Ca^{2+} sensors or Ca^{2+} binding proteins which relayed the signals into downstream response processes such as regulation of gene expression and phosphorylation cascades [5,6].

There are four major classes of Ca^{2+} binding proteins characterized in plants, including calmodulins (CaM), calmodulin-like proteins (CaML), calcineurin B-like proteins (CBL) and calcium-dependent protein kinases (CDPK) [7,8,9]. Among them, CDPKs are the best characterized and are of particular interest, which constitute a large multigene family and are reported to be found throughout the plant kingdom from algae to angiosperms [10]. The CDPK protein possesses four characterized domains, a variable N-terminal domain, a catalytic Ser/Thr protein kinase domain, an autoinhibitory region, and a calmodulin-like domain [9,11,12,13]. The N-terminal domain is highly variable and contains myristoylation or palmotylation sites which may contribute to membrane localization [12,14]. And the calmodulin-like domain contains EF-hands structure for binding of Ca^{2+} [11,12].

It was confirmed that CDPK genes were involved in regulating plant response to various stimuli, including abiotic and biotic stresses and hormones. Overexpression of the rice CDPK gene *OsCDPK7* enhanced the tolerance of the transgenic rice plants to cold, salt, and drought stress [15,16], and *OsCPK12* overexpression plants increased the tolerance to salt stress but to blast fungus [17]. The low temperatures induced *ZmCDPK1* expression in maize [18], and *ZmCPK11* participated in touch- and wound-induced pathways [19]. In tobacco, *NtCPK4* expression was increased in response to the treatment of gibberellin or NaCl [20]. In Arabidopsis, *AtCPK10* was reported to participate in ABA- and Ca^{2+}-medicated stomatal regulation in response to drought stress [21], and *AtCPK3* and *AtCPK6* were also shown as positive transducers of stomatal ABA signaling in guard cells [22]. *AtCPK23* was demonstrated to function in response to drought and salt stresses [23]. In addition, some CDPK genes were also involved in pollen tube growth [24], root development [25], cell division and differentiation, and cell death [26,27].

So far, many CDPK genes have been identified from numerous plant species. In Arabidopsis, 34 CDPK genes were revealed [9,12]. And similarly, 31 CDPK genes in rice genome has identified by a genome-wide analysis [28]. In wheat, 20 CDPK genes including 14 full-length cDNA sequences were comprehensively studied [29]. And in poplar, 30 CDPK genes were identified [30], and 35 CDPK genes were initially revealed in maize genome [31]. Other higher plants such as tomato [32], potato [33,34], tobacco [35,36] and grapevine [37] also have multiple CDPK genes characterized in recent years. However, compared to the extensive studies of CDPK genes in many other plant species, little is known about this gene family in cotton. Till now, only one CDPK gene, *GhCPK1*, has been identified from upland cotton. And its transcripts accumulated primarily in the elongating fiber, which suggested that *GhCPK1* might play a vital role in the calcium signaling events associated with fiber elongation [38,39].

Cotton is the major source of natural fibers used in the textile industry and is cultivated worldwide. Cotton, which belongs to the genus of *Gossypium*, includes approximately 45 diploid and five tetraploid species and serves as an excellent model system for evolutionary studies of polyploidy plants [40]. With completion of the genome sequencing of the diploid cotton *Gossypium raimondii* [41], genome-wide analysis of all the genes belonging to specific gene families have been realized. Here, 41 CDPK genes from *G. raimondii* were identified by database searches and classified according to phylogenetic analysis. Furthermore, the expression profiles of CDPK genes in response to low temperature, which affects seed germination,plant development, and final yield in cotton, were investigated. The identification and comprehensive study for CDPK genes from *G. raimondii* will provide valuable information for breeding stress-resistant cotton and further studying of the biological function and evolutionary relationship of this family in cotton.

Materials and Methods

Gene retrieval and annotation

The *G. raimondii* genome database (release v2.1) [41] was downloaded from Phytozome (http://www.phytozome.net/cotton.php). The published CDPK protein information for Arabidopsis [12] and rice [28] were obtained from the Arabidopsis Information Resource (TAIR release 10, http://www.arabidopsis.org) and the Rice Genome Annotation Project Database (RGAP release 7, http://rice.plantbiology.msu.edu/index.shtml), respectively. To identify potential CDPK genes in *G. raimondii*, the Arabidopsis and rice CDPK proteins were used as queries by searching against the *G. raimondii* genome database using BlastP and tBlastN programs with default parameters. Subsequently, all hits were verified with the InterProScan program (http://www.ebi.ac.uk/Tools/pfa/iprscan/) [42] to confirm the presence of the protein kinase domain. Finally, the Pfam (http://pfam.sanger.ac.uk/) [43] and SMART (http://smart.embl-heidelberg.de/) [44] database were applied to manually determine each candidate member of the CDPK family. The EF hand was predicted by ScanProsite tool (http://prosite.expasy.org/scanprosite/) [45]. The N-myristoylation motif and the palmitoylation site were predicted by Myristoylator (http://web.expasy.org/myristoylator/) [46] and CSS-Plam program [47], respectively. The molecular weight (MW) of the full-length protein was calculated by Compute pI/Mw tool (http://web.expasy.org/cgi-bin/compute_pi/pi_tool) [48].

Phylogenetic analysis and Gene structure prediction

Multiple alignments of the full-length protein sequences were performed using Clustal X version 2.0 program [49] with default parameters. Phylogenetic trees were constructed using the method of Neighbor Joining MEGA 5.2 [50] with pairwise deletion option and poisson correction model. For statistical reliability, bootstrap tests were carried out with 1000 replicates.

The gene structures were obtained through comparing the genomic sequences and their predicted coding sequences of GSDS(http://gsds.cbi.pku.edu.cn/) [51].

Chromosomal locations and gene duplications

All CDPK genes were mapped on the *G. raimondii* chromosomes according to their starting positions given in the genome annotation document. The chromosome location image was generated by MapInspect software [52].

Gene duplication events of CDPK genes in *G. raimondii* were also investigated. The gene duplication was defined according to (1) the length of aligned sequence cover >80% of the longer gene, (2) the identity of the aligned regions >80%, and (3) only one duplication event was counted for tightly-linked genes [53,54]. With the chromosomal locations of CDPK genes, two types of gene duplications were recognized, i.e., tandem duplication and segmental duplication.

Estimating Ka/Ks ratios for duplicated gene pairs

The CDPK duplicated gene pairs of *G.raimondii* were firstly aligned by Clustal X version 2.0 program [49]. Then Ks (synonymous substitution rate) and Ka (non-synonymous substitution rate) were calculated using the DnaSP v5.0 software (DNA polymorphism analysis) [55]. Finally, the Ka/Ks ratio was analyzed to assess the selection pressure for each gene pair.

Cis-element analysis

For promoter analysis, 2000 bp genomic DNA sequences upstream of the initiation codon (ATG) were retrieved from the genome sequence. These sequences were subjected to search in the PLACE database (http://www.dna.affrc.go.jp/PLACE/signalscan.html) [56] to identify putative *cis*-elements in promoter regions.

Plant materials and low temperature stress treatment

All the plants of *G. raimondii* were grown in a temperature-controlled chamber at 28°C with a photoperiod of 16 hours light and 8 hours dark. After ten days, the leaves, stems, roots, and cotyledons of some seedlings were collected to analyze tissue-specific expression, and the rest seedlings were used to examine the expression patterns of CDPK genes under low-temperature stress. Plant leaves of the seedlings grown in the temperature-controlled chamber treated at 10°C were harvested at 0, 3, 6, and 12 hours, which represented normal plants, slight stress, moderate stress, and severe stress, respectively. All collected samples were immediately frozen in liquid nitrogen and stored at −80°C.

RNA isolation and quantitative real-time PCR (qRT-PCR)

Total RNA was extracted from all samples using EASYspin Plus RNAprep Kit, and the first-strand cDNAs were synthesized with PrimerScript 1st Strand cDNA synthesis kit (TaKaRa). All protocols followed the manufacturer's protocol. For quantitative real-time PCR (qRT-PCR) assay, gene-specific primers were designed for the 11 selected CDPK genes according to their CDSs (Table S1). The qRT-PCR was performed with SYBR premix Ex taq (TakaRa) and CFX96 Realtime System (BioRad) by strictly following the manufacturer's instructions. Each reaction was done in a final volume of 20 µl containing 10 µl of SYBR premix Ex taq, 1.0 µl of cDNA sample, and 0.5 µl of each gene-specific

Table 1. The information of 41 CDPK genes from *G. raimondii*.

Gene symbol	ID	Chromosome	Alternative splicing	CDS	Amino acid	MW (kDa)	No. of EF hands	N-terminal aa	N-Myristoylation	N-palmitoylation
GrCPK1	Gorai.001G135000	1	_	1542	513	57.3	4	MGNCCSRG	Yes	Yes
GrCPK2	Gorai.001G138000	1	2	1590	529	59.5	2	MGNCCATT	No	Yes
GrCPK3	Gorai.002G088800	2	4	1614	537	60.5	4	MGSCLTKS	Yes	Yes
GrCPK4	Gorai.002G153600	2	2	1776	591	66.3	4	MGNSCAKS	Yes	Yes
GrCPK5	Gorai.003G009500	3	8	1626	541	61.3	4	MGACLSAT	Yes	Yes
GrCPK6	Gorai.003G084000	3	_	1527	508	57.5	3	MGNCNGLP	No	Yes
GrCPK7	Gorai.003G092900	3	3	1608	535	60.6	3	MGNCCATP	Yes	Yes
GrCPK8	Gorai.004G015100	4	_	1617	538	60.5	4	MGNCCSRG	Yes	Yes
GrCPK9	Gorai.005G019800	5	_	1599	532	60.7	4	MGSCVARP	Yes	Yes
GrCPK10	Gorai.005G074300	5	_	1527	508	56.9	4	MNNQSSSI	No	No
GrCPK11	Gorai.005G216500	5	6	1707	568	63.6	4	MGNTCRGS	No	Yes
GrCPK12	Gorai.006G124800	6	_	1575	524	58.6	4	MGNCCSCG	Yes	Yes
GrCPK13	Gorai.006G128200	6	3	1596	531	59.8	3	MGNCCATP	No	Yes
GrCPK14	Gorai.006G137800	6	_	1596	531	60.2	3	MGNCCVTS	No	Yes
GrCPK15	Gorai.006G147600	6	_	1836	611	68.4	4	MGNNCFKT	No	Yes
GrCPK16	Gorai.007G025000	7	5	1593	530	60.4	4	MGNCNRPP	No	Yes
GrCPK17	Gorai.007G035100	7	3	1653	550	62.3	4	MGNCNACV	No	Yes
GrCPK18	Gorai.007G194500	7	3	1665	554	62.8	4	MGACLSTT	Yes	Yes
GrCPK19	Gorai.007G378700	7	_	1584	527	59.2	3	MGNCCRSP	No	Yes
GrCPK20	Gorai.008G013700	8	_	1728	575	64.6	4	MRLHYCMR	No	Yes
GrCPK21	Gorai.008G251000	8	2	1605	534	60.2	4	MGNCNSQP	Yes	Yes
GrCPK22	Gorai.009G078000	9	_	1599	532	59.5	4	MGNLCSRS	Yes	Yes
GrCPK23	Gorai.009G191500	9	_	2622	873	97.1	4	MGICQSLC	No	Yes
GrCPK24	Gorai.009G290200	9	5	1617	538	60.4	4	MNKKIAGS	No	No
GrCPK25	Gorai.009G351200	9	_	1605	534	60.7	4	MGSCISAP	Yes	Yes
GrCPK26	Gorai.009G394700	9	_	1947	648	71.6	4	MGNVCATL	No	Yes
GrCPK27	Gorai.009G395400	9	5	1719	572	63.6	4	MGNACAGP	No	Yes
GrCPK28	Gorai.009G438300	9	_	1575	524	58.7	4	MGGCLTKT	Yes	Yes
GrCPK29	Gorai.010G001300	10	_	1581	526	59.1	4	MGLCQSLG	Yes	Yes
GrCPK30	Gorai.010G252400	10	_	1584	527	59.3	4	MGCCSSKN	Yes	Yes
GrCPK31	Gorai.011G014200	11	_	1656	551	62.0	4	MGCFSSKH	Yes	Yes
GrCPK32	Gorai.011G098300	11	_	1635	544	62.0	4	MGICLSTT	Yes	Yes
GrCPK33	Gorai.011G228500	11	5	1764	587	65.4	4	MGNTCVGP	No	Yes
GrCPK34	Gorai.012G045700	12	2	1494	497	55.9	4	MSRTSSGT	No	No
GrCPK35	Gorai.012G114600	12	_	1584	527	59.4	3	MGNCCRSP	No	Yes

Table 1. Cont.

Gene symbol	ID	Chromosome	Alternative splicing	CDS	Amino acid	MW (kDa)	No. of EF hands	N-terminal aa	N-Myristoylation	N-palmitoylation
GrCPK36	Gorai.012G138900	12	2	1659	552	62.0	4	MGNTCRGP	No	Yes
GrCPK37	Gorai.013G003100	13	2	1740	579	64.6	4	MGNTCVGP	No	Yes
GrCPK38	Gorai.013G064400	13	5	1464	487	54.6	4	MRRAIDHQ	No	No
GrCPK39	Gorai.013G064500	13	2	1572	523	58.3	3	MGNTCLGS	No	Yes
GrCPK40	Gorai.013G159300	13	3	1611	536	60.4	4	MGGCLTKN	Yes	Yes
GrCPK41	Gorai.013G253100	13	–	1584	527	58.8	4	MGNCCTRG	No	Yes

– represents no alternative splice variants.

primer. The qRT-PCR cycles were conducted with 40 cycles and an annealing temperature of 60°C, the amplification programs were as follows: 95°C for 30 seconds, 40 cycles of 94°C for 10 seconds, 60°C for 10 seconds, and 72°C for 15 seconds. The cotton *UBQ7* gene was used as internal reference for all the qRT–PCR analyses. Each cDNA sample was tested in three replicates. The relative expression levels were calculated according to the $2^{-\triangle\triangle Ct}$ method [57]. The expression profiles were clustered using the Cluster 3.0 software [58].

Results and Discussion

Identification and annotation of CDPK genes in *G. raimondii*

The availability of the *G. raimondii* genome sequences [41] makes it possible to identify all CDPK gene family members in the diploid cotton. BLASTP and TBLASTN programs were performed to search the candidate CDPK genes from the *G. raimondii* genome with the query sequences of Arabidopsis and rice CDPK genes. Then Pfam and SMART analyses were used to verify the retrieved sequences, and finally a total of 41 non-redundant CDPK genes were confirmed and described (Table1). According to the proposed nomenclature for CDPK genes [59], we designated these 41 CDPK genes as *GrCPKs*, from *GrCPK1* to *GrCPK41*. The numbering of these *GrCPKs* was based on their position from top to bottom on corresponding chromosomes, from chromosome 1 to chromosome 13. Through analyzing the transcriptome sequencing data downloaded from the NCBI Sequence Read Archive (SRA), it was found that at least 37 *GrCPKs* were actively expressed in *G. raimondii* (Figure S1). The total number of CDPK genes identified from *G. raimondii* was greater than that in Arabidopsis [12] and rice [28]. The length and molecular weights (MW) of 41 GrCPK proteins were deduced from their predicted protein sequences. All identified GrCPK genes encoded proteins with amino acid numbers varying from 487 to 873 and molecular weight range between 54.6 kDa and 97.1 kDa, which were comparable with CDPK genes from other plant species [12,28,30,31].

All 41 CDPKs identified in our study possessed the typical CDPK structure with four characterized domains, including a variable N-terminal domain, a catalytic Ser/Thr protein kinase domain, an autoinhibitory region, and a calmodulin-like domain. The N-terminal domain of several CDPK proteins contained myristoylation motif with a Gly residue at the second position, which was thought to be critical for mediating protein-protein and protein-membrane interactions [60]. It was reported that CDPK genes which had an N-myristoylation motif tended to localize in the plasma membranes [14,21,29,61,62], demonstrating that myristoylation was required for membrane binding. In addition, palmitoylation, as a second type of lipid modification, was also necessary to stabilize for membrane association [14]. Among 41 CDPKs in our study, 18 CDPKs were predicted to have myristoylation motifs at the N-terminus, and all of them possessed at least one palmitoylation site. However, the membrane association of CDPKs was complex and might be affected by other motifs. For example, *TaCPK3* and *TaCPK15* lacking myristoylation motifs could also be associated with the membranes [29], while *ZmCPK1* which was predicted to have an N-myristoylation motif was found to localize into the cytoplasm and nucleus [63]. Therefore, the subcellular localizations of these CDPKs were still needed to be further characterized experimentally.

The calmodulin-like domain contained Ca^{2+} binding EF hand structure that allow CDPK proteins to function as a Ca^{2+} sensor.

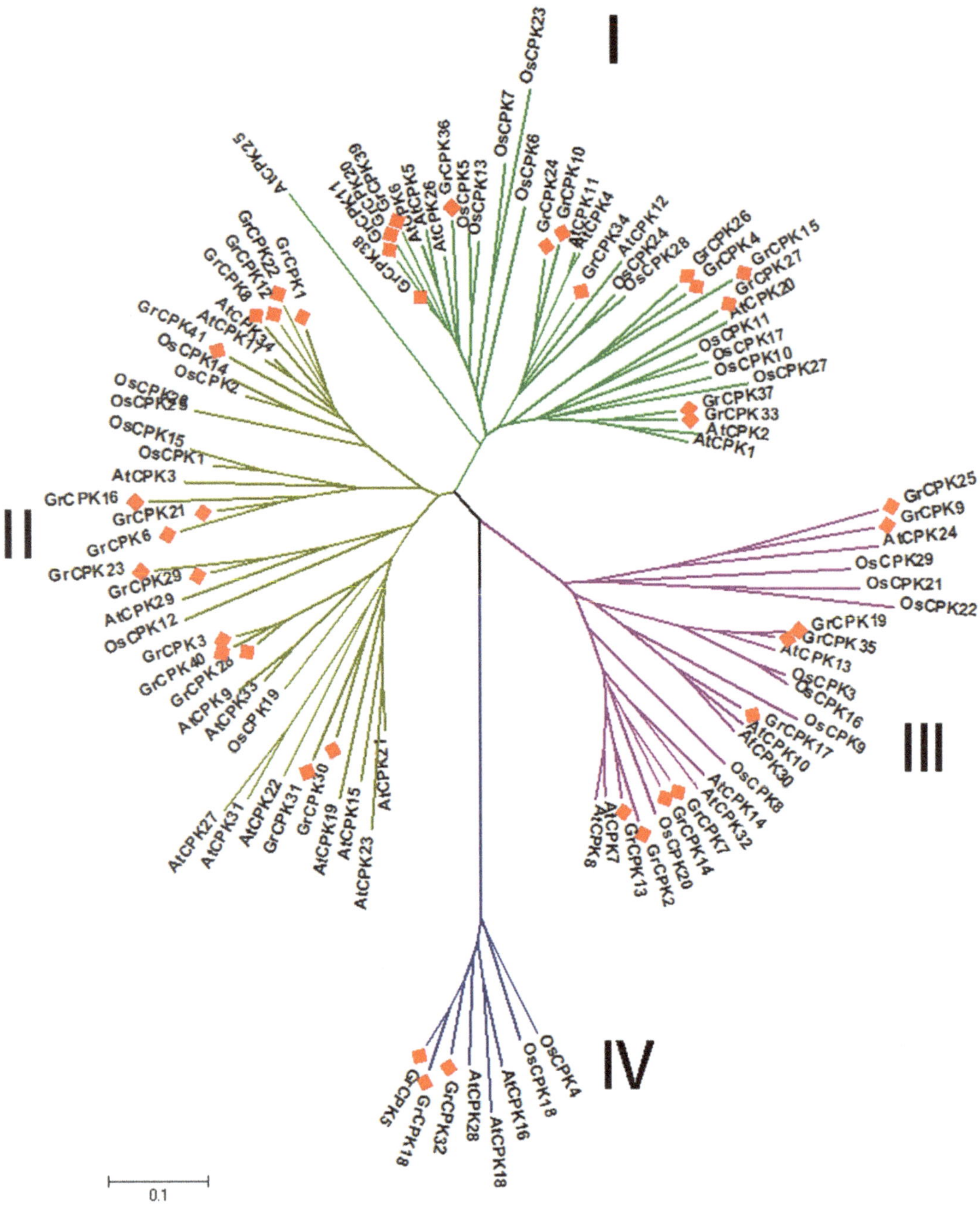

Figure 1. Unrooted phylogenetic tree of CDPK genes from *G. raimondii*, Arabidopsis and rice. Four subfamilies are labeled as I, II, III, and IV. And the branches of each subfamily are indicated in a specific color.

In *G. raimondii*, 33 CDPKs contained four Ca^{2+} binding EF hands, seven CDPKs, including GrCPK6, GrCPK7, GrCPK13, GrCPK14, GrCPK19, GrCPK35, and GrCPK39, had three EF hand motifs each, whereas GrCPK2 had two EF hand motifs only. This difference in EF hand was also found in the CDPK family of other plants [9,30,31,64]. Studies have demonstrated that the number and position of EF hands might be important for determining the Ca^{2+} regulation of CDPK activity [65,66], and compared to C-terminal EF3 and EF4 motifs, N-terminal EF1 and EF2 motifs with lower Ca^{2+}-binding affinities were more important for activating the kinases [67,68].

Phylogenetic analysis of the CDPK gene family

To detect the evolutionary relationships of CDPK genes, an unrooted phylogenetic tree was generated from alignments of the full-length sequences of *G. raimondii*, Arabidopsis and rice CDPK proteins. All the CDPK proteins were divided into four subgroups (Figure 1): CDPK I, CDPK II, CDPK III and CDPK IV. CDPK I contained 14 CDPKs from *G. raimondii*, 11 from rice, and ten from Arabidopsis. CDPK II contained 15 CDPKs from *G. raimondii*, eight from rice, and 13 from Arabidopsis. The CDPKs numbers in CDPK I and CDPK II from *G. raimondii* were greater than that from Arabidopsis and rice, mainly due to more CDPK genes in *G. raimondii*. In CDPK III, there were nine CDPKs from *G. raimondii*,

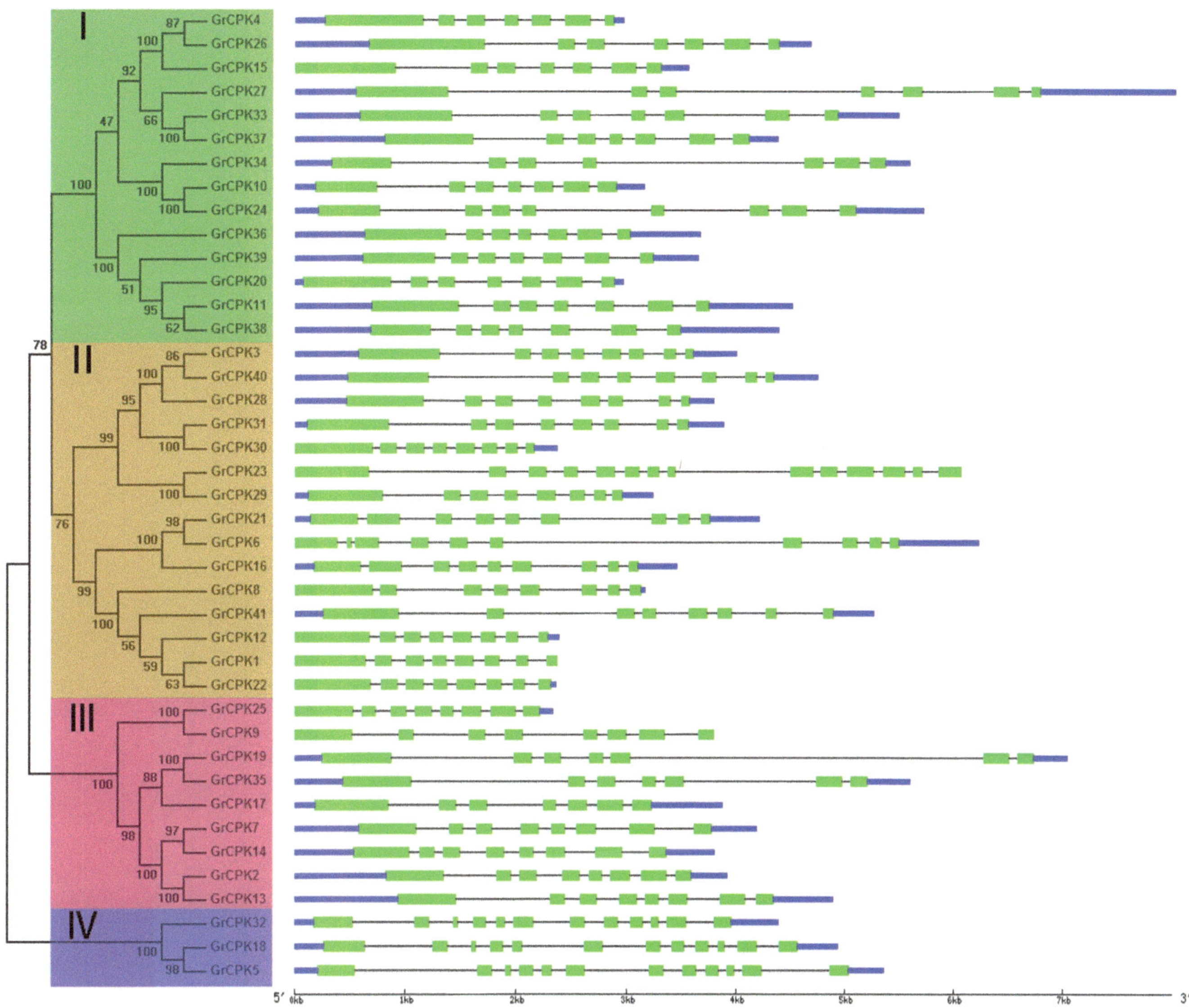

Figure 2. Phylogenetic relationship and gene structure of CDPK genes from *G. raimondii*. Four subfamilies labeled as I, II, III, and IV are marked with different color backgrounds. Exons are represented by green boxes and introns by black lines. The untranslated regions (UTRs) are indicated by thick blue lines.

eight from rice, and eight from Arabidopsis, while CDPK IV genes consisted of the smallest subfamilies in all of the three species, which contained three CDPKs from *G. raimondii*, two from rice, and three from Arabidopsis. And the number of genes in CDPK III and CDPK IV were approximately identical across these three species. Interestingly, among the four subfamilies, CDPK I in rice comprised the most members, while CDPK II in *G. raimondii* and Arabidopsis was the largest subfamily. Recent study indicated that poplar had total 30 CDPK genes with 11 genes in CDPK I, eight genes in CDPK II, nine genes in CDPK III and two genes in CDPK IV [30]. These results implied that the difference in the number of CDPK genes was mainly due to the occurrence of gene gain or loss in CDPK I or CDPK II subfamilies independently among the different organisms.

Phylogenetic analysis also showed that *GrCPKs* were more closely allied to *AtCPKs* than to *OsCPKs*, consistent with the evolutionary relationships among *G. raimondii*, Arabidopsis, and rice. Moreover, all the CDPK genes from the three species sorted into four distinct clades, implying that these four subfamilies existed before the divergence of monocots and dicots, which also supported the hypothesis that CDPK genes radiated into four subfamilies before algae and land plants split [69].

Structural analysis and chromosomal localization of GrCPKs

To get further insight into the possible structural evolution of *GrCPKs*, a separate unrooted phylogenetic tree was constructed using the protein sequences of all the CDPK genes from *G. raimondii*, and then the diverse exon-intron organizations of *GrCPKs* were compared. As shown in Figure 2, the *GrCPKs* clustered in the same subfamily shared very similar exon-intron structures. Most members in Group I possessed seven exons, except for *GrCPK24*, which contained eight exons. Most genes in Group II had eight exons, but *GrCPK16* and *GrCPK21* had nine exons each, *GrCPK6* contained ten exons, and *GrCPK23* contained 14 exons. Six genes in group III had eight exons, whereas three genes contained seven

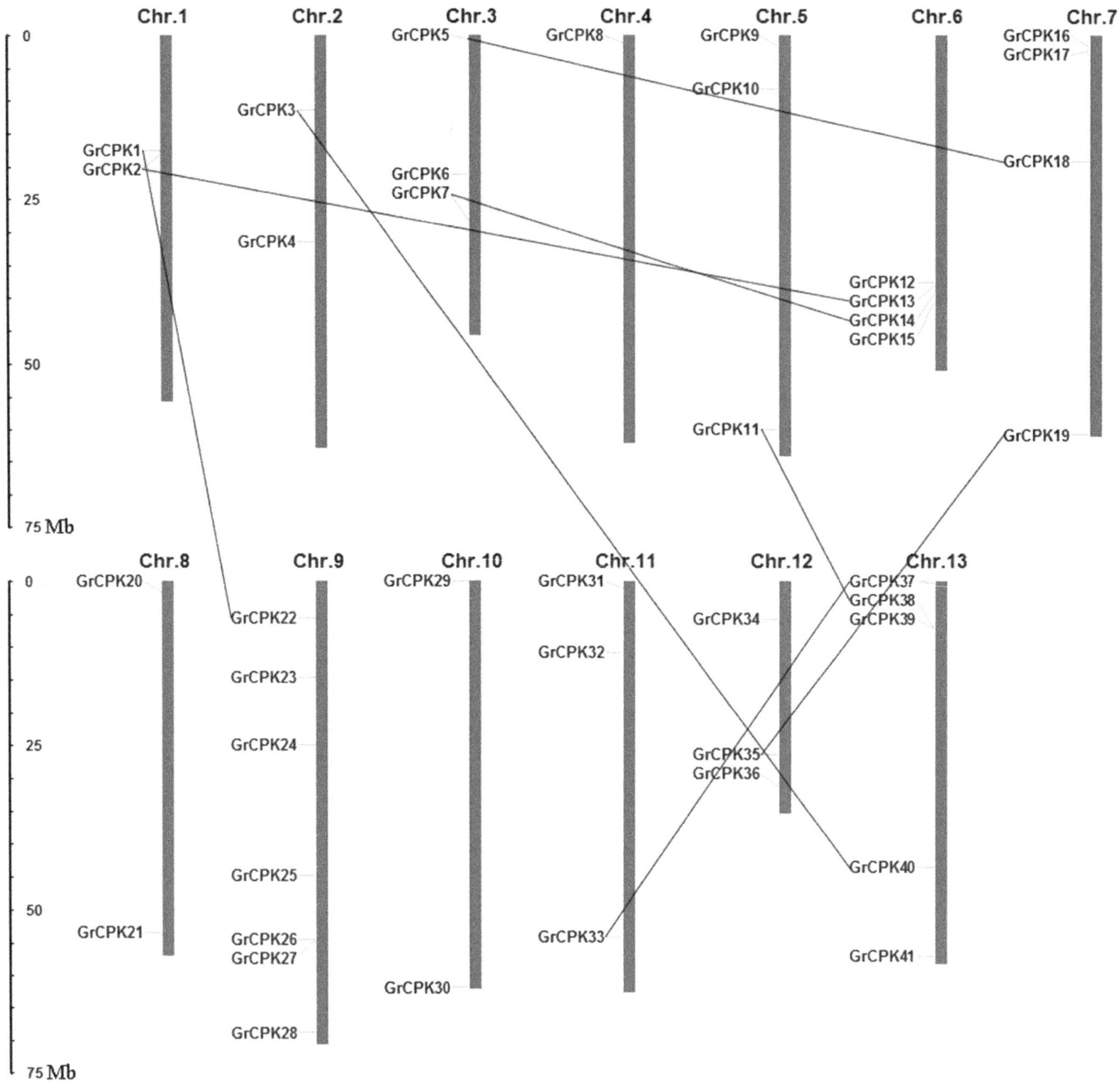

Figure 3. Distributions of CDPK genes from *G. raimondii* in 13 chromosomes. Chromosome numbers are indicated above each vertical bar. The scale represents megabases (Mb). The duplicated gene pairs are connected with black lines.

exons each. All genes in Group IV contained 12 exons. These conserved gene structure within each group supported their close evolutionary relationship.

The 41 nonredundant CDPK genes were mapped on the 13 *G. raimondii* chromosomes (Figure 3). They were found to be distributed unevenly among 13 chromosomes and the number of CDPK genes on each chromosome varied widely. Chromosome 9 which contained 7 CDPK genes was the highest one in gene numbers, followed by chromosome 13 on which five genes were found. Chromosome 6 and 7 both had four genes, and chromosome 3, 5, 11, and 12 contained three genes each. Chromosome 1, 2, 8, and 10 had two genes each only, whereas only single CDPK gene was localized on chromosome 4.

CDPK gene duplications and functional divergence

To elucidate the expanded mechanism of the CDPK gene family in *G. raimondii*, gene duplication events, including tandem and segmental duplications, were investigated, which were thought to play a significant role in the amplification of gene family members in the genome [70,71]. A total of eight duplication events, *GrCPK2/GrCPK13*, *GrCPK7/GrCPK14*, *GrCPK18/GrCPK5*, *GrCPK19/GrCPK35*, *GrCPK22/GrCPK1*, *GrCPK37/GrCPK33*, *GrCPK38/GrCPK11*, and *GrCPK40/GrCPK3*, were found in the *G. raimondii* genome and all of them were segmental duplication events based on the chromosomal distribution of the CDPK genes (Figure 3). This result suggested that the CDPK gene family

Table 2. Ka/Ks analysis for the duplicated gene pairs.

Duplicated gene 1	Duplicated gene 2	Ka	Ks	Ka/Ks	Purifying selection	Duplicate type
GrCPK2	*GrCPK13*	0.068	0.422	0.161	Yes	Segmental
GrCPK14	*GrCPK7*	0.083	0.484	0.171	Yes	Segmental
GrCPK18	*GrCPK5*	0.057	0.589	0.096	Yes	Segmental
GrCPK19	*GrCPK35*	0.028	0.399	0.071	Yes	Segmental
GrCPK22	*GrCPK1*	0.080	0.683	0.118	Yes	Segmental
GrCPK37	*GrCPK33*	0.062	0.511	0.121	Yes	Segmental
GrCPK38	*GrCPK11*	0.034	0.515	0.065	Yes	Segmental
GrCPK40	*GrCPK3*	0.038	0.528	0.071	Yes	Segmental

expansion in *G. raimondii* was mainly attributed to segmental duplication events.

The duplicated gene pairs might experience three alternative outcomes during the process of evolution, i.e., (i) one copy may simply become silenced and lost original functions (nonfunctionalization); (ii) one copy may acquire a novel, beneficial function, with the other copy retaining the original function (neofunctionalization); or (iii) both copies may become partition of original functions (subfunctionalization) [72]. We subsequently calculated the non-synonymous to synonymous substitution ratio (Ka/Ks) for each pair of duplicated CDPK genes, which showed the selective force acting on the protein, to reveal whether Darwinian positive selection was associated with functional divergence after gene duplications. Generally, Ka/Ks >1 indicates positive selection, Ka/Ks $= 1$ indicates neutral selection, while Ka/Ks <1 indicates negative or purifying selection. In this study, the Ka/Ks ratios for eight duplicated CDPK gene pairs were no larger than 0.2 (Table 2), which demonstrated that the CDPK genes from *G. raimondii* had mainly experienced strong purifying selection pressure with limited functional divergence after segmental duplications. These results suggested that functions of the duplicated CDPK genes did not diverge much during subsequent evolution.

Expression analysis of the CDPK genes in *G. raimondii*

Cold stress is one of the serious environmental stresses that most land plants might encounter during the process of their growth. And so far, numerous of CDPK genes identified in various plant species have been proven to play crucial roles in cold stress response. Cotton is a subtropical crop and its cultivation has been

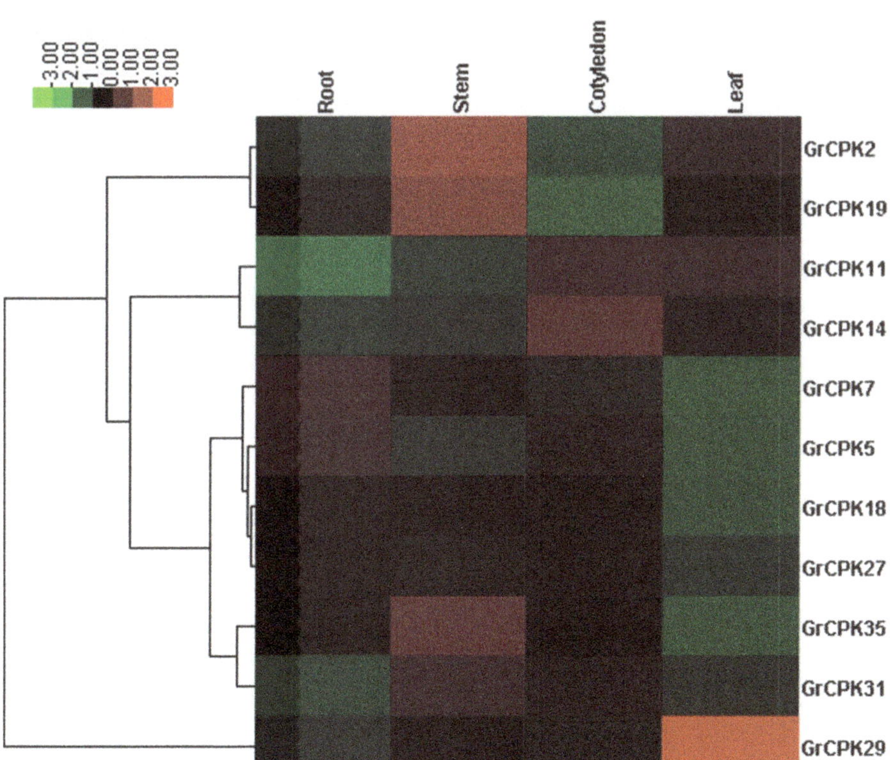

Figure 4. Expression patterns of 11 *GrCPKs* in four representative tissues of *G. raimondii* seedlings. The color bar represents the relative signal intensity values.

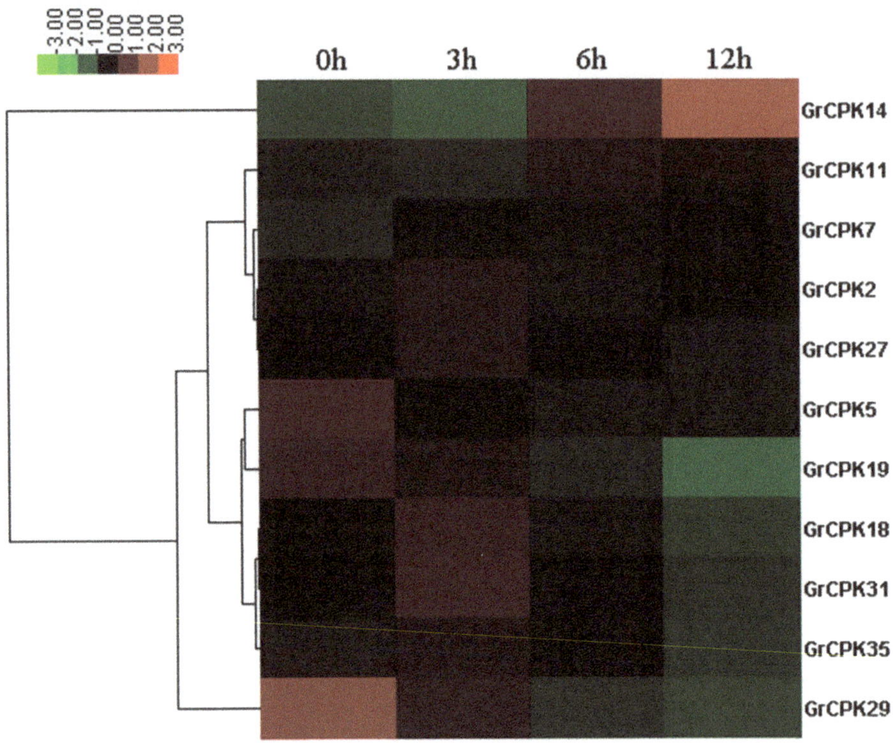

Figure 5. Expression profiling of 11 *GrCPKs* in leaves under 10°C treatment. The color bar represents the relative signal intensity values.

extended from tropical and subtropical to colder regions. Low temperature (<15°C) can adversely affect plant development, resulting in poor germination, infection by fungi and other disease-causing organisms, and yield loss [73]. There is little information about the functions of CDPK genes in the cotton response to low temperature.

The *cis*-element analysis in promoter regions might provide some indirect evidence for the functional dissection of CDPK genes in stress responses [74]. To investigate the potential functions of *GrCPKs* in cotton under cold stress, the possible presence of low temperature-responsive element (LTRE) in the promoter regions of all the *GrCPKs* was detected by searching against the PLACE database (Table S2). The results revealed that the majority, 27 out of 41, of *GrCPKs* contained the LTRE in promoter sequences, indicating that these *GrCPKs* might participate in the signal transduction of the plant response to cold stress. In order to verify the expression patterns of these genes, eight *GrCPKs* which had the LTRE in promoter regions were selected for qRT-PCR analysis. In addition, as the gene expression was a complex process, the existence of some *cis*-elements did not always correlate with the gene function [74], therefore, three *GrCPKs*, *GrCPK2*, *GrCPK11*, and *GrCPK27*, which had no LTRE in promoter regions were also selected in this expression analysis.

Firstly, qRT-PCR analysis was performed to examine expression patterns of 11 selected *GrCPKs* in roots, stems, cotyledons, and leaves of 10-day-old seedlings. As shown in Figure 4, *GrCPK5* and *GrCPK7* were expressed at high levels in roots, while *GrCPK2*, *GrCPK19*, and *GrCPK35* shared high expressions in young stems. In cotyledons, *GrCPK14* exhibited relatively high transcript abundance, whereas *GrCPK29* were predominantly expressed in leaves, which showed about 6-fold higher expression than all other tissues. The results demonstrated that most of these 11 *GrCPKs* had specific spatial expression patterns, which was similar with CDPK genes in maize [54] and rice [74] that also exhibited tissue-specific expression.

To further confirm the cold stress response of *GrCPKs*, the expression patterns of 11 *GrCPKs* in 10-day-old leaves under low temperature stress of 10°C were investigated. The expression levels of all the 11 CDPK genes responsive to slight cold stress (3 hours), moderate cold stress (6 hours), and severe cold stress (12 hours) were shown in Figure 5, comparing with normal plants. The result showed that the expression of all the 11 *GrCPKs* were induced or repressed by clod treatment. Among them, *GrCPK2*, *GrCPK7*, *GrCPK11*, *GrCPK14*, *GrCPK18*, *GrCPK27*, *GrCPK31*, and *GrCPK35* were positively regulated response to cold stress according to their transcription levels. For example, *GrCPK14* showed a significantly increase in its expression after long time cold stress treatment. Whereas, the transcription levels of the other three genes, *GrCPK5*, *GrCPK19*, and *GrCPK29*, were down-regulated in response to cold stress.

Conclusions

A complete set of 41 CDPK genes were identified in the *G. raimondii* genome, which were categorized into four subgroups and mapped on the 13 chromosomes. Segmental duplication contributed to the expansion of *GrCPKs*, and the duplicated genes mainly undergone strong purifying selection pressure with limited functional divergence. The majority of *GrCPKs* contained the low temperature-responsive elements in their promoter sequences. The expression of *GrCPKs* were also induced or repressed by the clod stress treatment. The results would be helpful for better understanding of the biological functions of the CDPK genes in cotton and the evolutionary relationship of this family in *Gossypium* species.

Supporting Information

Figure S1 Expression analysis of *GrCPKs* using the transcriptome sequencing data. The transcriptome sequencing datasets of *G. raimondii* for three tissue samples (mature leaves, 0DPA ovules, and 3DPA ovules) were downloaded from the NCBI Sequence Read Archive (SRA) with accession numbers SRX111367, SRX111365 and SRX111366. Then sequenced reads of these three datasets were mapped to the sequences of *GrCPKs*, respectively. And matches were converted to RPKM to estimate gene expression levels. The expression profiles were clustered using the Cluster 3.0 software. The color bar represents the relative signal intensity values. DPA: Days Post Anthesis.

Acknowledgments

We are grateful to Lei Mei, Ning Zhu, Cheng Li, Yue Chen, Yingchao Sun and Jingjing Pan (Zhejiang University, China) for their support in this study.

Author Contributions

Conceived and designed the experiments: SJZ JHC. Performed the experiments: W. Liu W. Li MKD JHC. Analyzed the data: W. Liu W. Li JHC QLH. Contributed reagents/materials/analysis tools: SJZ. Wrote the paper: W. Liu W. Li SJZ MKD.

References

1. Xiong L, Schumaker KS, Zhu JK (2002) Cell signaling during cold, drought, and salt stress. Plant Cell 14 Suppl: S165–S183.
2. Yamaguchi-Shinozaki K, Shinozaki K (2006) Transcriptional regulatory networks in cellular responses and tolerance to dehydration and cold stresses. Annu Rev Plant Biol 57: 781–803.
3. Valliyodan B, Nguyen HT (2006) Understanding regulatory networks and engineering for enhanced drought tolerance in plants. Curr Opin Plant Biol 9: 189–195.
4. Tran LS, Nakashima K, Shinozaki K, Yamaguchi-Shinozaki K (2007) Plant gene networks in osmotic stress response: from genes to regulatory networks. Methods Enzymol 428: 109–128.
5. Evans NH, McAinsh MR, Hetherington AM (2001) Calcium oscillations in higher plants. Curr Opin Plant Biol 4: 415–420.
6. Tuteja N, Mahajan S (2007) Calcium signaling network in plants: an overview. Plant Signal Behav 2: 79–85.
7. McCormack E, Braam J (2003) Calmodulins and related potential calcium sensors of Arabidopsis. New Phytologist 159: 585–598.
8. Kolukisaoglu U, Weinl S, Blazevic D, Batistic O, Kudla J (2004) Calcium sensors and their interacting protein kinases: genomics of the Arabidopsis and rice CBL-CIPK signaling networks. Plant Physiol 134: 43–58.
9. Cheng SH, Willmann MR, Chen HC, Sheen J (2002) Calcium signaling through protein kinases. The Arabidopsis calcium-dependent protein kinase gene family. Plant Physiol 129: 469–485.
10. Harmon AC, Gribskov M, Gubrium E, Harper JF (2001) The CDPK superfamily of protein kinases. New Phytologist 151: 175–183.
11. Harper JF, Sussman MR, Schaller GE, Putnam-Evans C, Charbonneau H, et al. (1991) A calcium-dependent protein kinase with a regulatory domain similar to calmodulin. Science 252: 951–954.
12. Hrabak EM, Chan CW, Gribskov M, Harper JF, Choi JH, et al. (2003) The Arabidopsis CDPK-SnRK superfamily of protein kinases. Plant Physiol 132: 666–680.
13. Klimecka M, Muszynska G (2007) Structure and functions of plant calcium-dependent protein kinases. Acta Biochim Pol 54: 219–233.
14. Martin ML, Busconi L (2000) Membrane localization of a rice calcium-dependent protein kinase (CDPK) is mediated by myristoylation and palmitoylation. Plant J 24: 429–435.
15. Saijo Y, Hata S, Kyozuka J, Shimamoto K, Izui K (2000) Over-expression of a single Ca^{2+}-dependent protein kinase confers both cold and salt/drought tolerance on rice plants. Plant J 23: 319–327.
16. Saijo Y, Kinoshita N, Ishiyama K, Hata S, Kyozuka J, et al. (2001) A Ca^{2+}-dependent protein kinase that endows rice plants with cold- and salt-stress tolerance functions in vascular bundles. Plant Cell Physiol 42: 1228–1233.
17. Asano T, Hayashi N, Kobayashi M, Aoki N, Miyao A, et al. (2012) A rice calcium-dependent protein kinase OsCPK12 oppositely modulates salt-stress tolerance and blast disease resistance. Plant J 69: 26–36.
18. Berberich T, Kusano T (1997) Cycloheximide induces a subset of low temperature-inducible genes in maize. Mol Gen Genet 254: 275–283.
19. Szczegielniak J, Borkiewicz L, Szurmak B, Lewandowska-Gnatowska E, Statkiewicz M, et al. (2012) Maize calcium-dependent protein kinase (ZmCPK11): local and systemic response to wounding, regulation by touch and components of jasmonate signaling. Physiol Plant 146: 1–14.
20. Zhang M, Liang S, Lu YT (2005) Cloning and functional characterization of NtCPK4, a new tobacco calcium-dependent protein kinase. Biochim Biophys Acta 1729: 174–185.
21. Zou JJ, Wei FJ, Wang C, Wu JJ, Ratnasekera D, et al. (2010) Arabidopsis calcium-dependent protein kinase CPK10 functions in abscisic acid- and Ca^{2+}-mediated stomatal regulation in response to drought stress. Plant Physiol 154: 1232–1243.
22. Mori IC, Murata Y, Yang Y, Munemasa S, Wang YF, et al. (2006) CDPKs CPK6 and CPK3 function in ABA regulation of guard cell S-type anion- and Ca^{2+}-permeable channels and stomatal closure. PLoS Biol 4: e327.
23. Ma SY, Wu WH (2007) AtCPK23 functions in *Arabidopsis* responses to drought and salt stresses. Plant Mol Biol 65: 511–518.
24. Estruch JJ, Kadwell S, Merlin E, Crossland L (1994) Cloning and characterization of a maize pollen-specific calcium-dependent calmodulin-independent protein kinase. Proc Natl Acad Sci U S A 91: 8837–8841.
25. Ivashuta S, Liu J, Liu J, Lohar DP, Haridas S, et al. (2005) RNA interference identifies a calcium-dependent protein kinase involved in *Medicago truncatula* root development. Plant Cell 17: 2911–2921.
26. Yoon GM, Cho HS, Ha HJ, Liu JR, Lee HS (1999) Characterization of NtCDPK1, a calcium-dependent protein kinase gene in *Nicotiana tabacum*, and the activity of its encoded protein. Plant Mol Biol 39: 991–1001.
27. Lee SS, Cho HS, Yoon GM, Ahn JW, Kim HH, et al. (2003) Interaction of NtCDPK1 calcium-dependent protein kinase with NtRpn3 regulatory subunit of the 26S proteasome in *Nicotiana tabacum*. Plant J 33: 825–840.
28. Ray S, Agarwal P, Arora R, Kapoor S, Tyagi AK (2007) Expression analysis of calcium-dependent protein kinase gene family during reproductive development and abiotic stress conditions in rice (*Oryza sativa* L. ssp. *indica*). Mol Genet Genomics 278: 493–505.
29. Li AL, Zhu YF, Tan XM, Wang X, Wei B, et al. (2008) Evolutionary and functional study of the CDPK gene family in wheat (*Triticum aestivum* L.). Plant Mol Biol 66: 429–443.
30. Zuo R, Hu R, Chai G, Xu M, Qi G, et al. (2013) Genome-wide identification, classification, and expression analysis of CDPK and its closely related gene families in poplar (*Populus trichocarpa*). Mol Biol Rep 40: 2645–2662.
31. Ma P, Liu J, Yang X, Ma R (2013) Genome-wide identification of the maize calcium-dependent protein kinase gene family. Appl Biochem Biotechnol 169: 2111–2125.
32. Chang WJ, Su HS, Li WJ, Zhang ZL (2009) Expression profiling of a novel calcium-dependent protein kinase gene, *LeCPK2*, from tomato (*Solanum lycopersicum*) under heat and pathogen-related hormones. Biosci Biotechnol Biochem 73: 2427–2431.
33. Gargantini PR, Giammaria V, Grandellis C, Feingold SE, Maldonado S, et al. (2009) Genomic and functional characterization of StCDPK1. Plant Mol Biol 70: 153–172.
34. Giammaria V, Grandellis C, Bachmann S, Gargantini PR, Feingold SE, et al. (2011) *StCDPK2* expression and activity reveal a highly responsive potato calcium-dependent protein kinase involved in light signalling. Planta 233: 593–609.
35. Witte CP, Keinath N, Dubiella U, Demouliere R, Seal A, et al. (2010) Tobacco calcium-dependent protein kinases are differentially phosphorylated *in vivo* as part of a kinase cascade that regulates stress response. J Biol Chem 285: 9740–9748.
36. Yang DH, Hettenhausen C, Baldwin IT, Wu J (2012) Silencing *Nicotiana attenuata* calcium-dependent protein kinases, *CDPK4* and *CDPK5*, strongly up-regulates wound- and herbivory-induced jasmonic acid accumulations. Plant Physiol 159: 1591–1607.
37. Dubrovina AS, Kiselev KV, Khristenko VS (2013) Expression of calcium-dependent protein kinase (*CDPK*) genes under abiotic stress conditions in wild-growing grapevine *Vitis amurensis*. J Plant Physiol 170: 1491–1500.
38. Huang QS, Wang HY, Gao P, Wang GY, Xia GX (2008) Cloning and characterization of a calcium dependent protein kinase gene associated with cotton fiber development. Plant Cell Rep 27: 1869–1875.
39. Wang H, Mei W, Qin Y, Zhu Y (2011) 1-Aminocyclopropane-1-carboxylic acid synthase 2 is phosphorylated by calcium-dependent protein kinase 1 during cotton fiber elongation. Acta Biochim Biophys Sin (Shanghai) 43: 654–661.

40. Chen ZJ, Scheffler BE, Dennis E, Triplett BA, Zhang T, et al. (2007) Toward sequencing cotton (*Gossypium*) genomes. Plant Physiol 145: 1303–1310.
41. Paterson AH, Wendel JF, Gundlach H, Guo H, Jenkins J, et al. (2012) Repeated polyploidization of *Gossypium* genomes and the evolution of spinnable cotton fibres. Nature 492: 423–427.
42. Quevillon E, Silventoinen V, Pillai S, Harte N, Mulder N, et al. (2005) InterProScan: protein domains identifier. Nucleic Acids Res 33: W116–W120.
43. Punta M, Coggill PC, Eberhardt RY, Mistry J, Tate J, et al. (2012) The Pfam protein families database. Nucleic Acids Res 40: D290–D301.
44. Letunic I, Doerks T, Bork P (2012) SMART 7: recent updates to the protein domain annotation resource. Nucleic Acids Res 40: D302–D305.
45. de Castro E, Sigrist CJ, Gattiker A, Bulliard V, Langendijk-Genevaux PS, et al. (2006) ScanProsite: detection of PROSITE signature matches and ProRule-associated functional and structural residues in proteins. Nucleic Acids Res 34: W362–W365.
46. Bologna G, Yvon C, Duvaud S, Veuthey AL (2004) N-Terminal myristoylation predictions by ensembles of neural networks. Proteomics 4: 1626–1632.
47. Ren J, Wen L, Gao X, Jin C, Xue Y, et al. (2008) CSS-Palm 2.0: an updated software for palmitoylation sites prediction. Protein Eng Des Sel 21: 639–644.
48. Bjellqvist B, Basse B, Olsen E, Celis JE (1994) Reference points for comparisons of two-dimensional maps of proteins from different human cell types defined in a pH scale where isoelectric points correlate with polypeptide compositions. Electrophoresis 15: 529–539.
49. Larkin MA, Blackshields G, Brown NP, Chenna R, McGettigan PA, et al. (2007) Clustal W and Clustal X version 2.0. Bioinformatics 23: 2947–2948.
50. Tamura K, Peterson D, Peterson N, Stecher G, Nei M, et al. (2011) MEGA5: molecular evolutionary genetics analysis using maximum likelihood, evolutionary distance, and maximum parsimony methods. Mol Biol Evol 28: 2731–2739.
51. Guo AY, Zhu QH, Chen X, Luo JC (2007) [GSDS: a gene structure display server]. Yi Chuan 29: 1023–1026.
52. Zhao Y, Zhou Y, Jiang H, Li X, Gan D, et al. (2011) Systematic analysis of sequences and expression patterns of drought-responsive members of the HD-Zip gene family in maize. PLoS One 6: e28488.
53. Wei H, Li W, Sun X, Zhu S, Zhu J (2013) Systematic analysis and comparison of nucleotide-binding site disease resistance genes in a diploid cotton *Gossypium raimondii*. PLoS One 8: e68435.
54. Kong X, Lv W, Jiang S, Zhang D, Cai G, et al. (2013) Genome-wide identification and expression analysis of calcium-dependent protein kinase in maize. BMC Genomics 14: 433.
55. Librado P, Rozas J (2009) DnaSP v5: a software for comprehensive analysis of DNA polymorphism data. Bioinformatics 25: 1451–1452.
56. Higo K, Ugawa Y, Iwamoto M, Korenaga T (1999) Plant cis-acting regulatory DNA elements (PLACE) database: 1999. Nucleic Acids Res 27: 297–300.
57. Livak KJ, Schmittgen TD (2001) Analysis of relative gene expression data using real-time quantitative PCR and the $2^{-\Delta\Delta CT}$ Method. Methods 25: 402–408.
58. de Hoon MJ, Imoto S, Nolan J, Miyano S (2004) Open source clustering software. Bioinformatics 20: 1453–1454.
59. Hrabak EM, Dickmann LJ, Satterlee JS, Sussman MR (1996) Characterization of eight new members of the calmodulin-like domain protein kinase gene family from *Arabidopsis thaliana*. Plant Mol Biol 31: 405–412.
60. Johnson DR, Bhatnagar RS, Knoll LJ, Gordon JI (1994) Genetic and biochemical studies of protein N-myristoylation. Annu Rev Biochem 63: 869–914.
61. Mehlmer N, Wurzinger B, Stael S, Hofmann-Rodrigues D, Csaszar E, et al. (2010) The Ca^{2+}-dependent protein kinase CPK3 is required for MAPK-independent salt-stress acclimation in Arabidopsis. Plant J.
62. Lu SX, Hrabak EM (2013) The myristoylated amino-terminus of an Arabidopsis calcium-dependent protein kinase mediates plasma membrane localization. Plant Mol Biol 82: 267–278.
63. Wang C, Shao J (2013) Characterization of the *ZmCK1* gene encoding a calcium-dependent protein kinase responsive to multiple abiotic stresses in maize. Plant Molecular Biology Reporter 31: 222–230.
64. Asano T, Tanaka N, Yang G, Hayashi N, Komatsu S (2005) Genome-wide identification of the rice calcium-dependent protein kinase and its closely related kinase gene families: comprehensive analysis of the *CDPKs* gene family in rice. Plant Cell Physiol 46: 356–366.
65. Hong Y, Takano M, Liu C, Gasch A, Chye M, et al. (1996) Expression of three members of the calcium-dependent protein kinase gene family in *Arabidopsis thaliana*. Plant molecular biology 30: 1259–1275.
66. Zhao Y, Pokutta S, Maurer P, Lindt M, Franklin RM, et al. (1994) Calcium-binding properties of a calcium-dependent protein kinase from *Plasmodium falciparum* and the significance of individual calcium-binding sites for kinase activation. Biochemistry 33: 3714–3721.
67. Christodoulou J, Malmendal A, Harper JF, Chazin WJ (2004) Evidence for differing roles for each lobe of the calmodulin-like domain in a calcium-dependent protein kinase. J Biol Chem 279: 29092–29100.
68. Franz S, Ehlert B, Liese A, Kurth J, Cazale AC, et al. (2011) Calcium-dependent protein kinase CPK21 functions in abiotic stress response in *Arabidopsis thaliana*. Mol Plant 4: 83–96.
69. Chen F, Fasoli M, Tornielli GB, Dal Santo S, Pezzotti M, et al. (2013) The evolutionary history and diverse physiological roles of the grapevine calcium-dependent protein kinase gene family. PLoS One 8: e80818.
70. Cannon SB, Mitra A, Baumgarten A, Young ND, May G (2004) The roles of segmental and tandem gene duplication in the evolution of large gene families in *Arabidopsis thaliana*. BMC Plant Biol 4: 10.
71. Maere S, De Bodt S, Raes J, Casneuf T, Van Montagu M, et al. (2005) Modeling gene and genome duplications in eukaryotes. Proc Natl Acad Sci U S A 102: 5454–5459.
72. Lynch M, Conery JS (2000) The evolutionary fate and consequences of duplicate genes. Science 290: 1151–1155.
73. Kargiotidou A, Deli D, Galanopoulou D, Tsaftaris A, Farmaki T (2008) Low temperature and light regulate delta 12 fatty acid desaturases (FAD2) at a transcriptional level in cotton (*Gossypium hirsutum*). J Exp Bot 59: 2043–2056.
74. Wan B, Lin Y, Mou T (2007) Expression of rice Ca^{2+}-dependent protein kinases (CDPKs) genes under different environmental stresses. FEBS Lett 581: 1179–1189.

3

Insights into Alpha-Hemolysin (Hla) Evolution and Expression among *Staphylococcus aureus* Clones with Hospital and Community Origin

Ana Tavares[1], Jesper B. Nielsen[2], Kit Boye[2], Susanne Rohde[2], Ana C. Paulo[3], Henrik Westh[2,4], Kristian Schønning[2,4], Hermínia de Lencastre[1,5], Maria Miragaia[1,6]*

1 Laboratory of Molecular Genetics, Instituto de Tecnologia Química e Biológica (ITQB), Oeiras, Portugal, **2** Dept. of Clinical Microbiology 445, Copenhagen University Hospital, Hvidovre, Denmark, **3** Molecular Microbiology of Human Pathogens, ITQB, Oeiras, Portugal, **4** Faculty of Health and Medical Sciences, University of Copenhagen, Copenhagen, Denmark, **5** Laboratory of Microbiology and Infectious Diseases, The Rockefeller University, New York, New York, United States of America, **6** Laboratory of Bacterial Evolution and Molecular Epidemiology, ITQB, Oeiras, Portugal

Abstract

Background: Alpha-hemolysin (Hla) is a major virulence factor in the pathogenesis of *Staphylococcus aureus* infection, being active against a wide range of host cells. Although *hla* is ubiquitous in *S. aureus*, its genetic diversity and variation in expression in different genetic backgrounds is not known. We evaluated nucleotide sequence variation and gene expression profiles of *hla* among representatives of hospital (HA) and community-associated (CA) *S. aureus* clones.

Methods: 51 methicillin-resistant *S. aureus* and 22 methicillin-susceptible *S. aureus* were characterized by PFGE, *spa* typing, MLST and SCC*mec* typing. The internal regions of *hla* and the *hla* promoter were sequenced and gene expression was assessed by RT-PCR.

Results: Alpha-hemolysin encoding- and promoter sequences were diverse, with 12 and 23 different alleles, respectively. Based on phylogenetic analysis, we suggest that *hla* may have evolved together with the *S. aureus* genetic background, except for ST22, ST121, ST59 and ST93. Conversely, the promoter region showed lack of co-evolution with the genetic backgrounds. Four non-synonymous amino acid changes were identified close to important regions of *hla* activity. Amino acid changes in the RNAIII binding site were not associated to *hla* expression. Although expression rates of *hla* were in general strain-specific, we observed CA clones showed significantly higher *hla* expression ($p = 0.003$) when compared with HA clones.

Conclusion: We propose that the *hla* gene has evolved together with the genetic background. Overall, CA genetic backgrounds showed higher levels of *hla* expression than HA, and a high strain-to-strain variation of gene expression was detected in closely related strains.

Editor: J. Ross Fitzgerald, University of Edinburgh, United Kingdom

Funding: This work was funded by project Ref. P-99911 from Fundação Calouste Gulbenkian (http://www.gulbenkian.pt/Institucional/pt/Homepage) and Project PTDC/BIA-MIC/3195/2012 from Fundação para a Ciência e Tecnologia (http://www.fct.pt/) awarded to HdL; Project PTDC/BIA-EVF/117507/2010 from Fundação para a Ciência e Tecnologia (http://www.fct.pt/) awarded to MM; and through grant Ref. Pest-OE/EQB/LA0004/2011 from Fundação para a Ciência e Tecnologia (FCT), Portugal. A. Tavares was supported by grant SFRH/BD/44220/2008 from FCT. The funders had no role in study design, data collection and analysis, decision to publish, or preparation of the manuscript.

Competing Interests: Co-author Herminia de Lencastre is a PLOS ONE Editorial Board member. This does not alter the authors' adherence to PLOS ONE Editorial policies and criteria.

* Email: miragaia@itqb.unl.pt

Introduction

Staphylococcus aureus is a human opportunistic pathogen responsible for a wide range of infections that can vary in its clinical presentation and severity. Methicillin-resistant *S. aureus* (MRSA) emerged in 1960 in the United Kingdom and has been a major problem in hospitals (HA-MRSA) worldwide during the last 40 years; however since the late 1990s, MRSA has been emerging as a leading cause of severe infection also in the community, in individuals without recent health-care contact (CA-MRSA) [1,2].

CA-MRSA present distinct genetic backgrounds from their hospital counterparts, are more susceptible to antibiotics other than beta-lactams, carry the smallest staphylococcal cassette chromosome *mec* types (SCC*mec* IV or V), and have higher virulence capacity [1,2,3]. The underlying reasons behind the enhanced virulence of CA-MRSA appear to be multiple including a different capacity to overcome host cell response [4], different distribution of mobile genetic elements carrying virulence determinants [5], allelic variation in virulence determinants located in the core genome and in mobile genetic elements [6], and different levels of expression and protein production of virulence determinants (alpha-hemolysin, collagen adhesin, staphylokinase, coagulase, lipase, enterotoxins C3 and Q, V8 protease and cysteine protease) [7,8,9].

The alpha-hemolysin or α-toxin (Hla), is one of the major virulence determinants implicated in the pathogenesis of *S. aureus*, associated to severe skin and soft tissue infections (SSTI), necrotizing pneumonia and even sepsis [10]. Hla is the most prominent *S. aureus* cytotoxin that can act against a wide range of host cells including erythrocytes, epithelial cells, endothelial cells, T cells, monocytes and macrophages [10,11,12]. The gene encoding Hla is located in the core genome and is expressed as a water-soluble monomer (33.2 kD) that assembles to form a membrane-bound heptameric β-barrel pore (232.4 kD) on susceptible cells leading to cell death and lysis [11]. The overall structure is mushroom-like, divided into three domains: 1) Cap domain: largely hydrophobic, defining the entry of the pore; 2) Rim domain: underside of the Cap, in close proximity to membrane bilayer; 3) Stem domain: part of the transmembrane channel, forming the membrane-perforating β-barrel pore (Figure 1) [10,11]. Hla expression is mainly controlled by the global toxin accessory gene regulator (*agr*), via the regulatory effector molecule RNAIII [13]. While *agr* provides the first and most important mechanism of up-regulation of *hla*, expression can also be modulated by other regulators, such as SaeR, SarZ, ArlS [14,15,16] (up-regulators) and Rot, SarT [17](down-regulators).

Although polymorphisms in the *hla* promoter region have been described [18], the range of genetic diversity and evolution of this toxin has never been assessed in a large representative *S. aureus* collection. Furthermore, although differences in *hla* expression have been described between community- and hospital-associated MRSA, these studies have been performed with a limited number of CA-MRSA epidemic clones [9], or almost exclusively with representatives of the USA300 clone [19,20,21]. To better understand the evolutionary history of *hla* and its importance as a virulence factor for CA-MRSA, in this study we compared the *hla* nucleotide sequence and expression among the major epidemic and minor CA and HA clones, including both MRSA and MSSA strains.

Materials and Methods

Ethics Statement

Isolates were obtained from routine diagnostic and were analyzed anonymously and only the isolates, not humans, were studied. All data was collected according to the European Parliament and Council decision for the epidemiological surveillance and control of communicable disease in the European community. Ethical approval and informed consent were for that reason not required.

Bacterial collection

A total of 73 *S. aureus*, including 51 MRSA and 22 MSSA were analyzed in this study. Strains were collected in 13 different countries (Belgium, Bulgaria, Czech Republic, Denmark, Greece, Netherlands, Portugal, Romania, Spain, Sweden, United Kingdom, USA and Brazil), between 1961 and 2009 from both community (n = 46) and hospital (n = 27). The strains comprised a total of 52 *spa* types and 23 sequence types (STs) (see Table S1).

Strains were defined as belonging to CA or HA clones if they contained the same or related genetic backgrounds as the reference CA-MRSA and HA-MRSA epidemic control strains, based on ST, *spa* type and SCC*mec* (in case of MRSA).

Media and bacterial growth conditions

Before RT-PCR analysis, strains were grown overnight at 37°C on tryptic soy agar plates (TSA). Bacterial growth experiments were performed by growing bacteria in tryptic Soy Broth (TSB) at 37°C with shake and measuring OD (600 nm) each hour in the follow up automatic incubator Bioscreen C (Oy Growth Curves AB, Helsinki, Finland). Plates of 100-well honeycomb (Oy Growth Curves AB, Helsinki, Finland) were filled with 300 µl/well of

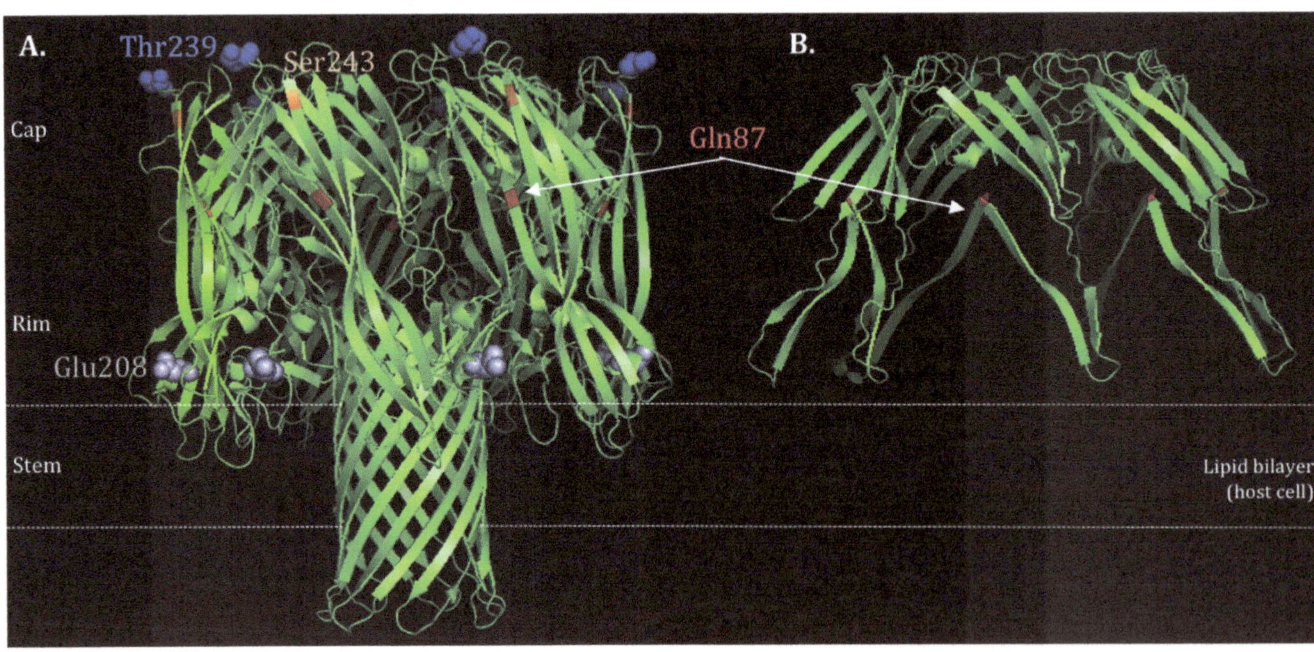

Figure 1. HLA protein structure. A) wildtype (highlighted the non-synonymous mutations Gln87, Glu208, Thr239 and Ser243) and B) truncated protein due to a stop codon at Gln87. Structure generated by the program PyMOL v.1.6.

overnight culture diluted to $OD_{600} = 0.05$ in TSB growth medium. Three individual growth experiments (SetC, SetD and SetE) were performed for each strain and named accordingly e.g. HLZ6C, HLZ6D and HLZ6E (see Figure S2.I to III).

Nucleotide sequence of *hla* and promoter region

Chromosomal DNA was extracted from overnight cultures, using the boiling method (100°C for 10 min followed by centrifuged at 13.000 g for 5 minutes). Two sets of primers were designed to span the most polymorphic regions within the *hla* gene and *hla* promoter (considered as the region located −600 bp from *hla* starting codon), after alignment of sequences available on NCBI for *S. aureus*. One set of primers (Forward: hla-F_CGAA-AGGTACCATTGCTGGT; Reverse: hla-R_CCAATCGATTT-TATATCTTTC) amplified an internal fragment of the *hla* gene (nt 1170419–1170982, CP000730.1) and the other set (Forward: hlaPro-F_CACTATATTAAAAATACATAC; Reverse: hlaPro-R_GTTGTTACTGAGCTGAC) amplified an internal fragment of the *hla* promoter region (nt 1171289–1171773, CP000730.1) (Figure S3). PCR products were sequenced (Macrogen Europe, Amsterdam, The Netherlands) and sequences were analyzed using SeqMan (DNAstar, Lasergene v9, Madison, WI, USA). To each unique *hla* promoter (P) and gene sequence (*hla*) - allotype - a single Arabic number was attributed (e.g. P1, P2; *hla*1, *hla*2). Gene and promoter sequences were deposited in GenBank (accession numbers KM019547–KM019606; KM019607–KM019674).

Phylogenetic analysis

Phylogenetic relatedness was analyzed using the MEGA5 v5.05 software (http://www.megasoftware.net/) for gene, promoter region and concatenated sequences obtained from 1) gene with promoter region and 2) seven MLST alleles from the 23 representative STs within the collection. Phylogenetic trees were constructed using the Neighbor-Joining clustering method, and 1000 bootstrap replicates, which assigns confidence values for the groupings in the tree.

Moreover, nucleotide diversity (ND) between the two clusters was calculated based on the estimation of the average evolutionary divergence over sequence pairs within the two groups, where the number of base substitutions per site from averaging over all sequence pairs within each group are compared using the maximum composite likelihood model [22].

Detection of recombination

Alignments from the *hla* gene, *hla* promoter and internal fragments of each of the seven MLST gene were screened for the occurrence of putative recombination events using Recombination Detection Program version 4 (RDP4) (http://web.cbio.uct.ac.za/) with the default settings (with highest acceptable probability value of 0.05). Identification of recombinant sequences recombination breakpoints and major parent was determined using simultaneously nine recombination detection methods (RDP, BOOT-SCAN, GENECONV, MAXCHI, CHIMAERA, SISCAN, PhylPro, LARD and 3SEQ. The "minor parent" is considered a sequence closely related to that from which sequences in the proposed recombinant region may have been derived (the presumed donor). The "major parent" was considered as a sequence closely related to that from which the greater part of the recombinant's sequence may have been derived.

RT-PCR analysis

Culture growth was stopped at late exponential phase, when alpha-toxin is described to have maximal activity [23], corresponding to the time-points 1) 3 hours 30 min in one group (65 strains) and 2) 4 hours 30 min in another (8 strains). Total RNA was extracted from three biological replicates. Cells were mechanically disrupted with FastPrep-24 Instrument (MP Biomedicals, Solon, OH, USA) and RNA was protected using RNA Protect (Qiagen, Valencia, USA). RNA was extracted automatically using the QIAsymphony platforms (Qiagen, Valencia, USA) with QIAsymphony RNA kit (Qiagen, Valencia, USA).

The RT-PCR assay was performed on a 7500 Real-Time PCR System (Applied Biosystems, Foster City, CA) using the following primers and TaqMan probes: Hla RT_F: TAATGAATCCTG-TCGCTAATGCC; HlaRT_R: CACCTGTTTTTACTGTAG-TATTGCTTCC; Hla RT Probe: 6FAM-AAACCGGTACTA-CAGATAT-MGBNFQ. The RT-PCR reaction was performed using the EZ RT-PCR Core Reagents (Applied Biosystems, Foster City, USA), in which RNA is reverse transcribed and amplified in a single reaction. The following PCR protocol was used: 50°C for 2 min, 60°C for 30 min, 95°C for 5 min, followed by 42 cycles of 95°C for 20 sec and 62°C for 1 min. The 16S gene was used as internal or reference control. The primers used for 16S RNA amplification were those previously described [24].

RT-PCR data analysis

The relative *hla* gene expression was calculated based on the C_t (RT-PCR output) of the gene of interest (C_t hla) as compared to the C_t of the internal control (C_t 16S) as follows: Delta $C_t = C_t$ hla-C_t 16S. The lower the Delta C_t the higher is the amount of *hla* mRNA and the more the gene is expressed. The reproducibility of the assay was evaluated by the calculation of the arithmetic mean of the relative expression of the three biological replicates (Mean Delta C_{t1-3} = Average (Delta C_{t1}; Delta C_{t2}; Delta C_{t3}). The reproducibility of RT-PCR reaction was evaluated by the calculation of the standard deviation (STDEV) of Delta C_t obtained for each biological replica (Delta C_{t1}; Delta C_{t2}; Delta C_{t3}). Values were considered valid when at least two C_t readings exist with STDEV<2.

Protein structure visualization (pyMOL)

The protein structure was modeled using PyMOL v.1.6 (http://www.pymol.org/) if a nucleotide mutation gave rise to a stop codon.

Statistical analysis

The statistical analysis was performed using the Graphpad Prism 6 (http://www.graphpad.com/scientific-software/prism/), with the two-tailed Student's t-test to determine whether the differences of mean expression rates (MSSA *versus* MSSA; HA backgrounds *versus* CA backgrounds) were statistically significant ($p \leq 0.05$).

Regression tree analysis was used to explore which variables could be related with the *hla* expression [25]. Trees explain the variation of a single response variable (in this study the *hla* mRNA expression) by repeatedly splitting the data into more homogeneous groups, using combinations of explanatory variables (in our case, the ST, *spa* type, MRSA, MSSA and the type of SCC*mec*).

Results

Analysis of polymorphisms in the *hla* gene and *hla* promoter

The sequence analysis of the internal region of *hla* and the *hla* promoter region among the 73 strains identified a total of 12 *hla* and 23 promoter region different sequences (allotypes) (Table 1). We obtained no amplification products for *hla* and *hla* promoter

Table 1. Summary of molecular characterization, sequence variation and relative expression rates of *S. aureus* strains collection.

Nº	Isolate ID	SCC*mec*	*spa* type	MLST	Branch[1]	Promotor Allotype	Gene Allotype	Nonsynonymous Mutation	Hla Expression (Mean Delta C_t)[2]	Stddev Delta C_t[3]	Expression (High/Low)
1	HLZ6	II	t002	**ST5**	L	P4	hla1	D208E	8.69	2*	Low
2	BK2464	II	t002	**ST5**	L	nt	hla1	D208E	5.37	1	**High**
3	HBR73	II	t067	**ST5**	L	P5	hla1	D208E	8.75	1	Low
4	C013	VI	t002	**ST5**	L	P3	hla1	D208E	6.84	1	Low
5	HDES26	VI	t062	**ST5**	L	P3	hla1	D208E	8.01	1	Low
6	HDE288	VI	t311	**ST5**	L	P3	hla1	D208E	6.67	1	Low
7	HSA29	–	t002	**ST5**	L	P3	hla1	D208E	Not Valid	–	Not Valid
8	HDE461	IV	t022	**ST22**	H	P10	hla12	S239T; T243S	6.60	1	Low
9	HAR22	IV	t022	**ST22**	H	P11	hla13	S239T; T243S	6.43	1	Low
10	HSMB280	IV	t032	**ST22**	H	P10	hla12	S239T; T243S	4.71	1	**High**
11	LBM12	IV	t747	**ST1806**	H	nt	hla12	S239T; T243S	9.28	1	Low
12	HSMB184	–	t5951	**ST1806**	H	P10	hla12	S239T; T243S	6.74	1	Low
13	HPH2	II	t018	**ST36**	H	P7	hla8	D208E; S239T; stop codon	8.02	2*	Low
14	HAR24	II	t018	**ST36**	H	nt	hla8	D208E; S239T; stop codon	9.62	2*	Low
15	DEN4415	II	t021	**ST36**	H	P7	hla8	D208E; S239T; stop codon	8.95	2*	Low
16	C563	IV	t015	**ST45**	H	nt	hla10	S239T	7.02	1	Low
17	C036	V	t015	**ST45**	H	nt	hla10	S239T	6.24	0	Low
18	HAR38	IV	t004	**ST45**	H	P7	hla10	S239T	10.38	1	Low
19	HFX77	III	t037	**ST239**	L	P1	hla4	–	8.74	2*	Low
20	HUC343	IIIA	t037	**ST239**	L	P1	hla4	–	8.27	0	Low
21	HU25	IIIA	t138	**ST239**	L	P1	hla4	–	8.17	1	Low
22	BK1953	IA	t051	**ST247**	L	P1	hla4	–	7.71	1	Low
23	HPV107	IA	t051	**ST247**	L	P1	hla4	–	7.56	0	Low
24	HSJ419	IA	t725	**ST247**	L	P1	hla4	–	8.23	1	Low
27	E2125	I	t051	**ST247**	L	P1	hla4	–	7.29	0	Low
25	10395	I	t008	**ST250**	L	P2	hla4	–	8.15	1	Low
26	COL	I	t008	**ST250**	L	P1	hla4	–	8.01	1	Low
28	HFX74	IV	t008	**ST8**	L	P1	hla4	–	6.46	1	Low
29	USA300	IV	t008	**ST8**	L	P1	hla4	–	6.19	3*	Low
30	C438	IV	t024	**ST8**	L	P1	hla4	–	6.07	1	Low
31	C574B	IV	t1257	**ST612**	L	P1	hla4	–	Not Valid	–	Not Valid
32	LBM27	–	t024	**ST8**	L	P1	hla4	–	8.12	0	Low
33	LBM74	–	t008	**ST8**	L	P1	hla4	–	5.87	1	Low
34	C270	IV	t1381	**ST1**	L	P17	hla2	–	8.81	1	Low
35	USA400	IV	t127	**ST1**	L	P17	hla2	–	6.01	2*	Low

Table 1. Cont.

Nº	Isolate ID	SCC*mec*	*spa* type	MLST	Branch[1]	Promotor Allotype	Gene Allotype	Nonsynonymous Mutation	Hla Expression (Mean Delta C_t)[2]	Stddev Delta C_t[3]	Expression (High/Low)
36	LBM36	–	t127	**ST1**	L	P18	hla2	–	11.09	1	Low
37	C577	IV	t216	**ST59**	L	P20	hla5	–	5.35	0	**High**
38	C583	IV	t437	**ST59**	L	P19	hla5	–	5.31	1	**High**
39	C434	V	t437	**ST59**	L	P19	hla5	–	9.14	1	Low
40	C018	IV	t1819	**ST93**	L	nt	hla7	–	5.16	1	**High**
41	C491	IV	t202	**ST93**	L	P21	hla7	–	5.45	0	**High**
42	LBM54	IV	t011	**ST398**	H	P12	hla11	–	4.46	2*	**High**
43	C482	IV	t011	**ST398**	H	P13	hla11	–	3.25	1	**High**
44	C496	VII	t108	**ST398**	H	nt	hla11	–	2.85	1	**High**
45	LBM40	–	t034	**ST398**	H	P12	hla11	–	5.37	1	**High**
46	C017	IV	t019	**ST30**	H	nt	hla9	D208E; S239T	4.53	0	**High**
47	C385	IV	t019	**ST30**	H	P7	hla9	D208E; S239T	7.25	1	Low
48	C479	IV	t019	**ST30**	H	nt	hla9	D208E; S239T	8.10	1	Low
71	HUC585	–	t342	**ST30**	H	P7	hla9	D208E; S239T	5.14	1	**High**
69	HFF204	–	t318	**ST30**	H	P9	hla9	D208E; S239T	6.23	1	Low
70	HFA30	–	t012	**ST30**	H	P8	hla8	D208E; S239T; stop codon	7.94	1	Low
49	HSJO7	IV	t148	**ST72**	L	P14	hla1	D208E	6.56	1	Low
50	USA700	IV	t148	**ST72**	L	P14	hla1	D208E	5.76	0	Low
51	COO3	IV	t791	**ST72**	L	P15	hla1	D208E	6.28	1	Low
52	SAMS1024	IV	t1346	**ST1810**	L	P14	hla1	D208E	4.78	1	**High**
53	HUC594	–	t148	**ST72**	L	P14	hla1	D208E	8.36	1	Low
54	HFA28	–	t126	**ST72**	L	P14	hla1	D208E	4.56	2*	**High**
55	C238	–	t3682	**ST72**	L	P14	hla1	D208E	4.64	1	**High**
56	C168	IV	t044	**ST80**	L	P16	hla1	D208E	8.20	0	Low
57	C485	IV	t044	**ST80**	L	P16	hla1	D208E	5.72	1	**High**
58	C014	IV	t131	**ST80**	L	P16	hla1	D208E	4.87	0	**High**
59	LBM25	–	t1509	**ST15**	L	P2	hla1	D208E	6.69	0	Low
60	C157	–	t084	**ST15**	L	P2	hla1	D208E	4.86	1	**High**
61	C230	–	t346	**ST15**	L	P2	hla1	D208E	9.03	2*	Low
62	HBA33	–	t258	**ST25**	L	P6	hla1	D208E	5.73	1	**High**
63	C095	–	t2909	**ST25**	L	P6	hla1	D208E	4.16	1	**High**
64	C141	–	t081	**ST25**	L	P6	hla1	D208E	4.50	2*	**High**
65	HBA34	IV	t308	**ST121**	L	nt	hla6	–	5.62	1	High
66	HUC574	–	t435	**ST121**	L	P1	hla6	–	5.19	1	**High**
67	HUC587	–	t159	**ST121**	L	P2	hla6	–	5.09	1	**High**

Table 1. Cont.

Nº	Isolate ID	SCC*mec*	*spa* type	MLST	Branch[1]	Promotor Allotype	Gene Allotype	Nonsynonymous Mutation	Hla Expression (Mean Delta C_t)[2]	Stddev Delta C_t[3]	Expression (High/Low)
68	HUC578	-	t284	**ST121**	L	P1	hla6	-	7.10	2*	Low
72	LBM23	-	t100	**ST9**	L	P22	hla1	D208E	5.48	2*	**High**
73	HFX84	-	t267	**ST97**	L	P23	hla3	-	9.03	1	Low

[1]H: High polymorphism; L: Low polymorphism;
[2]Mean Delta C_{t1-3} = Average (Delta C_{t1}; Delta C_{t2}; Delta C_{t3}), Delta $C_t = C_t$ hla$-C_t$ 16S; Not valid: only one C_t reading;
[3]*low reproducibility between three C_T values (Stddv≤2). nt: non typable; Stddv: standard deviation.

region in one and 13 strains, respectively, which probably result from misparing of the primers used.

From the 12 *hla* (*hla*1–12), we observed that only a single *hla*-allotype was found among representatives of a specific ST, except for ST22 (*hla*12; *hla*13) and ST30 (*hla*8; *hla*9) where two different alleles were identified. On the other hand, the most frequent alleles, *hla1* (33.3%, n = 24) and *hla4* (20.8%, n = 15), were identified in more than one ST.

Regarding the nucleotide changes identified in the *hla*, some correspond to non-synonymous mutations (E208, T239 and S243) and, in one particular case, to a stop codon (Table 1 and 2). The substitutions observed did not correspond to any difference in the charge or polarity of the amino acid (aa). However, changes in molecular weight were observed: i) changes from aa D208 to aa E208 (D208E) and from aa S239 to T239 (S239T) gave rise to a higher molecular weight aa; and ii) change from aa T243 to S243 (T243S) resulted in a lower molecular weight aa; of note all changes occurred in the Rim domain of the protein. In a particular case, the aa change gave rise to a stop codon located in the CAP domain, in strains of ST36. Protein structure modeling showed that a protein of about one third of its real size is produced, truncated at the Gln87 (Figure 1, A and B). The truncation is in the outside part of the domain, suggesting that this will affect the capacity of the Hla to form cell wall pores, and ultimately to induce hemolysis.

A high number of sequence variations were identified in the *hla* promoter region, (n = 23) (P1–23) (Table 1 and 2). Although we found that some STs were associated to a specific promoter allotype, and some promoters were identified in a single ST, we also identified cases where single STs were associated to different promoters (8 out of 23) and examples in which a single promoter allotype was associated to different STs (5 out of 23). This is the case of the most frequent promoter (P1) that was found in about one third of the strains analyzed (25.4%, n = 16), including several different STs.

A particular highly polymorphic region corresponding to nt −22 to −24 from the start codon, was found in the majority (16 out of 23) of the promoter allotypes (exceptions P1, P6, P13, P14, P15, P18 and P23). These polymorphisms are located in the vicinity of RNAIII binding site [26]; however, we could not find a direct correlation between a particular nucleotide sequence and a specific expression pattern (high or low expression). For example, the sequence **TTT**, observed in two strains belonging to ST398 that have a high level expression, was also observed in strains with low expression belonging to other genetic backgrounds (ST8, ST239, ST247, ST250, ST36, ST45 and ST22).

Alpha-hemolysin evolutionary history

In order to better understand the evolution of *hla* gene within the *S. aureus* population, we constructed phylogenetic trees from the *hla* and *hla* promoter sequences, separately or concatenated (Figure 2, A) and compared it with the tree constructed from the concatenated sequences of the seven housekeeping genes used in MLST, including all the STs represented in the strain collection described here (Figure 2, B).

The phylogenetic tree constructed for the *hla* gene showed two distinct major clusters with different evolutionary clocks that differed in their nucleotide diversity (ND, see Materials and Methods): cluster (L) with lower diversity (ND = 0.005), and cluster H with higher diversity (ND = 0.019). Cluster L included more than 70% of strains (71.2%, n = 52), and five sub-clusters; Cluster H contained about 29% of the strains (28.8%, n = 21), and comprised four minor sub-clusters including *hla8–hla12* alleles,

Table 2. Strains data distribution based on promoter allotypes.

		Promotor allotype	Gene allotype	Non Synonymous Mutation	Isolates Molecular Characterization	Expression Category
CA backgrounds	**ST398**	P13	hla11	–	ST398-IV, t011	**High expression**
		P12			ST398, t034	**High expression**
		NT			ST398-VII, t108	**High expression**
		P12			ST398-IV, t011	**High expression***
	ST25	P6	hla1	D208E	ST25, t258	**High expression**
					ST25, t081	**High expression***
					ST25, t2909	**High expression**
	ST9	P22	hla1	D208E	ST9, t100	**High expression***
	ST93	P21	hla7	–	ST93-IV, t202	**High expression**
		NT			ST93-IV, t1819	**High expression**
	ST121	P2	hla6	–	ST121, t159	**High expression**
		P1			ST121, t435	**High expression**
		NT			ST121-IV, t308	**High expression**
		P1			ST121, t284	Low expression*
	ST72	P14	hla1	D208E	ST72-IV, t148	**High expression**
		P14			ST72, t3682	**High expression**
		P14			ST1810-IV, t1346	**High expression**
		P14			ST72, t126	**High expression***
		P15			ST72-IV, t791	Low expression
		P14			ST72-IV, t148	Low expression
		P14			ST72, t148	Low expression
	ST80	P16	hla1	D208E	ST80-IcV, t131	**High expression**
					ST80-IV, t044	**High expression**
					ST80-IV, t044	Low expression
	ST30	P7	hla9	D208E; S239T	ST30, t342	**High expression**
		NT			ST30-IV, t019	**High expression**
		P7			ST30-IV, t019	Low expression
		P9			ST30, t318	Low expression
		NT			ST30-IV, t019	Low expression
		P8	hla8	D208E; S239T; stop codon	ST30, t012	Low expression
	ST15	P2	hla1	D208E	ST15, t084	**High expression**
					ST15, t346	Low expression*
					ST15, t1509	Low expression
	ST59	P20	hla5	–	ST59-IV, t216	**High expression**
		P19			ST59-IV, t437	Low expression
		P19			ST59-V, t437	Low expression
	ST1	P17	hla2	–	ST1-IV, t1381	Low expression
		P17			ST1-IV, t127	Low expression*
		P18			ST1, t127	Low expression
	ST8	P1	hla4	–	ST8-IV, t008	Low expression
					ST8-IV, t024	Low expression
					ST8-IV, t008	Low expression*
					ST8, t008	Low expression
					ST612-IV, t1257	Not valid**
					ST8, t024	Low expression
	ST97	P23	hla3	–	ST97, t267	Low expression
HA backgrounds	**ST22**	P10	hla13	S239T; T243S	ST22-IV, t032	**High expression**
		P10	hla12		ST22-IV, t022	Low expression
		P11			ST22-IV, t022	Low expression

Table 2. Cont.

	Promotor allotype	Gene allotype	Non Synonymous Mutation	Isolates Molecular Characterization	Expression Category
	P10			ST1806, t5951	Low expression
	NT			ST1806-IV, t747	Low expression
ST5	NT	hla1	D208E	ST5-II, t002	**High expression**
	P3			ST5-VI, t002	Low expression
	P3			ST5-VI, t062	Low expression
	P3			ST5-VI, t311	Low expression
	P4			ST5-II, t002	Low expression*
	P3			ST5, t002,	Not valid**
	P5			ST5-II, t067	Low expression
ST36	P7	hla8	D208E; S239T; stop codon	ST36-II, t018	Low expression*
	P7			ST36-II, t021	Low expression*
	NT			ST36-II, t01	Low expression*
ST45	NT	hla10	S239T	ST45-IV, t015	Low expression
	NT			ST45-V, t015	Low expression
	P7			ST45-IV, t004	Low expression
ST239	P1	hla4	-	ST239-IIIA, t037	Low expression
				ST239-III, t037	Low expression*
			-	ST239-IIIA, t138	Low expression
ST247	P1	hla4	-	ST247-I, t051	Low expression
				ST247-IA, 051	Low expression
				ST247-IA, t051	Low expression
				ST247-IA, t725	Low expression
ST250	P1	hla4	-	ST250-I, t008	Low expression
	P2			ST250-I, t008	Low expression

(*)(**) relative expression values not valid (SDV≤2 or only one C_T reading).

which were found in strains of ST30, ST36, ST45, ST398 and ST22.

As opposed to the phylogenetic tree constructed from *hla* gene, the one constructed from the promoter region did not show two distinct evolutionary branches (Figure S1). Moreover, dissimilar subgroup clustering was noticed in the tree constructed from the promoter gene sequence. For example, ST45, ST30 and ST36 backgrounds were clustered together in the promoter sequence-based tree whereas in the *hla* sequence-based tree ST45 was placed separately from ST30 and ST36 cluster (branch H). The same type of observations can be drawn for most of STs. Overall the promoter region showed to be more diverse than the *hla* gene sequence among the different backgrounds.

On the other hand, when we compared the phylogenetic tree constructed with the *hla* gene with that constructed from MLST concatenated genes, the same type of division into two distinct main clusters was observed (Figure 2). Moreover, the majority of STs were equally distributed between the two clusters in the two trees. The only exceptions were ST22, ST121, ST59 and ST93 that in the two trees have exchanged their positions from one cluster to the other (Figure 2, B-blue arrows).

Detection of recombination in *hla* gene, h*la* promoter and MLST genes

To understand if recombination could explain the incongruence found between the trees constructed from *hla* and MLST concatenated genes, we screened the *hla* gene, *hla* promoter and each MLST gene for recombination events using the RDP4 software.

The SiScan and 3Seq methods detected one recombination event in the *hla* gene. This event corresponded to a fragment ending in positions 385–410 of the *hla* alignment, however the beginning breakpoint was not possible to determine. In the collection analyzed this event was detected in five isolates belonging to ST22 or related STs (HSMB280, HDE461, HAR22 and LBM12 (TLV ST22) and HSMB184 (TLV ST22)) and four isolates of ST398 (LBM54, LBM40, C496, C482_ST398). The ST30 HFF204 strain was identified as the minor parent (97.8% identity with ST22 strains and −99.3% identity with ST398 strains) and ST121 strain HUC587 was identified as the major parent (with 100% identity to ST398 strains and 93.5–95.2% identity with ST22 strains) of the recombining fragment. A trace signal of recombination of this same event was also identified among ST45 isolates; however this signal was not statistically significant. Interestingly all the recombination events were detected in strains belonging to the high genetic diversity cluster in the tree constructed from *hla* gene. In the *hla* promoter region no recombination events were detected.

We have performed the same type of analysis using the internal sequences of each of the seven housekeeping used in MLST scheme, including the alleles present in all STs identified in this

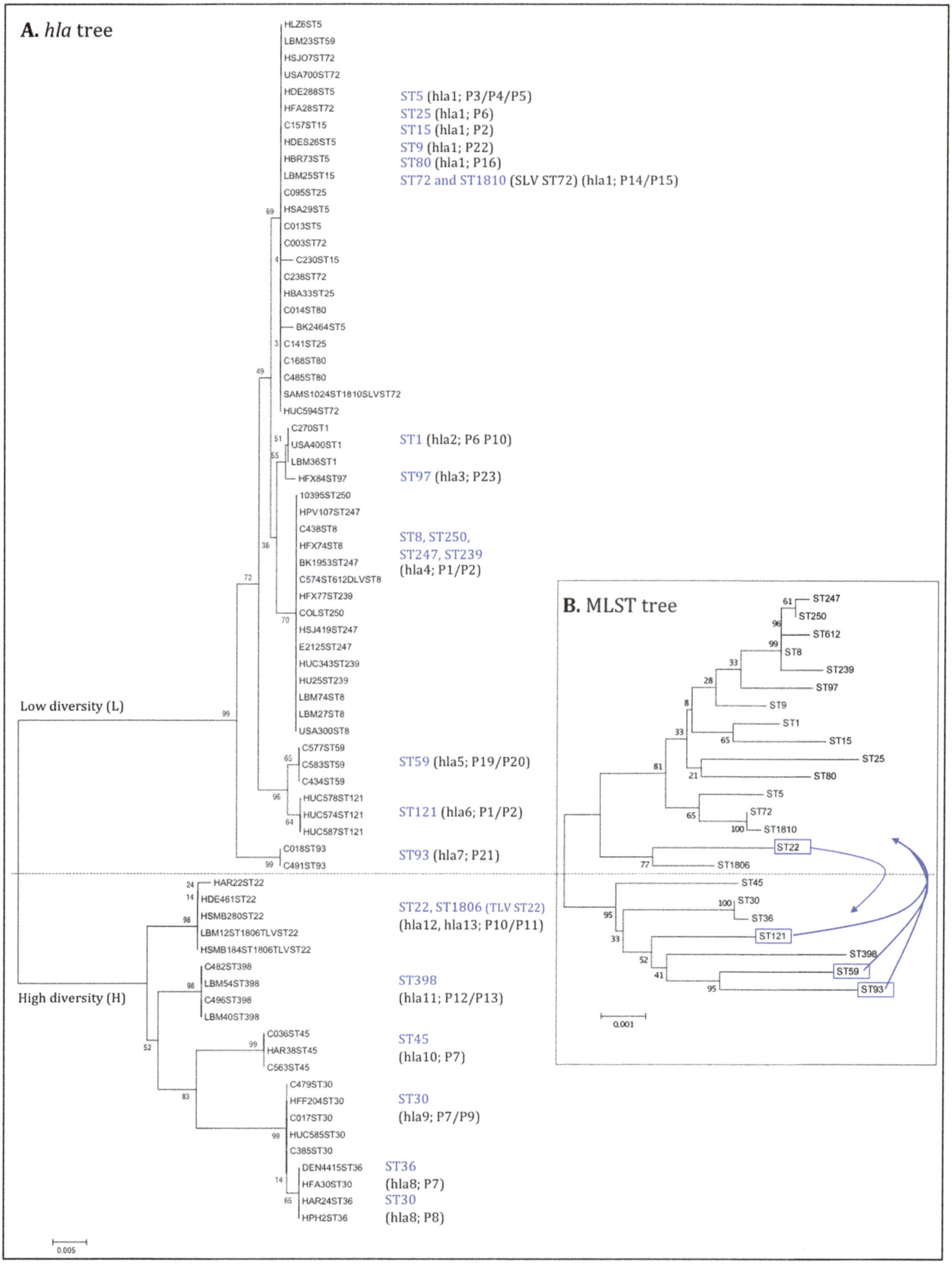
A. hla tree
ST5 (hla1; P3/P4/P5)
ST25 (hla1; P6)
ST15 (hla1; P2)
ST9 (hla1; P22)
ST80 (hla1; P16)
ST72 and ST1810 (SLV ST72) (hla1; P14/P15)
ST1 (hla2; P6 P10)
ST97 (hla3; P23)
ST8, ST250, ST247, ST239 (hla4; P1/P2)
ST59 (hla5; P19/P20)
ST121 (hla6; P1/P2)
ST93 (hla7; P21)
Low diversity (L)
High diversity (H)
ST22, ST1806 (TLV ST22) (hla12, hla13; P10/P11)
ST398 (hla11; P12/P13)
ST45 (hla10; P7)
ST30 (hla9; P7/P9)
ST36 (hla8; P7)
ST30 (hla8; P8)
0.005
B. MLST tree
0.001

Figure 2. Phylogenetic trees of *hla* gene (A) and concatenated sequences of MLST alleles (B) from 23 STs representatives of the strains collection. The tree was constructed using MEGA 5 with Neighbour-joining method and bootstrap values provided as percents over 1000 replications. Branch length values are indicated and the percentage of replicate trees (bootstrap test) are shown next to the branches. The dashed line indicates the separation of the two evolutionary branches.

study, however no recombination events were detected in any of the genes.

Altogether the data gathered suggest that for the majority of strains *hla* gene evolved together with the genetic background. The different clustering of ST22 and ST121 strains, in the trees constructed from MLST concatenated genes and *hla* gene, may derive from recombination events occurring in the *hla* gene. Similarly these type of events might explain the genetic diversity observed in cluster H in the *hla* tree in strains belonging to ST22, ST398, ST45, ST30 and ST36 (H cluster of *hla* tree).

Expression of alpha-hemolysin

The expression of alpha-hemolysin in the 73 strains was assessed by RT-PCR, in three biological replicates. Fifteen of the 73 strains (20.5%) were excluded from the final analysis, either because a single valid determination for Delta C_t (N = 2) was obtained or because C_T obtained from the different biological replicates were not reproducible (N = 13).

The analysis of the regression tree split the response variable into two distinct groups, according to the *spa* type of the strains. There was a group of strains with mean Delta $C_{t1-3}\leq 5.73$, that was classified as a high expression group and a second group with a mean Delta $C_{t1-3}>5.73$ classified as a low expression group (Table 1, Table 2 and Figure 3). Overall the regression tree explained 60% of the variance in the data. This is mostly because there were strains expressing a low or high mean Delta C_t that were classified in the same *spa* type; those were the cases of *spa* types t002, t019, t044 and t437.

Furthermore, we explored in each of the *spa* types what other explanatory variables (ST, MRSA, MSSA and type of SCC*mec*) could differentiate the inclusion of some strains in the low or high expression group, but we found no associations with the variables we measured in the study.

We observed that the *hla* expression level varied within strains of the same ST (Figure 3; Table 1 and 2). In fact, in some cases the same ST comprised strains with both high and low levels of expression (ST5, ST15, ST22, ST30, ST59, ST72 and ST80). Moreover, we found that the expression rates did not differ significantly (P = 0.665) between MRSA and MSSA strains. However, we did find a correlation between the *hla* expression and the origin of the genetic backgrounds. Actually, strains of CA genetic backgrounds showed, in general, higher mean expression rates than strains of HA backgrounds (p = 0.003) (Figure 4). Among the 21 strains (36.2%, 21 out of 58) with high expression level, only two (9.5%) belonged to HA backgrounds (ST22-IVh, t032 and ST5-II, t002) whereas the majority (90.5%, n = 19) were represented by CA backgrounds (Table 1 and Table 2). Moreover, two additional CA strains, ST72-IVa-t148 and ST8-MSSA-t008, showed expression rates near the cutoff value (5.73), with 5.76 and 5.87, respectively. These were considered as belonging to the low-level expression group.

The three strains with the highest expression rate were ST398-VII-t108 (2.85), ST398-IVa-t011 (3.25) and ST25-MSSA-t2909 (4.16) and strains with the lowest rate were ST1806 (TLV ST22)-IVh-t747 (9.28), ST45-IVa-t004 (10.38) and ST1-MSSA-t127 (11.09).

We observed that some promoters and gene alleles (P6, P12/P13, P21; and *hla7*, *hla9*, *hla11*) were exclusively associated to a high expression level profile, while others (P3/P4/P5, P7, P8/P9, P11, P15, P17/P18, P23; and *hla4*, *hla8*, *hla10*) were exclusively associated to a low expression level (Table 1 and 2). But we also found promoter and gene allotypes that were associated to both high and low expression levels.

Discussion

Although Hla is one of the most important *S. aureus* virulence factors [10], to the best of our knowledge, this is the first study in which the variation in *hla* nucleotide sequence and gene expression was assessed in such a large and representative collection.

We found that the nucleotide sequence of *hla* was highly diverse. The high degree of diversity found within *hla* is in accordance to results obtained for other exotoxins, which are generally highly polymorphic [27]. Four non-synonymous substitutions (Q87 stop codon, D208E, S239T and T243S) were identified, that are located in two structural protein domains which are essential for Hla oligomerization and pore formation (Rim and Cap) [11,28,29]. The impact of these amino acid (aa) changes on *hla* activity is uncertain. If by one hand, the aa changes described implicate differences in the molecular weight of the aa, that can have influence in the three dimensional structure stability and activity of the protein; on the other hand these aa changes did not match any of the aa previously described to be essential for Hla pore formation.

Furthermore, Walker and Bayley showed that multiple mutations in this same region (residues spanning Hla235–250) did not alter Hla activity in terms of binding, oligomerization or lysis. Thus, it would not be expected that S239T or T243S had significant biological impact in terms of toxin function. The unique mutation with an identified role in Hla function is the stop codon found in the ST36 and ST30 strains that was previously described by DeLeo and co-authors [30] to hinder toxin production and to originate a less virulent strain in a murine infection model. The true effect of the non-synonymous substitutions identified in our study in the activity of the protein would have to be tested by the construction of site directed mutagenesis mutants and by performing binding, oligomerization, hemolysis and *in vivo* models assays.

The construction of phylogenetic trees from the *hla* defined the existence of two clusters with different levels of genetic diversity suggesting that *hla* is evolving at different rates in different genetic backgrounds. Interestingly, the most diverse cluster included the clonal types which are presently more disseminated or that emerged recently (like ST398). This might be related to the fact that these clones still need to evolve to evade the human immune system and not enough time as elapsed for the most adapted allele to have been selected [31]. On the other hand the recombination events detected in the *hla* gene in this study were all in strains belonging to the high genetic diversity cluster, suggesting that this mechanism might have been important in the most recent *hla* evolution and diversification.

Interestingly, the phylogenetic tree constructed from the *hla* gene was similar to that constructed from MLST genes, in the sense that both trees distributed the different STs similarly in two main clusters. This observation suggests that *hla* gene has evolved

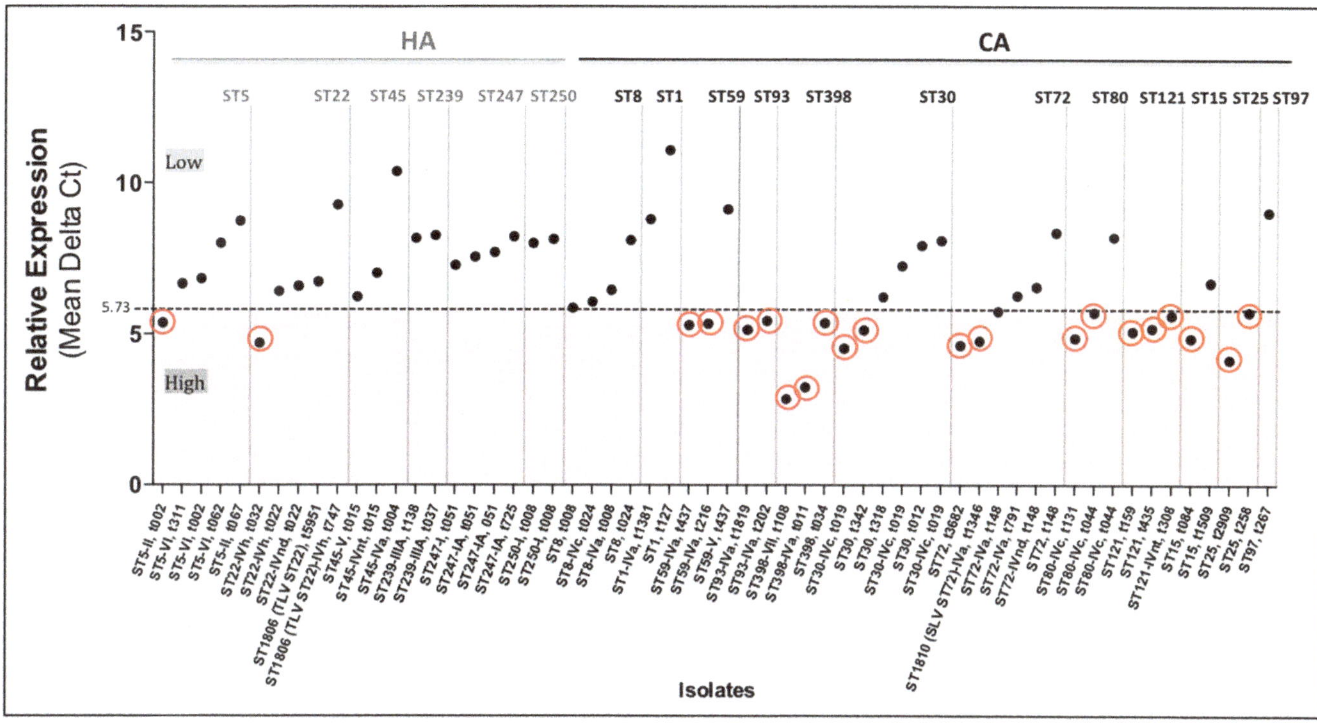

Figure 3. HA and CA strains relative expression distribution. Mean of expression rates from three biological replicates. Dashed line corresponding to the mean Ct value 5.73 results from the regression tree analysis which split strains in two distinct groups, at *spa* type level: a) high expression group - corresponding to strains with Mean Delta Ct≤5.73 and b) low expression group- corresponding to strains with Mean Delta Ct> 5.73). Highlighted in red are the high expressing strains.

together with the *S. aureus* genetic background. A similar type of correlation with the genetic background was previously described for adhesins, either located in the core genome (*clfA*, *clfB*, *fnbA*, *map*, *sdrC*, and *spa*) or accessory genome (*ebpS*, *fnbB*, *sdrD*, and *sdrE*) [32]. Although this was the case for the great majority of STs, we observed that four STs (ST22, ST121, ST59, ST93) were located in different clusters in the *hla* and MLST trees. Our results suggest that recombination occurring at the *hla* level, might explain the different clustering of strains belonging to ST22 and ST121. No recombination events were, however, detected in

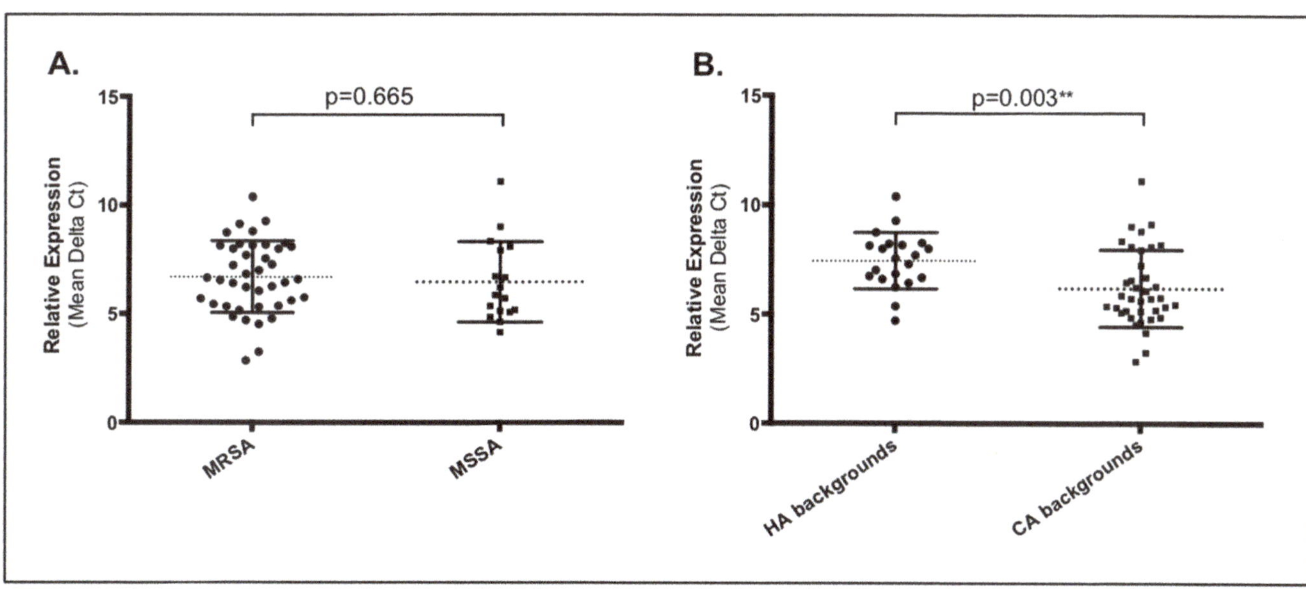

Figure 4. Distribution of the relative *hla* expression. Mean of relative expression of three independent readings. Expression comparison between a) MRSA and MSSA and b) HA and CA backgrounds using the Two-tailed Student's t-test. Statistically significance ($p \leq 0.05$) (**).

MLST genes or *hla* sequences of strains belonging to ST59 and ST93, suggesting that their displacement in the two trees could derive from different phenomena, like random mutation.

It was previously suggested that CA-MRSA expressed more *hla* than HA-MRSA [9]. Results from our study allowed us to extend this conclusion to virtually all epidemic CA, but also in two particular cases of HA genetic backgrounds. The CA strains belonging to ST398, ST25, ST121 and ST93 showed uniformly high relative expression rates and strains belonging to ST36, ST45, ST239, ST247 and ST250 showed uniformly low expression rates. To understand if in fact these patterns of expression are characteristic of these clones, more strains within each clone should be studied for *hla* expression. Nevertheless, we could not correlate the *hla* expression rate with any particular polymorphism within the promoter or any aa substitution in the *hla* gene. The results suggest that *hla* regulation is probably a result of combination of factors which are redundant, rather than associated to a single genetic event. In fact, it has been demonstrated by several authors that alpha-hemolysin is part of a complex regulatory network, that includes the main two-component systems (TCS) – Agr – that in turn is controlled by a diverse pool of regulatory networks that coordinately interact in response to external stimulus and cell signals, namely others TCS (SaeRS, ArlRS and SrrAB), alternative sigma factors (σ^B), and transcription factors (e.g. SarS, SarT, Rot, SarA, SarZ) [33,34].

We showed that *hla* evolved together with the genetic background. Moreover, the most epidemic CA-MRSA genetic backgrounds express more *hla* than the most epidemic HA-MRSA genetic backgrounds. However, the finding of frequent strain-to-strain variation in the expression level of *hla* within strains of the same clonal types suggests that *hla* polymorphisms cannot be used as genetic markers of virulence and investigators should remain cautious when inferring conclusions for the entire MRSA population from studies performed with a limited number of strains.

Supporting Information

Figure S1 Phylogenetic trees of the *hla* gene, promoter gene and concatenated sequences of both. The tree was constructed using MEGA 5 with Neighbour-joining method and bootstrap values provided as percents over 1000 replications. Branch length values are indicated and the percentage of replicate trees (bootstrap test) are shown next to the branches. The dashed line indicates the separation of the two evolutionary branches (L and H).

Figure S2 I. Growth curves for triplicates of each *S. aureus* strain – Set C. **II.** Growth curves for triplicates of each *S. aureus* strain – Set D. **III.** Growth curves for triplicates of each *S. aureus* strain – Set E.

Figure S3 Internal sequences of *hla* promoter (highlighted blue) and *hla* gene (highlighted orange) used for analysis in this study. Primers used are highlighted. The sequence shown corresponds to the promoter and *hla* regions of USA300 strain from our collection blasted against USA300_TCH1516.

Table S1 Molecular characterization of the 73 MRSA and MSSA strains included in this study [35–50].

Author Contributions

Conceived and designed the experiments: HdL MM. Performed the experiments: AT. Analyzed the data: AT MM ACP JBN KS. Contributed reagents/materials/analysis tools: HdL MM HW. Wrote the paper: AT MM. Manuscript revision: HdL HW KS JBN KB SR ACP.

References

1. Deurenberg RH, Stobberingh EE (2009) The molecular evolution of hospital- and community-associated methicillin-resistant *Staphylococcus aureus*. Curr Mol Med 9: 100–115.
2. David MZ, Daum RS (2010) Community-associated methicillin-resistant *Staphylococcus aureus*: epidemiology and clinical consequences of an emerging epidemic. Clin Microbiol Rev 23: 616–687.
3. Otto M (2013) Community-associated MRSA: What makes them special? Int J Med Microbiol.
4. Kobayashi SD, Voyich JM, Burlak C, DeLeo FR (2005) Neutrophils in the innate immune response. Arch Immunol Ther Exp (Warsz) 53: 505–517.
5. Baba T, Takeuchi F, Kuroda M, Yuzawa H, Aoki K, et al. (2002) Genome and virulence determinants of high virulence community-acquired MRSA. Lancet 359: 1819–1827.
6. Diep BA, Gill SR, Chang RF, Phan TH, Chen JH, et al. (2006) Complete genome sequence of USA300, an epidemic clone of community-acquired meticillin-resistant *Staphylococcus aureus*. Lancet 367: 731–739.
7. Burlak C, Hammer CH, Robinson MA, Whitney AR, McGavin MJ, et al. (2007) Global analysis of community-associated methicillin-resistant *Staphylococcus aureus* exoproteins reveals molecules produced in vitro and during infection. Cell Microbiol 9: 1172–1190.
8. Loughman JA, Fritz SA, Storch GA, Hunstad DA (2009) Virulence gene expression in human community-acquired *Staphylococcus aureus* infection. J Infect Dis 199: 294–301.
9. Li M, Cheung GY, Hu J, Wang D, Joo HS, et al. (2010) Comparative analysis of virulence and toxin expression of global community-associated methicillin-resistant *Staphylococcus aureus* strains. J Infect Dis 202: 1866–1876.
10. Berube BJ, Bubeck Wardenburg J (2013) *Staphylococcus aureus* alpha-Toxin: Nearly a Century of Intrigue. Toxins (Basel) 5: 1140–1166.
11. Song L, Hobaugh MR, Shustak C, Cheley S, Bayley H, et al. (1996) Structure of staphylococcal alpha-hemolysin, a heptameric transmembrane pore. Science 274: 1859–1866.
12. Valeva A, Palmer M, Bhakdi S (1997) Staphylococcal alpha-toxin: formation of the heptameric pore is partially cooperative and proceeds through multiple intermediate stages. Biochemistry 36: 13298–13304.
13. Novick RP, Ross HF, Projan SJ, Kornblum J, Kreiswirth B, et al. (1993) Synthesis of staphylococcal virulence factors is controlled by a regulatory RNA molecule. EMBO J 12: 3967–3975.
14. Ballal A, Ray B, Manna AC (2009) *sarZ*, a *sarA* family gene, is transcriptionally activated by MgrA and is involved in the regulation of genes encoding exoproteins in *Staphylococcus aureus*. J Bacteriol 191: 1656–1665.
15. Liang X, Yu C, Sun J, Liu H, Landwehr C, et al. (2006) Inactivation of a two-component signal transduction system, SaeRS, eliminates adherence and attenuates virulence of *Staphylococcus aureus*. Infect Immun 74: 4655–4665.
16. Liang X, Zheng L, Landwehr C, Lunsford D, Holmes D, et al. (2005) Global regulation of gene expression by ArlRS, a two-component signal transduction regulatory system of *Staphylococcus aureus*. J Bacteriol 187: 5486–5492.
17. Schmidt KA, Manna AC, Gill S, Cheung AL (2001) SarT, a repressor of alpha-hemolysin in *Staphylococcus aureus*. Infect Immun 69: 4749–4758.
18. Liang X, Hall JW, Yang J, Yan M, Doll K, et al. (2011) Identification of single nucleotide polymorphisms associated with hyperproduction of alpha-toxin in *Staphylococcus aureus*. PLoS One 6: e18428.
19. Bubeck Wardenburg J, Bae T, Otto M, Deleo FR, Schneewind O (2007) Poring over pores: alpha-hemolysin and Panton-Valentine leukocidin in *Staphylococcus aureus* pneumonia. Nat Med 13: 1405–1406.
20. Bubeck Wardenburg J, Patel RJ, Schneewind O (2007) Surface proteins and exotoxins are required for the pathogenesis of *Staphylococcus aureus* pneumonia. Infect Immun 75: 1040–1044.
21. Inoshima I, Inoshima N, Wilke GA, Powers ME, Frank KM, et al. (2011) A *Staphylococcus aureus* pore-forming toxin subverts the activity of ADAM10 to cause lethal infection in mice. Nat Med 17: 1310–1314.
22. Tamura K, Nei M, Kumar S (2004) Prospects for inferring very large phylogenies by using the neighbor-joining method. Proc Natl Acad Sci U S A 101: 11030–11035.
23. Vandenesch F, Kornblum J, Novick RP (1991) A temporal signal, independent of *agr*, is required for hla but not *spa* transcription in *Staphylococcus aureus*. J Bacteriol 173: 6313–6320.
24. Zielinska AK, Beenken KE, Joo HS, Mrak LN, Griffin LM, et al. (2011) Defining the strain-dependent impact of the Staphylococcal accessory regulator

(*sarA*) on the alpha-toxin phenotype of *Staphylococcus aureus*. J Bacteriol 193: 2948–2958.
25. De'ath G, Fabricius KE (2000) Classification and regression trees: a powerful yet simple thechnique for ecological data analysis. Ecology. Ecology 81: 3178–3192.
26. Morfeldt E, Taylor D, von Gabain A, Arvidson S (1995) Activation of alpha-toxin translation in *Staphylococcus aureus* by the trans-encoded antisense RNA, RNAIII. EMBO J 14: 4569–4577.
27. Wilson GJ, Seo KS, Cartwright RA, Connelley T, Chuang-Smith ON, et al. (2011) A novel core genome-encoded superantigen contributes to lethality of community-associated MRSA necrotizing pneumonia. PLoS Pathog 7: e1002271.
28. Montoya M, Gouaux E (2003) Beta-barrel membrane protein folding and structure viewed through the lens of alpha-hemolysin. Biochim Biophys Acta 1609: 19–27.
29. Walker B, Bayley H (1995) Key residues for membrane binding, oligomerization, and pore forming activity of staphylococcal alpha-hemolysin identified by cysteine scanning mutagenesis and targeted chemical modification. J Biol Chem 270: 23065–23071.
30. DeLeo FR, Kennedy AD, Chen L, Bubeck Wardenburg J, Kobayashi SD, et al. (2011) Molecular differentiation of historic phage-type 80/81 and contemporary epidemic *Staphylococcus aureus*. Proc Natl Acad Sci U S A 108: 18091–18096.
31. Castillo-Ramirez S, Harris SR, Holden MT, He M, Parkhill J, et al. (2011) The impact of recombination on dN/dS within recently emerged bacterial clones. PLoS Pathog 7: e1002129.
32. Kuhn G, Francioli P, Blanc DS (2006) Evidence for clonal evolution among highly polymorphic genes in methicillin-resistant *Staphylococcus aureus*. J Bacteriol 188: 169–178.
33. Novick RP (2003) Autoinduction and signal transduction in the regulation of staphylococcal virulence. Mol Microbiol 48: 1429–1449.
34. Thoendel M, Kavanaugh JS, Flack CE, Horswill AR (2011) Peptide signaling in the staphylococci. Chem Rev 111: 117–151.
35. Tavares A, Miragaia M, Rolo J, Coelho C, de Lencastre H (2013) High prevalence of hospital-associated methicillin-resistant *Staphylococcus aureus* in the community in Portugal: evidence for the blurring of community-hospital boundaries. Eur J Clin Microbiol Infect Dis 32: 1269–1283.
36. Oliveira DC, Tomasz A, de Lencastre H (2001) The evolution of pandemic clones of methicillin-resistant *Staphylococcus aureus*: identification of two ancestral genetic backgrounds and the associated *mec* elements. Microb Drug Resist 7: 349–361.
37. Roberts RB, de Lencastre A, Eisner W, Severina EP, Shopsin B, et al. (1998) Molecular epidemiology of methicillin-resistant *Staphylococcus aureus* in 12 New York hospitals. MRSA Collaborative Study Group. J Infect Dis 178: 164–171.
38. Aires-de-Sousa M, Correia B, de Lencastre H (2008) Changing patterns in frequency of recovery of five methicillin-resistant *Staphylococcus aureus* clones in Portuguese hospitals: surveillance over a 16-year period. J Clin Microbiol 46: 2912–2917.
39. Rolo J, Miragaia M, Turlej-Rogacka A, Empel J, Bouchami O, et al. (2012) High Genetic Diversity among Community-Associated *Staphylococcus aureus* in Europe: Results from a Multicenter Study. PLoS One 7: e34768.
40. Conceicao T, Tavares A, Miragaia M, Hyde K, Aires-de-Sousa M, et al. (2010) Prevalence and clonality of methicillin-resistant *Staphylococcus aureus* (MRSA) in the Atlantic Azores islands: predominance of SCC*mec* types IV, V and VI. Eur J Clin Microbiol Infect Dis 29: 543–550.
41. Sa-Leao R, Santos Sanches I, Dias D, Peres I, Barros RM, et al. (1999) Detection of an archaic clone of *Staphylococcus aureus* with low-level resistance to methicillin in a pediatric hospital in Portugal and in international samples: relics of a formerly widely disseminated strain? J Clin Microbiol 37: 1913–1920.
42. Amorim ML, Aires de Sousa M, Sanches IS, Sa-Leao R, Cabeda JM, et al. (2002) Clonal and antibiotic resistance profiles of methicillin-resistant *Staphylococcus aureus* (MRSA) from a Portuguese hospital over time. Microb Drug Resist 8: 301–309.
43. Milheirico C, Oliveira DC, de Lencastre H (2007) Multiplex PCR strategy for subtyping the staphylococcal cassette chromosome *mec* type IV in methicillin-resistant *Staphylococcus aureus*: 'SCC*mec* IV multiplex'. J Antimicrob Chemother 60: 42–48.
44. Richardson JF, Reith S (1993) Characterization of a strain of methicillin-resistant *Staphylococcus aureus* (EMRSA-15) by conventional and molecular methods. J Hosp Infect 25: 45–52.
45. Oliveira DC, Milheirico C, Vinga S, de Lencastre H (2006) Assessment of allelic variation in the *ccrAB* locus in methicillin-resistant *Staphylococcus aureus* clones. J Antimicrob Chemother 58: 23–30.
46. Faria NA, Oliveira DC, Westh H, Monnet DL, Larsen AR, et al. (2005) Epidemiology of emerging methicillin-resistant *Staphylococcus aureus* (MRSA) in Denmark: a nationwide study in a country with low prevalence of MRSA infection. J Clin Microbiol 43: 1836–1842.
47. Sanches IS, Ramirez M, Troni H, Abecassis M, Padua M, et al. (1995) Evidence for the geographic spread of a methicillin-resistant *Staphylococcus aureus* clone between Portugal and Spain. J Clin Microbiol 33: 1243–1246.
48. de Lencastre H, Chung M, Westh H (2000) Archaic strains of methicillin-resistant *Staphylococcus aureus*: molecular and microbiological properties of isolates from the 1960s in Denmark. Microb Drug Resist 6: 1–10.
49. Crisostomo MI, Westh H, Tomasz A, Chung M, Oliveira DC, et al. (2001) The evolution of methicillin resistance in *Staphylococcus aureus*: similarity of genetic backgrounds in historically early methicillin-susceptible and -resistant isolates and contemporary epidemic clones. Proc Natl Acad Sci U S A 98: 9865–9870.
50. McDougal LK, Steward CD, Killgore GE, Chaitram JM, McAllister SK, et al. (2003) Pulsed-field gel electrophoresis typing of oxacillin-resistant *Staphylococcus aureus* isolates from the United States: establishing a national database. J Clin Microbiol 41: 5113–5120.

4

Genome-Wide Analysis of the NADK Gene Family in Plants

Wen-Yan Li, Xiang Wang, Ri Li, Wen-Qiang Li, Kun-Ming Chen*

State Key Laboratory of Crop Stress Biology in Arid Areas, College of Life Sciences, Northwest A&F University, Yangling, Shaanxi, China

Abstract

Background: NAD(H) kinase (NADK) is the key enzyme that catalyzes *de novo* synthesis of NADP(H) from NAD(H) for NADP(H)-based metabolic pathways. In plants, NADKs form functional subfamilies. Studies of these families in *Arabidopsis thaliana* indicate that they have undergone considerable evolutionary selection; however, the detailed evolutionary history and functions of the various NADKs in plants are not clearly understood.

Principal Findings: We performed a comparative genomic analysis that identified 74 NADK gene homologs from 24 species representing the eight major plant lineages within the supergroup Plantae: glaucophytes, rhodophytes, chlorophytes, bryophytes, lycophytes, gymnosperms, monocots and eudicots. Phylogenetic and structural analysis classified these NADK genes into four well-conserved subfamilies with considerable variety in the domain organization and gene structure among subfamily members. In addition to the typical NAD_kinase domain, additional domains, such as adenylate kinase, dual-specificity phosphatase, and protein tyrosine phosphatase catalytic domains, were found in subfamily II. Interestingly, NADKs in subfamily III exhibited low sequence similarity (~30%) in the kinase domain within the subfamily and with the other subfamilies. These observations suggest that gene fusion and exon shuffling may have occurred after gene duplication, leading to specific domain organization seen in subfamilies II and III, respectively. Further analysis of the exon/intron structures showed that single intron loss and gain had occurred, yielding the diversified gene structures, during the process of structural evolution of NADK family genes. Finally, both available global microarray data analysis and qRT-RCR experiments revealed that the NADK genes in *Arabidopsis* and *Oryza sativa* show different expression patterns in different developmental stages and under several different abiotic/biotic stresses and hormone treatments, underscoring the functional diversity and functional divergence of the NADK family in plants.

Conclusions: These findings will facilitate further studies of the NADK family and provide valuable information for functional validation of this family in plants.

Editor: Manoj Prasad, National Institute of Plant Genome Research, India

Funding: This research was supported by grants from the National Natural Science Foundation of China (grant number 31270299), the Basic Scientific Fund of The State Key Laboratory of Crop Stress Biology for Arid Areas, NWAFU (CSBAAZD1301), the Basic Scientific Fund of NWAFU (2014YB033), and the Program for New Century Excellent Talents in University of China (NCET-11-0440). The funders had no role in study design, data collection and analysis, decision to publish, or preparation of the manuscript.

Competing Interests: The authors have declared that no competing interests exist.

* Email: kunmingchen@nwsuaf.edu.cn

Introduction

NAD(H) and NADP(H) are crucial coenzymes and play important and distinguishable roles in all living organisms [1]. NAD(H) is primarily involved in catabolic reactions, whereas NADP(H) participates in anabolic reactions, such as NADP(H)-dependent reductive anabolic pathways, signal transduction, and defense against oxidative stress [1,2]. Hence, regulation of the intracellular balance of NAD(H) and NADP(H) is critical. NAD(H) kinase (NADK) is the key enzyme in the *de novo* biosynthesis of NADP(H), catalyzing the transfer of a phosphoryl group from ATP to NAD(H), and thus plays an important role in the regulation of intracellular NAD(H)/NADP(H) balance for NADP(H)-based metabolic pathways.

NADK genes have been found in nearly all living organisms, including Archaea, eubacteria and eukaryotes, except for the intracellular parasite *Chlamydia trachomatis* [3]. NADK genes have been cloned from a wide variety of species, including Archaea (*Methanococcus jannaschii* [4]), eubacteria (*Mycobacterium tuberculosis* [5]), *Escherichia coli* [6], yeast (*Saccharomyces cerevisiae* [7,8]), humans (*Homo sapiens* [9]) and plants (*Arabidopsis thaliana* [10,11,12]). Moreover, the number of NADKs in different organisms varies, with most prokaryotic organisms, such as Archaea and eubacteria, having only one NADK, whereas most eukaryotic organisms, such as yeast and plants, have several NADKs. For example, *Euglena gracilis* [13] has two NADKs and *S. cerevisiae* [14,15] and *Arabidopsis* have three [10,11,16].

The enzymatic properties of natural or recombinant NADKs from several organisms, including their substrate specificity and structural properties, especially of the active site, have been well characterized [2,9,17,18,19]. All characterized NADKs are homomultimers and exhibit species-dependent phosphoryl donor and acceptor specificity. Hence, these enzymes are classified into two groups: ATP-NAD kinases and inorganic polyphosphate poly(P)/ATP-NAD kinases, based on the phosphoryl donor specificity [2]. For example, NADKs from gram-positive bacteria (*M. tuberculosis*) and Archaea (*M. jannaschii*) utilize both ATP and poly(P) as phosphoryl donors and they phosphorylate NAD^+/NADH, whereas NADKs from gram-negative bacteria (*E. coli*) and eukaryotes (*S. cerevisiae*, *Arabidopsis* and human) utilize ATP but not poly(P), and phosphorylate NAD^+ and NAD^+/NADH, respectively [2]. In addition, analysis of the sequences and crystal structures of NADKs identified three highly conserved and functionally important motifs within the NADK family (a GGDG motif, an NE/D motif and a Gly-rich motif) and a possible phosphate transfer mechanism [20,21,22,23,24].

The biological significance and physiological functions of NADKs have been characterized from studies of several organisms. For example, mutation of the single NADK gene in *M. tuberculosis* and *Salmonella enterica* is lethal [3,4]. Similarly, mutation of all three NADK genes in *S. cerevisiae* (*utr1/yef1/pos5*) or two of the three genes (*utr1/pos5*) also causes lethality, whereas the respective NADK gene mutations are not lethal in *Arabidopsis* [11,16,25,26]. In addition, studies of the physiological functions of NADKs in *S. cerevisiae* and humans revealed that NADKs play a major role in protecting living cells against oxidative stress [8,15,26,27] because NADPH is vital in the intracellular anti-oxidative defense system of most organisms [28].

In plants, NADKs are involved in regulating redox balance, biotic and abiotic stress responses and various developmental processes. Notably, NADK was the first protein in plants demonstrated to be regulated by the calcium-sensing protein calmodulin (CaM) [29], and plant NADKs are divided into CaM-independent and CaM-regulated isoforms. CaM-dependent NADK is essential for survival of plants under difficult conditions and for protecting plants against invading pathogens by helping to provide reductants for the NADPH-dependent oxidative burst [29,30,31]. For example, CaM-dependent NADK activity, but not CaM-independent NADK activity, increases under cold stress in green bean [32], and decreased in response to high salinity and drought in tomato [33] and wheat [34]. In *Arabidopsis*, three genes encoding NADK (*NADK1/NADK2/NADK3*) have been identified and their physiological functions have been characterized [2,35]. *AtNADK1* is located in the cytosol and expressed mainly in roots, *AtNADK2* is located in the chloroplasts and expressed mainly in leaves, whereas *AtNADK3* is located in the peroxisome and is strongly expressed in reproductive tissues, such as the stigma, pollen and carpel vasculature [10,11,36]. The *AtNADK1*-deficient mutant exhibits sensitivity to γ-irradiation and paraquat-induced oxidative stress [16]. The *AtNADK2*-deletion mutant displays hypersensitivity to environmental stresses that induce oxidative stress, such as UVB-irradiation, drought, heat shock and high salinity [11]. Similarly, the *AtNADK3*-null mutant shows hyper-sensitivity to oxidative stress, including methyl viologen, high salinity and osmotic shock [10]. Moreover, AtNADK2 plays a vital role in chlorophyll synthesis and protects chloroplasts against oxidative damage [11]. In addition, plants growth and fertility are affected in the NADK-deficient mutants [2,11,37].

Although these studies in bacteria, yeast, human and *Arabidopsis* have led to an understanding of the biochemical properties and physiological functions of NADKs, there has been no systematic study of the evolution and functional divergence of the NADK gene family, especially in Plantae. Here, we performed a comprehensive analysis of the NADK gene family in 24 species, representing the eight major plant lineages within the supergroup Plantae. Phylogenetic analysis was performed to delineate the evolutionary history of the NADK family in Plantae, and exon/intron structure analysis was performed to gain insight into the possible mechanisms of the structural diversity of NADK gene family. Finally, the tissue-specificity and inducibility of NADK gene expression in *Arabidopsis* and rice (*Oryza sativa*) were characterized by examining publicly available microarray datasets and by qRT-PCR experiments. The results obtained here will broaden our understanding of the roles of plant NADKs and provide a framework for further functional investigations of these genes in plants.

Results

Identification of NADK family members in plants

To comprehensively investigate and characterize the NADK gene family in plants, 24 species representing the eight major plant lineages within the supergroup Plantae, were selected for analysis (Figure 1). A hidden Markov model (HMM) search was performed with the obtained sequences and 74 NADK homologs were identified (Figure 1, Table S1). Except for the gymnosperm *Picea sitchensis*, for which the full genome sequence was not yet available, two or more NADK genes were identified in each genome of the selected species (Figure 1). Most of the aquatic algae, including glaucophytes, rhodophytes and chlorophytes, contained two NADK genes per genome and only three species in the chlorophytes, *Micromonas pusilla* RCC299, *Chlamydomonas reinhardtii* and *Volvox carteri*, carried three NADK genes per genome (Figure 1). In contrast, three or more NADK genes were found in all land plants, including bryophytes, lycophytes, monocots and eudicots (Figure 1). In addition, only one NADK gene was identified per genome in the cyanophytes (*Acaryochloris marina* MBIC11017 and *Prochlorococcus marinus* MIT 9301) and bacteria (*E. coli* K-12), outgroup (Figure 1).

The Pfam and SMART databases were used to analysis the functional domains of the identified NADK candidates. All of the putative NADKs possess a typical NAD_kinase domain (Pfam accession number PF01513) and some also contain other functional domains such as an adenylate kinase domain (ADK; PF00406), a dual-specificity phosphatase, catalytic domain (DSPc; PF00782), and/or a protein tyrosine phosphatase, catalytic domain/DSPc domain (PTPc/DSPc; SMART accession number SM000012). Only one NADK candidate, CpNADK2, which did not contain a complete NAD_kinase domain, was excluded from the following analysis.

Phylogenetic analysis and classification of the NADK family

To explore the phylogenetic relationships among NADK family members in plants, we first generated a rooted maximum-likelihood phylogenetic tree with the 73 NADKs from the 24 species (Figure 2A), which was inferred from the amino acid sequences of their NAD_kinase domains (Figure S1). Furthermore, phylogenetic trees reconstructed by the neighbor joining, minimum evolution and maximum parsimony methods showed very similar topologies to the maximum-likelihood tree (data not shown). Using *EcNADK1*, *AmNADK1* and *PmNADK1* from bacteria and cyanophytes as the outgroup, all NADK homologs in plants can be classified into four well-conserved subfamilies (I–IV; Figure 2A) with high statistical support, according to the topology and the deep duplication nodes of NADK paralogues in the maximum-likelihood tree. Interestingly, the topological relationship of members within subfamilies was highly consistent with the evolutionary relationships between species in Plantae (Figures 1 and 2A). The majority of aquatic algae contained two NADK genes per genome and grouped into subfamilies II and IV, whereas the majority of land plants, except for the gymnosperm

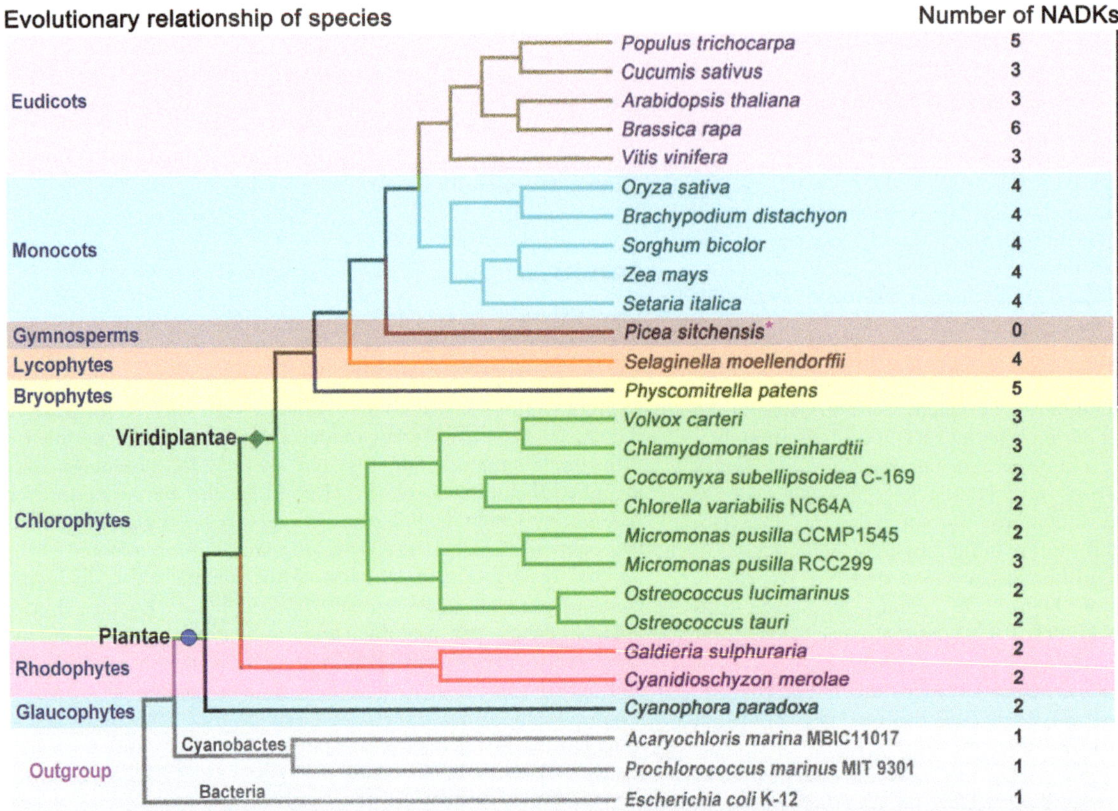

Figure 1. Systematic evolutionary relationships of 24 species among eight lineages within the supergroup Plantae. The numbers of NADK homologs in each species are listed next to the tree. *, the genome sequencing of *Picea sitchensis* is not complete.

Picea sitchensis, carried ≥ three NADK genes per genome and were clustered into subfamilies I, II and III (Figure 2A). Moreover, only one of the NADK genes from green algae (*CrNADK3*) fell into subfamily I and only two (*VcNADK3* and *MpNADK3*) fell into subfamily III (Figure 2A).

Examination of the chromosomal locations of the NADK family genes in the genomes of algae (*Ostreococcus lucimarinus*, *Ostreococcus tauri*, *C. reinhardtii* and *V. carteri*), bryophytes (*Physcomitrella patens*), lycophytes (*Selaginella moellendorffii*), monocots (*Brachypodium distachyon*, rice, *Sorghum bicolor* and *Zea mays*) and eudicots (*Arabidopsis*, *Brassica rapa*, *Populus trichocarpa* and *Vitis vinifera*) showed that NADK genes are randomly distributed (Figure S3). Moreover, a search for NADK paralogs using the Plant Genome Duplication Database (PGDD; http://chibba.agtec.uga.edu/duplication/) [38] revealed eight paralogous gene pairs in *P. patens*, rice, *Z. mays*, *B. rapa* and *P. trichocarpa* (Figure S4A), but none in the other species.

To further explore association of positive selection with duplication and divergence of NADK family genes, the rate of non-synonymous substitution (Ka), synonymous substitution (Ks) and the Ka/Ks ratios were calculated for the eight paralogous gene pairs and used to estimate duplication and divergence times. The Ka/Ks ratios varied from 0.18 to 0.39 among the five different species (Figure S4B). The fact that the Ka/Ks ratios were <1 suggests that the NADK family genes have undergone strong negative selection pressure, and the duplication event was estimated to have occurred ~13.8–67.6 million years ago. Divergence of *P. patens*, rice, *Z. mays*, *P. trichocarpa* and *B. rapa* was estimated to have occurred 67.6, 61.5, 13.8, 20.8 and 23.1–33.1 million years ago, respectively (Figure S4B).

Further analysis of the functional domains showed that domain organization of NADKs in the different subfamilies varied considerably (Figure 2B). All of the identified NADKs contained a typical NAD_kinase domain at the C terminus, but the proteins in subfamily II also carried the additional N-terminal catalytic domains noted earlier (DSPc, PTPc/DSPc or ADK; Figure 2B). Interestingly, the NAD_kinase domain of proteins in subfamily III was not only divided into two parts, but also exhibited low sequence similarity (~30%) between subfamily III members and the other subfamilies (Figures 2C and S1). In addition, based on the NAD_kinase domain sequence similarity, the NADK gene subfamilies could be further divided into two clusters: cluster I (subfamilies I, II and IV) and cluster II (subfamily III).

More detailed analysis of the NAD_kinase domains of the 73 NADKs revealed three highly conserved and functionally important motifs: a GGDG motif, an NE/D motif and a Gly-rich motif (Figures 2C and S1). The GGDG motif is involved in ATP-binding, whereas the NE/D and Gly-rich motifs are involved in NAD(H) binding [2,20,21,23,24].

Structure analysis of NADK family genes

Intron position is generally very well-conserved in orthologous genes over long evolutionary time intervals, whereas exon/intron structure is slightly less, but sufficiently, conserved in paralogous genes to reveal evolutionary relationships between introns [39,40,41]. To investigate the gene structural diversity and possible mechanisms for the structural evolution of NADK homologs in green plants (Viridiplantae, excluding glaucophytes and rhodophytes), we analyzed the exon/intron organization in the coding sequence. Overall, there was considerable diversity in

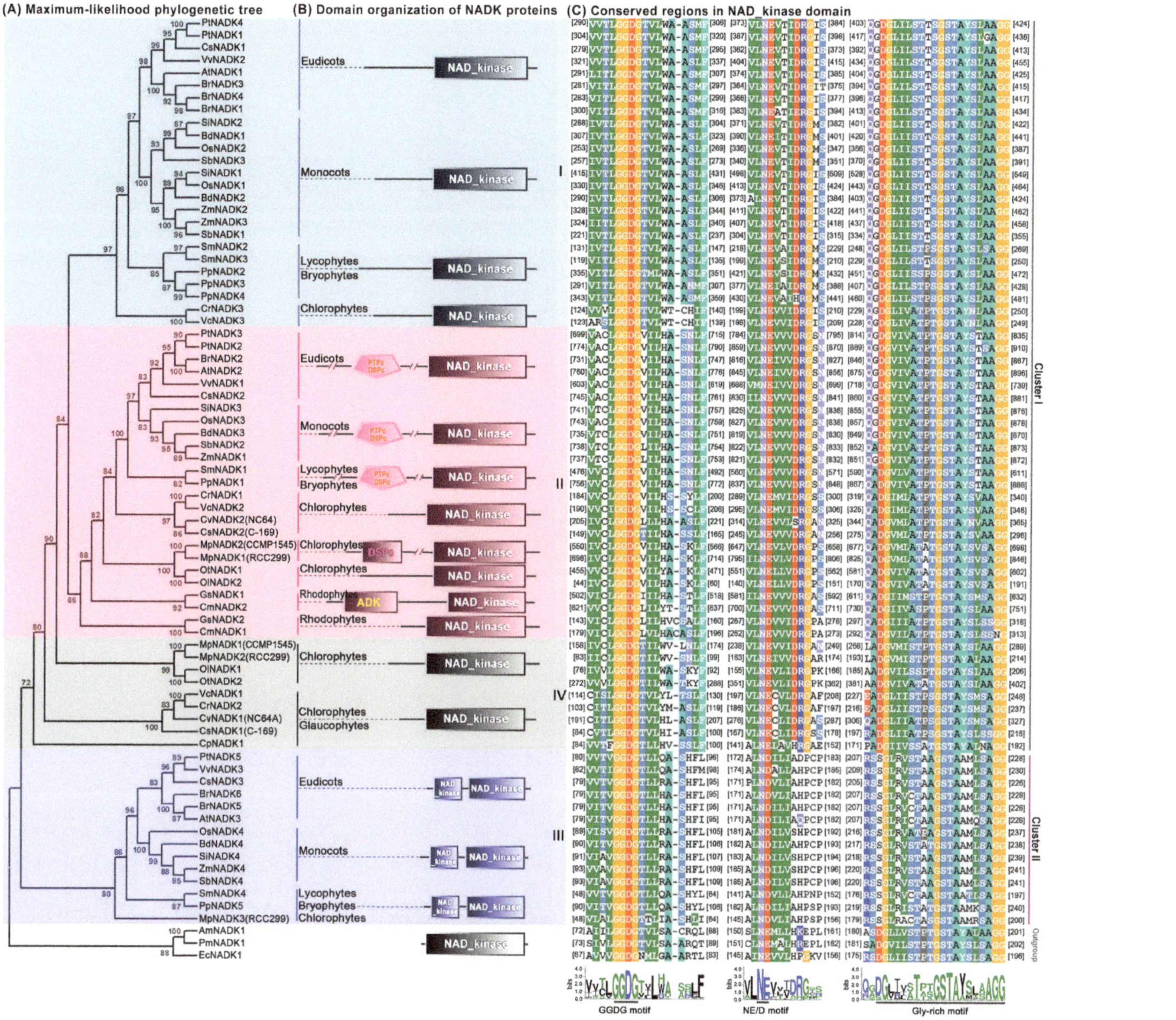

Figure 2. Phylogenetic relationships and domain organization of NADK genes in plants. (A) The rooted maximum-likelihood phylogenetic tree of NADK family members was inferred from the amino acid sequence alignment of the NAD_kinase domain. Numbers above the nodes represent bootstrap values from 1000 replications. **(B)** Domain organization of the NADKs. **(C)** Amino acid sequence alignment of conserved motifs within the NAD_kinase domain.

the number of introns (0–11) and the length of introns (50–7452 bp) in the NADK family genes (Figure 3). Interestingly, NADK family members within the same subfamily shared similar gene structure in terms of intron number, exon length, and/or intron phases, with the exception of subfamily IV (Figure 3). For instance, NADK genes in subfamily I had 6–11 introns, 48% (12/25) and 84% (21/25) of which contained 10 and 9–11 introns, respectively. We also investigated intron phases with respect to codons in the NADK genes. The intron phases were remarkably well conserved among subfamily members, whereas the intron arrangement and phases were strikingly distinct between subfamilies (Figure 3B).

To further explore intron loss or gain within the NADK family genes, we next examined the exon/intron organization of paralogs and orthologs in the land plants (Figure 4A). This analysis revealed that single intron loss and gain likely occurred during the structural evolution of NADK family genes in land plants. For example, The paralogous genes *PtNADK1/4* and *SmNADK2/3* showed conserved exon/intron structure in terms of the number of introns and exon length, whereas a single intron appears to have been lost during the evolution of the *BrNADK1/4* and *PpNADK3/4* paralogs (Figures 3B and 4A). By contrast, a single intron gain occurred between the paralogs *PtNADK2* and *3* (Figures 3B and 4A). Among the orthologous NADK genes in land plants, the majority of subfamilies I, II and III members contained 10, 7 and 4 conserved common introns, respectively (Figure 4A), whereas some subfamily I and II NADK orthologs contained fewer introns. For example, *OsNADK1*, *SiNADK1*, *BdNADK1/2*, *ZmNADK2/3* (subfamily I) have only nine conserved common introns. Similar intron loses were seen with *BrNADK1/3/4*, *SbNADK3* and

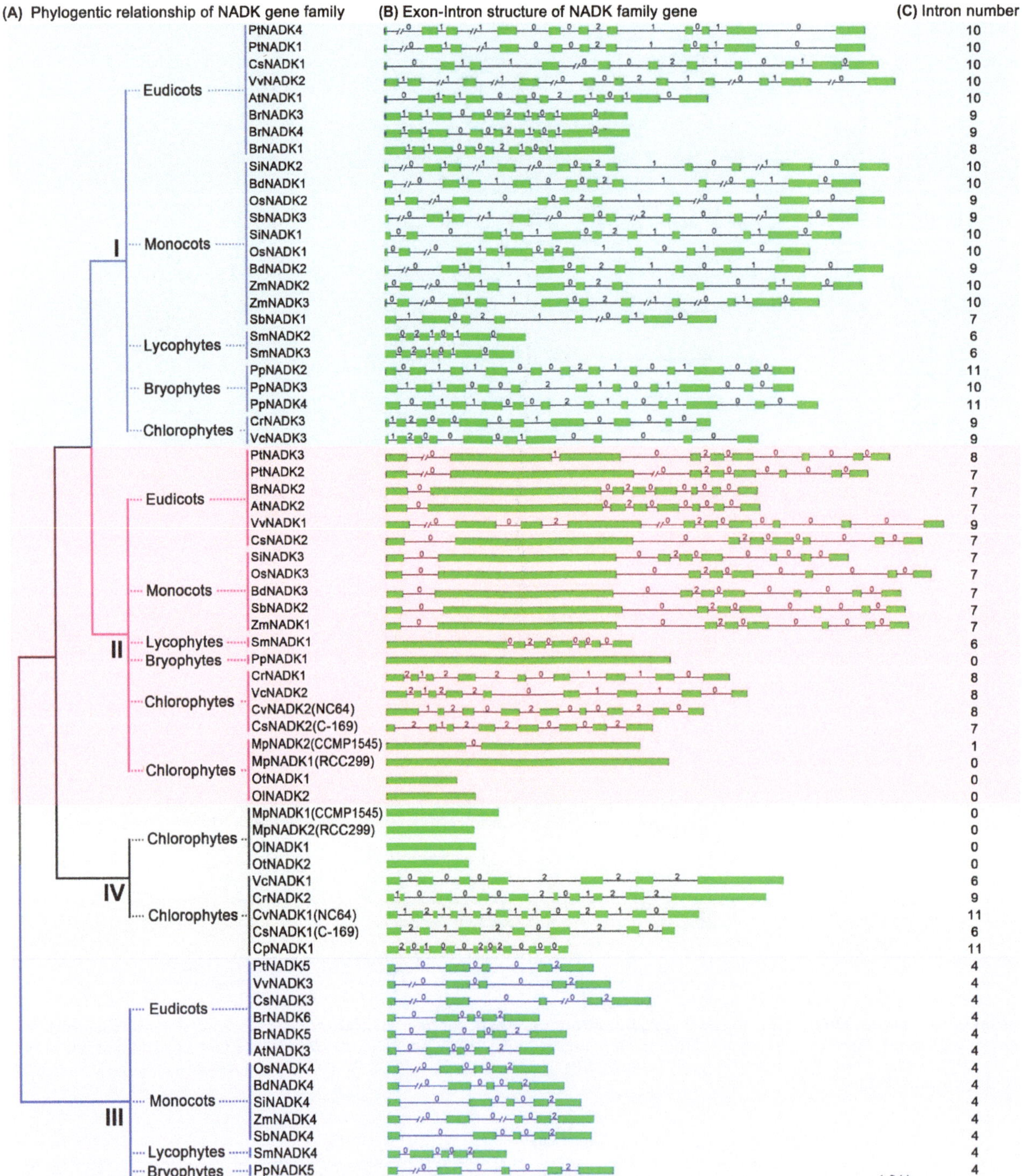

Figure 3. Exon/intron structure of NADK family genes of green plants. Green boxes represent exons; black lines represent introns; numbers 0, 1 and 2 are intron phases. The length of the boxes and lines are scaled relative to the length of the gene, and longer introns are denoted by a double slash.

SmNADK2/3 (subfamily I), *PpNADK1* (subfamily II), and *SmNADK4* (subfamily III; Figure 4A). Single intron gain was also observed in the NADK genes of land plants. For instance, *OsNADK1*, *SiNADK1*, and *PpNADK2/3/4* in subfamily I and *PtNADK3* in subfamily II appear to have gained an intron, whereas *VvNADK1* in subfamily II appears to have gained two introns (Figure 4A).

Tissue-specific expression patterns of NADK genes in *Arabidopsis* and rice

To investigate differences in expression of NADK genes in *Arabidopsis* and rice during plant development, we first analyzed the expression profiles of AtNADK and OsNADK genes available in expressed sequence tag (EST) databases from vegetative and reproductive development stages (Figures 5A and 5B and Tables

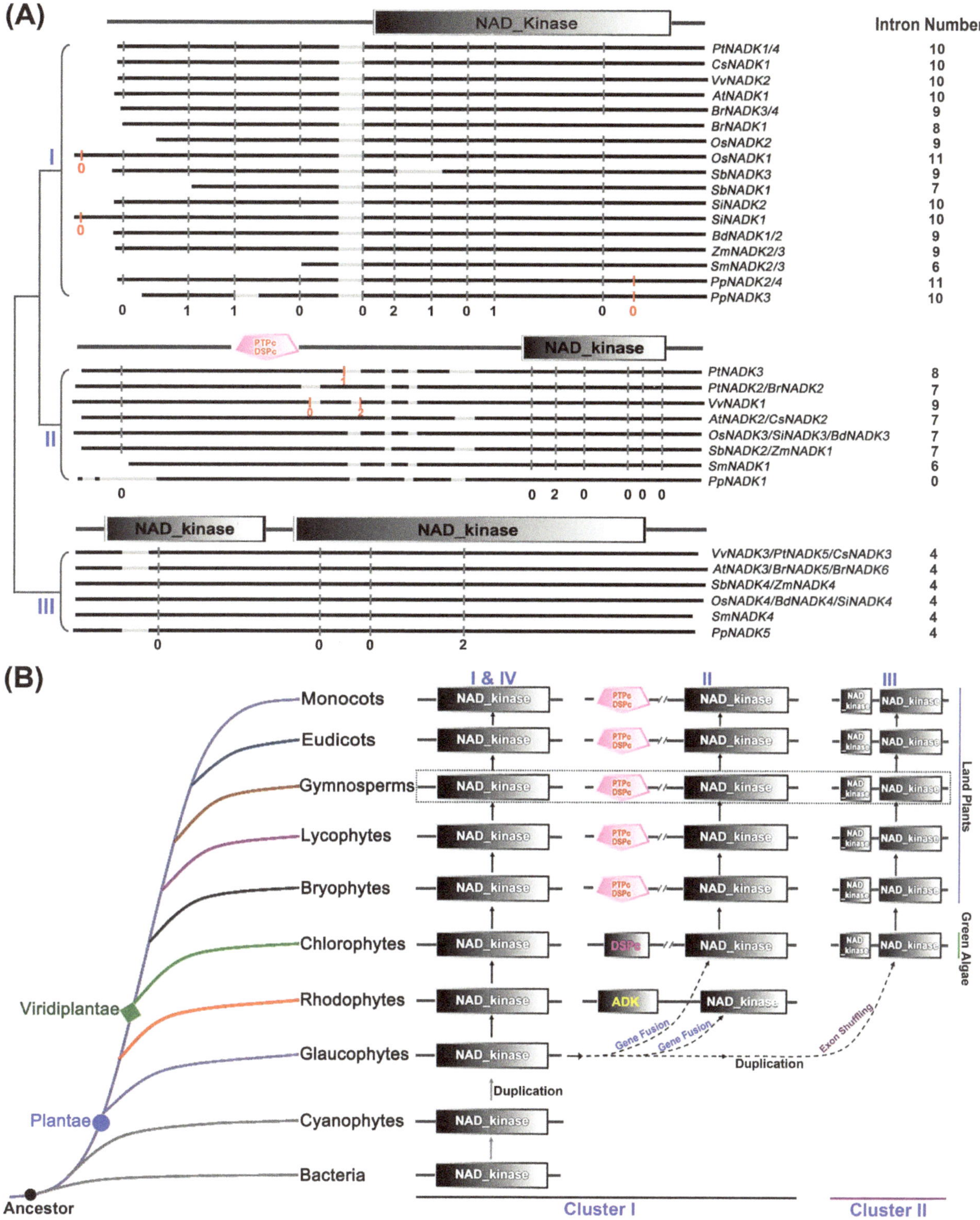

Figure 4. The expansion and evolution of the NADK gene family in plants. (**A**) Schematic comparison of intron distribution in NADK orthologs of land plants generated with the CIWOG software. Black horizontal lines are aligned sequences; gray horizontal lines are gaps in the alignment; gray vertical bars are conserved common introns; red vertical bars are gained introns. The numbers 0, 1 and 2 are intron phases. (**B**) A model for the expansion and evolution of the NADK gene family in Plantae.

S3 and S4, respectively). Varying levels of ESTs for each of the AtNADK and OsNADK genes were found the database, indicating that all of the NADK genes in *Arabidopsis* and rice are expressed, and that they are likely differential expressed in different tissues or developmental stages. We then used *Arabidopsis* (ATH1, 22 k array) and rice (Os 51 k array) microarray data in

Genevestigator to analyze the expression patterns of the NADK genes in 10 and 9 developmental stages/tissues, respectively (Figure 5C, D and S5). Essentially identical developmental expression profiles for the AtNADK and OsNADK genes were also obtained with data from the *Arabidopsis* and rice eFP browsers in the Bio-Analytic Resource (http://bar.utoronto.ca/welcome.htm) database [42]. *AtNADK3* was not included in the comparison of tissue/developmental expression patterns because it was not included in the microarray data. The combined expression patterns of *AtNADKs* and *OsNADKs* from the EST and microarray data analyses are summarized in Figure 5E.

Overall, all of the AtNADK and OsNADK genes were expressed during the vegetative and reproductive development stages, and they displayed strong tissue specificity. In *Arabidopsis*, *AtNADK2* showed higher expression in most tissues and developmental stages than *AtNADK1* (Figures 5C and S5). The highest *AtNADK2* expression is seen in leaves and flowers (Figures 5A, C and S5), consistent with Waller et al. (2010) [36]; whereas *AtNADK1* expression is highest in the dry seed and senescence stages (Figures 5C and S5). *AtNADK3* appears to be mainly expressed in the reproductive tissues such as flowers and siliques (Figure 5A), also consistent with the findings of Waller et al. (2010) [36]. In rice, the *OsNADK1*, *OsNADK2* and *OsNADK3* transcripts tend to accumulate to higher levels than the *OsNADK4* transcript in most tissues/developmental stages (Figures 5B, D and S5), but their individual expression patterns differ. For instance, during the germinate seed stage, *OsNADK1* transcript levels are low compared to *OsNADK2* (Figure 5D), whereas the opposite is observed in callus (Figure 5B). The highest levels of *OsNADK3* transcript are seen in leaf and stem (Figures 5D and S5); whereas relatively low levels of the Os*NADK4* transcript are seen in most tissues/developmental stages except for stem and inflorescence (Figures 5D and S5).

Response profiles of NADK genes under abiotic/biotic stresses and hormone treatments in *Arabidopsis* and rice

To understand the molecular mechanism of *AtNADKs* and *OsNADKs* transcriptional regulation under abiotic/biotic stresses and hormone treatments, we first identified potential *cis*-elements in the promoter regions of each NADK gene using the

(A)

Pool Name	Total ESTs	*AtNADK1*		*AtNADK2*		*AtNADK3*	
		Gene ESTs	Spot Intensity*	Gene ESTs	Spot Intensity*	Gene ESTs	Spot Intensity*
Cell culture	1100	0		0		0	
Root	49512	1		1		2	
Leaf	12182	0		3		0	
Vegetative tissue	2284	2		0		0	
Floral meristem	4229	0		0		0	
Bud	6299	0		0		0	
Flower	62048	0		1		1	
Inflorescence	2246	0		0		0	
Silique	13014	0		0		3	
Seed	16148	0		1		1	

(B)

Pool Name	Total ESTs	*OsNADK1*		*OsNADK2*		*OsNADK3*		*OsNADK4*	
		Gene ESTs	Spot Intensity*	Gene ESTs	Spot Intensity*	Gene ESTs	Spot Intensity*	Gene ESTs	Spot Intensity*
Callus	162069	29		4		4		4	
Vegetative meristem	4537	0		0		0		0	
Root	67194	9		4		1		0	
Leaf	175658	4		6		22		0	
Stem	124773	14		0		27		3	
Flower	134391	13		6		10		0	
Panicle	136249	8		2		0		0	
Seed	32409	3		2		0		0	

*, Spot Intensity means the transcripts per million (TPM), according to the EST database definition.

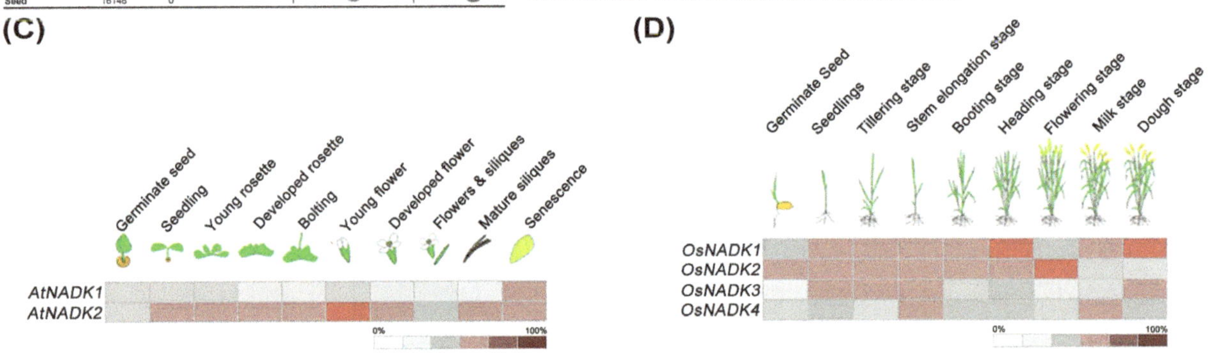

(E)

Gene Name	EST counts	FL_cDNA	Germinate seed	Seedling	Root	Leave	Stem	Flower	Inflorescence	Silique	Seed	Callus
AtNADK1	19	AY383545 AK318804	•	•	• •	•	•	•	•	•	•	NA
AtNADK2	30	AF337912 AY099794	•	• •	• •	• • •	• •	• • •	• •	• •	•	NA
AtNADK3	25	BT022078 BT033136	•	•	•	•	•	•	•	• •	○	NA
OsNADK1	86	AK099730 AK061785	•	• •	• •	• •	• •	• •	• •	NA	• •	• •
OsNADK2	30	AK289181 AK243031	• •	• •	• •	• •	• •	• •	• •	NA	•	• •
OsNADK3	69	AK065215 AK101259 AK060751	•	• •	• •	• •	• •	• •	• •	NA	• •	• •
OsNADK4	10	AK073719	•	•	•	•	• •	•	•	NA	•	•

•, weak expression; • •, moderate expression; • • •, strong expression; ○, without expression; NA, not available.

Figure 5. Developmental expression patterns of NADK family genes in *Arabidopsis* and rice. Expression profiles of (**A**) *AtNADKs* and (**B**) *OsNADKs* inferred from public EST data. Expression profiles of (**C**) *AtNADKs* (except *AtNADK3*) and (**D**) *OsNADKs* in different developmental stages obtained from microarray data reported in Genevestigator. Results are shown as heat maps in white/gray/red (low to high) that reflect the percent of expression. (**E**) Summary of the relative expression patterns of AtNADK and OsNADK genes inferred from the combined EST (A, B; Tables S3 and S4) and microarray (C, D) profiles.

PlantCARE program (Tables 1 and S2). We found that the promoter regions of AtNADK and OsNADK genes contain response elements for several abiotic/biotic stresses, such as low temperature, heat, drought (MYB binding sites), anaerobic conditions, and pathogens (TC-rich repeats, W box, GCC box, Box S, Box-W1 and EIRE) (Tables 1 and S2). In addition, the AtNADK and OsNADK gene promoters contain *cis*-elements for responding to several hormones, such as auxin, gibberellin, abscisic acid (ABA), ethylene, salicylic acid (SA) and methyl jasmonic acid (MeJA) (Tables 1 and S2). Further analysis showed that all of the *AtNADK* and *OsNADK* promoters contain anaerobic- and ABA-responsive elements; and the majority of them also carry heat stress, drought (MYB binding sites) and MeJA response elements (Tables 1 and S2).

To further demonstrate that the expression of AtNADK and OsNADK genes is induced by abiotic and biotic stress, we again examined *Arabidopsis* and rice microarray data in the Genevestigator database, as well as by qRT-PCR experiments. As shown in Figures 6, 7 and S6, the expression of AtNADK and OsNADK genes is induced to varying degrees by abiotic stress such as cold, heat, drought (PEG), salt (NaCl) and oxidative (methyl viologen, MV), as well as by biotic stress such as the pathogens *Botrytis cinerea, Blumeria graminis, Pseudomonas syringae, Agrobacterium tumefaciens, Mycosphaerella graminicola, Magnaporthe oryzae*, and *Magnaporthe grisea*. In *Arabidopsis*, *AtNADK1* is up-regulated in whole plants under heat and MV stresses, whereas it is down-regulated and does not change under PEG and cold treatment, respectively (Figure 6). By contrast, *AtNADK2* is down-regulated or does not change under these conditions (Figure 6). *AtNADK3* is up-regulated under cold stress, while it is down-regulated under other abiotic stresses including heat, PEG, NaCl, and MV (Figure 6). Additionally, both *AtNADK1* and *AtNADK2* can be up-regulated by some biotic stresses, although with different expression profiles under the same stress (Figure S6). In rice, *OsNADK1/2* are up-regulated in shoot and root under cold stress, in root under heat stress and in shoot under MV stress, respectively; while they are slightly down-regulated or does not change under these conditions (Figure 7). *OsNADK3* shows up-regulated under PEG in shoot and down-regulated or does not change under other abiotic stresses such as cold, heat, NaCl and MV (Figure 7). Similarly, *OsNADK4* is up-regulated in root under cold and heat and down-regulated or does not change under other abiotic stresses including PEG, NaCl, and MV (Figure 7). Additionally, *OsNADK1–4* can be induced by biotic stresses. For instance, *OsNADK1* is up-regulated in callus after infection by *A. tumefaciens* and root after infection by *M. oryzae*, whereas they are down-regulated in root and leaf after infection by *M. graminicola* and *M. grisea*, respectively (Figure S6).

As several hormone response elements were identified in the promoter regions of the AtNADK and OsNADK genes, we also examined their expression profiles under phytohormone treatments in the public *Arabidopsis* and rice microarray data (Figure S7), as well as by qRT-PCR experiments (Figures 6 and 7). *AtNADK1* and *AtNADK2* are differentially expressed in seedlings and cells treated with phytohormones, such as the auxins indole-3-acetic acid (IAA) and naphthaleneacetic acid (NAA), zeatin, SA, ABA, and ABA + SA (Figure S7). Interestingly, *AtNADK1* is up-regulated by ABA, ABA + SA, and ABA + MeJA treatment; whereas *AtNADK2* is down-regulation by ABA and ABA + SA treatment (Figure S7). The expressions of AtNADK genes by qRT-PCR showed *AtNADK1–2* are up-regulated by MeJA treatment and down-regulated or do not change by ABA treatment; *AtNADK3* is down-regulated by ABA treatment and does not change by MeJA treatment (Figure 6). In rice, *OsNADK1–4* are differentially expressed in seedlings and leaves treated with IAA, NAA, zeatin, gibberellin, kinetin, ABA, SA and jasmonic acid (JA) (Figure S7). For instance, *OsNADK3* is up-regulated in roots under *trans*-zeatin treatment, whereas *OsNADK4* is down-regulated in leaves under the same treatment (Figure S7). It is also notable that *OsNADK1–3* can be induced in shoot by ABA treatment and do not change in root, but *OsNADK4* is not obviously changed with the same treatment. All *OsNADK1–4* were up-regulated or not obviously changed by MeJA treatment.

Discussion

Gene fusion and exon shuffling after gene duplication contributed to the expansion and evolution of the NADK family in plants

Gene duplication is a common phenomenon in eukaryotes and contributes to biological diversity during evolution [43,44,45]. The NADKs are represented by at least one gene in nearly every living organism. In this study, we found that the 24 representative plant species from the supergroup Plantae examined contain 2–6 NADK homologs, whereas cyanophytes, considered to be the ancestor of Plantae, had only one NADK homolog per genome, suggesting that a single gene duplication leading to the expansion of the NADK gene family, occurred during the divergence of ancestral cyanophytes from Plantae. The scattered distribution of NADK family genes on chromosomes (Figure S3) and the eight paralogous NADK gene pairs found in five land plants (Figure S4) suggest that segmental duplications may have been involved in the expansion of the NADK gene family and caused differences in the number of NADK genes within subfamilies and species of land plants after divergence from aquatic algae.

Gene fusion and exon shuffling after gene duplication are mechanisms that can enhance the functional divergence of duplicated genes by creating additional domains or rearranging the original functional domains [46,47,48,49]. In this study, phylogenetic analysis, together with the domain organizations and gene structures of NADK family genes, showed distinct evolutionary differences among subfamilies (Figure 2). Therefore, a model was constructed to account for the expansion and evolution of the NADK gene family in plants (Figure 4B). In this model, all NADK family members originated from a common ancestor, which contained only the typical NAD_kinase domain and existed in all living organisms from prokaryotic bacteria to eukaryotic angiosperms; cyanophytes were considered to be the ancestor of Plantae, and at least one gene duplication occurred to yield two NADK genes in eukaryotic glaucophytes (Figures 1 and 4B). Moreover, during the evolutionary diversification from glaucophytes to rhodophytes and Viridiplantae, gene fusion events occurred and additional catalytic domains (ADK, DSPc or PTPc/DSPc) were acquired in the N-terminus, leading to the diversified domain organization seen in subfamily II (Figures 2B and 4B). In addition, exon shuffling after gene duplication contributed to the formation of the bifurcated NAD_kinase domain in subfamily III (Figures 2B and 4B).

Single intron loss and gain lead to the diversified gene structures

Gene structural diversity within gene families is another evolutionary mechanism that promotes variability, and intron loss or gain is important in generating structural diversity and complexity [41,50]. Analyzing the exon/intron structures of NADK genes, we found that the number of introns and intron phases among subfamily members are remarkably conserved, whereas the intron arrangement and intron phases are strikingly distinct between subfamilies (Figures 3 and 4). Further analysis of

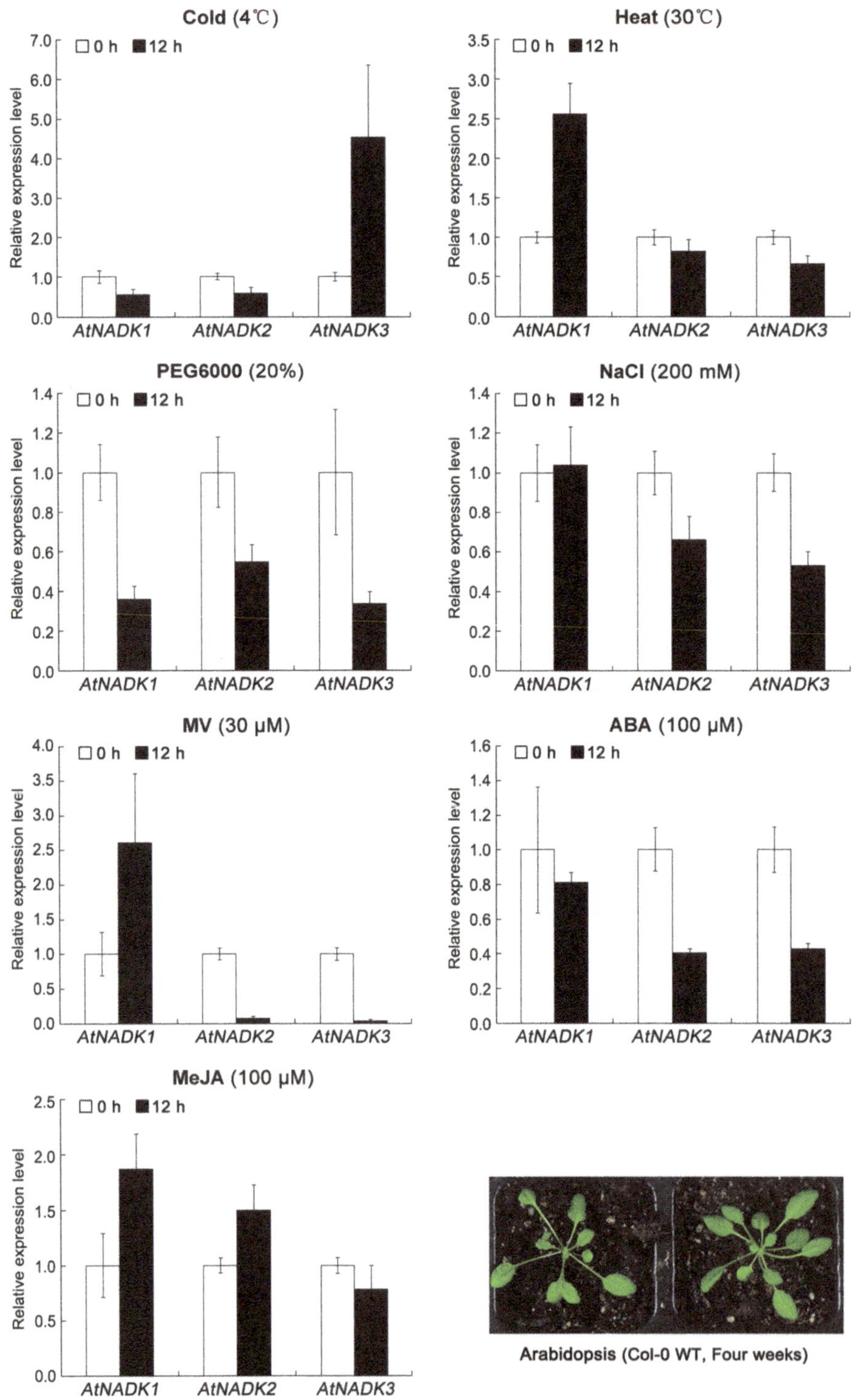

Figure 6. Expression patterns of NADK family genes in *Arabidopsis* under abiotic stress and hormone treatments. Expression levels of *AtNADK1–3* assayed by qRT-PCR under cold (4°C), heat (30°C), drought (20% PEG6000), salt (200 mM NaCl), oxidative (30 μM MV) stresses and MeJA (100 μM), ABA (100 μM) hormone treatments. Data are means ± SD (n = 3) and are representative of similar results from three independent experiments.

the orthologous and paralogous genes in land plants (Figure 4) suggests that single intron loss as well as intron gain likely occurred and contributed to the diversification of gene structure, and consequent functional diversity and divergence, during the evolution of the NADK family in plants.

NADKs are involved in plant responses to several abiotic and biotic stresses

The three NADK genes of *Arabidopsis* belong to three different subfamilies according to our phylogenetic analysis (Figures 1 and 2). The three *AtNADKs* are known to have anti-oxidative functions [10,11,12,16,36], but they have distinct mechanisms by which

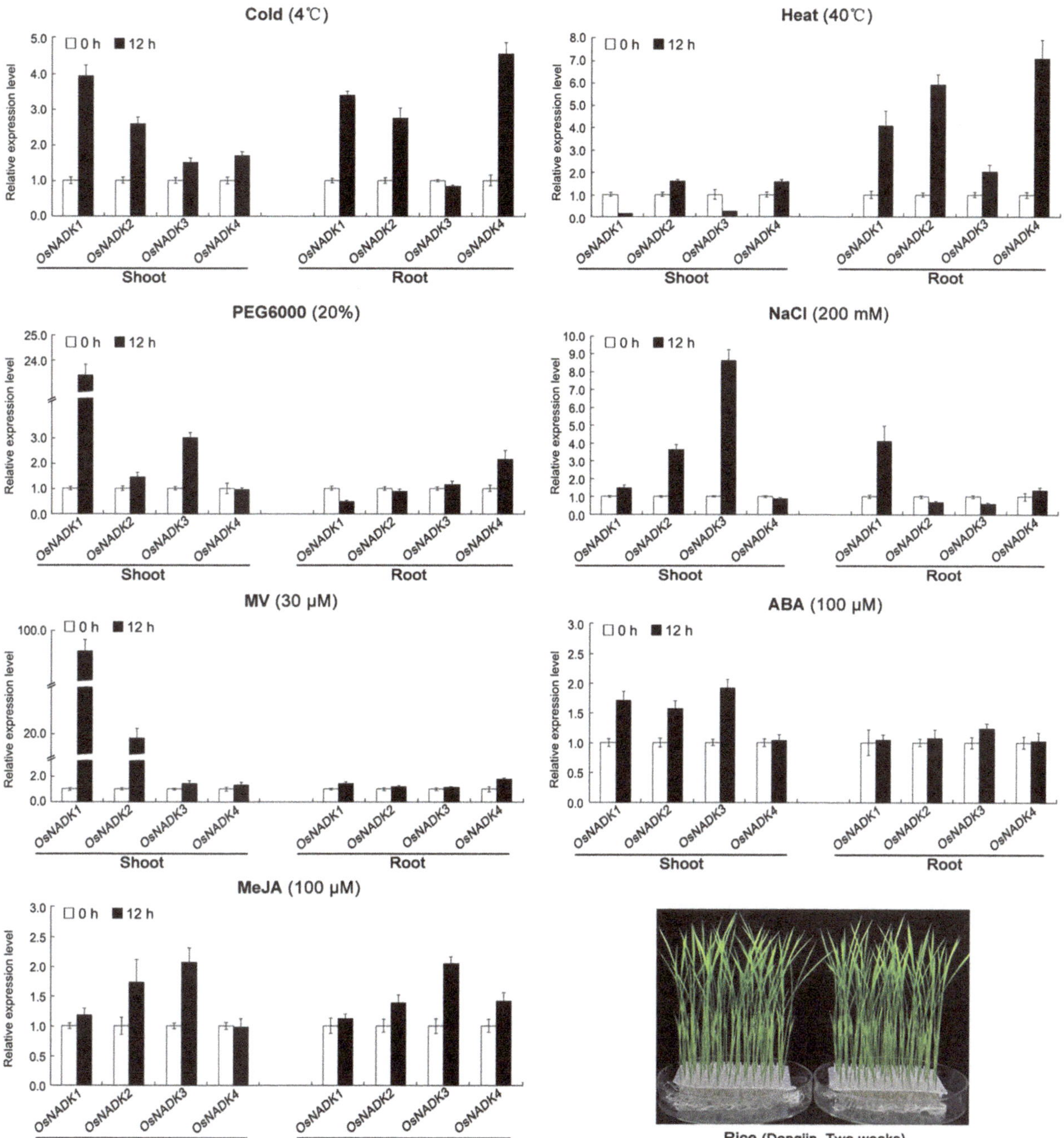

Figure 7. Expression patterns of NADK family genes in rice under abiotic stress and hormone treatments. Expression levels of *OsNADK1–4* assayed by qRT-PCR under cold (4°C), heat (30°C), drought (20% PEG6000), salt (200 mM NaCl), oxidative (30 μM MV) stresses and MeJA (100 μM), ABA (100 μM) hormone treatments. Data are means ± SD (n = 3) and are representative of similar results from three independent experiments.

they facilitate plant resistance to oxidative stress. AtNADK1 (subfamily I) is a cytosolic enzyme that indirectly provides cytosolic NADP for plasma membrane NADPH oxidases, which are the key producers of reactive oxygen species under both normal and stress conditions in plants [36,51,52,53]. AtNADK2 (subfamily II) is a chloroplastic enzyme that plays a vital role in chlorophyll synthesis and chloroplast protection against oxidative damage by regulating plastidic NADP-biosynthesis [11,36]. AtNADK3 (subfamily III) is a peroxisomal enzyme that plays a prominent anti-oxidation role by providing the peroxisomal reductant NADPH [10,36]. NADK genes in subfamily IV were only present in aquatic algae and grouped with *S. cerevisiae NADK1* (Pos5) in our broader phylogenetic analysis (Figure S2). Pos5 is located in the mitochondria and its deletion causes slow growth, sensitivity to oxidative stress, such as paraquat, hyperoxia, H_2O_2, and Cu^{2+}, and biosynthesis deficiency on iron-sulfur clusters [25,54,55]. These observations suggest that plant NADKs in subfamily IV may also localize to mitochondria and protect against oxidative damage by regulating mitochondrial NADP-biosynthesis.

We found that plant NADKs are also involved in responding to several abiotic/biotic stresses, including cold, heat, drought, salt, oxidative and pathogens. *Arabidopsis NADK1* expression is up-regulated by H_2O_2, irradiation and the bacterial pathogen *P. syringae* pv. *tomato*; and the *AtNADK1* deficient mutant is sensitive to γ-irradiation and paraquat-induced oxidative stress [16]. The *AtNADK2* deletion mutant is also hypersensitive to environmental stresses that trigger oxidative stress, such as UVB-irradiation, drought, heat, and high salinity [11]; similarly *AtNADK3* transcription can be induced by MV, high salinity and osmotic shock [10,36]. In this study, we found that *AtNADK1* and *OsNADK1/2* belongs to subfamily I and can be induced to varying degrees by cold, drought (PEG), salt (NaCl), oxidative (MV) and pathogens (Figures 6, 7, S6 and S7). We also found that *AtNADK2*, which belongs to subfamily II, is up-regulated by pathogens such as *P. syringae* pv. *phaseolicola* and *Hyaloperonospora arabidopsidis* (Figure S6); *OsNADK3*, which also belongs to subfamily II, is up-regulated by *Nilaparvata lugens* and down-regulated under anaerobic conditions (Figure S6). *OsNADK4* belongs to subfamily III, and is up-regulated under anaerobic conditions, *N. lugens* and heat in root, and slightly up-regulated by drought or PEG in root (Figures 7 and S6). These observations, together with our analysis of *cis*-elements (Table 1), suggest that NADK genes in subfamilies I, II and III play an important role in plant responses to invading pathogens and abiotic stresses. The functions of NADKs in subfamily IV in abiotic/biotic-stress responses remains to be examined.

NADKs may participate in regulation of hormone signaling

AtNADK3 transcription can be slightly induced by ABA and the mutant is hypersensitive to ABA [10]. In this study, we found that all *AtNADK1–3* showed down-regulated or slightly down-regulated in response to ABA (Figure 6), while *OsNADK1–3* were up-regulated by ABA treatment (Figure 7). It's important to note that response profiles of NADK genes by qRT-PCR are not completely consisted with microarray data or previous studies, because of the different experimental material, processing and analyzing methods used in this study. Moreover, seven AtNADK and OsNADK genes (*AtNADK1–3* and *OsNADK1–4*) contain *cis*-acting ABA responsive elements (Table 1), suggesting that NADKs probably participate in regulation of ABA signaling in plants.

JA and its derivative MeJA are important signaling molecules in plant responses to many abiotic and biotic stresses such as wounding and pathogens [56,57]. Moreover, JA biosynthesis occurs in peroxisomes [57,58]. AtNADK3 is also a peroxisomal enzyme and may phosphorylate the NADH derived from β-oxidation to yield NADPH needed for anti-oxidant defense [36]. *AtNADK1/2* and *OsNADK1–4* were up-regulated or slightly up-regulated under MeJA treatment (Figures 6 and 7). Moreover, several of these genes (*AtNADK1/2/3* and *OsNADK1/4*) contain CGTCA-MeJA-responsiveness motifs in their promoter regions (Table 1). These observations suggest that NADKs may also participate in regulation of JA or MeJA signaling in plants.

It should be pointed out that although transcription factors and *cis*-regulatory elements play important roles in regulating gene expression, the relationship is not always direct and the inducibility of expression is often affected by many factors and on multiple levels [59,60]. For example, ABA-responsive *cis*-regulatory elements were identified in the promoter regions of all seven NADK genes in *Arabidopsis* and rice, but only *OsNADK1–3* were up-regulated or slightly up-regulated by ABA treatment (Figures 7 and S7). Moreover, considering the limitations of the PlantCARE program in predicting *cis*-regulatory elements, further experiments are necessary to verify the relationship between the inducible expression profiles of NADK genes and the *cis*-elements within their promoters.

Materials and Methods

Data retrieval and identification of NADK genes

The protein sequences of 24 completely or partially sequenced plant genomes representing the eight major plant lineages were retrieved from public databases. All of the protein sequences were the most current non-redundant sequences from the following sources (Table S1): the glaucophyte *Cyanophora paradoxa* from the *Cyanophora* Genome Project (http://cyanophora.rutgers.edu/cyanophora/home.php) [61]; the rhodophytes *Cyanidioschyzon merolae* and *Galdieria sulphuraria* from the *Cyanidioschyzon* Genome Project (http://merolae.biol.s.u-tokyo.ac.jp/) [62] and the *Galdieria* Genome Project (http://genomics.msu.edu/galdieria/) [63], respectively; the chlorophytes *O. tauri* (version 2.0) [64], *O. lucimarinus* (version 2.0) [65], *M. pusilla* strain RCC299 (version 3.0) and strain CCMP1545 (version 3.0) [66], *Chlorella variabilis* NC64A (version 1.0) [67], *Coccomyxa subellipsoidea* C-169 (version 2.0) [68], *C. reinhardtii* (version 4.0) [69] and *V. carteri* (version 2.0) and the bryophyte *P. patens* (version 3.0) [70] and lycophyte *S. moellendorffii* (version 1.1) [71] from the Joint Genome Institute (JGI, http://genome.jgi-psf.org/); for the gymnosperm *P. sitchensis* [72] from NCBI (http://www.ncbi.nlm.nih.gov/, partial sequences only because of its genome is only partially sequenced); the monocots *B. distachyon* (version 1.2) [73], *Setaria italica* (version 2.1) [74] and *S. bicolor* (version 2.1) [75] from JGI, rice (version 7.0) and *Z. mays* (version 5.6) from the Institute for Genomic Research Rice Genome Annotation Project (http://rice.plantbiology.msu.edu/index.shtml) [76] and MaizeSequence (http://www.maizesequence.org/index.html) [77], respectively; the eudicots *P. trichocarpa* (version 2.2) [78] and *Cucumis sativus* (version 1.0) [79] from JGI, *Arabidopsis* (version 10.0) from the Arabidopsis Information Resource (http://www.arabidopsis.org/) [80], *B. rapa* (version 1.5) from the *Brassica* Database (http://brassicadb.org/brad/) [81] and *V. vinifera* (version 1.0) from GenoScope (http://www.genoscope.cns.fr/externe/GenomeBrowser/Vitis/) [82]. All of the above protein sequences were integrated into a local protein database for the subsequent identification of NADK homologs.

To identify the NADK genes and their homologs in the Plantae supergroup, HMMER v3.0 [83,84] was used to perform an HMM search against the local protein database, using the family-specific NAD_kinase domain (PF01513) HMM profile obtained from the Pfam database [85,86]. The HMM search was performed with the default parameters and an E-value cutoff of $1e^{-5}$. If a candidate gene had multiple alternative splice variants, the longest variant was used to represent the candidate protein. The Pfam and SMART [87,88] databases were employed to detect conserved domains in the candidate proteins, and the search results were refined manually to eliminate partial NADK domains and other potential false positives.

Sequence alignment and phylogenetic analysis

The NAD_kinase domain sequences of the candidate proteins were aligned using the MUSCLE v3.8 program with the default parameters [89,90] and the alignments were manually edited using the BioEdit v7.0 program [91]. The rooted maximum-likelihood tree was inferred from the resulting alignments using the Phylip v3.68 package [92] under the γ-corrected Jones–Taylor–Thornton model [93] with default parameters, and the reliability of interior branches was assessed with 1000 bootstrap resamplings. In addition, the MEGA v5.0 program [94] was also used to reconstruct the phylogenetic trees by the neighbor joining,

Table 1. Abiotic/biotic stress and hormone response elements in *AtNADK* and *OsNADK* promoters*.

Cis-elements	Core sequences	Functions of the cis-elements	*OsNADK1*	*OsNADK2*	*OsNADK3*	*OsNADK4*	*AtNADK1*	*AtNADK2*	*AtNADK3*
Abiotic/biotic stress									
LTR	CCGAAA	low-temperature responsiveness		√		√			
HSE	AAAAAATTTC	heat-stress responsiveness	√	√	√	√	√		√
MBS	TAACTG	MYB binding site, drought inducibility	√	√	√	√			√
ARE	TGGTTT	anaerobic induction	√	√	√	√	√	√	√
TC-rich repeats	ATTCTCTAAC	defense and stress responsiveness		√	√			√	√
W box	TTGACC	wound and pathogen responsiveness			√		√		
Box S	AGCCACC	wound and pathogen responsiveness	√						
GCC box	AGCCGCC	wound and pathogen responsiveness				√			
EIRE	TTCGACC	Elicitor responsiveness						√	
Box-W1	TTGACC	fungal elicitor responsiveness			√		√		√
Hormone response elements									
AuxRR-core	GGTCCAT	auxin responsiveness			√				
TGA-element	AACGAC	Auxin responsiveness					√	√	
GARE-motif	AAACAGA	Gibberellin responsiveness		√			√		
ABRE	CACGTG	abscisic acid responsiveness	√	√	√	√	√	√	√
ERE	ATTTCAAA	Ethylene responsiveness	√	√	√		√		
TCA-element	GAGAAGAATA	salicylic acid responsiveness					√		√
CGTCA-motif	CGTCA	MeJA responsiveness	√			√	√	√	√

*the *cis*-elements were identified with the PlantCARE program (http://bioinformatics.psb.ugent.be/webtools/plantcare/html/) using the sequences 1500 bp upstream from the transcription start site of each NADK gene. The "√" means the NADK gene contains this *cis*-element in the promoter region.

minimal evolution and maximum parsimony methods, and to display the phylogenetic trees.

The rates of non-synonymous substitution (Ka) and synonymous substitution (Ks) were estimated for the orthologous and paralogous gene pairs of NADK family genes using the Codeml program in PAML v4.3 [95] interface tool of PAL2NAL [96], based on the aligned amino acid sequences and the corresponding nucleotide sequences. The duplication and divergence times of each gene pairs were estimated from the Ks of λ substitutions per synonymous site per year as $T = Ks/2\lambda$ ($\lambda = 6.5\times10^{-9}$) [97,98].

Exon/intron structure and conserved motif analysis

The exon/intron structures of individual NADK genes were obtained through the Gene Structure Display Server (http://gsds.cbi.pku.edu.cn) [99] by aligning the coding or cDNA sequences with their corresponding genomic DNA sequences from Phytozome v9.1 (http://www.phytozome.net/) [100]. To illustrate the evolution of introns, gene models were inspected for annotation of introns, and exon/intron boundaries were manually checked. For a subset of genes, predictions pertaining to the types of introns were independently checked using Common Introns Within Orthologous Genes software (http://ciwog.gdcb.iastate.edu/) [101].

The conserved functional motifs within NAD_kinase domains were identified using the MEME v4.9 program (http://meme.sdsc.edu) [102] with the default parameters, and the sequence logos (graphical representations) of these motifs or domains were generated with the WebLogo v3.3 server (http://weblogo.threeplusone.com/create.cgi) [103] based on the results of protein sequence alignments.

Cis-regulatory elements and expression profile analysis

The 1500 bp upstream of the transcription start site of all NADK genes in *Arabidopsis* and rice were obtained from Phytozome v9.1 (http://www.phytozome.net/), and the *cis*-regulatory elements were identified using the PlantCARE program (http://bioinformatics.psb.ugent.be/webtools/plantcare/html/) [104,105].

ATH1 22 k and Os 51 k microarray data in the Genevestigator V3 database were used to analyze the tissue-specific and inducible expression profiles of NADK genes in *Arabidopsis* and rice [106], respectively. In addition, EST profiles of each NADK gene in *Arabidopsis* and rice were also analyzed by BLASTN search against the corresponding NCBI (http://www.ncbi.nlm.nih.gov/) EST database, with the coding sequences of the individual NADK gene as query. The BLASTN searches were performed with the following criteria: E-value$<1e^{-10}$ and nucleotide identities >95% over 150 bp.

Plant materials, treatments and quantitative real-time PCR (qRT-PCR) analysis

The *Arabidopsis* (Col-0) seeds were pretreated at 4°C for 3 days and directly sown in Sunshine MVP potting soil, and then grown in growth chambers under 16 h light/8 h dark at 22±1°C; while rice (*Oryza sativa* ssp. *japonica* cv. Dongjin) seeds were pretreated germination at 26±1°C for 4 days and transplanted to 1/2 Hogland nutrient solution, and then grown in growth chambers under 16 h light/8 h dark at 26±1°C. Following two (rice) or four (*Arabidopsis*) weeks of growth, the seedlings were grown under 4°C or 30°C (*Arabidopsis*)/40°C (rice) for 12 h for cold and heat treatments, respectively; submerged in 200 mM NaCl or 20% PEG6000 solutions for 12 h for drought and salt treatments, respectively. For oxidative stress and hormone treatments, solutions of 30 μM MV, 100 μM MeJA and ABA were separately sprayed on seedlings for 12 h. All the MV, MeJA and ABA used for treatments were purchased from Sigma-Aldrich. Samples were collected following treatment and immediately frozen at −80°C.

Total RNA was extracted by using Trizol reagent (Takara, Japan) according to the manufacturer's instructions and treated with RNase-free DNase I (Invitrogen, USA) for 15 min to remove any DNA contamination. RNA concentration and quality were verified by using the NanoDrop 1000 Spectrophotometers (Thermo, USA), and cDNAs were synthesized by using oligo $d(T)_{18}$ reverse primer from 5 μg of total RNA in a total volume of 20 μL by using EasyScript First-Strand cDNA Synthesis SuperMix (TransGen Biotech, China). qRT-PCR reactions were carried out in 96-well (20 μL) format by using the FastStart Essential DNA Green Master (Roche, Switzerland), and were performed in an CFX96 Touch Real-Time PCR Detection System (BIO-RAD, USA). The reactions were repeated three times and the quantitative analysis used the $2^{-\Delta\Delta CT}$ method. All gene-specific primers were designed to avoid the conserved region and span introns or cross an exon-exon junction. The detailed primer sequences are shown in Table S5. The *AtTub6* (AT5G12250) and *OsActin1* (accession ID KC140126) were chosen as the internal control in *Arabidopsis* and rice, respectively.

Supporting Information

Figure S1 Amino acid sequence alignment of NAD_kinase domains in NADKs.

Figure S2 Phylogenetic relationship and domain organization of NADK genes in plants, yeast and humans.

Figure S3 Chromosomal locations of NADK family genes in plants.

Figure S4 Segmental duplications and duplication and divergence times of NADK family genes in land plants.

Figure S5 Developmental expression patterns of NADK family genes in *Arabidopsis* and rice.

Figure S6 Expression patterns of NADK family genes in *Arabidopsis* and rice under abiotic and biotic stresses.

Figure S7 Expression patterns of NADK family genes in *Arabidopsis* and rice with various hormone treatments.

Table S1 The NADK family genes in plants.

Table S2 *Cis*-element analysis of the *AtNADK* and *OsNADK* promoters.

Table S3 EST profiles of NADK genes in *Arabidopsis*.

Table S4 EST profiles of NADK genes in rice.

Table S5 The gene-specific primers used for qRT-PCR in this study.

Author Contributions

Conceived and designed the experiments: WYL KMC. Performed the experiments: XW RL. Analyzed the data: WYL WQL. Contributed reagents/materials/analysis tools: WYL XW RL. Wrote the paper: WYL WQL. Provided funding and critically revised the manuscript: KMC.

References

1. Ying W (2008) NAD+/NADH and NADP+/NADPH in cellular functions and cell death: regulation and biological consequences. Antioxid Redox Signal 10: 179–206.
2. Kawai S, Murata K (2008) Structure and function of NAD kinase and NADP phosphatase: key enzymes that regulate the intracellular balance of NAD(H) and NADP(H). Biosci Biotechnol Biochem 72: 919–930.
3. Grose JH, Joss L, Velick SF, Roth JR (2006) Evidence that feedback inhibition of NAD kinase controls responses to oxidative stress. Proc Natl Acad Sci U S A 103: 7601–7606.
4. Kawai S, Fukuda C, Mukai T, Murata K (2005) MJ0917 in archaeon *Methanococcus jannaschii* is a novel NADP phosphatase/NAD kinase. J Biol Chem 280: 39200–39207.
5. Kawai S, Mori S, Mukai T, Suzuki S, Yamada T, et al. (2000) Inorganic Polyphosphate/ATP-NAD kinase of Micrococcus flavus and *Mycobacterium tuberculosis* H37Rv. Biochem Biophys Res Commun 276: 57–63.
6. Kawai S, Mori S, Mukai T, Hashimoto W, Murata K (2001) Molecular characterization of *Escherichia coli* NAD kinase. Eur J Biochem 268: 4359–4365.
7. Kawai S, Suzuki S, Mori S, Murata K (2001) Molecular cloning and identification of UTR1 of a yeast *Saccharomyces cerevisiae* as a gene encoding an NAD kinase. FEMS Microbiol Lett 200: 181–184.
8. Outten CE, Culotta VC (2003) A novel NADH kinase is the mitochondrial source of NADPH in *Saccharomyces cerevisiae*. EMBO J 22: 2015–2024.
9. Lerner F, Niere M, Ludwig A, Ziegler M (2001) Structural and functional characterization of human NAD kinase. Biochem Biophys Res Commun 288: 69–74.
10. Chai MF, Wei PC, Chen QJ, An R, Chen J, et al. (2006) NADK3, a novel cytoplasmic source of NADPH, is required under conditions of oxidative stress and modulates abscisic acid responses in *Arabidopsis*. Plant J 47: 665–674.
11. Chai MF, Chen QJ, An R, Chen YM, Chen J, et al. (2005) NADK2, an *Arabidopsis* chloroplastic NAD kinase, plays a vital role in both chlorophyll synthesis and chloroplast protection. Plant Mol Biol 59: 553–564.
12. Turner WL, Waller JC, Vanderbeld B, Snedden WA (2004) Cloning and characterization of two NAD kinases from *Arabidopsis*. identification of a calmodulin binding isoform. Plant Physiol 135: 1243–1255.
13. Stephan C, Renard M, Montrichard F (2000) Evidence for the existence of two soluble NAD(+) kinase isoenzymes in *Euglena gracilis* Z. Int J Biochem Cell Biol 32: 855–863.
14. Li YF, Shi F (2006) Partial rescue of pos5 mutants by YEF1 and UTR1 genes in *Saccharomyces cerevisiae*. Acta Biochim Biophys Sin (Shanghai) 38: 293–298.
15. Shi F, Kawai S, Mori S, Kono E, Murata K (2005) Identification of ATP-NADH kinase isozymes and their contribution to supply of NADP(H) in *Saccharomyces cerevisiae*. FEBS J 272: 3337–3349.
16. Berrin JG, Pierrugues O, Brutesco C, Alonso B, Montillet JL, et al. (2005) Stress induces the expression of AtNADK-1, a gene encoding a NAD(H) kinase in *Arabidopsis thaliana*. Mol Genet Genomics 273: 10–19.
17. Oganesyan V, Huang C, Adams PD, Jancarik J, Yokota HA, et al. (2005) Structure of a NAD kinase from *Thermotoga maritima* at 2.3 A resolution. Acta Crystallogr Sect F Struct Biol Cryst Commun 61: 640–646.
18. Mori S, Kawai S, Mikami B, Murata K (2001) Crystallization and preliminary X-ray analysis of NAD kinase from *Mycobacterium tuberculosis* H37Rv. Acta Crystallogr D Biol Crystallogr 57: 1319–1320.
19. Ando T, Ohashi K, Ochiai A, Mikami B, Kawai S, et al. (2011) Structural Determinants of Discrimination of NAD+ from NADH in Yeast Mitochondrial NADH Kinase Pos5. J Biol Chem 286: 29984–29992.
20. Raffaelli N, Finaurini L, Mazzola F, Pucci L, Sorci L, et al. (2004) Characterization of *Mycobacterium tuberculosis* NAD kinase: functional analysis of the full-length enzyme by site-directed mutagenesis. Biochemistry 43: 7610–7617.
21. Garavaglia S, Raffaelli N, Finaurini L, Magni G, Rizzi M (2004) A novel fold revealed by *Mycobacterium tuberculosis* NAD kinase, a key allosteric enzyme in NADP biosynthesis. J Biol Chem 279: 40980–40986.
22. Labesse G, Douguet D, Assairi L, Gilles AM (2002) Diacylglyceride kinases, sphingosine kinases and NAD kinases: distant relatives of 6-phosphofructokinases. Trends Biochem Sci 27: 273–275.
23. Mori S, Yamasaki M, Maruyama Y, Momma K, Kawai S, et al. (2005) NAD-binding mode and the significance of intersubunit contact revealed by the crystal structure of *Mycobacterium tuberculosis* NAD kinase-NAD complex. Biochem Biophys Res Commun 327: 500–508.
24. Liu J, Lou Y, Yokota H, Adams PD, Kim R, et al. (2005) Crystal structures of an NAD kinase from *Archaeoglobus fulgidus* in complex with ATP, NAD, or NADP. J Mol Biol 354: 289–303.
25. Shianna KV, Marchuk DA, Strand MK (2006) Genomic characterization of POS5, the *Saccharomyces cerevisiae* mitochondrial NADH kinase. Mitochondrion 6: 94–101.
26. Bieganowski P, Seidle HF, Wojcik M, Brenner C (2006) Synthetic lethal and biochemical analyses of NAD and NADH kinases in *Saccharomyces cerevisiae* establish separation of cellular functions. J Biol Chem 281: 22439–22445.
27. Pollak N, Niere M, Ziegler M (2007) NAD kinase levels control the NADPH concentration in human cells. J Biol Chem 282: 33562–33571.
28. Singh R, Mailloux RJ, Puiseux-Dao S, Appanna VD (2007) Oxidative stress evokes a metabolic adaptation that favors increased NADPH synthesis and decreased NADH production in *Pseudomonas fluorescens*. J Bacteriol 189: 6665–6675.
29. Anderson JM, Charbonneau H, Jones HP, McCann RO, Cormier MJ (1980) Characterization of the plant nicotinamide adenine dinucleotide kinase activator protein and its identification as calmodulin. Biochemistry 19: 3113–3120.
30. Harding SA, Oh SH, Roberts DM (1997) Transgenic tobacco expressing a foreign calmodulin gene shows an enhanced production of active oxygen species. EMBO J 16: 1137–1144.
31. Zhou CY, Wu GL, Duan ZQ, Wu LL, Gao YS, et al. (2010) H_2O_2-NOX system: an important mechanism for developmental regulation and stress response in plants. Chinese Bulletin of Botany 45: 615–631.
32. Ruiz JM, Sanchez E, Garcia PC, Lopez-Lefebre LR, Rivero RM, et al. (2002) Proline metabolism and NAD kinase activity in green bean plants subjected to cold-shock. Phytochemistry 59: 473–478.
33. Delumeau O, Paven M-CM-L, Montrichard F, Laval-Martin DL (2000) Effects of short-term NaCl stress on calmodulin transcript levels and calmodulin-dependent NAD kinase activity in two species of tomato. Plant, Cell & Environment 23: 329–336.
34. Zagdanska B (1990) NAD kinase activity in wheat leaves under water deficit. Acta Biochim Pol 1990;37(3): 385–9.
35. Shi F, Li Y, Wang X (2009) Molecular properties, functions, and potential applications of NAD kinases. Acta Biochim Biophys Sin (Shanghai) 41: 352–361.
36. Waller JC, Dhanoa PK, Schumann U, Mullen RT, Snedden WA (2010) Subcellular and tissue localization of NAD kinases from *Arabidopsis*: compartmentalization of de novo NADP biosynthesis. Planta 231: 305–317.
37. Wu LL, Zhou CY, Gao YS, Cong YX, Chen KM, et al. (2011) Cloning and genetic transformation of *OsNADK3* gene in rice. Journal of Nuclear Agricultural Sciences 25: 0863–0870.
38. Lee TH, Tang H, Wang X, Paterson AH (2013) PGDD: a database of gene and genome duplication in plants. Nucleic Acids Res 41: D1152–1158.
39. Hardison RC (1996) A brief history of hemoglobins: plant, animal, protist, and bacteria. Proc Natl Acad Sci U S A 93: 5675–5679.
40. Rogozin IB, Wolf YI, Sorokin AV, Mirkin BG, Koonin EV (2003) Remarkable interkingdom conservation of intron positions and massive, lineage-specific intron loss and gain in eukaryotic evolution. Curr Biol 13: 1512–1517.
41. Li W, Liu B, Yu L, Feng D, Wang H, et al. (2009) Phylogenetic analysis, structural evolution and functional divergence of the 12-oxo-phytodienoate acid reductase gene family in plants. BMC Evol Biol 9: 90.
42. Winter D, Vinegar B, Nahal H, Ammar R, Wilson GV, et al. (2007) An "Electronic Fluorescent Pictograph" browser for exploring and analyzing large-scale biological data sets. PLoS One 2: e718.
43. Bowers JE, Chapman BA, Rong J, Paterson AH (2003) Unravelling angiosperm genome evolution by phylogenetic analysis of chromosomal duplication events. Nature 422: 433–438.
44. Van de Peer Y, Fawcett JA, Proost S, Sterck L, Vandepoele K (2009) The flowering world: a tale of duplications. Trends Plant Sci 14: 680–688.
45. Magadum S, Banerjee U, Murugan P, Gangapur D, Ravikesavan R (2013) Gene duplication as a major force in evolution. J Genet 92: 155–161.
46. Kolkman JA, Stemmer WP (2001) Directed evolution of proteins by exon shuffling. Nat Biotechnol 19: 423–428.
47. Jones CD, Begun DJ (2005) Parallel evolution of chimeric fusion genes. Proc Natl Acad Sci U S A 102: 11373–11378.
48. Kaessmann H (2010) Origins, evolution, and phenotypic impact of new genes. Genome Res 20: 1313–1326.
49. Morgante M, Brunner S, Pea G, Fengler K, Zuccolo A, et al. (2005) Gene duplication and exon shuffling by helitron-like transposons generate intraspecies diversity in maize. Nat Genet 37: 997–1002.
50. Zhang Z, Kishino H (2004) Genomic background predicts the fate of duplicated genes: evidence from the yeast genome. Genetics 166: 1995–1999.
51. Wang GF, Li WQ, Li WY, Wu GL, Zhou CY, et al. (2013) Characterization of Rice NADPH Oxidase Genes and Their Expression under Various Environmental Conditions. Int J Mol Sci 14: 9440–9458.
52. Sagi M, Fluhr R (2006) Production of reactive oxygen species by plant NADPH oxidases. Plant Physiol 141: 336–340.
53. Foreman J, Demidchik V, Bothwell JH, Mylona P, Miedema H, et al. (2003) Reactive oxygen species produced by NADPH oxidase regulate plant cell growth. Nature 422: 442–446.

54. Strand MK, Stuart GR, Longley MJ, Graziewicz MA, Dominick OC, et al. (2003) POS5 gene of *Saccharomyces cerevisiae* encodes a mitochondrial NADH kinase required for stability of mitochondrial DNA. Eukaryot Cell 2: 809–820.
55. Pain J, Balamurali MM, Dancis A, Pain D (2010) Mitochondrial NADH kinase, Pos5p, is required for efficient iron-sulfur cluster biogenesis in *Saccharomyces cerevisiae*. J Biol Chem 285: 39409–39424.
56. Chehab EW, Kaspi R, Savchenko T, Rowe H, Negre-Zakharov F, et al. (2008) Distinct roles of jasmonates and aldehydes in plant-defense responses. PLoS One 3: e1904.
57. Wasternack C, Hause B (2013) Jasmonates: biosynthesis, perception, signal transduction and action in plant stress response, growth and development. An update to the 2007 review in Annals of Botany. Ann Bot 111: 1021–1058.
58. Schaller A, Stintzi A (2009) Enzymes in jasmonate biosynthesis - structure, function, regulation. Phytochemistry 70: 1532–1538.
59. Ma Q, Chirn GW, Szustakowski JD, Bakhtiarova A, Kosinski PA, et al. (2008) Uncovering mechanisms of transcriptional regulations by systematic mining of *cis* regulatory elements with gene expression profiles. BioData Min 1: 4.
60. Jackson RJ, Hellen CU, Pestova TV (2010) The mechanism of eukaryotic translation initiation and principles of its regulation. Nat Rev Mol Cell Biol 11: 113–127.
61. Price DC, Chan CX, Yoon HS, Yang EC, Qiu H, et al. (2012) *Cyanophora paradoxa* genome elucidates origin of photosynthesis in algae and plants. Science 335: 843–847.
62. Matsuzaki M, Misumi O, Shin IT, Maruyama S, Takahara M, et al. (2004) Genome sequence of the ultrasmall unicellular red alga *Cyanidioschyzon merolae* 10D. Nature 428: 653–657.
63. Barbier G, Oesterhelt C, Larson MD, Halgren RG, Wilkerson C, et al. (2005) Comparative genomics of two closely related unicellular thermo-acidophilic red algae, *Galdieria sulphuraria* and *Cyanidioschyzon merolae*, reveals the molecular basis of the metabolic flexibility of *Galdieria sulphuraria* and significant differences in carbohydrate metabolism of both algae. Plant Physiol 137: 460–474.
64. Derelle E, Ferraz C, Rombauts S, Rouze P, Worden AZ, et al. (2006) Genome analysis of the smallest free-living eukaryote *Ostreococcus tauri* unveils many unique features. Proc Natl Acad Sci U S A 103: 11647–11652.
65. Palenik B, Grimwood J, Aerts A, Rouze P, Salamov A, et al. (2007) The tiny eukaryote *Ostreococcus* provides genomic insights into the paradox of plankton speciation. Proc Natl Acad Sci U S A 104: 7705–7710.
66. Worden AZ, Lee JH, Mock T, Rouze P, Simmons MP, et al. (2009) Green evolution and dynamic adaptations revealed by genomes of the marine picoeukaryotes *Micromonas*. Science 324: 268–272.
67. Blanc G, Duncan G, Agarkova I, Borodovsky M, Gurnon J, et al. (2010) The *Chlorella variabilis* NC64A genome reveals adaptation to photosymbiosis, coevolution with viruses, and cryptic sex. Plant Cell 22: 2943–2955.
68. Blanc G, Agarkova I, Grimwood J, Kuo A, Brueggeman A, et al. (2012) The genome of the polar eukaryotic microalga *Coccomyxa subellipsoidea* reveals traits of cold adaptation. Genome Biol 13: R39.
69. Merchant SS, Prochnik SE, Vallon O, Harris EH, Karpowicz SJ, et al. (2007) The *Chlamydomonas* genome reveals the evolution of key animal and plant functions. Science 318: 245–250.
70. Rensing SA, Lang D, Zimmer AD, Terry A, Salamov A, et al. (2008) The *Physcomitrella* genome reveals evolutionary insights into the conquest of land by plants. Science 319: 64–69.
71. Banks JA, Nishiyama T, Hasebe M, Bowman JL, Gribskov M, et al. (2011) The *Selaginella* genome identifies genetic changes associated with the evolution of vascular plants. Science 332: 960–963.
72. Ralph SG, Chun HJ, Kolosova N, Cooper D, Oddy C, et al. (2008) A conifer genomics resource of 200,000 spruce (*Picea spp.*) ESTs and 6,464 high-quality, sequence-finished full-length cDNAs for Sitka spruce (*Picea sitchensis*). BMC Genomics 9: 484.
73. International-Brachypodium-Initiative (2010) Genome sequencing and analysis of the model grass *Brachypodium distachyon*. Nature 463: 763–768.
74. Zhang G, Liu X, Quan Z, Cheng S, Xu X, et al. (2012) Genome sequence of foxtail millet (*Setaria italica*) provides insights into grass evolution and biofuel potential. Nat Biotechnol 30: 549–554.
75. Paterson AH, Bowers JE, Bruggmann R, Dubchak I, Grimwood J, et al. (2009) The *Sorghum bicolor* genome and the diversification of grasses. Nature 457: 551–556.
76. Goff SA, Ricke D, Lan TH, Presting G, Wang R, et al. (2002) A draft sequence of the rice genome (*Oryza sativa* L. *ssp. japonica*). Science 296: 92–100.
77. Schnable PS, Ware D, Fulton RS, Stein JC, Wei F, et al. (2009) The B73 maize genome: complexity, diversity, and dynamics. Science 326: 1112–1115.
78. Tuskan GA, Difazio S, Jansson S, Bohlmann J, Grigoriev I, et al. (2006) The genome of black cottonwood, *Populus trichocarpa* (Torr. & Gray). Science 313: 1596–1604.
79. Huang S, Li R, Zhang Z, Li L, Gu X, et al. (2009) The genome of the cucumber, *Cucumis sativus* L. Nat Genet 41: 1275–1281.
80. Arabidopsis-Genome-Initiative (2000) Analysis of the genome sequence of the flowering plant *Arabidopsis thaliana*. Nature 408: 796–815.
81. Wang X, Wang H, Wang J, Sun R, Wu J, et al. (2011) The genome of the mesopolyploid crop species *Brassica rapa*. Nat Genet 43: 1035–1039.
82. Jaillon O, Aury JM, Noel B, Policriti A, Clepet C, et al. (2007) The grapevine genome sequence suggests ancestral hexaploidization in major angiosperm phyla. Nature 449: 463–467.
83. Zhang Z, Wood WI (2003) A profile hidden Markov model for signal peptides generated by HMMER. Bioinformatics 19: 307–308.
84. Finn RD, Clements J, Eddy SR (2011) HMMER web server: interactive sequence similarity searching. Nucleic Acids Res 39: W29–37.
85. Finn RD, Mistry J, Schuster-Bockler B, Griffiths-Jones S, Hollich V, et al. (2006) Pfam: clans, web tools and services. Nucleic Acids Res 34: D247–251.
86. Punta M, Coggill PC, Eberhardt RY, Mistry J, Tate J, et al. (2012) The Pfam protein families database. Nucleic Acids Res 40: D290–301.
87. Schultz J, Milpetz F, Bork P, Ponting CP (1998) SMART, a simple modular architecture research tool: identification of signaling domains. Proc Natl Acad Sci U S A 95: 5857–5864.
88. Letunic I, Doerks T, Bork P (2012) SMART 7: recent updates to the protein domain annotation resource. Nucleic Acids Res 40: D302–305.
89. Edgar RC (2004) MUSCLE: multiple sequence alignment with high accuracy and high throughput. Nucleic Acids Res 32: 1792–1797.
90. Edgar RC (2004) MUSCLE: a multiple sequence alignment method with reduced time and space complexity. BMC Bioinformatics 5: 113.
91. Hall TA (1999) BioEdit: a user-friendly biological sequence alignment editor and analysis program for Windows 95/98/NT. Nucleic Acids Symposium Series 41: 95–98.
92. Krawetz S, Retief J (1999) Phylogenetic Analysis Using PHYLIP. Bioinformatics Methods and Protocols: Humana Press. 243–258.
93. Jones DT, Taylor WR, Thornton JM (1992) The rapid generation of mutation data matrices from protein sequences. Comput Appl Biosci 8: 275–282.
94. Tamura K, Peterson D, Peterson N, Stecher G, Nei M, et al. (2011) MEGA5: molecular evolutionary genetics analysis using maximum likelihood, evolutionary distance, and maximum parsimony methods. Mol Biol Evol 28: 2731–2739.
95. Yang Z (2007) PAML 4: phylogenetic analysis by maximum likelihood. Mol Biol Evol 24: 1586–1591.
96. Suyama M, Torrents D, Bork P (2006) PAL2NAL: robust conversion of protein sequence alignments into the corresponding codon alignments. Nucleic Acids Res 34: W609–612.
97. Puranik S, Sahu PP, Mandal SN, B VS, Parida SK, et al. (2013) Comprehensive genome-wide survey, genomic constitution and expression profiling of the NAC transcription factor family in foxtail millet (*Setaria italica* L.). PLoS One 8: e64594.
98. Lynch M, Conery JS (2000) The evolutionary fate and consequences of duplicate genes. Science 290: 1151–1155.
99. Guo AY, Zhu QH, Chen X, Luo JC (2007) GSDS: a gene structure display server. Yi Chuan 29(8): 1023–6.
100. Goodstein DM, Shu S, Howson R, Neupane R, Hayes RD, et al. (2012) Phytozome: a comparative platform for green plant genomics. Nucleic Acids Res 40: D1178–1186.
101. Wilkerson MD, Ru Y, Brendel VP (2009) Common introns within orthologous genes: software and application to plants. Briefings in Bioinformatics 10: 631–644.
102. Bailey TL, Boden M, Buske FA, Frith M, Grant CE, et al. (2009) MEME SUITE: tools for motif discovery and searching. Nucleic Acids Res 37: W202–208.
103. Crooks GE, Hon G, Chandonia JM, Brenner SE (2004) WebLogo: a sequence logo generator. Genome Res 14: 1188–1190.
104. Lescot M, Dehais P, Thijs G, Marchal K, Moreau Y, et al. (2002) PlantCARE, a database of plant *cis*-acting regulatory elements and a portal to tools for in silico analysis of promoter sequences. Nucleic Acids Res 30: 325–327.
105. Rombauts S, Dehais P, Van Montagu M, Rouze P (1999) PlantCARE, a plant *cis*-acting regulatory element database. Nucleic Acids Res 27: 295–296.
106. Hruz T, Laule O, Szabo G, Wessendorp F, Bleuler S, et al. (2008) Genevestigator v3: a reference expression database for the meta-analysis of transcriptomes. Adv Bioinformatics 2008: 420747.

5

New Insights into 1-Aminocyclopropane-1-Carboxylate (ACC) Deaminase Phylogeny, Evolution and Ecological Significance

Francisco X. Nascimento[1]*, Márcio J. Rossi[1], Cláudio R. F. S. Soares[1], Brendan J. McConkey[2], Bernard R. Glick[2]

1 Universidade Federal de Santa Catarina, Departamento de Microbiologia, Laboratório de Microbiologia do Solo, Florianópolis SC, Brazil, **2** University of Waterloo, Department of Biology, Waterloo, Ontario, Canada

Abstract

The main objective of this work is the study of the phylogeny, evolution and ecological importance of the enzyme 1-aminocyclopropane-1-carboxylate (ACC) deaminase, the activity of which represents one of the most important and studied mechanisms used by plant growth–promoting microorganisms. The ACC deaminase gene and its regulatory elements presence in completely sequenced organisms was verified by multiple searches in diverse databases, and based on the data obtained a comprehensive analysis was conducted. Strain habitat, origin and ACC deaminase activity were taken into account when analyzing the results. In order to unveil ACC deaminase origin, evolution and relationships with other closely related pyridoxal phosphate (PLP) dependent enzymes a phylogenetic analysis was also performed. The data obtained show that ACC deaminase is mostly prevalent in some Bacteria, Fungi and members of Stramenopiles. Contrary to previous reports, we show that ACC deaminase genes are predominantly vertically inherited in various bacterial and fungal classes. Still, results suggest a considerable degree of horizontal gene transfer events, including interkingdom transfer events. A model for ACC deaminase origin and evolution is also proposed. This study also confirms the previous reports suggesting that the *Lrp-like regulatory protein AcdR* is a common mechanism regulating ACC deaminase expression in Proteobacteria, however, we also show that other regulatory mechanisms may be present in some Proteobacteria and other bacterial phyla. In this study we provide a more complete view of the role for ACC deaminase than was previously available. The results show that ACC deaminase may not only be related to plant growth promotion abilities, but may also play multiple roles in microorganism's developmental processes. Hence, exploring the origin and functioning of this enzyme may be the key in a variety of important agricultural and biotechnological applications.

Editor: Bas E. Dutilh, Radboud University Medical Centre, NCMLS, Netherlands

Funding: The authors have no support or funding to report.

Competing Interests: The authors have declared that no competing interests exist.

* E-mail: fxnascimento@gmail.com

Introduction

One of the key bacterial traits in facilitating plant growth is the production of the enzyme 1-aminocyclopropane-1-carboxylate (ACC) deaminase (EC 3.5.99.7). This enzyme is responsible for the cleavage of the ethylene precursor, ACC, into ammonia and α-ketobutyrate [1]. By decreasing ACC levels in plants, ACC deaminase-producing organisms decrease plant ethylene levels [2,3], which when present in high concentrations can lead to a reduced plant growth and ultimately, plant death [4].

ACC deaminase was initially identified in the yeast *Hansenula saturnus* (now re-classified as *Cyberlindnera saturnus*) and the bacterium *Pseudomonas* sp. ACP [1]. Since then, many groups have reported the isolation and sometimes the manipulation of *acdS* genes (i.e. the structural gene encoding ACC deaminase) from a wide range of different organisms, mostly bacteria and fungi [5]. Moreover, several studies have addressed the detailed biochemistry of ACC deaminase and the atypical and important reaction mechanism of ACC breakdown [6]. Data obtained in these studies show that ACC deaminase is a multimeric enzyme (homodimer or homotrimer) with a subunit molecular mass of approximately 35–42 kDa and it uses one molecule of pyridoxal phosphate (PLP) per subunit. Based on its protein fold, ACC deaminase has been classified as belonging to the tryptophan synthase beta superfamily (fold type II) of PLP binding proteins [6]. In this family are also included the ACC deaminase homolog from *Pyrococcus horikoshii* [7] and the D-cysteine desulfhydrase from *E.coli* and *Salmonella typhymurium* [8,9].

ACC deaminase is central to the functional interactions of various plant associated bacteria and fungi. The root colonizing bacteria *Pseudomonas putida* GR12-2 and *Pseudomonas* sp. UW4 no longer promote canola root elongation after its *acdS* gene is knocked out [10,11]. The symbiotic efficiency of the root nodule forming bacteria, *Rhizobium leguminosarum* bv. *viciae* and *Mesorhizobium loti* MAFF303099, is decreased upon *acdS* gene deletion [12,13]. The endophytic plant growth-promoting bacteria *Burkholderia phytophirmans* PsJN, *Pseudomonas fluorescens* YsS6 and *Pseudomonas migulae* 8R6 are less effective when their *acdS* gene is deleted [14,15]. Similarly, when ACC deaminase expression is impaired in the fungus *Trichoderma asperellum* T203, the plant growth promotion abilities of this organism are also decreased [16,17].

Bacteria and fungi that express ACC deaminase can lower the impact of a range of different stresses that affect plant growth and development [3,17]. Using ACC deaminase-producing bacteria in association with plants subjected to different kinds of biotic and abiotic stresses resulted in enhanced plant tolerance [18–25]. The use of ACC deaminase-producing bacteria in association with plants for purposes of soil decontamination is also documented [26–28]. Increased phytoremediation potential and resistance to biotic and abiotic stresses are observed in transgenic plants expressing a bacterial ACC deaminase [29–32]. The expression of an exogenous ACC deaminase gene increases the symbiotic performance of many rhizobial strains [33–36].

Studies regarding the mechanisms regulating ACC deaminase expression have been reported for some Proteobacteria. Binding sites for CRP (cAMP receptor protein), FNR (fumarate-nitrate reduction regulatory protein) and LRP (leucine responsive regulatory protein) were present in the promoter region of the *Pseudomonas* sp. UW4 *acdS* gene and were shown to function in regulating *acdS* expression [37–39]. In addition, an LRP-like protein-coding region has been found in the immediate upstream region of many *acdS* genes. This gene was termed *acdR* (ACC deaminase regulatory protein), based on the evidence that it is necessary for optimum ACC deaminase expression in the presence of ACC. The *acdR* gene has also been demonstrated to participate in the regulation of ACC deaminase expression in *Rhizobium leguminosarum* bv. *viciae* 128C53K and *Azospirillum lipoferum* 4B [12,40]. Most other Proteobacteria that have been examined for ACC deaminase activity or *acdS* gene presence, possess an *acdR* gene in the vicinity of *acdS*, suggesting that this regulatory mechanism is widespread in *acdS*+ Proteobacteria [40].

Despite the fact that many biochemical and biological features of ACC deaminase are now understood, not much is known about the origin and phylogeny of the *acdS* gene and its regulatory elements. Based upon a phylogenic analysis of a limited number of *acdS* genes partially characterized and their comparison to the phylogeny of 16S rRNA genes from the same bacteria, Hontzeas et al. [41] proposed that some ACC deaminase genes have been transmitted through horizontal gene transfer (HGT). Using the same criteria, Blaha et al. [42] suggested that ACC deaminase genes in Proteobacteria were extensively subjected to HGT. In addition, Nascimento et al. [43] suggested that in many *Mesorhizobium* spp. the *acdS* gene is transferred between strains through symbiotic island exchange.

The phylogeny in Proteobacteria of *acdR* has also been investigated. Prigent-Combaret et al., [40] suggested that *acdR*, like *acdS*, may have evolved through HGT. This conclusion notwithstanding, these authors suggest that the evolution of *acdS* and *acdR* genes might not be coupled.

While phylogenetic studies of *acdS* and *acdR* genes have been focused primarily on Proteobacteria, other studies have demonstrated the presence of ACC deaminase activity in Actinobacteria [41,44–48], Firmicutes [44,48–51] and Bacteroidetes [52–55]. Furthermore, the presence of a putative functional ACC deaminase in *Phytophthora sojae* [56] further emphasizes the notion that the current view of *acdS* phylogeny and evolution is somewhat incomplete. To address this, here we have undertaken a comprehensive study of the phylogeny of *acdS* and *acdR* and the results are discussed in terms of evolutionary and ecological implications of ACC deaminase production by diverse microorganisms.

Methods

Obtaining the sequences

To obtain bacterial ACC deaminase (AcdS) and ACC deaminase regulatory protein (AcdR) sequences, BLAST searches were performed in the NCBI databases (www.ncbi.nlm.nih.gov/) using *Pseudomonas sp.* UW4 *acdS* and *acdR* gene, as well as AcdS and AcdR protein sequences as the queries. For fungal ACC deaminase sequence retrieval, BLAST searches were performed in the NCBI database using the *Penicillium citrinum* AcdS protein sequence as the query. Default BLAST parameters were used when obtaining the sequences.

An NCBI genomic BLAST search (www.ncbi.nlm.nih.gov/sutils/genom_table.cgi) was also performed using *Pseudomonas* sp. UW4 *acdS* and AcdS sequences in order to evaluate the presence of ACC deaminase in other completely sequenced organisms. An additional BLAST search was performed in the nematode genomic database (www.nematodes.org) using *Pseudomonas* sp. UW4 or *Penicillium citrinum acdS* gene as query.

Moreover, all putative AcdS sequences were analyzed for key protein residues known to be important for ACC deaminase activity, namely Lys51, Ser78, Tyr295, Glu296 and Leu322 [7,57,58] using *Pseudomonas* sp. UW4 as a reference. The AcdS sequences were aligned using MUSCLE [59] and the presence of key amino acid positions were verified. Sequences presenting different amino acids in the above mentioned positions were discarded, as they are likely to represent related PLP dependent enzymes, such as D-cysteine desulfhydrase [58].

Sequence identities and similarities were analyzed using SIAS (http://imed.med.ucm.es/Tools/sias.html) with default parameters.

When available, the genomic regions containing the *acdS* gene were analyzed in order to identify any patterns present in the *acdS* gene neighborhood.

Strain information and 16S rRNA gene sequences were obtained via NCBI (http://www.ncbi.nlm.nih.gov), Goldcard (http://www.genomesonline.org/cgi-bin/GOLD/index.cgi) and SILVA (http://www.arb-silva.de), where available.

The accession numbers for sequences used in this study as well as strains descriptions are presented in Tables S1 (Actinobacteria, Deinococcus-Thermus and Firmicutes), S2 (α-Proteobacteria), S3 (β-Proteobacteria), S4 (γ-Proteobacteria), and S5 (Eukaryotes).

ACC deaminase protein sequence analysis and comparison to closely related enzymes

Protein sequence analysis was conducted on AcdS proteins found in completely sequenced representative bacteria. The functional AcdS protein sequences of the Proteobacteria *Agrobacterium tumefaciens* D3 [60], *Azospirillum lipoferum* 4B [40], *Bradyrhizobium japonicum* USDA110 [61], *Mesorhizobium loti* MAFF303099 [13], *Phyllobacterium brassicacearum* STM196 [62], *Rhizobium leguminosarum* 128C53K [12], *Sinorhizobium meliloti* SM11 [63], *Burkholderia phytofirmans* PsJN [14], *Burkholderia graminis* C4D1M [64], *Ralstonia solanacearum* GMI1000 [42], *Variovorax paradoxus* 5C2 [22], *Pseudomonas* sp. UW4 [38], *Pseudomonas* sp. ACP [1] and the Fungi, *Cyberlidnera saturnus* [1], *Penicillium citrinum* [65], *Trichoderma asperellum* T203 [16], together with the AcdS from *Herbaspirillum frisingense* GSF30 [66] and the putative AcdS sequences from *Agreia* sp. PHSC20C1, *Rhodococcus* sp. R04 (Actinobacteria), *Meiothermus ruber* DSM1279 (Deinococcus-Thermus) were used. Sequences were aligned using MUSCLE and the presence of conserved and variable sites was analyzed.

Sequence comparisons were also performed with closely related enzymes. Therefore, D-cysteine desulfhydrase sequences from *E.*

coli [8], as well as the ACC deaminase homologs from *Pyrococcus horikoshi* [7] and *Solanum lycopersicum* [58] were used and compared to the various ACC deaminase proteins.

Phylogenetic analysis

The sequences were aligned using MUSCLE and phylograms were constructed in Seaview v.4.2.12 [67] using PhyML [68].

In order to obtain the best substitution model for the construction of the phylogenetic trees, the resulting alignments were analyzed with jModeltest2 [69] and ProtTest [70]. The substitution models were chosen based on minimum BIC (Bayesian Information Criteria) values.

The *acdS*, *acdR* and 16S rRNA gene evolutionary history was inferred by using the Maximum Likelihood method based on the GTR model with a discrete Gamma distribution (4 Gamma categories). The AcdS and AcdR phylograms were constructed using the Maximum likelihood method based on the WAG model with a discrete Gamma distribution (4 Gamma categories). Branch support was evaluated using both aLRT (SH like) [71] and bootstrap analysis (100 replicates). Only bootstrap values above 0.75 (75%) are included in the phylograms. The resulting phylogenetic trees were plotted using FigTree v.1.4.1 (http://tree.bio.ed.ac.uk/software/figtree).

Estimates of evolutionary divergence between *acdS* sequences or 16S rRNA sequences in groups of bacterial strains were computed using MEGA software 6.06 [72]. The number of base substitutions per site from between sequences was calculated and analyses were conducted using the Maximum Composite Likelihood model with 1000 bootstrap replications. The analysis involved 3 nucleotide sequences per group of bacterial species, previously aligned using MUSCLE. Codon positions included were 1st+2nd+3rd+Non-coding. All positions containing gaps and missing data were removed.

Results and Discussion

ACC deaminase prevalence in completely sequenced organisms

After performing multiple searches in the NCBI database (http://www.ncbi.nlm.nih.gov/sutils/genom_table.cgi) using *Pseudomonas* sp. UW4 *acdS* gene as query, it was observed that the *acdS* gene is not commonly seen in most sequenced organisms. The *acdS* gene is mainly found in Actinobacteria, members from the Deinococcus-Thermus phylum (*Meiothermus*), three classes from Proteobacteria (α, β and γ), in various Fungi classes belonging to Ascomycota and Basidiomycota, and in Stramenopiles members.

These results are in agreement with previous reports, which have demonstrated ACC deaminase activity in many Actinobacteria, α, β and γ-Proteobacteria. Remarkably, putative *acdS* genes were found in *Meiothermus*, yet, there is no record of ACC deaminase activity in these thermophile strains. Putative *acdS* genes were also found many in members of Stramenopiles, mostly in *Phytophthora*. By computational analysis, Singh and Kashyap, [56] suggest that the *acdS* gene found in *Phytophthora sojae* encodes a functional ACC deaminase.

Interestingly, despite the known ACC deaminase activity display by bacteria belonging to the Bacteroidetes/Chlorobi or Firmicutes, it was not possible to identify *acdS* genes in the completely sequenced bacteria belonging to these phyla. In 478 completely sequenced bacteria (accessed in July, 2013) belonging to the Bacteroidetes/Chlorobi, including many *Flavobacterium* and *Chryseobacterium* species, the *acdS* gene is not found. Although candidate *acdS* genes are identified via BLAST, the active sites contain residues more consistent with D-cysteine desulfhydrase or a related PLP dependent enzyme [58], such as YP_001296100 in which threonines replace residues corresponding to active site residues E296 and L322. ACC deaminase activity has been previously reported to be present in *Flavobacterium* and *Chryseobacterium* species, although at very low levels [52,55], which may represent non-specific activity of D-cysteine desulfhydrase-like enzymes. Similarly, although ACC deaminase activity has been described in many *Bacillus* and *Paenibacillus* strains [48–51], it was not possible to identify the *acdS* gene in 271 completely sequenced strains belonging to the Bacilli class (Firmicutes phylum), including many soil and plant associated *Bacillus* and *Paenibacillus* species.

It is possible that in these and many other bacterial strains the presence of an *acdS* gene may be related to a strain's specific feature in which *acdS* acquisition happened by HGT by result of a co-existence with other ACC deaminase-producing bacteria in environments where ACC deaminase production provides the bacteria with some important advantages. Other possible explanations for this inconsistency may relate to the fact that genome sequencing is biased and the sequenced strains may not be representative of bacteria that interact extensively with plants.

Analysis of ACC deaminase (putative and functional) protein sequences

In the first instance, every sequence used in this study (Table S1–S5) contains the previously described AcdS conserved regions that have been found to be necessary for ACC deaminase activity. Moreover, all bacterial AcdS sequences shared high sequence identity (60 to 100%) to AcdS from *Pseudomonas* sp. UW4.

When comparing the putative AcdS sequences from Fungi with the functional ACC deaminase from *Penicillium citrinum*, sequence identities ranged between ~70 and 99% for the majority of fungal AcdS sequences. Exceptionally, some AcdS sequences from yeasts and some other fungi share only ~52–55% identity to *Penicillium citrinum* AcdS. Also, the Stramenopiles members share approximately 60% identity to the *Penicillium citrinum* AcdS. Interestingly, the AcdS sequences from yeasts, some other Fungi and Stramenopiles share higher identity to *Pseudomonas* sp. UW4 AcdS sequence (~70 to 85%), consistent with a relationship with Proteobacteria and the possibility of past horizontal gene transfers. A more detailed description of this issue is presented below.

Protein sequence analysis suggests that the putative *acdS* genes found in *Rhodococcus* sp. R04, *Agreia* sp., PHSC20C1 and *Meiothermus ruber* DSM1279, encode a true ACC deaminase. By sequence comparison, it was observed that the putative AcdS contain all the conserved features present in all known functional ACC deaminases and not present in the related enzymes (Figure S1). For instance, the putative AcdS sequences contain the important residues E295 and L322 known to be required for ACC deaminase activity [58] and not present in other related enzymes. These results are also supported by the fact that these Actinobacteria and *Meiothermus* AcdS protein sequences share high identity (70 to 82%) to the functional ACC deaminase from *Rhodococcus* sp. 4N-4 (partially characterized) [41]. In addition, these sequences show similar sequence identities to other β and γ-Proteobacteria AcdS sequences (~70%).

ACC deaminase phylogeny: Horizontal gene transfer or vertical transmission?

The comparison between the *acdS* phylogenetic tree (Fig. 1) and the 16S rRNA based phylogeny (Fig. 2), suggests that ACC deaminase has evolved mainly through vertical transmission with occasional horizontal gene transfer. In the *acdS* phylogram (Fig. 1), it is observed that closely related strains typically have similar *acdS*

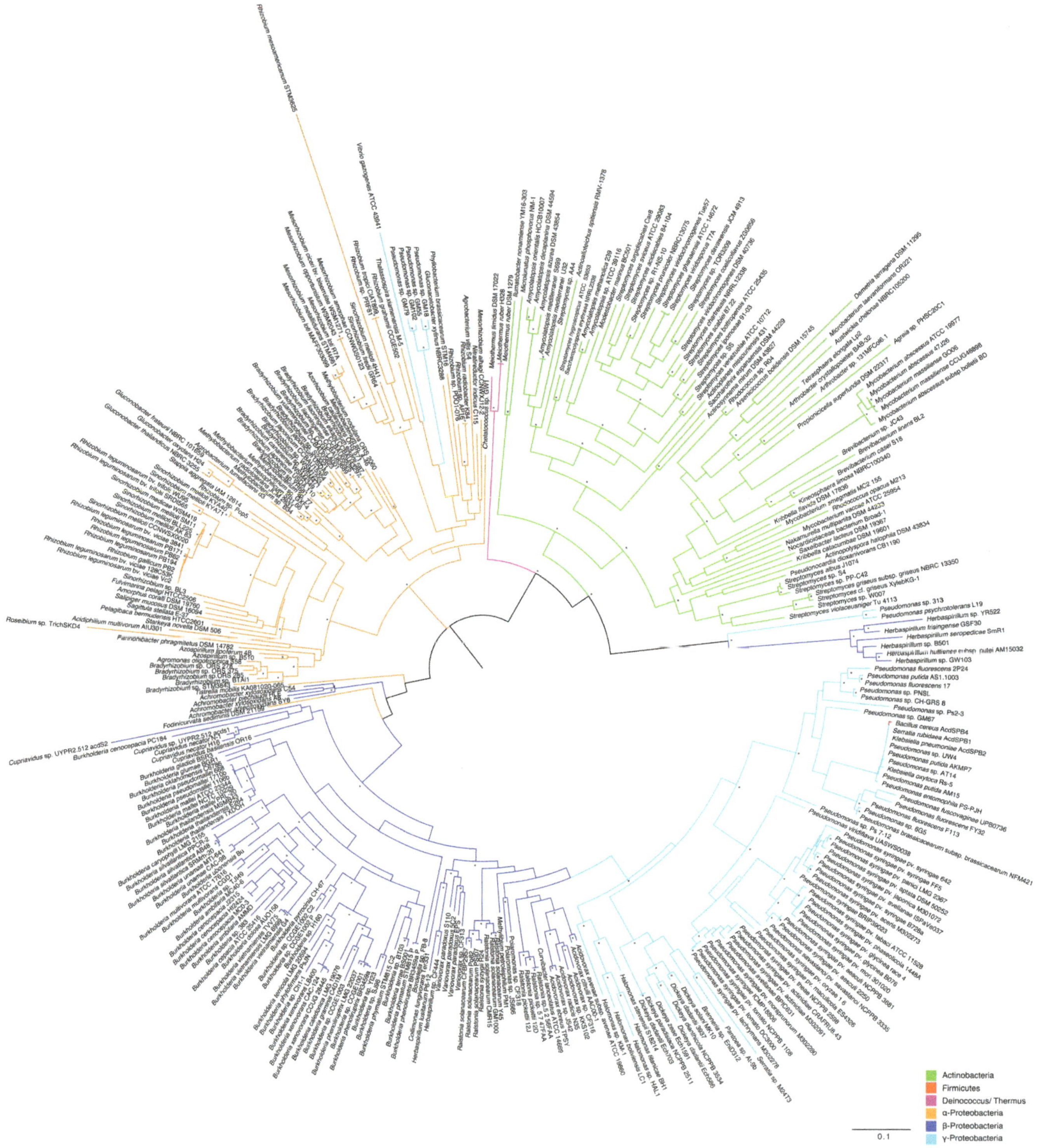

Figure 1. Phylogram based on the *acdS* gene. The evolutionary history was inferred by using the Maximum Likelihood method based on the GTR model. A discrete Gamma distribution was used to model evolutionary rate differences among sites (4 categories). Branch support was evaluated using both aLRT (SH like) and bootstrap analysis (100 replicates). Bootstrap values above 0.75 (75%) are displayed in the phylograms shown next to the branches as *. The analysis involved 335 nucleotide sequences and 931 patterns were found (out of a total of 1155 sites).

gene sequences. Furthermore, many strains with different origins and isolated from different habitats (Tables S1–S5), but belonging to the same species tend to have similar *acdS* genes.

The presence of the *acdS* gene in an organism like *Meiothermus ruber* is also consistent with the vertical transmission of this gene. It is unlikely that this bacterial thermophile (optimum growth at 60°C) isolated from a hot spring has acquired an *acdS* gene

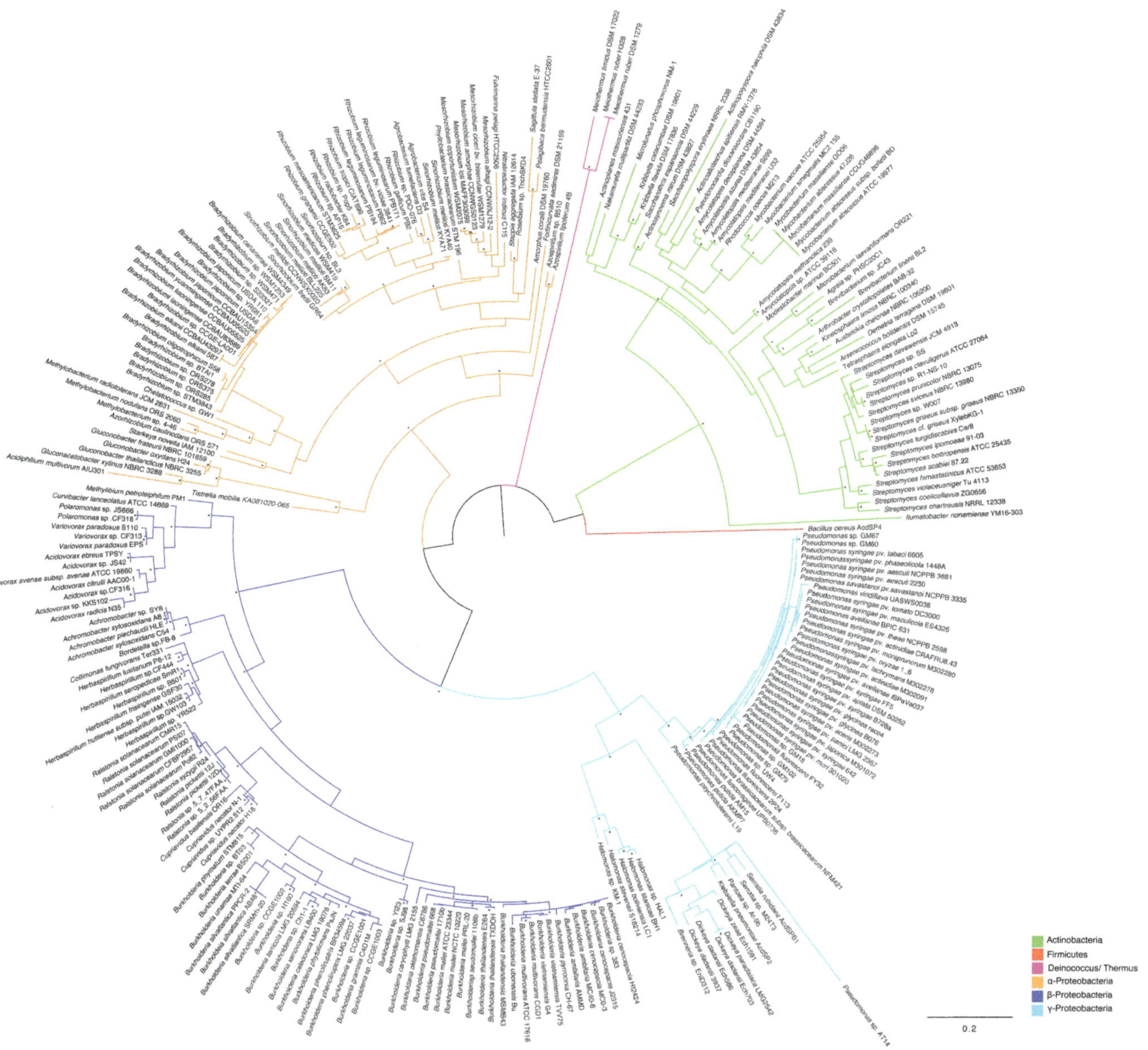

Figure 2. Phylogram based on 16S rDNA sequences. The evolutionary history was inferred by using the Maximum Likelihood method based on the GTR model. A discrete Gamma distribution was used to model evolutionary rate differences among sites (4 categories). Branch support was evaluated using both aLRT (SH like) and bootstrap analysis (100 replicates). Bootstrap values above 0.75 (75%) are displayed in the phylograms shown next to the branches as *. The analysis involved 272 nucleotide sequences and 768 patterns were found (out of a total of 1334 sites).

through HGT. This is strongly supported by the *acdS* gene phylogram (Fig. 1) showing a well bootstrap-supported and unique cluster grouping all *Meiothermus acdS* sequences distantly from all other *acdS* genes obtained from different bacterial phyla.

The presence of an *acdS* gene in the chromosome of the psychrophile marine actinobacterium, *Agreia* sp. PHSC20C1, (isolated in the Antarctic) and other soil Actinobacteria, is also consistent with the vertical transmission and ancient origin of the *acdS* gene. In *Azorhizobium* and *Bradyrhizobium* strains, the *acdS* gene is located far away from the "plastic" chromosomal symbiotic island containing the symbiotic genes. If these strains had acquired the *acdS* gene by HGT it might be expected that it would be present in a region that is more prone to such transfers, such as a symbiotic island or a plasmid.

Blaha et al. [42] and Glick et al. [3] have suggested that that the presence of *acdS* on plasmids may facilitate the lateral transfer of this gene. On the other hand, the presence of the *acdS* gene on a plasmid can also account for a different sequence divergence rate. Mobile elements and smaller replicons show higher evolutionary rates when compared to primary chromosomes [73,74]. By being present on smaller replicons, *acdS* genes may be subject to different evolutionary rates compared to genes present in primary chromosomes. This may help to explain the *acdS* gene phylogeny of *Burkholderia* and *Cupriavidus*. Instead of clustering together with the other β-Proteobacteria, strains belonging to the *Burkholderia* and *Cupriavidus* genus form a separate cluster (Fig. 1). From the available data, most *Burkholderia* and *Cupriavidus* strains have the *acdS* gene present in a second smaller chromosome. Other β-

Proteobacteria possesses an *acdS* gene in the primary chromosome or in plasmids (Table S3). This phenomenon is also observed in *Agrobacterium* and *Rhizobium* strains, *A. vitis* S4 and *R. radiobacter* K84 which have the *acdS* gene located in a second chromosome, and therefore, cluster distantly from their *Agrobacterium tumefaciens* D3 (*acdS* in plasmid) and *Rhizobium* (*acdS* in plasmid) relatives. Thus, there seems to be a connection between *acdS* phylogenetic distribution, evolution and *acdS* location in the replicon.

Environmental cues can also lead to different gene mutation rates [75]. Gene loss, acquisition, mutational rates and genome rearrangements may play a crucial role in bacterial adaptation and survival [76,77]. This is particularly important in organisms living in adverse environments like many of the organisms described here (Tables S1–S5). It is possible that bacteria adapted to different environments may present different *acdS* divergence rates, thus being responsible for some of the variance in *acdS* genes in bacteria from the same species. When calculating the 16S rRNA and *acdS* gene evolutionary distance estimates in specific bacterial species groups it was found that the ratio between 16S rRNA and *acdS* sequence divergence is not always identical between strains and groups (file S1). For instance, three *Burkholderia mallei* strains isolated from three different countries show identical 16S rRNA (1200 bp) ($d = 0$) and identical *acdS* gene (1019 bp) ($d = 0$) sequences. In three *Burkholderia silvatlantica* strains obtained from Brazil, this is not observed; all strains present identical 16S rRNA sequences (1200 bp) ($d = 0$) but show intraspecific differences in the *acdS* gene sequences (1019 bp) ($d = 0.0059 \pm 0.0020$), sometimes accounting to up to 5 different nucleotides. Interestingly, all three *B. mallei* were obtained from human and animal blood and are known pathogens, while the three *B. silvatlantica* strains were obtained from the rhizosphere of different plants where they act like plant growth-promoting bacteria [64].

Several authors suggested HGT for *acdS* genes based on results showing a specific relative position of some *Pseudomonas* (γ-Proteobacteria) strains in the *acdS* phylogenetic tree [41,42,6]. Instead of forming a separate cluster, some *Pseudomonas* strains clustered together with β-Proteobacteria. In this work, we obtained somewhat similar results. Although members of γ-Proteobacteria group very close to β-Proteobacteria, they form a unique cluster and are not scattered through the phylogenetic tree as observed in previous studies. A very close evolutionary relationship between these two classes has been reported [78–81]. In fact, some bacterial strains that belonged to the *Pseudomonas* genus (γ-Proteobacteria) have been reassigned to the *Burkholderia* genus (β-Proteobacteria) [82,83].

Interestingly, the *Pseudomonas* sp. ACP AcdS sequence shares higher identity (96.7%) with *Burkholderia xenovorans* LB400 functional ACC deaminase than with *Pseudomonas* sp. UW4 AcdS (85.3%). This is also observed in the AcdS phylogram, where *Pseudomonas* sp. ACP groups closer to *Burkholderia xenovorans* LB400 (Fig. 3). While, Honma and Shimomura [1] tentatively identified *Pseudomonas* sp. ACP bacterium by phenotypic methods, it is conceivable that *Pseudomonas* sp. ACP is in fact a *Burkholderia* strain [64]. If this is in fact the case, then previous studies regarding the phylogeny of *acdS* may also have been influenced by the confusing relationship between *Pseudomonas* and *Burkholderia*. Furthermore, due to the recent divergent evolution and close relationship between γ and β-Proteobacteria, it is very difficult to prove *acdS* HGT in these classes.

While less prevalent than previously thought, HGT likely does occur and accounts for a portion of *acdS* gene evolution. For example, it has been shown that some *Mesorhizobium* strains may acquire a specific *acdS* gene by the means of symbiotic island transfer [43]. Nandasena et al., [84] demonstrated that *Mesorhizobium opportunistum* WSM 2073 acquired a specific symbiotic island when it came in contact with non-endemic populations of *M. ciceri* bv. *biserrulae*, thus, gaining the ability to nodulate *Biserrula pelecinus*. The *acdS* gene was present within that symbiotic island and was therefore transferred between these strains. Moreover, the *acdS* gene sequences from those two strains share 100% identity, strongly supporting the idea of a recent transfer event.

Curiously, there are some cases where *acdS* horizontal transfers seem to have occurred between strains with a more distant evolutionary relationship. This is the case of *Pseudomonas* isolates GM 18, GM 55, GM 79 and GM 102, which are found to possess *acdS* genes like those of α-Proteobacteria (Fig. 1). Despite belonging to the γ-Proteobacteria, *Vibrio gazogenes* ATCC43941 has an *acdS* gene resembling those of α-Proteobacteria (Fig. 1). Chen et al., (2013) showed that *Bacillus cereus* AcdSPB4 isolated from the casing soil of *Agaricus bisporus* possesses an *acdS* gene highly similar to those of *Pseudomonas* (Fig. 1) thus, strengthening the idea of *acdS* horizontal transfer between distantly related strains.

Interestingly, *Herbaspirillum seropedicae* SmR1, *H. frisingense* GFS30, *H. huttiense* subsp. *putei* AM15032, *H.* sp. B501, *H.* sp. GW103, *H.* sp. YR522, *Pseudomonas psychrotolerans* L19 and *Pseudomonas* sp. 313 strains possess *acdS* genes that are not similar to those found in other bacteria from the same Class or even to other *Herbaspirillum* and *Pseudomonas* strains. Instead they form a unique group in the phylogenetic tree (Fig. 1, 3). Furthermore, these strains also possess *acdR* genes that are frequently found in AcdS+ Proteobacteria. In this scenario it is possible that these strains have horizontally acquired *acdS* and *acdR* genes from a different class of bacteria yet to be determined. One may also assume that the putative *acdS* genes in these strains encode a different type of deaminase, however, ACC deaminase activity has been detected in *Herbaspirillum frisingense* GFS30 [66]. Moreover, the putative AcdS from *H. frisingense* GFS30 shows the conserved regions known to be important in functional ACC deaminases (i.e E295, L322) (Fig. S1). Curiously, *Herbaspirillum frisingense* GFS30 and *Herbaspirillum* sp. YR522 are both endophytes isolated from *Miscanthus* and *Populus deltoides*, respectively.

Similar to what is observed in Bacteria, the AcdS phylogeny in Fungi indicates that closely related strains possess a similar ACC deaminase (Fig. 3). This is consistent with the notion that *acdS* genes are vertically transmitted in Fungi. However, some fungal strains like *Penicillium marneffei* and *Talaromyces stipitatus* (Ascomycota/Eurotiomycetes) are likely to have acquired the *acdS* from other Fungi belonging to the Sordariamycetes class, suggesting that like in some bacteria, fungal *acdS* genes may also be acquired by HGT.

In addition, the yeasts *Cyberlindnera saturnus*, *Cyberlindnera jadinii* NBRC 0988, *Clavispora lusitaniae* ATCC 42720 (Ascomycota/Saccharomycetes) and *Schizosaccharomyces pombe* 972h- (Ascomycota/Schizosaccharomycetes) seem to have acquired an ACC deaminase gene separately from most Ascomycota and presumably from Proteobacteria. ACC deaminase genes like those of Proteobacteria have also been detected in Fungi belonging to different classes such as *Punctularia strigosozonata* HHB-11173 (Basidiomycota/Agaricomycetes), *Fomitopsis pinicola* FP-58527 (Basidiomycota/Agaricomycetes), *Aureobasidium pullulans* AY4 (Ascomycota/Dothideomycetes), *Macrophomina phaseolina* MS6 (Ascomycota/Dothideomycetes) and *Guignardia citricarpa* CGMCC3.14348 (Ascomycota/Dothideomycetes).

The Stramenopiles, Phytophthora infestans T30-4, P. ramorum Pr102, P. sojae P6497, P. lateralis, P. kernoviae, P. parasitica, Pseudoperonospora cubensis and Hyaloperonospora arabidopsidis Emoy2 also have ACC deaminase genes most similar to those of Proteobacteria (Fig. 3).

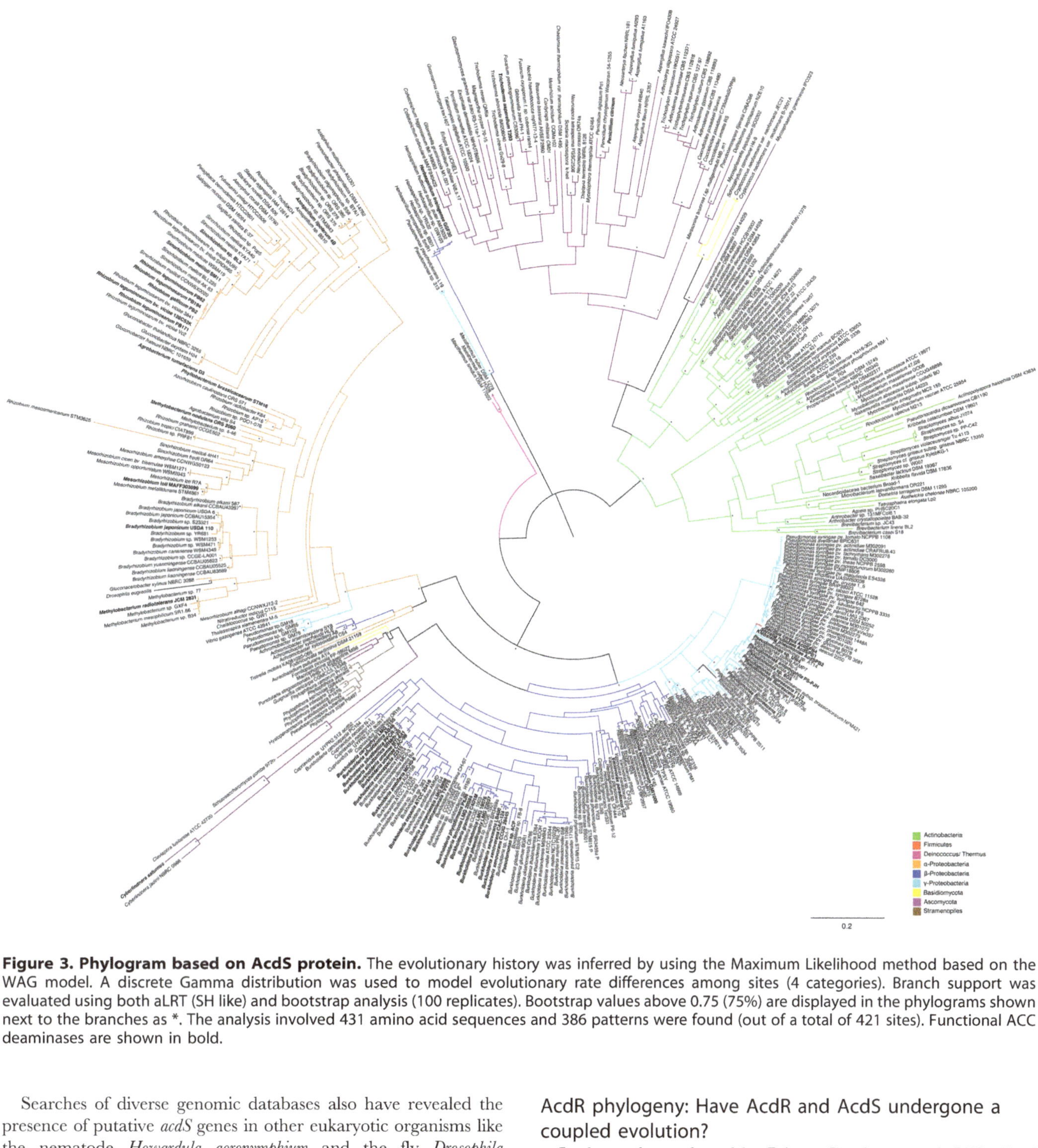

Figure 3. Phylogram based on AcdS protein. The evolutionary history was inferred by using the Maximum Likelihood method based on the WAG model. A discrete Gamma distribution was used to model evolutionary rate differences among sites (4 categories). Branch support was evaluated using both aLRT (SH like) and bootstrap analysis (100 replicates). Bootstrap values above 0.75 (75%) are displayed in the phylograms shown next to the branches as *. The analysis involved 431 amino acid sequences and 386 patterns were found (out of a total of 421 sites). Functional ACC deaminases are shown in bold.

Searches of diverse genomic databases also have revealed the presence of putative *acdS* genes in other eukaryotic organisms like the nematode *Howardula aoronymphium* and the fly *Drosophila eugracilis*. Furthermore, these genes show high similarity to *acdS* from Proteobacteria (Fig. 3). Some *acdS* genes are found in bacteria known to be associated with Eukaryotic organisms, for example, *Serratia* sp. M24T3, isolated from the nematode *Bursaphelenchus xylophilus*, and *Pantoea* sp. At-9b, the leaf cutter ant symbiont (Table S4). While it is possible that *Howardula aoronymphium* and *Drosophila eugracilis* may have acquired *acdS* genes from associated bacteria, it is most likely that the presence of *acdS* in these organisms results from contamination of genomic DNA.

AcdR phylogeny: Have AcdR and AcdS undergone a coupled evolution?

In the study conducted by Prigent-Combaret et al. [40], 45 of 48 studied Proteobacteria were found to possess an LRP homolog (*acdR*) near the *acdS* gene. Here, we report the presence of *acdR* in, at least, 166 of 261 Proteobacteria possessing an *acdS* gene. Still, it was not possible to obtain the *acdR* sequence in many (78) organisms and others only have their *acdS* gene described. The *acdR* gene was not found at least in 17 *acdS*+ completely sequenced strains (6 *Mesorhizobium* strains, 2 *Rhizobium* strains, *Fulvimarina pelagi*, 3 root nodule *Burkholderia* strains containing the 2nd *acdS* copy in a plasmid, *Halomonas titanicae* BH1 and 4 *Pseudomonas* strains).

Moreover, the *acdR* gene is found in the opposite direction of the *acdS* gene in most studied Proteobacteria (data not shown). This is consistent with the previous reports of Grichko and Glick [37], Ma et al. [12] and Prigent-Combaret et al. [40]. This data suggest that *acdR* is a common mechanism regulating ACC deaminase expression in most Proteobacteria. Moreover, the phylogeny of *acdR* (Fig. 4) is related to the *acdS* gene phylogeny (Fig. 1), suggesting that these genes evolved in a similar and dependent manner. Closely related strains have similar *acdR* genes, as also observed in the *acdS* phylogram, suggesting that *acdR* is primarily vertically inherited. In the phylogram based on the *acdR* gene it is also observed a grouping according to the bacterial Class (taxonomy) and the gene location in the replicon (Example: 2nd chromosome location in *Burkholderia* and *Cupriavidus* vs. primary chromosome/plasmid location in other β and γ-Proteobacteria).

Interestingly, there are few cases where it seems that the *acdR* and *acdS* are not inherited together or may have undergone genomic rearrangements. While some strains don't have *acdR* genes in the vicinity of the *acdS* gene others don't have an *acdR* gene at all. In *Gluconacetobacter xylinus* NBRC 3288 there are various regions coding putative LRP in the upstream region of the *acdS* gene. However, they are not true *acdR* genes. A sequence sharing high homology to the *acdR* gene is found far away (aprox. 9 kb) from the *acdS* gene. This is also observed in *Burkholderia xenovorans* LB400. In this case, despite the fact that this strain has an *acdR* gene located far from the *acdS* gene, it is still able to express ACC deaminase [64]. In *Rhizobium leguminosarum* bv. *viciae* 3841 an *acdR* gene is not found. In *Mesorhizobium loti* MAFF303099 the *acdR* gene is also not present, but in this case, the *acdS* gene transcription is regulated by NifA [13].

It is possible that genome rearrangements or gene insertions in smaller replicons can account for the absence of *acdR* genes in some *acdS*+ bacterial strains. The strains *Burkholderia* sp. CCGE1002, *B. phymatum* STM815 and *B. phenoliruptrix* BR3459a (isolated from root nodules) have two copies of the *acdS* gene, one on the second chromosome and the other on a megaplasmid. The *acdS* gene copy present on the megaplasmid seems to be the result of *acdS* gene duplication and later insertion into this smaller replicon. This is consistent with the high identity between the two *acdS* copies and also the presence of transposase genes in the immediate upstream and downstream regions of the *acdS* gene. In this case, the *acdR* gene is not present and may have been lost in this process.

The exceptions notwithstanding, in the majority of cases in Proteobacteria it appears that the evolution of *acdS* and *acdR* is coupled. This result is in agreement with previous reports showing that *acdR* is necessary for optimum ACC deaminase expression [38,39].

Despite being mostly inherited together, it is observed that these genes may have different evolutionary rates. Thus, for example, compared to the *Pseudomonas* sp. UW4 AcdR sequence, other AcdR sequences from Proteobacteria show identities ranging from 51% to 87%. This degree of variability is not observed in Proteobacteria AcdS sequences. By coding a regulatory protein it is most likely that the *acdR* gene is more prone to modifications, thus, allowing fine-tuning of *acdS* transcription and expression.

Previously, Nikolic et al. [85] stated that "the *acdR–acdS* gene cluster is rather rare and typically occurs in few α and β-Proteobacterial genera" based on finding the *acdR–acdS* gene cluster in four α, six β-Proteobacteria and in only three *Pseudomonas syringae* strains. They concluded that the operon is rather uncommon among γ-Proteobacteria. However, more detailed data presented by Prigent-Combaret et al. [40] and also in this study, supports a widespread occurrence of *acdR–acdS* gene cluster. The apparent lack of *acdR-acdS* clusters in Nikolic et al. [85] may be due to the inclusion of putative ACC deaminase sequences that were not confirmed by comparison with conserved protein domains. Thus, sequences coding for D-cysteine desulfhydrases, and possibly other deaminases and aminotransferases, were considered as ACC deaminases, leading to a confusing relationship between *acdS* and *acdR* and also the presence of *acdS* in some bacterial groups.

Other mechanisms regulating ACC deaminase transcription

The expression of ACC deaminase by organisms that don't possess *acdR* genes indicates that the presence of this regulator is not absolutely necessary for *acdS* transcription. The presence of CRP and FNR binding sites in the immediate upstream region of the *acdS* gene in many Proteobacteria [37,40] suggests that these elements can also account for ACC deaminase expression regulation in some Proteobacteria. It has been demonstrated directly in some instances that FNR as well as CRP regulate *acdS* transcription [37,38,40]. The NifA protein is also a known regulator of ACC deaminase expression in *Mesorhizobium loti* MAFF303099. In this strain ACC deaminase expression occurs only inside of formed nodules [13,86]. In addition, a NifA binding site is found in the immediate upstream region of the *acdS* gene in this and many other *Mesorhizobium* strains, suggesting that this regulatory mechanism is widespread in this genus [43]. Interestingly, the NifA binding site (5′-TGT-N_{9-11}-ACA-3′) is quite similar to the CRP binding site (5′-TGTGA-N_6-TCACA-3′).

In many Actinobacteria and in *Meiothermus*, a gene encoding a protein from the GntR family of transcriptional regulators is found next to the *acdS* gene. We putatively termed it *acd-AR* (Actinobacteria) and *acd-MR* (*Meiothermus*). When performing BLAST searches using one Acd-AR protein sequence as query, the main hits are always related to other Acd-AR protein sequences found in *acdS*+ Actinobacteria, suggesting a close relationship between *acdS* and *acd-AR*. The same trend is observed in *Meiothermus* despite the fact Acd-AR shares low identity to Acd-MR. There are no sequences in the database that share a high degree of similarity to Acd-MR. These results are consistent with the possibility that both *acd-AR* and *acd-MR* might be involved in the regulation of ACC deaminase expression in these organisms. Curiously, when analyzing the immediate upstream region of the *acdS* gene in various Actinobacteria it is observed that some strains appear to have no promoter regions (Fig. S2-A). The same is observed in *Meiothermus*. In these strains the *acdS* gene forms an operon together with the *acd-AR* gene and *acd-MR* gene, respectively.

Interestingly, in Nocardioidaceae Broad-1 a leucine responsive protein is found in the vicinity of the *acdS* gene, however, it is quite different from the Proteobacteria AcdR protein. Also, in some Actinobacteria and Proteobacteria strains the *acdS* gene is located near a transcriptional regulator belonging to the LysR family. Moreover, in *Saccharopolyspora erythraea* NRRL 233 and *Streptomyces hygroscopicus* ATCC 53653 strains the *acdS* gene also appears to be part of an operon consisting of a gene encoding a MFS family protein and another gene encoding a M20 peptidase (Fig. S2-B). Interestingly, a LysR transcriptional regulator is also found in the vicinity of the *acdS* gene in some Proteobacteria like *Brenneria* sp. EniD312, *Burkholderia xenovorans* LB400, *Dickeya* spp. and *Pantoea* sp. At-9b (Fig. S2-C-H). The presence of peptidase M20 in the vicinity of *acdS* is also observed in some of these strains.

Further studies are necessary in order to characterize the importance of these regulators in ACC deaminase expression in different organisms.

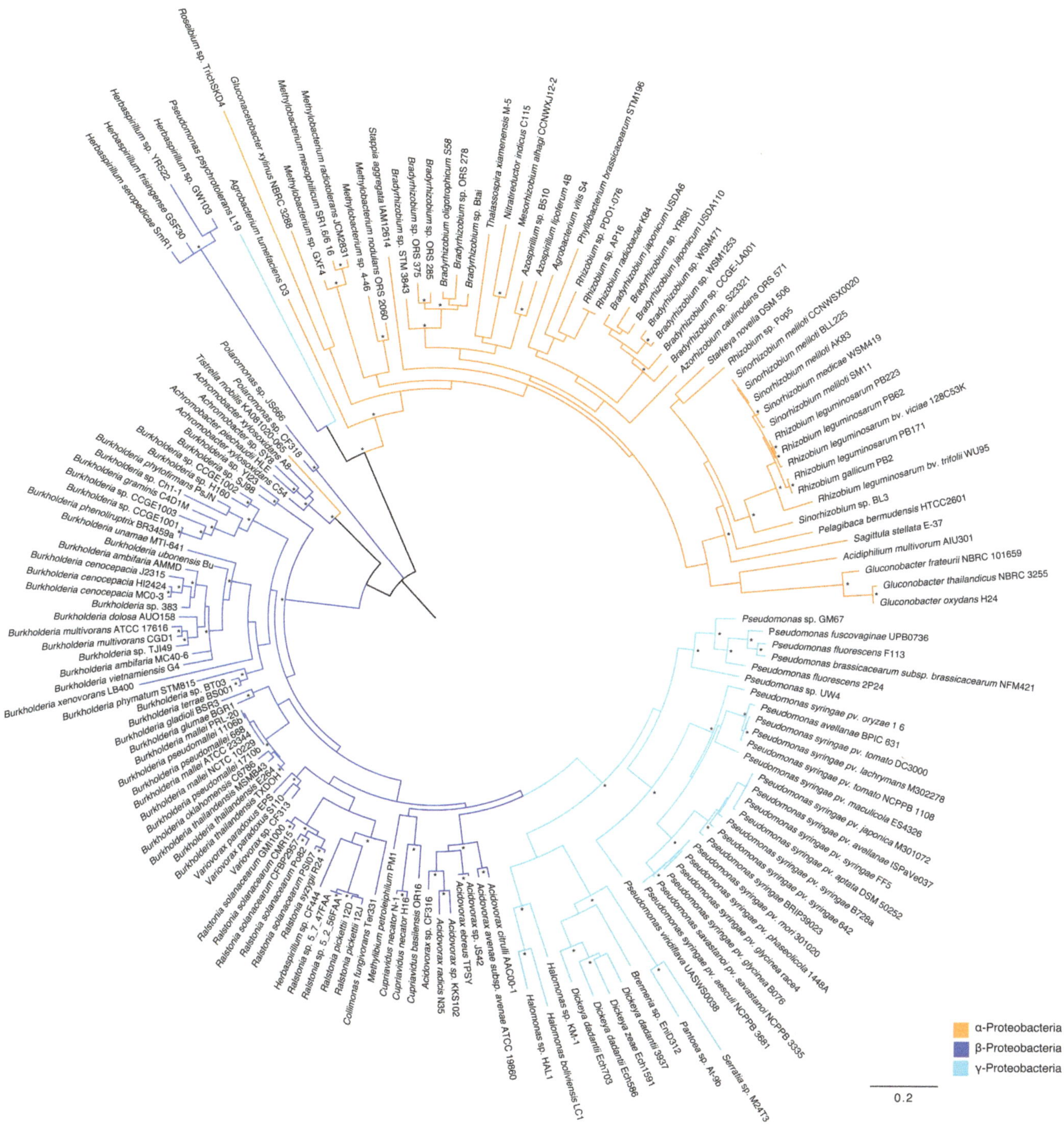

Figure 4. Phylogram based on *acdR* gene. The evolutionary history was inferred by using the Maximum Likelihood method based on the GTR model. A discrete Gamma distribution was used to model evolutionary rate differences among sites (4 categories). Branch support was evaluated using both aLRT (SH like) and bootstrap analysis (100 replicates). Bootstrap values above 0.75 (75%) are displayed in the phylograms shown next to the branches as *. The analysis involved 166 nucleotide sequences and 509 patterns were found (out of a total of 594 sites).

ACC deaminase origin

To gain additional knowledge regarding the origin and evolution of ACC deaminase multiple searches of the database were conducted; sequences showing a high similarity to different deaminases were obtained and a phylogram was constructed (Fig. 5). In this instance it was observed that ACC deaminase forms a distinct and unique group, where ACC deaminases from different organisms like Bacteria and Fungi cluster together. This is also observed with D-cysteine desulfhydrase, however only a few representatives of the considered "true" D-cysteine desulfhydrases (*E.coli* D-cysteine desulfhydrase) were obtained. When searching in the database, it was observed that D-cysteine desulfhydrase is an enzyme whose distribution is not widespread and it may also be not nearly as conserved as ACC deaminase. Its presence has been

Phytophthora infestans T30-4
Phytophthora sojae P6497
Emiliania huxleyi CCMP1516
Thalassiosira oceanica
Selaginella moellendorffii
Triticum urartu
Zea mays
Arabidopsis thaliana
Arabidopsis lyrata subsp. lyrata
Prunus persica
Fragaria vesca subsp. vesca
Solanum lycopersicum
Vitis vinifera
Cicer arietinum
Glycine max
Volvox carteri f. nagariensis
Chlamydomonas reinhardtii
Chlorella variabilis
Coccomyxa subellipsoidea C-169
Micromonas sp. RCC299
Oikopleura dioica
Amphimedon queenslandica
Nematostella vectensis
Branchiostoma floridae
Capitella teleta
Strongylocentrotus purpuratus
Trichoplax adhaerens
Ciona intestinalis
Haliangium ochraceum DSM14365
Unidentified deaminase (s) /D-cysteine desulfhydrase/ ACC deaminase?
Caldithrix abyssi
Saccharophagus degradans 2-40
Teredinibacter turnerae T7901
Simiduia agarivorans SA1
Luminiphilus syltensis NOR5-1B
Bacillus bataviensis
Staphylococcus pettenkoferi
Johnsonella ignava
Coprobacillus sp. 8_2_54BFAA
Clostridium symbiosum
Alkaliphilus metalliredigens QYMF
Syntrophobotulus glycolicus DSM8271
Fusobacterium ulcerans
Sorangium cellulosum Soce56
Thermotoga maritima MSB8
Unidentified deaminase(s)
Pyrococcus horikoshii OT3
Thermococcus sp. AM4
Deaminase (D-serine)
Lactobacillus parafarraginis
Acaryochloris sp. CCMEE5410
Psychrobacter arcticus 273-4
Roseobacter sp. SK209-2-6
Methylobacterium populi BJ001
Pseudomonas stutzeri
Unidentified deaminase (s)
Herbaspirillum frisingense GSF30
Pseudomonas psychrotolerans L19
ACC deaminase/ unidentified deaminase?
Azospirillum lipoferum 4B
Agrobacterium tumefaciens D3
Rhizobium leguminosarum 128C53K
Sinorhizobium meliloti SM11
Mesorhizobium loti MAFF303099
Phyllobacterium brassicacearum STM196
Bradyrhizobium japonicum USDA110
Cyberlidnera saturnus
Pseudomonas sp. ACP
Burkholderia phytofirmans PsJN
Burkholderia graminis C4D1M
Variovorax paradoxus 5C2
Ralstonia solanacearum GMI1000
Phytophthora infestans T30-4
Phytophthora sojae P6497
Pseudomonas sp. UW4
Rhodococcus sp. R04
Agreia sp. PHSC20C1
Penicillium citrinum
Trichoderma asperellum T203
Meiothermus ruber DSM1279
ACC deaminase
Pseudomonas fluorescens F113
Pseudomonas sp. UW4
Burkholderia pseudomallei 305
Escherichia coli K-12
Salmonella typhimurium LT2
Cronobacter sakazakii
Erwinia tasmaniensis Et1/99
D-cysteine desulfhydrase
Bacillus megaterium DSM319
Bacillus cereus
Bacillus thuringiensis
Deinococcus deserti VCD115
Pseudoalteromonas atlantica T6c
Vibrio tubiashii
Dinoroseobacter shibae DFL12
Bordetella bronchiseptica MO149
Pectobacterium carotovorum
Ruegeria pomeroyi DSS-3
Hydrogenophaga sp. PBC
Sagittula stellata E-37
Roseovarius sp. 217
Vibrio splendidus
Oceanibaculum indicum
Caulobacter crescentus CB15
Ramlibacter tataouinensis TTB310
D-cysteine desulfhydrase / unidentified deaminase (s)?
0.2

Figure 5. Phylogram constructed based on ACC deaminase and related PLP enzymes protein sequences. The evolutionary history was inferred by using the Maximum Likelihood method based on the WAG model. A discrete Gamma distribution was used to model evolutionary rate differences among sites (4 categories). Branch support was evaluated using both aLRT (SH like) and bootstrap analysis (100 replicates). Bootstrap values above 0.75 (75%) are displayed in the phylograms shown next to the branches as *. The analysis involved 99 aminoacid sequences and 570 patterns were found (out of a total of 594 sites). Sequences used for the construction of this phylogram are described in Table S6.

verified mainly in γ-Proteobacteria. Other proteins showing some homology were found in Firmicutes and other α-Proteobacteria, but in those instances showing low identity scores (39%).

Interestingly, D-cysteine desulfhydrase activity has been demonstrated for *Solanum lycopersicum* [58] and *Arabidopsis thaliana* [87]. However, these enzymes form a distinct phylogenetic cluster, far away from *E. coli* and other γ-Proteobacteria D-cysteine desulfhydrase. Controversially, ACC deaminase activity has also been demonstrated for *Arabidopsis thaliana*. Although Riemenschneider et al. [87] did not detect ACC deaminase activity from the product of *Arabidopsis thaliana* gene "At1g48420", McDonell et al. [88] showed that this gene encoded a protein with the ability to breakdown ACC. Moreover, McDonnel et al. [88] suggest that the gene is responsible for regulation of *Arabidopsis thaliana* endogenous ACC levels. The same authors also suggest that this enzyme may be present in many other plant species. Curiously, Todorovic and Glick [58] did not find ACC deaminase activity in the *Solanum lycopersicum* At1g48420 protein homolog, sharing 70% identity and clustering together with *Arabidopsis thaliana* At1g48420 protein (Fig. 5).

Despite showing D-cysteine desulfhydrase and ACC deaminase activity *in vitro*, it is conceivable that the at1g48420 gene product does not represent a "true" D-cysteine desulfhydrase or ACC deaminase, or at least, is only distantly related to bacterial D-cysteine desulfhydrase and ACC deaminase. The grouping that is observed in Fig. 5 supports this latter conclusion. Instead of clustering with bacterial ACC deaminase or D-cysteine desulfhydrase, the plant protein homologs form a distant and unique cluster with a different phylogenetic background within the broader family of these PLP-dependent enzymes.

Proteins with similar origin and function often tend to be conserved. Thus, if the at1g48420 gene encoded a true ACC deaminase (similar phylogenetic background and specialization towards ACC breakdown), it would likely cluster together with bacterial and fungal ACC deaminases and have similar amino acid residues in specific sites as are present in functional ACC deaminases that are important for ACC breakdown (i.e E295, L322). Nevertheless, it is possible that proteins like At1g48420 evolved and specialized in ACC degradation in a different route than those bacterial and fungal ACC deaminases. In this case, additional studies are necessary to further unveil the characteristics of At1g48420 like proteins.

It is most likely that the ability of at1g48420 gene product to use D-cysteine or ACC as substrates results from the high promiscuity that many deaminases show in cleaving multiple substrates that share similar characteristics. For example, it has been shown that ACC deaminase from *Pseudomonas* sp. ACP is able to use multiple substrates like D-cysteine and also other D-aminoacids. *E. coli* and *S. thyphimurium* D-cysteine desulfhydrases are able to efficiently use β-chloro-D-alanine (β-CDA) and other substrates (Table 1).

Intriguingly, Todorovic and Glick [58] demonstrated that mutations in amino acids (E295S/L322T) in *Pseudomonas* sp. UW4 ACC deaminase lead to the loss of ACC deaminase activity, yet, these mutations conferred an increased D-cysteine desulfhydrase activity to the mutant enzyme. The Km of the double mutant for D-cysteine was much lower than the Km of the native ACC deaminase towards ACC. Moreover, the Km of the double mutant enzyme towards D-cysteine is equivalent to that of a true D-cysteine desulfhydrase. Still, this mutant enzyme shows great inefficiency (K cat = 10.9 min-1) in D-cysteine cleavage. This data shows that small amino acid changes can confer different substrate usage abilities in closely related deaminases.

If ACC deaminase can use multiple substrates, it is possible that in some organisms the production of ACC deaminase can be important for cleavage of such substances, thus, giving these organisms the ability to use other nutrient sources, or to grow under otherwise toxic conditions. This can have major implications in a microorganism's fitness, especially in organisms living under limiting conditions. For example, Soutourina et al. [89] demonstrated that the expression of D-cysteine desulfhydrase by *E.coli* relieved some of the toxic effects of D-cysteine in bacterial growth. Also, D-cysteine desulfhydrase production allowed the growth of *E. coli* in a minimal medium containing D-cysteine as the sole sulfur source, demonstrating the importance of D-cysteine desulfhydrase in bacterial growth under sulfate limitation. It is possible that by maintaining a broad ability to cleave ACC-like substrates and some D-aminoacids, ACC deaminase genes were maintained in organisms that live in environments where ACC is not present. On the other hand, the presence of ACC deaminase in organisms that are associated with plants or other ACC-producing organisms, gave them a significant advantage in their ecology so that *acdS* genes were maintained. This may have led to the significant *acdS* gene presence in plant associated microorganisms, especially plants grown under perennially stressful conditions [51], and increased *acdS* gene loss in microorganisms living in environments where ACC is nonexistent.

Overall, it seems that bacterial and fungal ACC deaminases (here considered to be the representatives of true ACC deaminases) belong to a large group of PLP dependent deaminases (including bacterial D-cysteine desulfhydrase) related to tryptophan synthase beta subunit and sharing a common origin. Further, mutations and other evolutionary forces may have lead to some level of substrate specialization. Yet, some conserved features appear to allow these enzymes to be able to utilize a wide range of related substrates. This is exemplified by the data presented in Table 1.

ACC deaminase phylogenetic distribution and evolution

A model for ACC deaminase evolution and phylogenetic distribution is proposed based on the AcdS phylogenetic analysis, AcdS protein sequence analysis, *acdS* gene location, organism habitat and origin. The evolutionary relationship among Archaea, Bacteria and Eukaryotes was also taken into account when attempting to resolve the evolution of ACC deaminase [90,91,78].

From the available sequence data, it would appear that the most ancient point for the origin of ACC deaminase in Bacteria dates to the Actinobacteria or Deinococcus-Thermus. Most Actinobacteria strains investigated (Table S1) possess an *acdS* gene in their primary and unique chromosome. In another ancient bacterial lineage, the Deinococcus-Thermus, the *acdS* gene is also found in the chromosome of its representatives *Meiothermus*, suggesting an *acdS* chromosomal location in a common ancestor for Bacteria.

In many α-Proteobacteria, including *Azorhizobium* and *Bradyrhizobium*, the *acdS* gene is found in the primary chromosome of these

Table 1. Substrate cleavage abilities of studied ACC deaminase, D-cysteine desulfhydrase and other PLP dependent (ACC deaminase or D-cysteine desulfhydrase homologs) enzymes.

Enzyme	Tested substrates	Functional substrates	K_m (mM)	Reference
Arabidopsis thaliana **"At1g48420" ACC deaminase homolog**	ACC, D-cys	ACC, D-cys	0.25 (D-cys)	[87,88]
Bradyrhizobium japonicum **USDA110 ACC deaminase**	ACC, D-ser, β-CDA	ACC, D-ser	n.a	[61]
Cyberlidnera saturnus **ACC deaminase**	ACC, DCA, D-cys, D-ser, β-CDA, OAD-ser	ACC, DCA, D-cys, D-ser	2.6 (ACC)	[1,102]
E. coli **D-cysteine desulfhydrase**	D-cys, 3-CDA, D-cyst, DLAC, DLSC, DLSCyst, Dlan, D-ala, L-ala, D-ser, L-ser, D-phen, L-phen, D-tryp, and others.	D-cys, 3-CDA, D-cyst, DLac, DLsc, DLSCyst, DLan.	0.15 (D-cys), 0.91 (3-CDA), 0.27 (D-cyst), 0.29 (DLac), 0.04 (DLsc), 0.11 (Dlan)	[8]
Methylobacterium nodulans **ACC deaminase**	ACC, D-cys, L-cys	ACC	0.8 (ACC)	[103]
Methylobacterium radiotolerans **ACC deaminase**	ACC, D-cys, L-cys	ACC	1.8 (ACC)	[103]
Penicillium citrinum **ACC deaminase**	ACC, DCA, L-ser, D-ser, DACC, DACA	ACC, DCA, D-ser	4.8 (ACC)	[65]
Pirococcus horikoshi **OT3 ACC deaminase homolog**	ACC, D-Ala, L-Ala, D- Ser, and L-Ser, D-cys	D-ser, L-ser	n.a	[7]
Pseudomonas putida **UW4 ACC deaminase**	ACC, D-cys	ACC, D-cys	3.4 (ACC)	[104]
Pseudomonas putida **UW4 ACC deaminase mutant E295S/L322T**	ACC, D-cys	D-cys	0.34 (D-cys)	[104]
Pseudomonas **sp. ACP ACC deaminase**	ACC, D-VG, β-CDA, β-FDA, D-ser, V-ACC, APC, Cys, L-hom, L-thr, L-try, L-met, L-tyr, L-cys, L-aba, DCA, D-EAC, D-TAF	ACC, D-VG, β-CDA, D-ser, β-FDA, V-ACC, OAD-ser, β-2CDA, β-2FDA, DCA, D-EAC, D-TAF	1.5 and 9.2 (ACC), 1.1 (β-FDA), 4.4 (V-ACC), 5.4 (β-CDA), 36.2 (DCA), 56 (OAD-ser), 97 (D-VG)	[1,7,105]
Salmonella typhimurium **D-cysteine desulfhydrase**	D-Cys, β-CDA, D-Ser, L-Ser, ACC, D-Ala	D-Cys, β-CDA, D-ser	0.34 (D-cys)	[9]
Solanum lycopersicum **ACC deaminase homolog**	ACC, D-cys, L-cys	D-cys	0.21 (D-cys)	[58]
Solanum lycopersicum **ACC deaminase homolog S358E/T386L**	ACC, D-cys	ACC	n.a	[58]

1-amino-2-vinylcyclopropane-l-carboxylic acid (**V-ACC**), 1-aminocyclopentane-1-carboxylate (**APC**), 1-aminocyclopropane-1-carboxylic acid (**ACC**), 3-chloro-D-alanine (**3-CDA**), Cystathionine (**Cyst**), D-alanine (**D-ala**), D-cysteine (**D-Cys**), D-cystine (**D-cyst**), D-erythro-2-amino-3-chlorobutyrate (**D-EAC**), D-methionine (**D-met**), D-phenylalanine (**D-phen**), D-serine (**D-ser**), D-threo-2-amine-3-fluorobutyrate (**D-TAF**), D-tryptophan (**D-tryp**), D-vinylglycine (**D-VG**), Dimethyl-ACC (**DACC**), DL-lanthionine (**Dlan**), DL-allocoronamic acid (**DACA**), DL-allocystathionine (**DLAC**), DL-coronamic acid (**DCA**), DL-selenocysteine (**DLSCyst**), DL-selenocystine (**DLSC**), L-alanine (**L-ala**), L-aminobutyric acid (**L-aba**), L-cysteine (**L-cys**), L-homoserine (**L-hom**), L-methione (**L-met**), L-phenylalanine (**L-phen**), L-serine (**L-ser**), L-threonine (**L-thr**), L-tryptophan (**L-try**), L-tyrosine (**L-tyr**), β -chloro-D-alanine (**β-CDA**), β-fluoro-D-alanine (**β-FDA**), β, β- dichloro-D-alanine (**β2CDA**), β, β- difluoro-D-alanine (**β2FDA**), O-acetyl-D-serine (**OAD-ser**).

strains. The *acdS* gene is also found in the primary chromosome of many marine α-Proteobacteria, and in the vinegar isolate *Gluconacetobacter xylinus* NBRC 3288. Despite the fact that ACC deaminase genes were not yet detected in any δ or ε - Proteobacteria (181 genome sequences available in the database), the presence of *acdS* genes in α-Proteobacterial strains that live in environments where ACC is not present, suggests that *acdS* was present in a Proteobacteria ancestor, located in its primary chromosome and likely was acquired by vertical transmission.

Many α-Proteobacteria have *acdS* genes located on plasmids, symbiotic islands or second chromosomes; this is the case of the studied Rhizobiaceae (*Rhizobium*, *Sinorhizobium*, and *Agrobacterium*), Phyllobacteriaceae (*Phyllobacterium* and *Mesorhizobium*) and *Azospirillum* strains (Table S2). Extensive gene transfer analysis between completely sequenced α-Proteobacteria suggested that secondary chromosomes originated from intragenomic transfers from primary chromosomes to ancestral plasmids [92]. This mechanism may have not only led to the origin of a second chromosome in some α-Proteobacteria but also in other Proteobacteria. In this regard, it is possible that *acdS* was transferred from the primary chromosome to a plasmid in some α-Proteobacteria. This possibility is consistent with the presence of *acdS* genes in the plasmids of *Rhizobium* and *Sinorhizobium* species and in the second chromosome of *Rhizobium radiobacter* K84 and *Agrobacterium vitis* S4. Slater et al. [92] also suggested that some strains like *Mesorhizobium* could have evolved by plasmid gene integration into the main chromosome. This suggestion is consistent with the observation that the same gene arrangement is found between *Mesorhizobium* symbiotic islands and some *Rhizobium* and *Sinorhizobium* symbiotic plasmids, where the *acdS* gene is located.

It is likely that intragenomic transfers of *acdS* genes from primary chromosomes to plasmids may have occurred in members of α-Proteobacteria as well as in β and γ-Proteobacteria. The presence of *acdS* genes in the second chromosome of *Burkholderia* and megaplasmids in *Ralstonia* and some strains of *Pseudomonas* is consistent with this idea. The occurrence of such phenomena may have led to a dispersal of *acdS* genes through plasmids that are readily transmissible between closely and more distant related strains. This leads to the puzzling phylogeny of the *acdS* gene that is observed in bacteria belonging to the same Order or Family (taxonomy) (Fig. 1).

In most Fungi, AcdS sequences share an average sequence identity of ~50% with Bacterial AcdS. An exception to this case is the AcdS from yeasts, *Punctularia strigosozonata* HHB-11173, *Fomitopsis pinicola* FP-58527 (Basidiomycota/Agaricomycetes), *Aureobasidium pullulans* AY4, *Macrophomina phaseolina* MS6 and *Guignardia citricarpa* CGMCC3.14348 (Ascomycota/Dothideomycetes), and Stramenopiles, which appear to have a Bacterial origin. As observed in the AcdS based phylogram (Fig. 3), it seems that Fungal (excluding the above mentioned exceptions) and Bacterial AcdS diverged long ago. At this point it's not possible to corroborate both hypothesis of AcdS monophyletic or paraphyletic origin. Still, protein sequence analysis show some conserved amino acid regions (e.g. His80 and Ala161) in ACC deaminase from Fungi, Actinobacteria, Deinococcus-Thermus, and α-Proteobacteria, suggesting a common origin for *acdS* in these organisms. Organisms belonging to β- and γ-Proteobacterial classes show different amino acids in the referred positions, suggesting a later divergence from the α-Proteobacteria and the rest of ancient classes. The Fungi grouping closer to Actinobacteria is also observed in the phylogram (Fig. 3) suggesting a common origin for ACC deaminase in these organisms.

Based on the currently available data, we suggest that *acdS* genes had an ancient origin that may date to a Eukaryote and Bacterial common ancestor that possessed this gene in its chromosome. Furthermore, it is most likely that ACC deaminase originated as a consequence of specific mutations in an already existing PLP dependent enzyme showing high similarity to tryptophan synthase beta subunit. This is consistent with the results obtained by Todorovic and Glick [58] showing that small amino acid changes in related enzymes can be responsible for the ability to use a specific substrate.

Through time, it is probable that the *acdS* gene evolved by continuous vertical transmission, in which different constraints like habitat adaptation led to *acdS* divergence and sometimes gene loss. Intragenomic transfers of *acdS* genes from primary chromosomes to plasmids may have been selected for as a consequence of the advantage of ACC deaminase production, and this probably led to HGT events and increased divergence of *acdS* genes. These intragenomic transfer events and the presence of *acdS* on plasmids may have also lead to gene loss in many organisms. This is consistent with the results obtained by Prigent-Combaret et al. [40] showing that *Azospirillum lipoferum* 4B loses the plasmid containing an *acdS* gene during phenotypic variation events.

The role of ACC deaminase production in microorganism's ecology and fitness

From the available information, it is observed that many of the *acdS*+ organisms here described were isolated from heavily contaminated soils or otherwise stressed environments (Tables S1–S5), suggesting that ACC deaminase-producing microorganisms are more prevalent and better able to live in such conditions.

Organisms that produce ACC deaminase normally bind to plant tissues, and take up ACC to convert into ammonia and α-ketobutyrate [2]. The products of ACC cleavage are potential nitrogen and carbon sources [2,33] that can play a role in the microorganism's fitness under stressful situations. Under stress conditions plants produce higher levels of the phytohormone ethylene, which means that the plants also produce higher levels of ACC [3]. Microorganisms that bind to plant tissues typically utilize plant exudates as a nutrient source. Under stress conditions, not only is the amount of ACC produced by the plant increased, the vast majority of rhizosphere microorganisms produce the phytohormone indoleacetic acid (IAA) which acts to loosen plant cell walls thereby facilitating root exudation. Bacterial IAA production has also been shown to increase ACC synthase expression in plants [93]. Thus, microorganisms that can both produce IAA and utilize ACC may have a competitive advantage over other soil microorganisms [2,94].

Importantly, a recent study by Timmusk et al. [51], showed that ACC deaminase-producing organisms were more much abundant in the rhizosphere of wild barley (*Hordeum spontaneum*) growing in a stressed environment than they were in a similar (nearby) less stressed environment. This result was obtained despite the fact that both environments had similar soil, rock and topology characteristics. In addition, ACC deaminase-producing bacteria were abundant in plant rhizosphere samples and almost nonexistent in bulk soil samples. This suggests that organisms that produce ACC deaminase more readily survive in stressed environments by the mutualistic interaction with a plant host.

By degrading ACC, microorganisms decrease plant ethylene levels that under stress conditions are responsible for plant senescence and ultimately plant death [3]. Therefore, these organisms facilitate plant health under stress conditions. In turn, healthier plants provide their associated microorganisms with more nutrients thereby increasing the proliferation of these microorganisms.

Chen et al. [95] demonstrated that ACC deaminase-producing bacteria are also present in the casing soil of the ethylene-producing fungi *Agaricus bisporus*. The authors proposed a new model for the interaction between fungi and ACC deaminase producing bacteria. Bacteria possessing an *acdS* gene were able to increase fungal primordium initiation and proliferation by reducing endogenous ACC levels and consequently the inhibitory ethylene levels known to affect fungal development. These results show that ACC deaminase-producing bacteria might not only associate with plants but also with fungi, bringing significant advantages to fungal colonization in soil. On the other hand, bacteria producing ACC deaminase gain significant advantages by associating with extreme soil and plant colonizers like fungi. Being that these organisms constantly produce ACC, bacteria able to degrade ACC may gain extra nutrient sources as previously suggested.

ACC deaminase in Fungi: Relationship with plants or regulation of endogenous ACC levels?

The production of ACC deaminase by *Trichoderma asperellum* T230 has been shown to be an important mechanism for the plant growth promotion abilities of this fungal strain [16]. When ACC deaminase production is impaired, the fungal ability to promote canola root elongation is decreased, therefore, suggesting that ACC deaminase may act in a similar way as previously described by Glick et al. [2] for plant growth-promoting bacteria.

Nonetheless, it has been shown by Jia et al. [65] that ACC deaminase in *Penicillium citrinum* is produced independently of a relationship with a plant host. This happens because *Pennicillium citrinum* is capable of producing and accumulating ACC in its tissues. That is, *Penicillium citrinum* possesses not only an *acdS* gene but also an ACC synthase gene. Jia et al. [96] found that the ACC deaminase was induced by the presence of accumulated ACC in the intracellular spaces of *Penicilium citrinum*, indicating that ACC deaminase may participate in the regulation of ACC levels in this strain. As a consequence, ethylene production by *P. citrinum* can also be regulated by ACC deaminase. In fact, our search of the database revealed the presence of ACC synthase homologs in most fungal strains that possess an ACC deaminase (data not shown). Together these results suggest that ACC deaminase production by Fungi can account for the regulation of endogenous ACC concentrations, and therefore regulation of ethylene levels which can inhibit primordium initiation and formation.

What is the role of ACC deaminase in pathogenic microorganisms?

Surprisingly, *acdS* and *acdR* genes are found in a wide range of plant and human pathogenic microorganisms (Tables S1-S5), suggesting that ACC deaminase may play a role in these microrganisms' ecology. For example, the production of ACC deaminase has been reported in the human pathogenic *Burkholderia cenocepacia* J2315 [64]. However, this bacterial strain, like other pathogenic *Burkholderia* strains, is predominant in soils where it normally associates with plants [97–99]. The *acdS* gene is also found in pathogenic fungi like *Aspergillus* spp. and *Myceliophthora thermophila*. Despite causing severe diseases in immunocompromised humans, these strains are mainly found in soil [100,101]. This data suggests that the presence of *acdS* genes in human pathogenic organisms may not be related to their human pathogenesis mechanisms but rather to their possible ecological role in soil. Also, it is possible that the presence of *acdS* gene in these strains and in plant pathogenic bacteria is related to the continuous *acdS* vertical transmission and not to any beneficial effects of ACC deaminase production.

Nevertheless, ACC deaminase production by pathogenic microorganisms may ultimately play a role in: (a) obtaining extra nutrients sources from ACC or ACC like substrate degradation, (b) the plant or fungi growth promoting abilities of these organisms when they are not acting as human or plant pathogens ("opportunistic" pathogens), (c) augmenting the ability to overcome ethylene or ACC mediated plant response systems, (d) regulation of endogenous ACC levels, or a combination of these factors.

Conclusions

The results obtained in this study provide a more complete view of the role for ACC deaminase-producing organisms then was previously available. ACC deaminase genes are not only found in plant-associated microorganisms but also in other bacterial and fungal strains isolated from a wide range of different sources (i.e. hot springs, industrial sludge, sea), hence, challenging the notion that ACC deaminase-producing organisms only interact with plants, or more interestingly, that ACC deaminase can only use ACC as a substrate.

Based on multiple parameters like protein sequence analysis and phylogenetic studies we suggest that ACC deaminase belongs to a broad group of promiscuous PLP dependent enzymes (tryptophan synthase beta subunit family) sharing a common ancestor. It is most likely that ACC deaminase originated as a consequence of specific mutations in its ancestral enzyme gene. Small amino acid mutations conferred changes in substrate specificity, however, the ability to degrade similar substrates was somehow maintained. This can account for the presence of *acdS* genes in bacteria that don't associate with ACC producing organisms. The continuous vertical transmission of *acdS* genes may also be responsible for the presence of *acdS* in these organisms. Furthermore, contrary to previous reports, here we demonstrate that the *acdS* gene is mostly vertically inherited in various bacterial and fungal classes. An ancient origin dating a Bacterial/Eukaryote ancestor is also proposed for the *acdS* gene.

Nonetheless, horizontal gene transfer does account for a wide portion of ACC deaminase evolution. For instance, some fungal classes and some members of Stramenopiles may have acquired *acdS* genes from Bacteria, suggesting that HGT events not only occur between bacteria but also may occur between distantly related organisms.

The presence of the *acdR* gene is observed in most Proteobacteria possessing an *acdS* gene, suggesting a coupled evolution for these genes. In other microorganisms like Actinobacteria and Deinococcus-Thermus (*Meiothermus*) the presence of genes encoding a GntR family protein are observed in the vicinity of the *acdS* gene, suggesting a different mechanism of ACC deaminase regulation. Moreover, these regulatory genes (here termed *acd-AR* and *acd-MR*) are mostly found in these *acdS*+ bacteria groups, reinforcing the idea that specific regulatory elements can be found in different Bacteria classes.

Additional genetic and biochemical studies are needed to gain some additional understanding of ACC deaminase functioning and its possible role(s) in the ecology of various organisms. Also, exploring the origin of ACC deaminase and related enzymes may bring new insights into the functioning of this PLP family of enzymes that may be the key to their use in a variety of important biotechnological applications.

Supporting Information

Figure S1 Multiple sequence alignment based on functional ACC deaminases, putative ACC deaminase sequences from *Agreia* sp. PHSC20C1, *Rhodococcus* sp. R04 (Actinobacteria) and *Meiothermus ruber* DSM1279 (Deinococcus-Thermus). D-cysteine desulfhydrase from *E-coli*, PLP dependent deaminase from *Pyrococcus horikoshii* and PLP dependent deaminase from *Solanum lycopersicum* are highlighted in grey. Conserved residues between all protein groups are shown in blue. ACC deaminase conserved residues are shown in green.

Figure S2 Putative regulators, *acdS* and neighborhood genes organization in some Actinobacteria, Deinococcus-Thermus and Proteobacteria.

File S1 Evolutionary distances estimates of *acdS* and 16S rRNA genes from several bacterial groups.

Table S1 Accession numbers for Actinobacteria, Deinococcus-Thermus and Firmicutes 16S rRNA, *acdS* and *acdR* genes and AcdS and AcdR protein sequences. Description of the *acdS* gene location, ACC deaminase (ACCD) activity, strains relative habitat and geographical origin.

Table S2 Accession numbers for α-Proteobacteria 16S rRNA, *acdS* and *acdR* genes and AcdS and AcdR proteins sequences. Description of the *acdS* gene location, ACC deaminase (ACCD) activity, strains relative habitat and origin.

Table S3 Accession numbers for β-Proteobacteria 16S rRNA, *acdS* and *acdR* genes and AcdS and AcdR proteins sequences and description of the *acdS* gene location, ACC deaminase (ACCD) activity, strains relative habitat and origin.

Table S4 Accession numbers for γ-Proteobacteria 16S rRNA, *acdS* and *acdR* genes and AcdS and AcdR proteins

sequences and description of the *acdS* gene location, ACC deaminase (ACCD) activity, strains relative habitat and origin.

Table S5 Accession numbers for Eukaryotes AcdS complete sequences and description of ACC deaminase (ACCD) activity, strains relative habitat and geographical origin.

Table S6 Acession numbers for the sequences used in Figure 5.

Acknowledgments

Francisco X. Nascimento acknowledges a PhD fellowship (SFRH/BD/86954/2012) from Fundação de Ciência e Tecnologia (FCT), Portugal. We thank Michael D.J. Lynch (University of Waterloo) for all the help in the phylogenetic analysis, and Clarisse Brígido (Universidade de Évora) for the suggestions provided for the elaboration of this manuscript.

Author Contributions

Conceived and designed the experiments: FXN MJR CRFSS BJM BRG. Performed the experiments: FXN. Analyzed the data: FXN. Contributed reagents/materials/analysis tools: MJR CRFSS BJM BRG. Wrote the paper: FXN MJR CRFSS BJM BRG.

References

1. Honma M, Shimomura T (1978) Metabolism of 1-aminocyclopropane-1-carboxylic acid. Agric Biol Chem 42: 1825–1831.
2. Glick B, Penrose D, Li J (1998) A model for the lowering of plant ethylene concentrations by plant growth-promoting bacteria. J Theor Biol 190: 63–68.
3. Glick BR, Cheng Z, Czarny J, Duan J (2007) Promotion of plant growth by ACC deaminase-producing soil bacteria. Eur J Plant Pathol 119: 329–339.
4. Hyodo H (1991) Stress/wound ethylene. In: Mattoo AK, Shuttle JC (eds) The Plant Hormone Ethylene. CRC Press, Boca Raton, 65–80.
5. Glick BR (2014) Bacteria with ACC deaminase can promote plant growth and help to feed the world. Microbiol Res. 169:30–39.
6. Glick BR, Todorovic B, Czarny J, Cheng Z, Duan J, et al. (2007) Promotion of plant growth by bacterial ACC deaminase. Crit Rev Plant Sci. 26: 227–242.
7. Fujino A, Ose T, Yao M, Tokiwano T, Honma M, et al. (2004) Structural and enzymatic properties of 1-aminocyclopropane-1-carboxylate deaminase homologue from *Pyrococcus horikoshii*. J Mol Biol 341: 999–1013.
8. Nagasawa T, Ishii T (1985) D-Cysteine desulfhydrase of *Escherichia coli*. Purification and characterization. Eur J Biochem 153: 541–551.
9. Bharath SR, Bisht S, Harijan RK, Savithri HS, Murthy MRN (2012) Structural and mutational studies on substrate specificity and catalysis of *Salmonella typhimurium* D-cysteine desulfhydrase. PLoS One 7: e36267
10. Glick BR, Jacobson CB, Schwarze MMK, Pasternak JJ (1994) 1-aminocyclopropane-1-carboxylic acid deaminase mutants of the plant growth-promoting rhizobacterium *Pseudomonas putida* GR12-2 do not stimulate canola root elongation. Can J Microbiol 40: 911–915.
11. Li J, Ovakim DH, Charles TC, Glick BR (2000) An ACC deaminase minus mutant of *Enterobacter cloacae* UW4 no longer promotes root elongation. Curr Microbiol 41: 101–105.
12. Ma W, Guinel F, Glick B (2003) *Rhizobium leguminosarum* biovar viciae 1-aminocyclopropane-1-carboxylate deaminase promotes nodulation of pea plants. Appl Environ Microbiol 69: 4396–4402.
13. Uchiumi T, Ohwada T, Itakura M, Mitsui H, Nukui N, et al. (2004) Expression islands clustered on the symbiosis island of the *Mesorhizobium loti* genome. J Bacteriol 186: 2439–2448.
14. Sun Y, Cheng Z, Glick BR (2009) The presence of a 1-aminocyclopropane-1-carboxylate (ACC) deaminase deletion mutation alters the physiology of the endophytic plant growth-promoting bacterium *Burkholderia phytofirmans* PsJN. FEMS Microbiol Lett 296: 131–136.
15. Ali S, Charles TC, Glick BR (2012) Delay of flower senescence by bacterial endophytes expressing 1-aminocyclopropane-1-carboxylate deaminase. J Appl Microbiol 113: 1139–1144.
16. Viterbo A, Landau U, Kim S, Chernin L, Chet I (2010) Characterization of ACC deaminase from the biocontrol and plant growth-promoting agent *Trichoderma asperellum* T203. FEMS Microbiol Lett 305: 42–48.
17. Brotman Y, Landau U, Cuadros-Inostroza Á, Tohge T, Takayuki T, et al. (2013) *Trichoderma*-plant root colonization: escaping early plant defense responses and activation of the antioxidant machinery for saline stress tolerance. PLoS Pathog 9: e1003221
18. Wang C, Knill E, Glick BR, Défago G (2000) Effect of transferring 1-aminocyclopropane-1-carboxylic acid (ACC) deaminase genes into *Pseudomonas fluorescens* strain CHA0 and its *gacA* derivative CHA96 on their growth-promoting and disease-suppressive capacities. Can J Microbiol 46: 898–907.
19. Grichko V, Glick B (2001) Amelioration of flooding stress by ACC deaminase-containingplant growth-promoting bacteria. Plant Physiol Biochem 39: 11–17.
20. Mayak S, Tirosh T, Glick BR (2004) Plant growth-promoting bacteria confer resistance in tomato plants to salt stress. Plant Physiol Biochem 42: 565–572.
21. Mayak S, Tirosh T, Glick BR (2004) Plant growth-promoting bacteria that confer resistance to water stress in tomatoes and peppers. Plant Sci 166: 525–530.
22. Belimov A, Hontzeas N, Safronova VI, Demchinskaya SV, Piluzza G, et al. (2005) Cadmium-tolerant plant growth-promoting bacteria associated with the roots of Indian mustard (*Brassica juncea* L. Czern.). Soil Biol Biochem 37: 241–250
23. Hao Y, Charles TC, Glick BR (2007) ACC deaminase from plant growth-promoting bacteria affects crown gall development. Can J Microbiol 53: 1291–1299
24. Toklikishvili N, Dandurishvili N, Vainstein A, Tediashvili M, Giorgobiani N, et al. (2010) Inhibitory effect of ACC deaminase-producing bacteria on crown gall formation in tomato plants infected by *Agrobacterium tumefaciens* or *A. vitis*. Plant Pathol 59: 1023–1030.
25. Nascimento FX, Vicente CSL, Barbosa P, Espada M, Glick BR, et al. (2013) Evidence for the involvement of ACC deaminase from *Pseudomonas putida* UW4 in the biocontrol of pine wilt disease caused by *Bursaphelenchus xylophilus*. BioControl 58: 427–433.
26. Burd G, Dixon D, Glick B (1998) A plant growth-promoting bacterium that decreases nickel toxicity in seedlings. Appl Environ Microbiol 64: 3663–3668.
27. Nie L, Shah S, Rashid A, Burd GI, George Dixon D, et al. (2002) Phytoremediation of arsenate contaminated soil by transgenic canola and the plant growth-promoting bacterium *Enterobacter cloacae* CAL2. Plant Physiol Biochem 40: 355–361.
28. Reed MLE, Warner BG, Glick BR (2005) Plant growth-promoting bacteria facilitate the growth of the common reed *Phragmites australisin* the presence of copper or polycyclic aromatic hydrocarbons. Curr Microbiol 51: 425–429.
29. Klee HJ, Hayford MB, Kretzmer KA, Barry GF, Kishore GM (1991) Control of ethylene synthesis by expression of a bacterial enzyme in transgenic tomato plants. Plant Cell 3: 1187–1193.
30. Grichko VP, Filby B, Glick BR (2000) Increased ability of transgenic plants expressing the bacterial enzyme ACC deaminase to accumulate Cd, Co, Cu, Ni, Pb, and Zn. J Biotechnol 81: 45–53. 1.
31. Robison MM, Shah S, Tamot B, Pauls KP, Moffatt BA, et al. (2001) Reduced symptoms of *Verticillium* wilt in transgenic tomato expressing a bacterial ACC deaminase. Mol Plant Pathol 2: 135–145.
32. Sergeeva E, Shah S, Glick BR (2005) Growth of transgenic canola (Brassica napus cv. Westar) expressing a bacterial 1-aminocyclopropane-1-carboxylate (ACC) deaminase gene on high concentrations of salt. World J Microbiol Biotechnol 22: 277–282.
33. Ma W, Charles T, Glick B (2004) Expression of an exogenous 1-aminocyclopropane-1-carboxylate deaminase gene in *Sinorhizobium meliloti* increases its ability to nodulate alfalfa. Appl Environ Microbiol 70: 5891–5897.
34. Conforte VP, Echeverria M, Sánchez C, Ugalde RA, Menéndez AB, et al. (2010) Engineered ACC deaminase-expressing free-living cells of *Mesorhizobium loti* show increased nodulation efficiency and competitiveness on *Lotus* spp. J Gen Appl Microbiol 56:331–338.
35. Nascimento F, Brígido C, Alho L, Glick BR, Oliveira S (2012) Enhanced chickpea growth-promotion ability of a *Mesorhizobium* strain expressing an exogenous ACC deaminase gene. Plant Soil 353: 221–230.
36. Nascimento FX, Brígido C, Glick BR, Oliveira S, Alho L (2012) *Mesorhizobium ciceri* LMS-1 expressing an exogenous 1-aminocyclopropane-1-carboxylate (ACC) deaminase increases its nodulation abilities and chickpea plant resistance to soil constraints. Lett Appl Microbiol 55: 15–21
37. Grichko VP, Glick BR (2000) Identification of DNA sequences that regulate the expression of the *Enterobacter cloacae* UW4 1-aminocyclopropane-1-carboxylic acid deaminase gene. Can J Microbiol 46: 1159–1165.
38. Li J, Glick BR (2001) Transcriptional regulation of the *Enterobacter cloacae* UW4 1-aminocyclopropane-1-carboxylate (ACC) deaminase gene (*acdS*). Can J Microbiol 47: 359–367.
39. Cheng Z, Duncker BP, McConkey BJ, Glick BR (2008) Transcriptional regulation of ACC deaminase gene expression in *Pseudomonas putida* UW4. Can J Microbiol 136: 128–136.
40. Prigent-Combaret C, Blaha D, Pothier JF, Vial L, Poirier M-A, et al. (2008) Physical organization and phylogenetic analysis of *acdR* as leucine-responsive regulator of the 1-aminocyclopropane-1-carboxylate deaminase gene *acdS* in phytobeneficial *Azospirillum lipoferum* 4B and other Proteobacteria. FEMS Microbiol Ecol 65: 202–219.
41. Hontzeas N, Richardson A (2005) Evidence for horizontal transfer of 1-aminocyclopropane-1-carboxylate deaminase genes. Appl Environ Microbiol 71: 7556–7558.

42. Blaha D, Prigent-Combaret C, Mirza MS, Moënne-Loccoz Y (2006) Phylogeny of the 1-aminocyclopropane-1-carboxylic acid deaminase-encoding gene *acdS* in phytobeneficial and pathogenic Proteobacteria and relation with strain biogeography. FEMS Microbiol Ecol 56: 455–470.
43. Nascimento FX, Brígido C, Glick BR, Oliveira S (2012) ACC deaminase genes are conserved among *Mesorhizobium* species able to nodulate the same host plant. FEMS Microbiol Lett 336: 26–37.
44. Sziderics H, Rasche F, Trognitz F, Sessitsch A, Wilhelm E (2007) Bacterial endophytes contribute to abiotic stress adaptation in pepper plants (*Capsicum annuum* L.). Can J Microbiol 53: 1195–1202.
45. Dell'Amico E, Cavalca L, Andreoni V (2008) Improvement of *Brassica napus* growth under cadmium stress by cadmium-resistant rhizobacteria. Soil Biol Biochem 40: 74–84.
46. El-Tarabily KA (2008) Promotion of tomato (*Lycopersicon esculentum* Mill.) plant growth by rhizosphere competent 1-aminocyclopropane-1-carboxylic acid deaminase-producing streptomycete actinomycetes. Plant Soil 308: 161–174.
47. Dastager SG, Deepa CK, Pandey A (2010) Isolation and characterization of novel plant growth promoting *Micrococcus* sp NII-0909 and its interaction with cowpea. Plant Physiol Biochem 48: 987–992.
48. Siddikee M, Chauan PS, Anandham R, Han G-Y, Sa T (2010) Isolation, characterization, and use for plant growth promotion under salt stress, of ACC deaminase-producing halotolerant bacteria derived from coastal soil. J Microbiol Biotechnol 20: 1577–1584.
49. Ghosh S, Penterman JN, Little RD, Chavez R, Glick BR (2003) Three newly isolated plant growth-promoting bacilli facilitate the growth of canola seedlings. Plant Physiol Biochem 41: 277–281.
50. Sharma M, Mishra V, Rau N, Sharma RS (2010) Functionally diverse rhizobacteria of *Saccharum munja* (a native wild grass) colonizing abandoned morrum mine in Aravalli hills (Delhi). Plant Soil 341: 447–459.
51. Timmusk S, Paalme V, Pavlicek T, Bergquist J, Vangala A, et al. (2011) Bacterial distribution in the rhizosphere of wild barley under contrasting microclimates. PLoS One 6: e17968.
52. Maimaiti J, Zhang Y, Yang J, Cen Y-P, Layzell DB, et al. (2007) Isolation and characterization of hydrogen-oxidizing bacteria induced following exposure of soil to hydrogen gas and their impact on plant growth. Environ Microbiol 9: 435–444.
53. Pragash M, Narayanan KB, Naik PR, Sakthivel N (2009) Characterization of *Chryseobacterium aquaticum* strain PUPC1 producing a novel antifungal protease from rice rhizosphere soil. J Microbiol Biotechnol 19: 99–107.
54. Nadeem SM, Zahir ZA, Naveed M, Arshad M (2009) Rhizobacteria containing ACC-deaminase confer salt tolerance in maize grown on salt-affected fields. Can J Microbiol 55: 1302–1309.
55. Marques APGC, Pires C, Moreira H, Rangel AOSS, Castro PML (2010) Assessment of the plant growth promotion abilities of six bacterial isolates using *Zea mays* as indicator plant. Soil Biol Biochem 42: 1229–1235.
56. Singh N, Kashyap S (2012) *In silico* identification and characterization of 1-aminocyclopropane-1-carboxylate deaminase from *Phytophthora sojae*. J Mol Model 18: 4101–4111.
57. Glick BR (2005) Modulation of plant ethylene levels by the bacterial enzyme ACC deaminase. FEMS Microbiol Lett 251: 1–7.
58. Todorovic B, Glick BR (2008) The interconversion of ACC deaminase and D-cysteine desulfhydrase by directed mutagenesis. Planta 229: 193–205.
59. Edgar RC (2004) MUSCLE: multiple sequence alignment with high accuracy and high throughput. Nucleic Acids Res 32: 1792–1797.
60. Hao Y, Charles T, Glick B (2011) ACC deaminase activity in avirulent *Agrobacterium tumefaciens* D3. Can J Microbiol 286: 278–286.
61. Murset V, Hennecke H, Pessi G (2012) Disparate role of rhizobial ACC deaminase in root-nodule symbioses. Symbiosis 57: 43–50.
62. Contesto C, Desbrosses G, Lefoulon C, Béna G, Borel F, et al. (2008) Effects of rhizobacterial ACC deaminase activity on *Arabidopsis* indicate that ethylene mediates local root responses to plant growth-promoting rhizobacteria. Plant Sci 175: 178–189.
63. Stiens M, Schneiker S, Keller M, Kuhn S, Pühler A, et al. (2006) Sequence analysis of the 144-kilobase accessory plasmid pSmeSM11a, isolated from a dominant *Sinorhizobium meliloti* strain identified during a long-term field release experiment. Appl Environ Microbiol 72: 3662–3672.
64. Onofre-Lemus J, Hernández-Lucas I, Girard L, Caballero-Mellado J (2009) ACC (1-aminocyclopropane-1-carboxylate) deaminase activity, a widespread trait in Burkholderia species, and its growth-promoting effect on tomato plants. Appl Environ Microbiol 75: 6581–6590.
65. Jia YJ, Kakuta Y, Sugawara M, Igarashi T, Oki N, et al. (1999) Synthesis and degradation of 1-aminocyclopropane-1-carboxylic acid by *Penicillium citrinum*. Biosci Biotechnol Biochem 63: 542–549.
66. Rothballer M, Eckert B, Schmid M, Fekete A, Schloter M, et al. (2008) Endophytic root colonization of gramineous plants by *Herbaspirillum frisingense*. FEMS Microbiol Ecol 66: 85–95.
67. Gouy M, Guindon S, Gascuel O (2010) SeaView version 4: A multiplatform graphical user interface for sequence alignment and phylogenetic tree building. Mol Biol Evol 27: 221–224.
68. Guindon S, Gascuel O (2003). A simple, fast, and accurate algorithm to estimate large phylogenies by maximum likelihood. Syst Biol 52:696–704.
69. Darriba D, Taboada GL, Doallo R, Posada D (2012) jModelTest 2: more models, new heuristics and parallel computing. Nature Methods 9:772.
70. Darriba D, Taboada GL, Doallo R, Posada D (2011) ProtTest 3: fast selection of best-fit models of protein evolution. Bioinformatics 27:1164–1165.
71. Anisimova M, Gascuel O (2006) Approximate likelihood-ratio test for branches: a fast, accurate, and powerful alternative. Syst Biol 55:539–552.
72. Tamura K, Stecher G, Peterson D, Filipski A, Kumar S (2013) MEGA6: Molecular Evolutionary Genetics Analysis version 6.0. Mol Biol Evol 30: 2725–2729.
73. Chain PSG, Denef VJ, Konstantinidis KT, Vergez LM, Agulló L, et al. (2006) *Burkholderia xenovorans* LB400 harbors a multi-replicon, 9.73-Mbp genome shaped for versatility. Proc Natl Acad Sci U S A 103: 15280–15287.
74. Cooper VS, Vohr SH, Wrocklage SC, Hatcher PJ (2010) Why genes evolve faster on secondary chromosomes in bacteria. PLoS Comput Biol 6: e1000732.
75. Massey RC, Buckling A (2002) Environmental regulation of mutation rates at specific sites. Trends Microbiol 10: 580–584.
76. Boussau B, Karlberg EO, Frank AC, Legault B-A, Andersson SGE (2004) Computational inference of scenarios for alpha-proteobacterial genome evolution. Proc Natl Acad Sci U S A 101: 9722–9727.
77. Dobrindt U, Chowdary MG, Krumbholz G, Hacker J (2010) Genome dynamics and its impact on evolution of *Escherichia coli*. Med Microbiol Immunol 199: 145–154.
78. Gupta RS (2000) The natural evolutionary relationships among prokaryotes. Crit Rev Microbiol 26: 111–131.
79. Gupta RS (2005) Protein signatures distinctive of alpha proteobacteria and its subgroups and a model for alpha-proteobacterial evolution. Crit Rev Microbiol 31: 101–135.
80. Gupta RS, Sneath PH (2007) Application of the character compatibility approach to generalized molecular sequence data: branching order of the proteobacterial subdivisions. J Mol Evol 64:90–100.
81. Ludwig W, Klenk H-P (2005) Overview: a phylogenetic backbone and taxonomic framework for prokaryotic systamatics. In: Boone DR, Castenholz RW (eds) Bergey's manual of systematic bacteriology. Springer-Verlag, Berlin, 49–65.
82. Urakami T, Ito-Yoshida C, Araki H, Kijima T, Suzuki K, et al. (1994) Transfer of *Pseudomonas plantarii* and *Pseudomonas glumae* to *Burkholderia* as *Burkholderia* spp. and description of *Burkholderia vandii* sp. nov. Int J Syst Bacteriol 44: 235–245.
83. Viallard V, Poirier I, Cournoyer B, Haurat J, Wiebkin S, et al. (1998) *Burkholderia graminis* sp. nov., a rhizospheric *Burkholderia* species, and reassessment of (*Pseudomonas*) *phenazinium*, (*Pseudomonas*) *pyrrocinia* and (*Pseudomonas*) *glathei* as *Burkholderia*. Int J Syst Bacteriol 48: 549–563.
84. Nandasena KG, O'Hara GW, Tiwari RP, Sezmiş E, Howieson JG (2007) In situ lateral transfer of symbiosis islands results in rapid evolution of diverse competitive strains of mesorhizobia suboptimal in symbiotic nitrogen fixation on the pasture legume *Biserrula pelecinus* L. Environ Microbiol 9: 2496–2511.
85. Nikolic B, Schwab H, Sessitsch A (2011) Metagenomic analysis of the 1-aminocyclopropane-1-carboxylate deaminase gene (*acdS*) operon of an uncultured bacterial endophyte colonizing *Solanum tuberosum* L. Arch Microbiol 193: 665–676.
86. Nukui N, Minamisawa K, Ayabe S-I, Aoki T (2006) Expression of the 1-aminocyclopropane-1-carboxylic acid deaminase gene requires symbiotic nitrogen-fixing regulator gene *nifA2* in *Mesorhizobium loti* MAFF303099. Appl Environ Microbiol 72: 4964–4969.
87. Riemenschneider BA, Wegele R, Schmidt A, Papenbrock J (2005) Isolation and characterization of a D-cysteine desulfhydrase protein from *Arabidopsis thaliana*. FEBS J 272: 1291–1304.
88. McDonnell L, Plett JM, Andersson-Gunnerås S, Kozela C, Dugardeyn J, et al. (2009) Ethylene levels are regulated by a plant encoded 1-aminocyclopropane-1-carboxylic acid deaminase. Physiol Plant 136: 94–109.
89. Soutourina J, Blanquet S, Plateau P (2001) Role of D-cysteine desulfhydrase in the adaptation of *Escherichia coli* to D-cysteine. J Biol Chem 276: 40864–40872.
90. Woese C (1998) The universal ancestor. Proc Natl Acad Sci U S A 95: 6854–6859.
91. Wu D, Hugenholtz P, Mavromatis K, Pukall R, Dalin E, et al. (2009) A phylogeny-driven genomic encyclopaedia of Bacteria and Archaea. Nature 462: 1056–1060.
92. Slater SC, Goldman BS, Goodner B, Setubal JC, Farrand SK, et al. (2009) Genome sequences of three *Agrobacterium* biovars help elucidate the evolution of multichromosome genomes in bacteria. J Bacteriol 191: 2501–2511.
93. Kende H (1993). Ethylene biosynthesis. Annu Rev Plant Physiol Plant Mol Biol 44: 283–307.
94. Stearns JC, Woody OZ, McConkey BJ, Glick BR (2012) Effects of bacterial ACC deaminase on *Brassica napus* gene expression measured with an *Arabidopsis thaliana* microarray. Mol Plant Microbe Interact 25: 668–676.
95. Chen S, Qiu C, Huang T, Zhou W, Qi Y, et al. (2013) Effect of 1-aminocyclopropane-1-carboxylic acid deaminase producing bacteria on the hyphal growth and primordium initiation of *Agaricus bisporus*. Fungal Ecol 6: 110–118.
96. Jia Y, Ito H, Matsui H, Honma M (2000) 1-aminocyclopropane-1-carboxylate (ACC) deaminase induced by ACC synthesized and accumulated in *Penicillium citrinum* intracellular spaces. Biosci Biotechnol Biochem 64: 299–305.
97. Fiore A, Laevens S, Bevivino A, Dalmastri C, Tabbacchioni S, et al. (2001) *Burkholderia cepacia* complex: distribution of genomovars among isolates from the maize rhizosphere in Italy. Environ Microbiol 3: 137–143.

98. Baldwin A, Mahenthiralingam E, Drevinek P, Vandamme P, Govan JR, et al. (2007) Environmental *cepacia* complex isolates in human infections. Emerg Infect Dis 13: 458–461.
99. Suárez-Moreno ZR, Caballero-Mellado J, Coutinho BG, Mendonça-Previato L, James EK, et al. (2012) Common features of environmental and potentially beneficial plant-associated *Burkholderia*. Microb Ecol 63:249–266.
100. Klich MA (2002) Biogeography of *Aspergillus* species in soil and litter. Mycologia 94: 21–27.
101. Destino L, Sutton DA, Helon AL, Havens PL, Thometz JG, et al. (2006) Severe osteomyelitis caused by *Myceliophthora thermophila* after a pitchfork injury. Ann Clin Microbiol Antimicrob 5: 21.
102. Minami R, Uchiyama K, Murakami T, Kawai J, Mikami K, et al. (1998) Properties, sequence, and synthesis in *Escherichia coli* of 1-aminocyclopropane-1-carboxylate deaminase from *Hansenula saturnus*. J Biochem 123: 1112–1118.
103. Fedorov DN, Ekimova GA, Doronina NV, Trotsenko YA (2013) 1-Aminocyclopropane-1-carboxylate (ACC) deaminases from *Methylobacterium radiotolerans* and *Methylobacterium nodulans* with higher specificity for ACC. FEMS Microbiol Lett 343: 70–76.
104. Hontzeas N, Zoidakis J, Glick BR, Abu-Omar MM (2004) Expression and characterization of 1-aminocyclopropane-1-carboxylate deaminase from the rhizobacterium *Pseudomonas putida* UW4: a key enzyme in bacterial plant growth promotion. Biochim Biophys Acta 1703: 11–19.
105. Walsh C, Pascal RA, Johnston M, Raines R, Dikshit D, et al. (1981) Mechanistic studies on the pyridoxal phosphate enzyme 1-aminocyclopropane-1-carboxylate deaminase from *Pseudomonas* sp. Biochemistry 20: 7509–7519.

6

Adaptation of a Cyanobacterium to a Biochemically Rich Environment in Experimental Evolution as an Initial Step toward a Chloroplast-Like State

Kazufumi Hosoda[1☯], Masumi Habuchi[2☯], Shingo Suzuki[3], Mikako Miyazaki[4], Go Takikawa[5], Takahiro Sakurai[5], Akiko Kashiwagi[6], Makoto Sueyoshi[5], Yusuke Matsumoto[5], Ayako Kiuchi[4], Kotaro Mori[2], Tetsuya Yomo[2,5,7]*

1 Institute for Academic Initiatives, Osaka University, Suita, Osaka, Japan, **2** Department of Frontier Biosciences, Graduate School of Frontier Biosciences, Osaka University, Suita, Osaka, Japan, **3** Quantitative Biology Center, RIKEN, Suita, Osaka, Japan, **4** Department of Biotechnology, Graduate School of Engineering, Osaka University, Suita, Osaka, Japan, **5** Department of Bioinformatic Engineering, Graduate School of Information Science and Technology, Osaka University, Suita, Osaka, Japan, **6** Faculty of Agriculture and Life Science, Hirosaki University, Hirosaki, Aomori, Japan, **7** Exploratory Research for Advanced Technology (ERATO), Japan Science and Technology Agency, Suita, Osaka, Japan

Abstract

Chloroplasts originated from cyanobacteria through endosymbiosis. The original cyanobacterial endosymbiont evolved to adapt to the biochemically rich intracellular environment of the host cell while maintaining its photosynthetic function; however, no such process has been experimentally demonstrated. Here, we show the adaptation of a model cyanobacterium, *Synechocystis* sp. PCC 6803, to a biochemically rich environment by experimental evolution. *Synechocystis* sp. PCC 6803 does not grow in a biochemically rich, chemically defined medium because several amino acids are toxic to the cells at approximately 1 mM. We cultured the cyanobacteria in media with the toxic amino acids at 0.1 mM, then serially transferred the culture, gradually increasing the concentration of the toxic amino acids. The cells evolved to show approximately the same specific growth rate in media with 0 and 1 mM of the toxic amino acid in approximately 84 generations and evolved to grow faster in the media with 1 mM than in the media with 0 mM in approximately 181 generations. We did not detect a statistically significant decrease in the autotrophic growth of the evolved strain in an inorganic medium, indicating the maintenance of the photosynthetic function. Whole-genome resequencing revealed changes in the genes related to the cell membrane and the carboxysome. Moreover, we quantitatively analyzed the evolutionary changes by using simple mathematical models, which evaluated the evolution as an increase in the half-maximal inhibitory concentration (IC_{50}) and estimated quantitative characteristics of the evolutionary process. Our results clearly demonstrate not only the potential of a model cyanobacterium to adapt to a biochemically rich environment without a significant decrease in photosynthetic function but also the properties of its evolutionary process, which sheds light of the evolution of chloroplasts at the initial stage.

Editor: Valerie de Crécy-Lagard, University of Florida, United States of America

Funding: This research was supported in part by JSPS KAKENHI grant number 25650147, and the "Global COE Program" of MEXT, Japan. The funders had no role in study design, data collection and analysis, decision to publish, or preparation of the manuscript.

Competing Interests: The authors have declared that no competing interests exist.

* E-mail: yomo@ist.osaka-u.ac.jp

☯ These authors contributed equally to this work.

Introduction

Chloroplasts originated from cyanobacteria through endosymbiosis [1–4]. Chloroplasts conduct photosynthesis and synthesize most fatty acids and some amino acids and hormones [5]. Although chloroplasts play many important roles in plant cells, they lack many genes and functions compared to cyanobacteria, depending instead on the host plant cells. These data suggest that the original cyanobacterial endosymbiont adapted to the environment in plant cells, which is biochemically richer than the environment of free-living cyanobacteria [6]. Thus, the original cyanobacterial endosymbiont evolved to depend on biochemical substances provided by the host plant cell, maintaining their photosynthetic functions to provide that cell with biochemical substances in return.

It is not clear how cyanobacteria can adapt to a biochemically rich environment while maintaining their photosynthetic function. Cyanobacteria inhabit almost every niche on earth, including freshwater and salty, hot, and arid environments [6], and employ various stress-management mechanisms [7]. Cyanobacteria also establish long-lived symbiotic associations with fungi and within the plant kingdom [8]; however, cyanobacteria are sensitive to many important biochemical substances. For example, in the case of *Synechocystis* sp. PCC 6803, a prominent model cyanobacteria, the original strain (stocked in the Pasteur Culture Collection) is sensitive to glucose in light [9]. While glucose-tolerant lines exist, they are still sensitive to several amino acids [10–12], possibly because their autotrophic ability opposes the ability to use nutrient biochemical substances. Indeed, Bell and Reboud [13–15] experimentally evolved *Chlamydomonas reinhardtii* in light (photoautotrophic) and dark (heterotrophic) environments and showed that some lines in the dark environment evolved to have higher specific growth rates in the dark environment but lower specific growth rates in the light environment, although *C. reinhardtii* is not a cyanobacterium. The experimental demonstration showing that cyanobacteria adapt to a biochemically rich environment sheds light on the mechanism of adaptation and how cyanobacteria evolved to become chloroplasts.

In this study, we experimentally evolved *Synechocystis* sp. PCC 6803 to adapt to a biochemically rich environment without loss of autotrophic ability. We used a synthetic medium (TCM1) that is sufficiently nutrient-rich to allow the growth of a heterotrophic model protozoan, *Tetrahymena thermophila*, and to inhibit growth of the cyanobacteria. We first identified the six amino acids in TCM1 that are toxic to the cyanobacteria (Figure S1 in File S1). We then cultivated the cyanobacteria in the medium with reduced concentrations of the toxic amino acids and serially transferred the culture, gradually increasing the amino acids concentrations. We found that the cyanobacterium adapted to the rich environment in approximately 100 generations, and we detected no significant difference in autotrophic growth in an inorganic medium between the ancestral and evolved populations. Whole-genome resequencing revealed changes in genes related to the cell membrane and to the carboxysome, which is involved in photosynthesis. Our results clearly show that the model cyanobacterium can evolve to adapt to a rich environment with no significant decrease in photosynthetic function and experimentally show the properties of the evolutionary process of the cyanobacteria to adapt to a biochemically rich environment.

Materials and Methods

Strains and culture conditions

The *Synechocystis* sp. PCC 6803 (GT) strain, originally obtained from Dr. J. G. K. Williams (Dupont de Nemours, Wilmington, DE, U.S.A.), was used as the ancestral strain. This is a glucose-tolerant strain and not the original strain (stocked in the Pasteur Culture Collection). All cells were grown at 30°C in 10 mL liquid culture in a 50 mL Erlenmeyer flask rotated at 80 rpm using NR-3 (Taitec, Saitama, Japan), with 10 $\mu mol \cdot m^{-2} \cdot s^{-1}$ light using an LED system (MIL-C1000T, MIL-U200, MIL-R18A, and MIL-B18A; Sanyo, Osaka, Japan). The LED board ratio of red (MIL-R18A) to blue (MIL-B18A) was 9. We used TCM1, TCM0, BG-11, and BG-11+AA for the culture media. TCM1 was derived from a chemically defined medium for a heterotrophic model protozoan, *Tetrahymena thermophila* CDM15 [16], with a few modifications (see Table S1 in File S1 for the detailed components). TCM1 includes the following 6 amino acids that are toxic to cyanobacteria: Arg, His, Lys, Met, Phe, and Thr in concentrations of approximately 1 mM. TCM0 does not include the 6 toxic amino acids and the other components are the same as the components in TCM1. We also used TCMx, where x is the concentration of the 6 toxic amino acids relative to TCM1 (*e.g.*, TCM0.1 when $x = 0.1$). BG-11 is an inorganic medium that is standard for *Synechocystis* sp. PCC 6803 culture [17]. BG-11+AA includes the 6 toxic amino acids in the same concentrations as the TCM1 and the other components are the same as BG-11.

Evolution Experiment

We first cultivated the cells (pre-cultured in BG-11) in TCM0, transferred them 4 times (as an initial adaptation, see Figure S2 in File S1), freeze-stocked them once by adding DMSO to the culture (final 7%), and recovered them in TCM0 before beginning the experimental evolution. The culture conditions were the same as in the evolution experiments except for the culture media. Simply, the difference between BG-11 and TCM0 is that TCM0 includes amino acids, vitamin, nucleosides, and glucose but BG-11 does not. We then serially transferred the cell culture following the protocol described below, always under 10 $\mu mol \cdot m^{-2} \cdot s^{-1}$ of light. At every transfer until day 79, we split the cell culture into three or four TCMx media that differed in x. After the cultivation, we identified the culture with the highest x that showed a population increase of more than the square root of 10, and we used it as the origin for the next transfer. The initial cell concentration of each transfer was set to be approximately 2×10^6 cells/mL (thus the dilution rate for transfers was different for each round and the bottleneck size of the population was 2×10^7 cells in 10 mL), and the actual cell concentrations in the evolution experiment are plotted in Figure 1A. The incubation time for each culture was 3 or 4 days so that we maintained almost exponential phase for the transferred culture (under approximately 2×10^8 cells/mL) as semi-continuous culture. After day 79, we serially transferred the cell culture to TCM1, examining the growth in TCM0 in parallel until day 271 (see Figure S4 in File S1). The generation during evolution was calculated as $\ln(\Delta_{rpop})/\ln 2$, where Δ_{rpop} is the total fold change in the population.

Measurement of cell concentrations

We measured the cell concentrations relative to a known concentration of fluorescent beads (Fluoresbrite YG Microspheres, 6 μm; Polysciences Inc., Warrington, PA, USA) using a Cytomics TM FC500 flow cytometer (Beckman Coulter, Inc., CA, USA) by

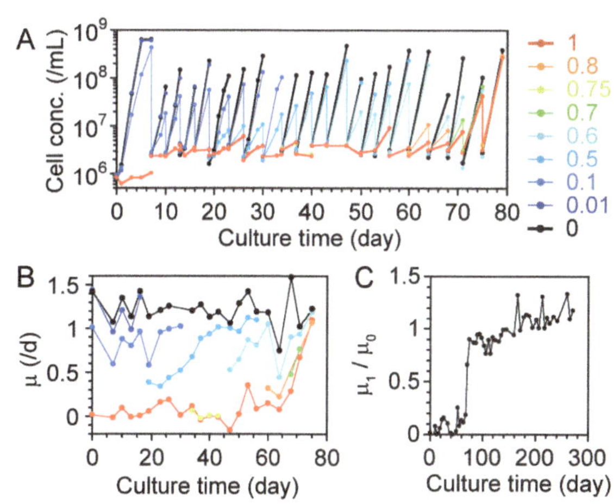

Figure 1. Evolutionary changes during adaptation to a nutrient-rich environment. We cultivated the cells in media with various concentrations of toxic amino acids until culture day 79, then used only TCM1 and TCM0. (A) Population dynamics in the evolution experiment until day 79. The colors show the culture media used (TCMx media; x is shown on the right). The transfers of the culture by dilution are depicted as the vertical decrease in the cell concentration. (B) Evolutionary changes in the specific growth rate (μ) in media with different toxic amino acid concentrations (the colors are same as in A). The specific growth rate was determined by the slope of the linear regression of the natural log of cell concentration per day, using first three (or two, if only two points were measured in the culture) data points. (C) The evolutionary changes in the relative specific growth rates of the cells in TCM1 (μ_1) and in TCM0 (μ_0). The raw data of the population dynamics are shown in Figure S4 in File S1.

loading culture samples mixed with the beads as described previously [18]. A 488-nm argon excitation laser was employed, and bandpass filters of 515–535 and 610–630 nm were used to measure green and red fluorescence, respectively. The cyanobacterial cells were detected as red fluorescent particles due to the presence of chlorophyll, and the cell concentration was calculated from the counts of these particles relative to the fluorescent beads, which were detected as green fluorescent particles.

Whole-genome resequencing

We analyzed 3 genome sequences of (i) ancestral cells, (ii) initially adaptive cells in TCM0 (Figure S2 in File S1), and (iii) evolved cells (at 408 days, 85 transfers, 548 generations, in TCM1). We prepared precultures by shaking stocked populations in media of BG-11, TCM0, and TCM1 for (i), (ii), and (iii), respectively, under the same culture conditions as described above, until the cell concentrations reached approximately 1×10^8 cells/mL. The cyanobacterial cells were collected by centrifugation at $14{,}170\times g$ for 2 min at 25°C, and the pelleted cells were stored at −80°C prior to genomic DNA purification. Genomic DNA was isolated and purified in accordance with a standard method [19]. Genome sequence analyses were performed with an Illumina MiSeq sequencer (Illumina, CA, USA) by Eurofins MWG Operon (AL, USA). Three hundred basepair insert libraries were generated according the Illumina protocol and sequenced in an Illumina MiSeq. In this study, 3 samples with different indexes were mixed and then sequenced in 2×150 bp sequencing mode. We obtained 1079, 917, and 1110 Mbp sequences for (i), (ii), and (iii), respectively. These numbers of base calls corresponded to ×301.9, ×256.6, and ×310.6 redundancy, respectively. We used Illumina's CASAVA pipeline version 1.8.2 software (Illumina, CA, USA) for data analyses including trimming and alignment to the reference genome sequence NC_000911.1 (*Synechocystis* sp. PCC 6803 chromosome, complete genome). We used VarScan 2 software [20] for variant analysis.

Modeling and simulations

Simulations of the evolutionary change in $\ln[IC_{50}]$ were performed using Matlab software (MathWorks, MA, USA). The Matlab code used is presented in Appendix S1 in File S1.

Results and Discussion

Evolutionary changes in adaptation to a nutrient-rich environment

We serially transferred the cyanobacterial cell culture, gradually increasing the concentrations of the 6 amino acids that were toxic to the cells and always maintaining the cultures under light to maintain their photosynthetic function. Figure 1A shows cell growth during evolution; the cells evolved the ability to grow in media with higher concentrations of the toxic amino acids through serial transfer. Figure 1B shows the specific growth rates (μ) in media with different toxic amino acid concentrations, TCMx, where x is the relative amino acid concentration (plots of μ as a function of x are shown in Figure S3 in File S1). We found a statistically significant increase in μ in TCM1 (μ_1, red line) while we did not find statistically significant increases in μ in TCM0 (μ_0, black line) by a Student's t-test to compare the slope of the linear regression of μ on the culture time with zero slope using a significance level of 0.05; the slopes were 0.0069 ($R^2=0.32$, the degree of freedom $DF=20$, $p=0.0057$) and −0.0014 ($R^2=0.032$, $DF=19$, $p=0.44$)/day^2 for TCM1 and TCM0, respectively. μ in TCMx for $0<x<1$ (colored lines) were in between μ_1 and μ_0. These results suggest that the key change was related not to global growth but to tolerance of the toxic amino acids. Then, we continued the transfer in TCM1 (the population dynamics are shown in Figure S4 in File S1). Figure 1C shows the evolutionary changes in μ_1 relative to μ_0. μ_1 approached 90% of μ_0 at approximately 75 days (21 transfers, 84 generations) and exceeded μ_0 at approximately 160 days (40 transfers, 181 generations). These results indicate that the model cyanobacteria had adapted to a biologically rich environment, increasing their tolerance to the previously toxic amino acids.

Autotrophic ability of the evolved strain

We found no statistically significant decrease in autotrophic ability (photosynthetic function) in the evolved cells. Figure 2 shows the μ of the ancestral and evolved cells in various culture media. First of all, we found that the evolved cells actually increased their growth ability in TCM1 because there was a statistically significant difference in the mean μ in TCM1 between the evolved and the ancestral cells as determined by a Welch's t-test using a significance level of 0.05 ($DF=7$, $p=1.9\times10^{-9}$). Importantly, there were no statistically significant differences between the evolved and the ancestral cells in the inorganic medium BG-11 ($DF=5$, $p=0.67$), suggesting that the photosynthetic function was maintained in the evolved cells. Furthermore, the evolved cells showed amino acid tolerance even in BG-11 (there were no statistically significant differences between BG-11 and BG-11+AA; $DF=2$, $p=0.11$), while the ancestral strain was sensitive (there was a statistically significant difference between BG-11 and BG-11+AA; $DF=4$, $p=5.9\times10^{-6}$). We did not observe decreased means of the specific growth rates in the evolved cells (trade-offs) in any of the tested media.

Whole-genome resequencing

The evolved changes in phenotypes were corroborated by the genetic changes detected by whole-genome resequencing. We analyzed 3 genomic DNA samples of ancestral cells, initially adaptive cells in TCM0 (Figure S2 in File S1), and evolved cells in TCM1. All mutations in the chromosomes found in the variant analysis are listed in Table S2 in File S1. We found 20 single-base substitutions and 8 small indels in ancestral cells compared with the reference sequence (NC_000911.1 *Synechocystis* sp. PCC 6803 chromosome, complete genome). These mutations were also detected in the other two types (initially adaptive and evolved) of cells. The detected mutations were identical for the ancestral and initially adaptive cells (see Table S2 in File S1).

The mutations detected only in the evolved strains are summarized in Table 1. All of the single-base substitutions

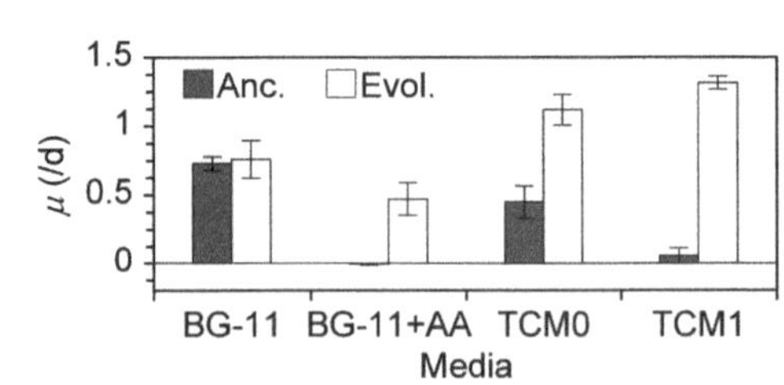

Figure 2. Specific growth rates of the ancestral and evolved cells in various environments. The black and white bars indicate the specific growth rates of the ancestral and evolved (at 408 days, 85 transfers, 548 generations) cells, respectively, in the culture media shown at the bottom. The error bars show the SD of five (for BG-11 and TCM1) or two (for the other media) independent cultures. The ancestral culture in BG11+AA showed almost zero growth.

Table 1. Genetic changes detected in the evolved cells relative to the ancestral cells.

#	Position (nt)	Nucleotide changes	Gene	AA changes in the Gene
1	335496	C to A	Carboxysome-formation protein CcmA; complementary	G to E
2	829508	C to T	RNA polymerase alpha subunit; complementary	R to Q
3	1114921	G to C	Periplasmic substrate-binding and integral membrane protein of the ABC-type Bgt permease for basic amino acids and glutamine BgtB; complementary	F to L
4	1128135	C to G	Unknown protein (sll1265); complementary	A to P
5	Next to 1905171	Del. GCCTCG	Penicillin-binding protein (cell-wall biogenesis); complementary	Del. AE
6	3203715	G to A	Probable cation transporter; complementary	P to S

All detected mutations in the chromosome of the ancestral cells, initially adaptive cells, and the evolved cells, relative to the reference (NC_000911.1; Synechocystis sp. PCC 6803 chromosome, complete genome), are shown in Table S2 in File S1.

identified in annotated genes were non-synonymous, suggesting that they were adaptive. Mutations in genes related to membrane biogenesis and transportations were detected (#3, 5 and 6 in Table 1), suggesting that the amino acid tolerance arose from changes in membrane functions, such as increased membrane protection from penetration by amino acids. Although gene #4 was annotated as an unknown protein, it has been reported that this gene is related to stress responses, such as butanol, hexane, salt stresses and nitrogen starvation [21]. These results are consistent with the increase in tolerance. A mutation in the carboxysome-formation protein CcmA [22] (#1) might indicate an optimization of photosynthetic activity in a biochemically rich environment because the CO_2-concentration mechanism (CCM) of the carboxysome is important to meeting habitat requirements [7]. A mutation was also identified in the RNA polymerase alpha subunit (#2). Although it is difficult to estimate the direct involvement of this mutation in the evolved changes in the phenotypes, the mutation in the RNA polymerase alpha subunit could potentially provide some benefit because the initiation of transcription is the main determinant for the regulation of gene expression in bacteria. Note that those individual mutations would not be directly correlated with acquired phenotype such as by the existence of epistasis [23,24] and epigenetic changes [25]. Further work, including construction of a mutant containing these mutations and analysis of phenotypes of these mutants, is needed to elucidate the relationship between genotype and phenotype. Also note that we analyzed mixed (not cloned) populations that were stocked directly from the evolution experiment, and some heterogeneous mutations were found (see Table S2 in File S1). However, it is possible that this heterogeneity did not arise from the heterogeneity of different cell groups because it is known that *Synechocystis* sp. PCC 6803 has a multicopy genome [26], and heterozygous mutations are commonly found in the genome of the *Synechocystis* sp. PCC 6803 [27]. Actually, the heterogeneic frequencies were maintained during the course of evolution, and the specific mutations in evolved cells shown in Table 1 were almost homogeneous (see Table S2 in File S1).

Quantitative analyses of the evolutionary changes using simple mathematical models

We analyzed the experimental results of the evolutionary changes using simple mathematical models to extract the quantitative properties of the evolution. Specifically, we estimated a key trait (IC_{50}, see below) in the evolution, the trait variance in the population, and some properties of the fitness landscape regarding the key trait, such as the probability of the beneficial mutation per replication and the extent of the change in the key trait per beneficial mutation. We used the data through day 75 (77 generations) where the growth rate of the cells in the transferred cultures was less than μ_0. First, we estimated the maximum specific growth rate (μm_{ax}) and the half-maximal inhibitory concentration (IC_{50}) of the cell population for every transfer round using the experimental data of the relationship between the specific growth rate and the amino acid concentration (Figure 1B, and Figure S3 in File S1). Specifically, we estimated those two parameters by fitting the experimental data to a basic noncompetitive-inhibition model $\mu = \mu m_{ax}/(1+x/IC_{50})$ [28] (the fitting results are shown in Figure S3 in File S1). Figure 3A-i summarizes the fitting results and clearly shows that the principal increase was in IC_{50}, not μm_{ax}; there was a statistically significant increase in $\ln[IC_{50}]$ (red square) while not in $\ln[\mu m_{ax}]$ (black circle) as determined by a Student's *t*-test to compare the slope of the linear regression of the parameters on generation with zero slope using a significance level of 0.05; the slope of $\ln[IC_{50}]$ and $\ln[\mu m_{ax}]$ was 0.035 ($R^2 = 0.67$, $DF = 20$, $p = 3.6\times10^{-6}$) and -0.0014 ($R^2 = 0.037$, $DF = 20$, $p = 0.39$) per generation, respectively. Thus, it is possible to quantitatively understand this evolutionary change as the change in a trait, IC_{50}, based on quantitative genetics as described below.

We analyzed the evolutionary rate of the IC_{50}, specifically the slope of $\ln[IC_{50}]$ (designated as S_{IC50}), to provide an interpretation of the experimental results, using a simple model of the directional selection based on quantitative genetics [29–32]. Let us assume a cell population in which each cell has a trait z ($=\ln[IC_{50}]$) and a specific growth rate $\mu(z)$, with the frequency $f(z,t)$ at the total N (*i.e.*, $\sum f(z,t) = N$). When a cell with z reproduces an offspring through replication, the trait of the progeny becomes $z+d_z$ by a beneficial mutation with a constant probability r, or else keeps z (probability $1-r$), which is the same as the parent (Figure 3B), *i.e.*, r is a beneficial mutation rate per generation and d_z is an effect size of a single beneficial mutation on the trait z. Constant r and d_z indicate that this model assumes a non-epistatic landscape [33]. The rate equation of $f(z,t)$ is described as

$$\frac{\partial f(z,t)}{\partial t} = (1-r)\mu(z)f(z,t) + r\mu(z-d_z)f(z-d_z,t) - \bar{\mu}(t)f(z,t) \quad (1)$$

where $\bar{\mu}(t)$ is the mean specific growth rate of the population (see Text S1 in File S1 for the derivation). The first and second terms on the right-hand side represent the increase in the population frequency of z by reproduction from the population of z without any beneficial mutation (a gray square fraction at z in Figure 3B) and by reproduction from the population of $z-d_z$ with a beneficial mutation (a red square fraction at $z-d_z$ in Figure 3B), respectively. The third term represents the decrease in the population by a non-selective dilution to keep the total population size (N) constant. The rate of the evolutionary change in the mean of z over generations in this model has the same meaning as S_{IC50} if constant. To obtain a simple understanding of the value of S_{IC50}, we used a set of rough approximations to obtain an approximate analytical solution of S_{IC50} from the model (see Text S1 in File S1). We found that the evolutionary rate became constant in this model and obtained its approximate analytical solution as

$$S_{IC50} \approx 2(\ln 2)\sigma_{st}^2 = \ln 2 \frac{d_z^{\,2} \ln N}{(\ln r)^2} \qquad (2)$$

where σ_{st} is a standard deviation of the trait z (the width of the trait frequency distribution) when the evolutionary rate is constant. Equation 2 indicates that S_{IC50} is a monotonically increasing function of d_z, r, and N for the ranges $d_z>0$, $0<r<1$, and $N>1$. The difference between this approximate analytical solution Equation 2 and the numerical simulation according to Equation 1 is shown in Figure S5 in File S1. Equation 2 means that the evolutionary rate is proportional to the variance ($\sigma_{st}{}^2$), known as Fisher's fundamental theorem of natural selection [34], and $\sigma_{st}\approx 0.16$ was obtained from the experimental value of S_{IC50} ($=0.035$) independently of the model parameters (see Figures 3A-ii and 3A-iii for graphical presentations of the frequency distribution).

Undetermined variables were only d_z and r because N was given in the experimental condition. The dependency of S_{IC50} on d_z and r is shown in Figure 3C. Therefore, if one of the two variables is determined, the other can also be determined from the experimental results, and the evolution can be visually reconstituted by the model, as shown by the blue line in Figure 3A-i and the corresponding population changes in Figure 3A-ii using (d_z, r) = (0.4, 0.00023) as an example. This value of $d_z = 0.4$ (*i.e.*, S_{IC50} increases 1.5-fold per beneficial mutation) is possible because the extent of the total increase in ln[IC_{50}] that is required to satisfy $\mu_0 \approx \mu_1$ (Δ_z) seems to be approximately 2.5 (−2 to 0.5) based on Figure 3A-i, and the total number of beneficial mutations (n_m) can be 6 based on the number of detected mutations, resulting in $d_z = \Delta_z / n_m \approx 0.4$. Then, r should be approximately 0.0002 to explain the experimental results by this model with $d_z = 0.4$ as above. With $r\approx 0.0002$ (*i.e.*, 0.02% of replications are beneficial), the fraction of beneficial mutation among all possible mutations can be estimated as 0.07 (*i.e.*, 7% of mutations are beneficial) from the value of r divided by the mutation rate per genome per replication m_{rep}, which is known to be approximately 0.003 in microbes [35]. Note that r is more sensitive than d_z in those range (Equation 2 and Figure 3C), *e.g.*, if the range of d_z is 0.3~0.5, the range of r is approximately 10^{-6}~10^{-3}. Also note that above analyses are rough estimation and validity is still unclear. For example, the total number of beneficial mutations n_m can be less than 6, for reasons such as non-beneficial mutations in Table 1 or few mutations until day 75. In addition, n_m also can be greater than 6 because heritable trait changes can occur not only through mutations but also by other epigenetic factors [25]. Although the validity of the values and the model itself are still unclear in this study, the above information provides important insights into the evolutionary process of the cyanobacteria to adapt to a biochemically rich environment.

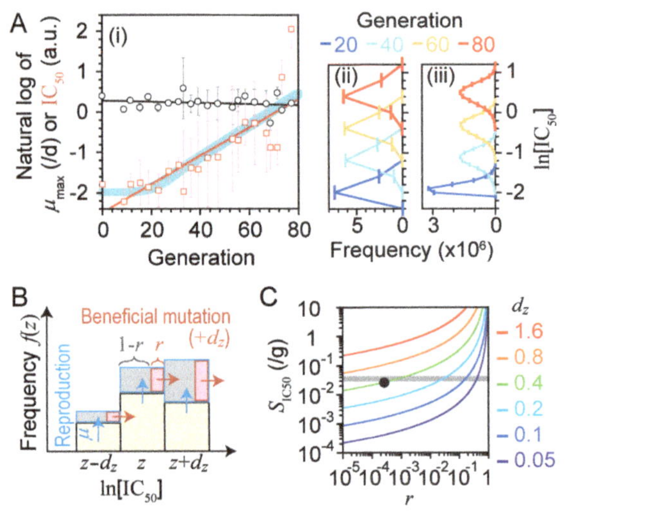

Figure 3. Quantitative analyses of the evolutionary changes by simple mathematical models. (A) Evaluation of the evolutionary changes in the maximum specific growth rate (μm_{ax}) and the half-maximal inhibitory concentration (IC_{50}) until day 79. (i) Evolutionary changes in ln[μm_{ax}] (black circle) and ln[IC_{50}] (red square), determined by the fitting shown in Fure S3 in File S1, with respect to the generation (*g*) of each round. The error bars show the standard error of the fitting estimates. The black and red lines show the linear regression of ln[μm_{ax}] and ln[IC_{50}], respectively ($-0.0014g+0.28$ and $0.035g-2.5$, respectively). The blue bold line is the change in the mean of ln[IC_{50}] calculated by the numerical simulation of the mathematical model of Eq. 1, using (d_z, r, N) = (0.4, 0.00023, 10^7), assuming a homogeneous population for the initial state. The source code of the simulation is shown in Appendix S1 in File S1. (ii and iii) Simulated evolutionary changes in frequency distributions of the cell population calculated by the numerical simulation using (d_z, r, N) of (0.4, 0.00023, 10^7: the same as the blue line in A-i) and (0.1, 0.04, 10^7), respectively. The vertical axis is ln[IC_{50}], as in A-i. (B) A conceptual scheme of the mathematical model of Eq. 1. See the text for explanation. Simply, the population shifts toward greater ln[IC_{50}] with a beneficial mutation rate per generation (r) and an effect size of a single beneficial mutation on ln[IC_{50}] (d_z). The axes correspond to those of A-ii and A-iii. (C) The dependency of the evolutionary rate of ln[IC_{50}], *i.e.*, the slope of the red line in A-i (S_{IC50}), on the model parameters d_z and r calculated from Eq. 2 using $N = 10^7$. The gray bold line shows the experimentally determined value ($S_{IC50} = 0.035/g$). The black point shows the value used for the simulation shown in A-ii, which is discussed in the text.

Conclusions

In this study, we show that a model cyanobacterium, *Synechocystis* sp. PCC 6803, can evolve to adapt to a biochemically rich environment without statistically significant losses of autotrophic ability. Our results thus show the potential of a cyanobacterium to evolve into a chloroplast-like endosymbiont. Furthermore, we showed some properties of the evolutionary changes such as an increase in tolerance to toxic amino acids rather than the global growth rate itself, the quantitative characteristics of the evolutionary rate, which provides estimated variance of the tolerance in the population and clues regarding the evolutionary fitness landscape, and genomic analyses that identified the mutated genes and were consistent with the increase in tolerance. Additional generations would be needed to show more qualitatively distinct changes toward a chloroplast-like phenotype, such as loss of growth ability in inorganic media while maintaining photosynthetic function and optimization of photosynthesis for a biochemically rich environment. Community-level function (shown in another cyanobacterium [36]) must be a key factor for loss of functions in the evolution [37]. Therefore, ultimately, the cyanobacteria cells would have to be cultivated with or within another eukaryotic cell [38–40] for the complete experimental evolution toward chloroplast-like endosymbiont.

Supporting Information

File S1 Supporting information which contains the following. Figure S1: Toxic components in TCM1. Figure S2: Preculture for the evolution experiment. Figure S3: The specific growth rates as a function of the relative concentration of the toxic amino acids in each transfer round. Figure S4: Population dynamics in the evolution experiment. Figure S5: The difference between the approximate analytical solution (Equation 2) and the numerical simulation according to Equation 1. Table S1: Components of TCM1. Table S2: All detected mutations in the chromosome, relative to the reference (NC_000911.1; Synechocystis sp. PCC 6803 chromosome, complete genome), by the genomic analysis. Text S1: The derivation of Equations 1 and 2. Appendix S1: The Matlab source code of the simulation shown in Figure 3A-i (blue bold line).

Acknowledgments

We thank Drs. Katsunori Yoshikawa, Takuyo Aita, and Saburo Tsuru, and Mr. Tsubasa Nakajima and Ms. Natsuki Hiasa for helpful discussions. We are also grateful to Ms. Naomi Tago and Ms. Ryoko Otsuki for their valuable technical assistance.

Author Contributions

Conceived and designed the experiments: KH MH SS MM GT TS A. Kashiwagi MS YM A. Kiuchi KM. Performed the experiments: KH MH SS MM GT TS A. Kashiwagi MS YM A. Kiuchi. Analyzed the data: KH MH SS. Wrote the paper: KH SS A. Kashiwagi TY.

References

1. Raven JA, Allen JF (2003) Genomics and chloroplast evolution: what did cyanobacteria do for plants? Genome Biol 4: 209.
2. Martin W, Kowallik KV (1999) Annotated English translation of Mereschkowsky's 1905 paper 'Uber Natur und Ursprung der Chromatophoren im Pflanzenreiche. Eur J Phycol 34: 287–295.
3. Mereschkowsky C (1905) Über Natur und Ursprung der Chromatophoren im Pflanzenreiche. Biol Centralbl 25: 593–604.
4. Margulis L (1981) Symbiosis in cell evolution: life and its environment on the early Earth. San Francisco: W. H. Freeman.
5. Taiz L, Zeiger E (2006) Plant physiology. Sunderland, Mass: Sinauer Associates.
6. Whitton BA (2012) Ecology of cyanobacteria II: their diversity in space and time. New York: Springer.
7. Srivastava AK, Rai AN, Neilan BA (2013) Stress biology of cyanobacteria: molecular mechanisms to cellular responses: CRC Press.
8. Bergman B, Zheng WW, Klint J, Ran L (2008) On the origin of plants and relations to contemporary cyanobacterial-plant symbioses. Plant Biotechnol-Nar 25: 213–220.
9. Williams JGK (1988) Construction of Specific Mutations in Photosystem-Ii Photosynthetic Reaction Center by Genetic-Engineering Methods in *Synechocystis*-6803. Method Enzymol 167: 766–778.
10. Labarre J, Thuriaux P, Chauvat F (1987) Genetic-Analysis of Amino-Acid-Transport in the Facultatively Heterotrophic Cyanobacterium *Synechocystis* sp Strain 6803. J Bacteriol 169: 4668–4673.
11. Flores E, Muropastor AM (1990) Mutational and Kinetic-Analysis of Basic-Amino-Acid Transport in the Cyanobacterium *Synechocystis* sp PCC-6803. Arch Microbiol 154: 521–527.
12. Eisenhut M, Bauwe H, Hagemann M (2007) Glycine accumulation is toxic for the cyanobacterium *Synechocystis* sp strain PCC 6803, but can be compensated by supplementation with magnesium ions. Fems Microbiol Lett 277: 232–237.
13. Bell G, Reboud X (1997) Experimental evolution in *Chlamydomonas* 2. Genetic variation in strongly contrasted environments. Heredity 78: 498–506.
14. Bell G (1997) Experimental evolution in *Chlamydomonas* 1. Short-term selection in uniform and diverse environments. Heredity 78: 490–497.
15. Reboud X, Bell G (1997) Experimental evolution in *Chlamydomonas* 3. Evolution of specialist and generalist types in environments that vary in space and time. Heredity 78: 507–514.
16. Mori K, Kashiwagi A, Yomo T (2011) Single-cell isolation and cloning of *Tetrahymena thermophila* cells with a fluorescence-activated cell sorter. J Eukaryot Microbiol 58: 37–42.
17. Stanier RY, Kunisawa R, Mandel M, Cohen-Bazire G (1971) Purification and properties of unicellular blue-green algae (order *Chroococcales*). Bacteriol Rev 35: 171–205.
18. Hosoda K, Suzuki S, Yamauchi Y, Shiroguchi Y, Kashiwagi A, et al. (2011) Cooperative adaptation to establishment of a synthetic bacterial mutualism. PLoS One 6: e17105.
19. Hihara Y, Ikeuchi M (1997) Mutation in a novel gene required for photomixotrophic growth leads to enhanced photoautotrophic growth of *Synechocystis* sp. PCC 6803. Photosynth Res 53: 243–252.
20. Koboldt DC, Zhang QY, Larson DE, Shen D, McLellan MD, et al. (2012) VarScan 2: Somatic mutation and copy number alteration discovery in cancer by exome sequencing. Genome Research 22: 568–576.
21. Qiao J, Shao M, Chen L, Wang J, Wu G, et al. (2013) Systematic characterization of hypothetical proteins in *Synechocystis* sp. PCC 6803 reveals proteins functionally relevant to stress responses. Gene 512: 6–15.
22. Ogawa T, Marco E, Orus MI (1994) A Gene (ccma) Required for Carboxysome Formation in the Cyanobacterium *Synechocystis* sp Strain PCC6803. J Bacteriol 176: 2374–2378.
23. Tenaillon O, Rodriguez-Verdugo A, Gaut RL, McDonald P, Bennett AF, et al. (2012) The Molecular Diversity of Adaptive Convergence. Science 335: 457–461.
24. Khan AI, Dinh DM, Schneider D, Lenski RE, Cooper TF (2011) Negative Epistasis Between Beneficial Mutations in an Evolving Bacterial Population. Science 332: 1193–1196.
25. Casadesus J, Low DA (2013) Programmed Heterogeneity: Epigenetic Mechanisms in Bacteria. Journal of Biological Chemistry 288: 13929–13935.
26. Chauvat F, Rouet P, Bottin H, Boussac A (1989) Mutagenesis by random cloning of an *Escherichia coli* kanamycin resistance gene into the genome of the cyanobacterium *Synechocystis* PCC 6803: selection of mutants defective in photosynthesis. Mol Gen Genet 216: 51–59.
27. Kanesaki Y, Shiwa Y, Tajima N, Suzuki M, Watanabe S, et al. (2012) Identification of substrain-specific mutations by massively parallel whole-genome resequencing of *Synechocystis* sp. PCC 6803. DNA Res 19: 67–79.
28. Han K, Levenspiel O (1988) Extended Monod Kinetics for Substrate, Product, and Cell-Inhibition. Biotechnol Bioeng 32: 430–437.
29. Iwasa Y (1988) Free Fitness That Always Increases in Evolution. Journal of Theoretical Biology 135: 265–281.
30. Kingman JFC (1978) Simple-Model for Balance between Selection and Mutation. Journal of Applied Probability 15: 1–12.
31. Brodie ED, Moore AJ, Janzen FJ (1995) Visualizing and Quantifying Natural-Selection. Trends in Ecology & Evolution 10: 313–318.
32. Ellegren H, Sheldon BC (2008) Genetic basis of fitness differences in natural populations. Nature 452: 169–175.
33. Kryazhimskiy S, Tkacik G, Plotkin JB (2009) The dynamics of adaptation on correlated fitness landscapes. Proceedings of the National Academy of Sciences of the United States of America 106: 18638–18643.
34. Fisher RA (1930) The genetical theory of natural selection. Oxford: The Clarendon press.
35. Drake JW, Charlesworth B, Charlesworth D, Crow JF (1998) Rates of spontaneous mutation. Genetics 148: 1667–1686.
36. Morris JJ, Johnson ZI, Szul MJ, Keller M, Zinser ER (2011) Dependence of the Cyanobacterium *Prochlorococcus* on Hydrogen Peroxide Scavenging Microbes for Growth at the Ocean's Surface. Plos One 6: e16805.
37. Morris JJ, Lenski RE, Zinser ER (2012) The Black Queen Hypothesis: evolution of dependencies through adaptive gene loss. MBio 3: e00036–00012.
38. Agapakis CM, Niederholtmeyer H, Noche RR, Lieberman TD, Megason SG, et al. (2011) Towards a synthetic chloroplast. PLoS One 6: e18877.
39. Gorelova OA, Kleimenov SY (2003) The accumulation and degradation dynamics of cyanophycin in cyanobacterial cells grown in symbiotic associations with plant tissues and cells. Microbiology 72: 318–326.
40. Nilsson M, Bhattacharya J, Rai AN, Bergman B (2002) Colonization of roots of rice (*Oryza sativa*) by symbiotic *Nostoc* strains. New Phytol 156: 517–525.

7

Adaptation to Fluctuating Temperatures in an RNA Virus Is Driven by the Most Stringent Selective Pressure

María Arribas[1], Kirina Kubota[2], Laura Cabanillas[1], Ester Lázaro[1]*

1 Centro de Astrobiología (INTA-CSIC), 28850 Torrejón de Ardoz, Madrid, Spain, 2 Facultad de Ciencias, Universidad Autónoma de Madrid, 28049 Cantoblanco, Madrid, Spain

Abstract

The frequency of change in the selective pressures is one of the main factors driving evolution. It is generally accepted that constant environments select specialist organisms whereas changing environments favour generalists. The particular outcome achieved in either case also depends on the relative strength of the selective pressures and on the fitness costs of mutations across environments. RNA viruses are characterized by their high genetic diversity, which provides fast adaptation to environmental changes and helps them evade most antiviral treatments. Therefore, the study of the adaptive possibilities of RNA viruses is highly relevant for both basic and applied research. In this study we have evolved an RNA virus, the bacteriophage Qβ, under three different temperatures that either were kept constant or alternated periodically. The populations obtained were analyzed at the phenotypic and the genotypic level to characterize the evolutionary process followed by the virus in each case and the amount of convergent genetic changes attained. Finally, we also investigated the influence of the pre-existent genetic diversity on adaptation to high temperature. The main conclusions that arise from our results are: i) under periodically changing temperature conditions, evolution of bacteriophage Qβ is driven by the most stringent selective pressure, ii) there is a high degree of evolutionary convergence between replicated populations and also among populations evolved at different temperatures, iii) there are mutations specific of a particular condition, and iv) adaptation to high temperatures in populations differing in their pre-existent genetic diversity takes place through the selection of a common set of mutations.

Editor: Valerie de Crécy-Lagard, University of Florida, United States of America

Funding: This work was supported by the Spanish Ministry of Science and Innovation (www.micinn.es) through grants FIS2011-27569 and BFU2011-25271. The funders had no role in study design, data collection and analysis, decision to publish, or preparation of the manuscript.

Competing Interests: The authors have declared that no competing interests exist.

* Email: lazarole@cab.inta-csic.es

Introduction

The frequency of change in the selective pressures together with the fitness effects of mutations across environments are major determinants of the evolutionary pathway followed by populations. When selective pressures remain constant for a long time, specialist organisms, which perform well in a particular environment and poorly elsewhere, are favoured [1–7]. In contrast to this, when the selective pressures change frequently, fitness costs of mutations across environments can limit adaptation [6], [8–10], promoting the selection of generalist organisms able to cope with a variety of conditions [8], [11–15]. Generalists frequently select genotypes containing mutations of beneficial effects under all the selective pressures that are to be encountered. Another possibility is to maintain as polymorphisms mutations advantageous in some environments but deleterious under alternative conditions. The existence of different targets of selection through time together with the high frequency of fitness trade-offs has led to the proposition that generalism could have an associated fitness cost [7], [11], [14], [16–18].

Many of the studies concerning the selection of generalism and specialism have been carried out through experimental evolution, which requires systems that evolve quickly and that allow to establish accurate correspondences between phenotypes and genotypes. Since RNA viruses are able to adapt in very short times to environmental changes [19], and also posses small genomes, they have been used oftenly [20–33]. The high adaptive capacity of RNA viruses lies on their elevated error rates that promote the generation of populations composed by a mutant spectrum with a very wide phenotypic and genetic heterogeneity [19], [34]. In these populations, mutants dominant under previous selective pressures can be kept as minority variants of the mutant spectrum and be rapidly selected when the virus is exposed again to the same condition. This reservoir of variants provides a molecular memory [35–37] which could be very useful during adaptation to fluctuating selective pressures. Typically, studies carried out with RNA viruses compare the results obtained when a selective pressure, commonly the virus host or some abiotic factor, either remains constant or alternates between different values. In most cases a pattern of specialization under constant conditions was found, which was occasionally accompanied by a cost in alternative environments [21], [22], [25], [26], [28], [30], [31]. Populations that alternated different environments in some cases presented a generalism-associated fitness cost [14], although in others they showed equal or superior improvement in each environment than the lineages evolved constantly in the same conditions [22], [24], [27–29], [32], [33].

In this study we have propagated a heterogeneous population of an RNA virus, the bacteriophage Qβ, at its optimal growth temperature (37°C) and at two new temperatures, one above the optimal value (43°C) and another one below (30°C). To compare the virus populations in constant versus deterministically fluctuating environments, we carried out evolution experiments in which the temperature was kept invariable throughout the process, together with other ones in which the temperature alternated periodically among the three values assayed, using two patterns of change. The analysis of the evolved populations allowed us to determine the evolutionary pathway followed by the virus under the different propagation conditions. We also analyzed the genomic changes candidate to be responsible for adaptation and the degree of evolutionary convergence between independent replicated experiments as well as among populations evolving under different conditions, including changing environments. Finally, we investigated whether the genetic changes that allow adaptation to a particular condition (in this case, high temperature) were the same or not in populations differing in their initial genetic diversity.

Materials and Methods

Viruses and bacteria: Standard procedures for infection

Bacteriophage Qβ was routinely propagated by infecting log-phase cultures of *Escherichia coli*, strain Hfr (Hayes) in NB medium (8 g/l Nutrient Broth from Merck and 5 g/l NaCl). The virus, which was provided by Prof. C. K. Biebricher, was adapted to replicate in liquid cultures in our laboratory by infecting 1 liter of an exponential phase *E. coli* culture at a multiplicity of infection (moi) of 1 plaque forming unit (pfu) per bacterium. The culture was incubated at 37°C with aeration during 90 min and subsequently an aliquot was used to infect another fresh exponential culture of *E. coli* at the same moi. This process was repeated for a total of 5 serial transfers [38]. The lysate at the fifth transfer was clarified by incubating with a few drops of chloroform for 30 min at room temperature with aeration, after which it was centrifuged for 40 min at 8000×g. Phages from the supernatant were purified by precipitation with 6% polyethylene glycol 6000 and differential sedimentation [38]. The virus population obtained in this way was called $Q\beta_0$, which was used as ancestor population for most of the evolution experiments described in this paper. We chose to use a heterogeneous population instead of a clone to better reproduce the action of natural selection in natural virus populations, which normally show high diversity.

Infections in liquid medium were always carried out using fresh exponential phase *E. coli* cultures (with an optical density at 550 nm between 0.6 and 0.8) that were infected with the virus at the moi indicated in each experiment. After 2 h of incubation at 37°C with aeration, cultures were treated with 1/20 vol of chloroform for 15 min at 37°C with shaking (300 rpm). Virus supernatants were harvested upon centrifugation at 13000×g for 10 min and maintained at 4°C for short-term use (less than 15 days) or at −80°C for long-term storage. Virus titres were determined by plaque assay and expressed as the number of pfu per ml of the phage suspension.

Evolution experiments

Adaptation to 30°C and 37°C was carried out by infecting 10^8 bacteria in exponential phase with 10^6 pfu of the population $Q\beta_0$ in a final volume of 1 ml in 10 ml polypropylene tubes. Since virus growing at 43°C was strongly inhibited, adaptation to this temperature was carried out under conditions that yielded higher virus titres (10^9 bacteria infected with 10^7 pfu in a final volume of 10 ml in 50 ml flasks). All experiments were carried out in duplicate. After 2 h of incubation at 37°C with aeration, the virus supernatants were collected as described above and the virus titres were estimated by plaque assay. The amount of phage suspensions containing 10^6 pfu (cultures performed at 30°C and 37°C) or 10^7 pfu (cultures carried out at 43°C) was used to infect a fresh *E. coli* culture. The procedure was repeated for a total of 15 transfers under constant conditions to obtain the virus populations $Q\beta_{37}$, $Q\beta_{30}$, and $Q\beta_{43}$ (Fig. 1). The approximate number of replication rounds experienced by the virus during the 15 transfers were 235 ($Q\beta_{37}$), 200 ($Q\beta_{30}$), and 120 ($Q\beta_{43}$) Adaptation to deterministically fluctuating temperatures took place by alternating 37°C, 43°C, and 30°C in this order, with the temperature change occurring either every transfer (populations $Q\beta_{F1}$) or every five transfers (populations $Q\beta_{F5}$) (Fig. 1). The culture conditions for each temperature were the same as described for experiments carried out at constant temperatures. In all cases replicate lineages were distinguished with numbers 1 and 2.

Experiments designed to study adaptation in populations differing in their pre-existent genetic diversity (Fig. 2) were carried out using as ancestor population a clonal population of bacteriophage Qβ (cQβ). The clonal population was obtained upon expression of the plasmid pBRT7Qβ which contains a cDNA of the wild type virus cloned in the plasmid pBR322 [39]. Briefly, we transformed *E. coli* DH5-α with the plasmid pBRT7Qβ, and used the supernatant from an overnight culture of a transformed colony to infect a lawn of *E. coli* Hfr in semisolid agar. A lytic plaque was isolated, and the viruses contained on it were extracted and adapted to replicate at 37°C or 43°C as described above (Fig. 2).

Fitness determinations

Growth rate values at 37°C, 30°C, or 43°C were used as a surrogate of fitness at each temperature. All the determinations were carried out by using liquid cultures that contained 10^8 bacteria which were inoculated with 10^4 pfu of the virus population indicated in a final volume of 1 ml. After two hours at 37°C with aeration, the virus supernatants were collected as described above and titrated to estimate the virus yield. Preliminary assays showed that bacteriophage Qβ grew exponentially under these conditions. Each determination was carried out in triplicate, and fitness was expressed as the $\log_2$ [$(virus_{end} - virus_0)/virus_0$], where $virus_0$ was the initial input of virus in each experiment and $virus_{end}$ was the number of pfu obtained at the end of the experiment.

RNA extraction, cDNA synthesis, PCR amplification and nucleotide sequencing

Virus RNA was prepared following standard procedures [38], [40]. RNAs were amplified by RT-PCR using Avian Myeloblastosis Virus RT (Promega) and Expand High Fidelity DNA polymerase (Roche). The cDNAs were purified with a Qiagen purification kit and subjected to cycle sequencing with Big Dye Chemistry (Applied Biosystems; Perkin Elmer). The following pairs of oligonucleotide primers were used for RT-PCR: 5′CGAATCTTCCGACACGCATCC3′ and 5′AAACGGTAA-CACGCTTCTCCAG3′ to amplify from nucleotide position 150 to 1497; 5′CTCAATCCGCGTGGGGTAAATCC3′ and 5′CA-GAAAATCGGCAGTGACGCAACA3′ to amplify from nucleotide position 1407 to 2817; and 5′GTGCCATACCGTTT-GACT3′with 5′GATCCCCCTCTCACTCGT3′ to amplify from nucleotide position 2254 to 4195. Sequences were aligned with the consensus sequence of the wild type phage with Clustal W. Mutations relative to the consensus sequence were identified using the program BioEdit. Nucleotides were numbered according

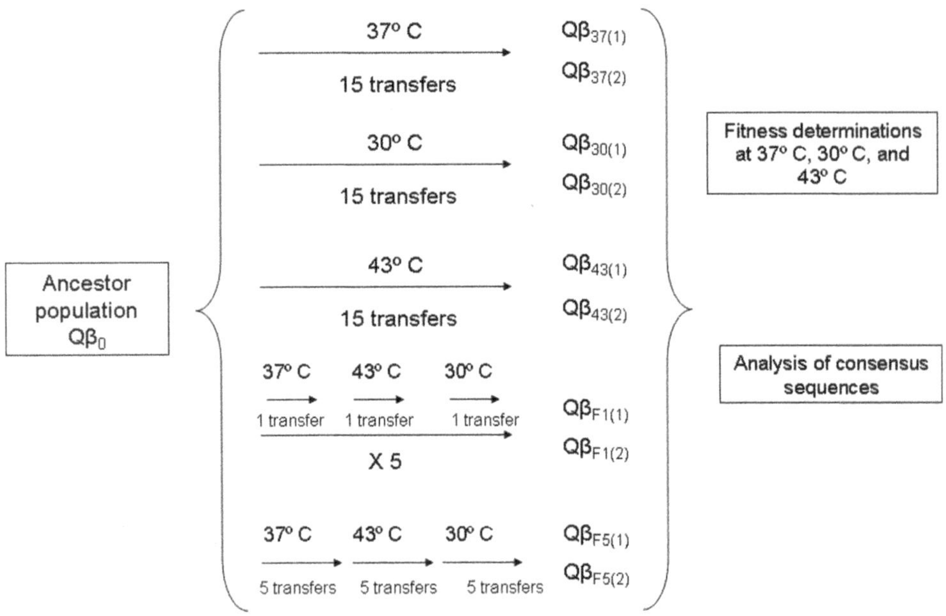

Figure 1. Description of the evolution experiment designed to study the adaptation of bacteriophage Qβ to constant or deterministically fluctuating temperatures. The same virus population ($Q\beta_0$) was used to found 5 replicated evolutionary lineages that were evolved in parallel during 15 transfers either at constant (populations $Q\beta_{37}$, $Q\beta_{30}$, and $Q\beta_{43}$), or fluctuating temperatures (populations $Q\beta_{F1}$, and $Q\beta_{F5}$) under the conditions described in Materials and Methods. Replicate populations were labelled with the subscripts (1) and (2). Each evolved population was analyzed at the phenotypic level (determination of fitness at 37°C, 30°C, and 43°C), and at the genotypic level (determination of the consensus sequence from nucleotide 250 to 4180).

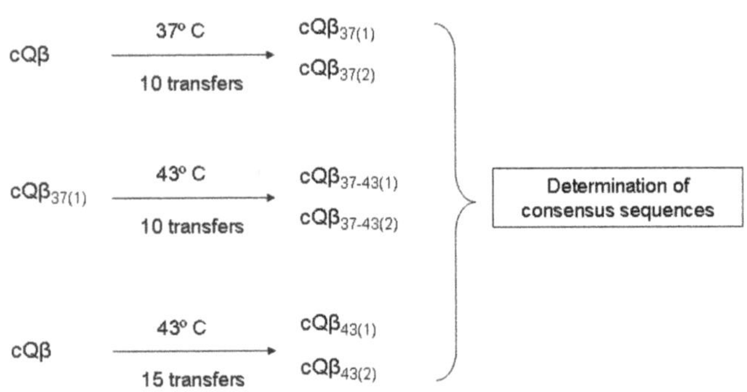

Figure 2. Description of the evolution experiment designed to study the influence of the pre-existent genetic diversity on the adaptation of bacteriophage Qβ to high temperature. The clonal population of the virus, cQβ, obtained upon expression of the plasmid pBRT7Qβ (see Materials and Methods), was propagated at 37°C during 10 serial transfers to yield populations $cQ\beta_{37(1)}$ and $cQ\beta_{37(2)}$. Population $cQ\beta_{37(1)}$ was subjected to 10 additional transfers at 43°C, to yield populations $cQ\beta_{37\text{-}43(1)}$ and $cQ\beta_{37\text{-}43(2)}$. The virus cQβ was also adapted to 43°C without previous optimization at 37°C. Adaptation took place through 15 serial transfers to render populations $cQ\beta_{43(1)}$ and $cQ\beta_{43(2)}$. All evolved populations were characterized at the genotypic level through their consensus sequences (from nucleotide 250 to 4180).

to the sequence of the cDNA of bacteriophage Qβ cloned in the plasmid pBR322 [39].

Statistics

The statistical significance of the fitness differences between the ancestor and the evolved populations was evaluated through the Student'*t* test. A nested ANOVA, including the temperature of evolution as main factor and the triplicate measures of fitness as random variables, was performed for each temperature at which fitness was determined. A two-way ANOVA was carried out to compare fluctuating populations with those evolved under each constant temperature condition. The factors considered were the temperature of evolution and the temperature of the fitness assay. Each replicate population was characterized by three fitness values, each one corresponding to the average of three determinations carried out at each temperature. Calculations were carried out using Mathematica 5 (Wolfram Research)

Results

Evolution of bacteriophage Qβ at constant temperatures

A heterogeneous population of bacteriophage Qβ ($Q\beta_0$), which had previously been adapted to replicate at 37°C (see Materials and Methods) was propagated during 15 serial transfers at 3 different temperatures (30°C, 37°C, and 43°C). All experiments were carried out in duplicate and the populations obtained at the last transfer were denoted Qβ$_{30}$, Qβ$_{37}$, and Qβ$_{43}$ respectively (Fig. 1). Fitness values were determined at the three temperatures assayed for both the ancestor population and the evolved populations. Population Qβ$_0$ had the highest fitness value at 37°C (20.50±0.19), which agrees with the optimal growth temperature of the bacterial host, an intermediate value at 30°C (17.92±1.26), and the lowest value at 43°C (9.80±0.58). Therefore, replication at 43°C was the selective pressure which produced a strongest decrease in the virus growth rate. All evolved populations experienced small, although significant, fitness gains at 37°C relative to the ancestor population ($P<0.05$, Student't test; Fig. 3, upper panel). There were no significant differences among populations that could be attributable to the temperature of evolution (nested ANOVA 37°C: $F_{2,3} = 2.0$, $P>0.05$). The fact that populations propagated at 30°C and 43°C also gained fitness at 37°C probably indicates that population Qβ$_0$ was not well adapted to the propagation conditions used in the adaptation experiment described in Fig. 1, and further optimization was still possible, independently of the temperature at which the virus was propagated. The generalized fitness gains at 37°C also imply that evolution at 30°C and 43°C does not have strong fitness costs on the replication of the virus at 37°C. In contrast to these results, fitness values determined at 30°C did not increase relative to the ancestor population in any of the evolved populations (Fig. 3, medium panel), and fitness at 43°C only increased significantly ($P<0.05$, Student't test) in the populations propagated at this temperature (Fig. 3, lower panel). Fitness determinations at both 30°C and 43°C were significantly affected by the temperature at which populations had evolved (nested ANOVA 30°C: $F_{2,3} = 25.5$, $P<0.025$; nested ANOVA 43°C: $F_{2,3} = 10.8$, $P<0.05$). The two populations evolved at 43°C had significantly lower fitness at 30°C than population Qβ$_0$ ($P<0.05$, Student't test), suggesting the existence of a fitness trade-off between adaptation to 43°C and replication at 30°C. Although evolution at 43°C was carried out by using larger culture volumes (10 ml) than evolution at 30°C or 37°C (1 ml), the fitness gains attained at 43°C (observed in an assay performed in 1 ml volume) indicate that the culture volume is not confounding the interpretation of the results.

Evolution of bacteriophage Qβ at deterministically fluctuating temperatures

Population Qβ$_0$ was also evolved at deterministically fluctuating temperatures that alternated 37°C, 43°C, and 30°C in that order, with the temperature change occurring either every transfer (populations Qβ$_{F1}$) or every five transfers (populations Qβ$_{F5}$) (Fig. 1). Determination of the fitness values of populations obtained at transfer number 15 allowed us to determine whether the fluctuating conditions assayed are compatible with adaptation. The results obtained (Fig. 4) show that both Qβ$_{F1}$ and Qβ$_{F5}$ resembled populations Qβ$_{43}$. They presented small though significant fitness gains at 37°C (with the only exception of Qβ$_{F5(2)}$), fitness losses at 30°C (although they were significant only in one of the replicated populations), and significant fitness gains at 43°C. In consonance with these observations, a two-way ANOVA showed that there was no significant interaction between the temperature of evolution and the temperature of the fitness assay when populations Qβ$_{F1}$ and Qβ$_{F5}$ were compared to populations Qβ$_{43}$ ($F_{4,9} = 1.1$, $P = 0.417$), whereas both factors interacted significantly when the same populations were compared to Qβ$_{37}$ ($F_{4,9} = 6.0$, $P = 0.013$) or Qβ$_{30}$ ($F_{4,9} = 7.1$, $P = 0.007$). It is remarkable that the fitness trade-off shown at 30°C by the populations that evolved at 43°C (Fig. 3) did not preclude adaptation to 43°C when the virus alternated these temperatures during its propagation. Evolution of bacteriophage Qβ at fluctuating temperatures shows a clear example in which the most stringent selective pressure drives adaptation. In general, fitness gains at 43°C were higher for populations evolved constantly at this temperature than for those that evolved under fluctuating temperature conditions. This could have been expected, since the former were adapting to 43°C for a larger number of transfers.

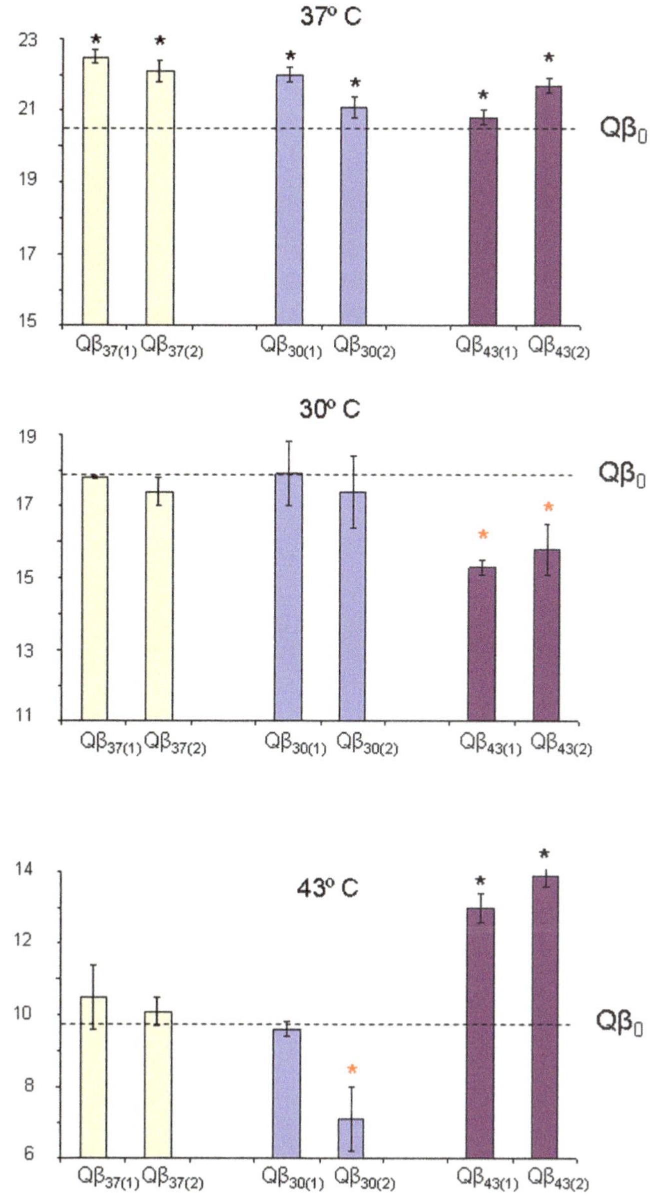

Figure 3. Fitness values of bacteriophage Qβ populations evolved at constant temperatures. Fitness values at 37°C (upper panel), 30°C (medium panel), and 43°C (lower panel) of the two replicas of populations Qβ$_{37}$, Qβ$_{30}$, and Qβ$_{43}$. Each bar represents the average of three parallel determinations carried out as described in Materials and Methods. The error bars represent the standard deviation. The fitness value obtained for the ancestor population, Qβ$_0$, at each temperature is indicated by a discontinuous line. Thus, bars above this line represent fitness increases relative to the ancestor and bars below the line correspond to fitness decreases. Asterisks mean that the average fitness for a given population is significantly higher (black ones) or lower (red ones) than the fitness value of population Qβ$_0$ ($P<0.05$, Student' t test).

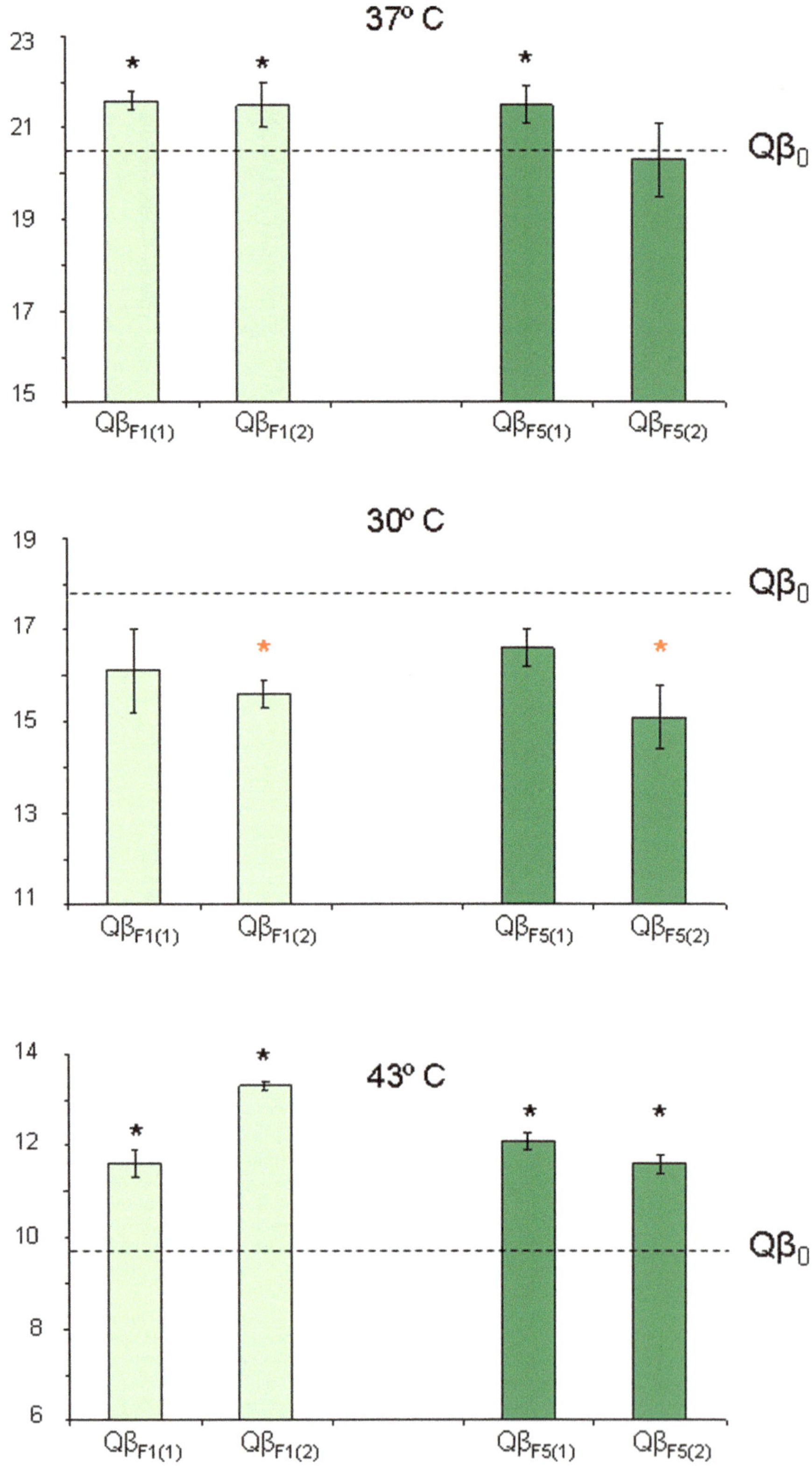

Figure 4. Fitness values of bacteriophage Qβ populations evolved at fluctuating temperatures. Fitness values at 37°C (upper panel), 30°C (medium panel), and 43°C (lower panel) of the two replicas of populations $Q\beta_{F1}$, and $Q\beta_{F5}$. Each bar represents the average of three parallel determinations carried out as described in Materials and Methods. The error bars represent the standard deviation. The fitness value obtained for the ancestor population $Q\beta_0$ at each temperature is indicated by a discontinuous line. Thus, bars above this line represent fitness increases relative to the ancestor and bars below the line correspond to fitness decreases. Asterisks mean that the average fitness for a given population is significantly higher (black ones) or lower (red ones) than the fitness value of population $Q\beta_0$ ($P<0.05$, Student' *t* test).

Table 1. Genomic substitutions present in the consensus sequence of the populations obtained upon evolution of $Q\beta_0$ at constant and deterministically fluctuating temperatures.

Substitution	Protein[1]	Amino acid change[2]	Repetitions[3]
U920A	A2	Phe286Tyr	1
A1088G	A2	Asp342Gly	3
U1198C	A2		1
U1763C	A1		1
G1773A	A1	Gly143Arg	2
C1806U	A1	Pro154Ser	4
G1817A	A1		3
A1823G	A1		1
A1930G	A1	Gln195Arg	4
U2006A	A1	Ser220Arg	1
A2187C	A1	Ser281Arg	7
U2297C	A1		1
U2776C	Rep	Val141Al	3
U3402C	Rep	Ser350Pro	6
U3731C	Rep		2
G3945A	Rep	Gly531Ser	1
C4031U	Rep		1

[1]Protein encoded by the gene where the nucleotide substitution indicated is located.
[2]Amino acid replacement produced by the nucleotide change. In the cases where no amino acid change is indicated, the substitution is synonymous.
[3]Number of populations where a given substitution appears either fixed or as a polymorphism.

Analysis of consensus sequences

To identify the genomic substitutions potentially responsible for the observed fitness changes we determined the consensus sequences of all evolved populations and compared them to that of population $Q\beta_0$. To make a more thorough analysis, the mutations, either fixed or polymorphic, detected in each population were visually inspected in the chromatograms of the remaining populations to evaluate their presence as polymorphisms not recognized by the sequence analysis programs. In total we detected 42 mutations, 35 of which were present as polymorphisms (Tables 1 and 2). There were 17 different substitutions, 8 of which were present in single populations whereas the rest were repeated in at least two populations (Table 1). The total amount of repeated mutations was 84%. The percentage of coincident changes between replicated populations was higher for populations evolved under constant conditions (67% for $Q\beta_{37}$, 60% for $Q\beta_{30}$, and 43% for $Q\beta_{43}$) than for those evolved under fluctuating conditions (29% for $Q\beta_{F1}$ and 11% for $Q\beta_{F5}$). When populations evolved under different conditions were compared, the percentages of coincident changes were 60% for $Q\beta_{37}$ with $Q\beta_{30}$, 11% for $Q\beta_{37}$ with $Q\beta_{43}$, and 9% for $Q\beta_{30}$ with $Q\beta_{43}$. In good agreement with the results obtained in the fitness assays, fluctuating populations showed a higher amount of coincident changes with populations evolved at 43°C (56% $Q\beta_{F1}$ and 33% $Q\beta_{F5}$) than with populations evolved at 37°C (25% $Q\beta_{F1}$ and 9% $Q\beta_{F5}$) or at 30°C (20% $Q\beta_{F1}$ and 8% $Q\beta_{F5}$). A high amount of mutations (76%) were non synonymous. The ratio dn/ds for the whole set of mutations was 1.1, which increased to 1.7 when the mutations represented in several populations were the only considered. The distribution of the substitutions along the $Q\beta$ genome (Table 2) shows a significant accumulation ($P<0.025$, chi squared test, 3 d.f.) in the region of the A1 protein that does not overlap with the coat protein (Fig. 5).

The analysis of the consensus sequences of the populations evolved under constant conditions shows that substitutions C1806U and A1930G were selected at 37°C and 30°C, and substitution A2187C was selected at the three temperatures (Table 2). There were also several substitutions that only appeared in the populations evolved at 30°C (U3731C and G3945A), or at 43°C (A1088G, G1773A, G1817A, U2297C, U2776C, and U3402C) (Table 2). In contrast to this, the populations evolved at 37°C did not show any specific mutations. From all these substitutions, the changes G1773A, C1806U, G1817A, A1930G, A2187C, and G3945A had previously been detected in our laboratory during the propagation of bacteriophage $Q\beta$, suggesting that they are beneficial mutations of general effect under the culture conditions used in our lab [40], [41]. Therefore, they are good candidates to be responsible for the small fitness gains observed at 37°C in populations evolved at 37°C, 30°C, and 43°C. Substitutions exclusive to a particular temperature, and not detected during the evolution of the virus under unrelated selective pressures, probably represent adaptive mutations for that particular condition. In this way, substitution U3731C, only detected in populations $Q\beta_{30}$, is a candidate to promote adaptation to 30°C. Similarly, substitutions A1088G, U2297C, U2776C, and U3402, could contribute to the fitness gains observed at 43°C.

The comparison of the substitutions found in the consensus sequences of the populations evolved under constant and fluctuating temperatures allowed us to determine whether the similarities shown at the phenotypic level by $Q\beta_{F1}$ and $Q\beta_{F5}$ with $Q\beta_{43}$ are also displayed at the genetic level. None of the populations $Q\beta_{F1}$ and $Q\beta_{F5}$ showed the substitutions exclusive to the population evolved at 30°C. In contrast to this, all substitutions but one (U2297C) of those exclusive to the populations evolved at 43°C were present at least in one of the populations evolved under fluctuating conditions (Table 2),

Table 2. Genomic location of the nucleotide positions mutated in the consensus sequences of the populations obtained upon evolution of $Q\beta_0$ at constant and deterministically fluctuating temperatures.

Population	A2			Coat	A1 (Non-overlapping region with the coat protein)								Replicase					
	920	1088	1198		1763	1773	1806	1817	1823	1930	2006	2187	2297	2776	3402	3731	3945	4031
$Q\beta_{37(1)}$										*		*						
$Q\beta_{37(2)}$							*			*		*						
$Q\beta_{30(1)}$							*			(*)						(*)	*	
$Q\beta_{30(2)}$							*			(*)		*				*		
$Q\beta_{43(1)}$		*						*				*	*		*			
$Q\beta_{43(2)}$		*				*		*						*	*			
$Q\beta_{F1(1)}$												*		*	*			
$Q\beta_{F1(2)}$						*	*	*				(*)			*			*
$Q\beta_{F5(1)}$		*			*						(*)			(*)	*			
$Q\beta_{F5(2)}$	*		*						*			(*)			*			

The asterisks indicate the presence of a mutation at the position indicated. All mutations were polymorphic at transfer number 15, except those shown in brackets which were fixed.

showing that this temperature also drives adaptation at the molecular level. One of the common mutations to populations $Q\beta_{30}$ and $Q\beta_{37}$ (C1806U) was present in one of the $Q\beta_{F1}$ populations, and the most widespread substitution, A2187C, was detected in the two $Q\beta_{F1}$ and in one $Q\beta_{F5}$ populations. Both $Q\beta_{F1}$ and $Q\beta_{F5}$ also have exclusive mutations. Substitution C4031U was unique to one of the $Q\beta_{F1}$ populations, whereas substitutions U920A, U1198C, U1763C, A1823G, and U2006A were only detected in the $Q\beta_{F5}$ populations (Table 2). Again, some of these substitutions (U1763C and A1823G) had previously been detected during the propagation of bacteriophage Qβ under different selective pressures in our lab [41].

Adaptation to 43°C of populations differing in the pre-existent genetic diversity

The high number of convergent mutations detected in the Qβ populations analyzed in this work (Tables 1 and 2) could be a consequence of their previous presence as minority variants in the mutant spectrum of the ancestor population $Q\beta_0$. We investigated whether the same set of mutations repetitively detected in populations exposed to 43°C would be selected when evolution starts with Qβ populations differing in their pre-existent genetic diversity. To this end, we prepared a clonal population of the virus (cQβ) and adapted it to replicate at 43°C either directly or after a pre-adaptation step to 37°C (Fig. 2). Upon propagation of cQβ at 37°C, both replicated populations $cQ\beta_{37(1)}$ and $_{(2)}$ fixed substitution A2187C, which was the most frequent in our previous experiments (Tables 1 and 2). There were no signs of polymorphism at any of the other mutated positions previously identified (Tables 1 and 2). Evolution of $cQ\beta_{37(1)}$ at 43°C rendered populations $cQ\beta_{37\text{-}43(1)}$ and $_{(2)}$: they maintained substitution A2187C and acquired the polymorphic substitutions A1088G, G1817A, U2776C, and U3402C, all exclusive to populations evolved at 43°C (Table 2). Although this experiment shows the selection of a common set of mutations when populations differing in their pre-existent genetic diversity ($Q\beta_0$ and $cQ\beta_{37(1)}$) were exposed to the same selective pressure (43°C), we cannot exclude the possibility that those substitutions were generated during the propagation of cQβ at 37°C, and, thus, were already present, although at undetectable frequency by the sequencing methods employed, in the mutant spectrum of population $cQ\beta_{37(1)}$. To determine whether the same mutations could also be selected when all the genetic diversity is generated *de novo* during adaptation, we sequenced the two replicated populations $cQ\beta_{43(1)}$ and $cQ\beta_{43(2)}$ generated from cQβ directly evolved to 43°C (Fig. 2). We detected the same set of polymorphic substitutions (A1088G, U2776C, U3402C, and U3784C) in both of them, confirming that substitutions A1088G, U2776C, and U3402C represent an adaptive pathway to high temperature that is accessible from different initial conditions. Substitution U3784C, which was not selected during the adaptation of $Q\beta_0$ at 43°C, could also be involved in adaptation to 43°C or be a substitution that increases the growth rate by a mechanism independent of temperature.

Discussion

Living organisms have to cope with frequent environmental changes that limit their survival if they are not able to adapt to the new conditions. When the selective pressures remain constant for a long time, populations usually evolve to perform optimally under those conditions, even if that entails a decrease of fitness in alternative environments [1–7]. On the contrary, when the selective pressures change frequently, populations do not have

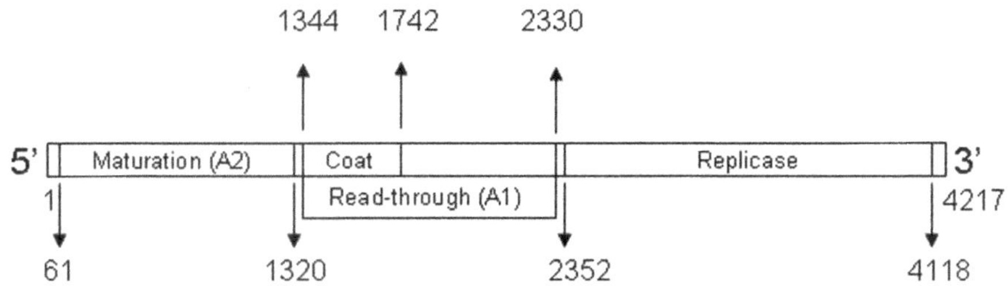

Figure 5. Genetic map of bacteriophage Qβ. The map shows the genes for the 4 proteins encoded by the virus [A2 (from nt 61 to nt 1320), coat (from nt 1344 to nt 1742), A1 (from nt 1344 to nt 2330, and replicase (from nt 2352 to nt 4118)], and the non-translated regions. The protein A1 synthesizes when the stop codon of the coat protein is read through by a trp-tRNA, giving rise to an elongated protein that incorporates at low amount in the capsid.

the time necessary to achieve the best adaptive solution for each condition [42]. The situation gets further complicated when mutations selected in a particular environment have deleterious effects in a different environment encountered by the population. The study of the evolutionary pathways followed by populations adapting to constant versus fluctuating environments is particularly interesting in the case of RNA viruses whose great capacity to adapt to new selective pressures hamper most antiviral treatments [43–45]. We have addressed this question by exposing an RNA virus, the bacteriophage Qβ to three constant environments that differed in the temperature (37°C, 43°C, and 30°C), and to two deterministically fluctuating environments in which the temperature alternated among 37°C, 43°C, and 30°C with two patterns of change.

Propagation of bacteriophage Qβ at different constant temperatures does not result in a common pattern of adaptation

We found that propagation of bacteriophage Qβ at any of the temperatures assayed increased the performance of the virus at 37°C, indicating that either the fitness effects of some of the mutations selected in lineages exposed to different treatments correlate across different temperatures or selection is acting on viral traits unrelated to this character. Since in all cases the virus was propagated according to a similar transmission regime, fitness increases at 37°C can arise as a consequence of adaptation to the culture conditions used. The coincidence of several mutations in populations evolved under different temperature conditions indicates that the common behaviour observed at 37°C has a genetic basis. The fact that several of those coincident mutations had also been detected in previous experiments in which bacteriophage Qβ was propagated in our lab under unrelated selective pressures also agrees with a general beneficial fitness effect on the adaptation of the virus to the culture conditions used [40], [41]. Fitness gains at a temperature different from that at which the virus was propagated have also been reported for vesicular stomatitis virus (VSV) adapted to 29°C and 37°C [33]. Another study that compared the adaptation to 44°C in the related viruses PhiX174 and G4 showed the existence of a fitness cost at 37°C in PhiX174 but not in G4 [46]. In our study, the general fitness gains observed at 37°C had no counterpart at 30°C or 43°C. In fact none of our populations increased fitness at 30°C, and only populations $Q\beta_{43}$ improved their performance at 43°C. This result suggests that the inhibition of virus growth that takes place at 30°C and 43°C masks the beneficial effect of the mutations that provide selective advantages unrelated to temperature. The observed difficulties for adaptation to 30°C agree with the hypothesis that argues that the rate-depressing effects of low temperature on biochemical reactions are difficult to overcome by genetic changes [47].

Populations adapted to 43°C performed worse at 30°C than the ancestor population, suggesting that some of the mutations responsible for adaptation to high temperature are deleterious when this parameter decreases below the optimal value. A possible explanation for this trade-off is that adaptation to high temperature selects for mutations that render more thermodynamically stable proteins, whose performance worsens when the temperature is lowered [48–51]. The main argument for the trade-off between function and stability lies in the belief that catalytic proteins must not be too stable to allow for function-related flexibility [52], a hypothesis supported by the fact that the stability of real proteins is not too high. Also there could be a trade-off between growth rate and survival, particularly at high temperatures, a possibility that deserves to be further analyzed.

Bacteriophage Qβ populations evolved under deterministically fluctuating temperatures perform similarly to populations propagated under the most restrictive condition

The fitness trade-off between adaptation to 43°C and virus performance at 30°C could pose a conflict when the virus has to thrive in environments in which the two temperatures alternate. However, populations evolved under fluctuating conditions showed striking similarities with populations evolved at 43°C. This result agrees with the selection of an evolutionary pathway that favours adaptation to the most restrictive condition, in this case 43°C, even if that means that the virus decreases its performance under other less stringent selective pressure (30°C). This pathway is probably favoured because the fitness cost at 30°C is lower than the growth inhibition that takes place at 43°C in non-adapted viruses. It is striking that we observe the same pattern of fitness across environments in the populations $Q\beta_{F1}$ and $Q\beta_{F5}$, which alternated temperatures differently. It could have been expected that mutations responsible for fitness gains at 43°C had been eliminated by negative selection during virus replication at 30°C, particularly in the population $Q\beta_{F5}$.

Our results differ from those obtained by Alto et al. during the adaptation of VSV to deterministically changing temperatures (29°C alternating with 37°C), where a generalist population with increased performance at each temperature was selected [33]. These results could have been facilitated by the absence in VSV of fitness trade-offs across the two temperatures assayed. Studies carried out with viruses challenged to alternate between different hosts sometimes also show equal or superior fitness gains on each host compared to viruses constantly evolved in the same host [22], [24], [27–29], [32], [33]. These fitness gains in the generalist lineages can even occur when trade-offs across different hosts exist [22].

The consensus sequences of the evolved bacteriophage Qβ populations show coincident mutations at different temperatures together with mutations exclusive to a particular condition

Most of the mutations that we found in the consensus sequences of the evolved bacteriophage Qβ populations were polymorphisms. A possible explanation is that not enough generations were allowed for mutations to reach fixation. An alternative interpretation could be that different beneficial mutations occurred in different lineages and competed with each other, such that interference would delay their fixation [53–55]. Antagonistic epistatic interactions among different beneficial mutations could also contribute to prolong the time that mutations remain as polymorphisms, especially if some of them cannot coexist in the same genome. This kind of interactions has been described in different viruses [56–58], and was shown to be responsible for the prolonged co-existence of two beneficial mutations during the adaptation of bacteriophage Qβ to a mutagen [41]. Investigating this question would require an analysis of the distribution of mutations in the individual components of the mutant spectra of the evolved Qβ populations. Propagation of the virus for a larger number of transfers would also help to determine whether a defined consensus sequence can be obtained.

The analysis of the consensus sequences of the evolved populations show a high number of coincident mutations. This fact, together with the high value of the ratio dn/ds calculated for the repeated mutations, is indicative of the action of natural selection. In the case of RNA, selection may also act at the level of synonymous mutations, whose fitness effect might appear through modifications of the RNA secondary and tertiary structures or by affecting genomic domains required for interaction with substrates. The high degree of convergent evolution here uncovered is similar to that found in other studies carried out with viruses adapting to a constant selective pressure [59–61].

In addition to the mutations repeated in populations propagated under different conditions, other mutations only appeared in the sequence of the virus exposed to a particular temperature. Some of these mutations also coincided in replicated populations, and thus are likely responsible for adaptation to a given temperature. Specifically, substitution U3731C could promote adaptation to 30°C, although this effect was not reflected in fitness increases at 30°C, which could mean that it acts on a trait with no influence on the growth rate. The other substitution exclusive to 30°C but only present in one of the replicated populations was G3945A, which was previously demonstrated to have a general beneficial fitness effect in experiments carried out in both our lab [41] and in others [62]. In the same way, substitutions A1088G, G1817A, U2776C, and U3402C, exclusive to populations evolved at 43°C, could be responsible for the fitness gains observed at this temperature. There were no mutations exclusive to 37°C, which agrees with the fact that the ancestor population $Q\beta_0$ had previously been propagated at 37°C and, therefore, the supply of beneficial mutations specific to this temperature could be exhausted. To establish unambiguous correspondences between particular mutations and their fitness effect, it would be desirable to build single site-directed mutants and determine their fitness at the temperature of interest, either in competition with the wild-type virus or through the estimation of the growth rate.

Populations evolved under fluctuating conditions, which were phenotypically similar to the constant ones evolved at 43°C, also resembled those populations at the genetic level. They presented some of the common mutations selected at different constant temperatures, did not have any of the mutations exclusive of 30°C, and showed most of the mutations exclusive of the populations evolved at 43°C, confirming their implication in the adaptation of bacteriophage Qβ to high temperature.

Evolution at 43°C of bacteriophage Qβ populations differing in their pre-existent genetic diversity leads to the selection of a common set of mutations

The high degree of parallel evolution shown by the populations analyzed in this work could be due to the fact that the ancestor population was heterogeneous, and, thus, it could contain in its mutant spectrum some of the mutations that consistently appeared during evolution in different lineages. To investigate this possibility we compared the mutations selected during adaptation to 43°C in populations that differed in their pre-existent genetic diversity. We observed that a clonal population and a heterogeneous population, obtained upon transmission at 37°C of the clonal population, coincided in the selection of the polymorphic substitutions A1088G, U2776C, and U3402C, which had previously been identified during the adaptation of population $Q\beta_0$ to 43°C. The result indicates that there are few pathways for adaptation of bacteriophage Qβ to high temperature. The population adapted directly to 43°C showed the additional selection of U3784C, which could also provide selective advantages. Pre-adaptation of the clonal population of bacteriophage Qβ to 37°C led to the fixation of a single substitution, A2187C, in two replicate experiments. Therefore, at least the most repeated substitution across all evolutionary lineages analyzed in this work could also be generated *de novo* during propagation of cQβ at 37°C. We did not detect any of the other substitutions to which we had attributed general beneficial effects, meaning that they were probably present in the mutant spectrum of population $Q\beta_0$ and reached high frequency upon its propagation.

Molecular mechanism of action of the mutations fixed during adaptation

Most mutations selected in lineages evolved at different temperatures were placed in the A1 protein and corresponded to the replacement of different amino acids by arginine residues. The A1 protein incorporates in 3–10 copies per virion and is essential for producing infectious virus particles, although its precise function remains unknown. It has recently been suggested that some domains of the A1 protein can be involved in protein interactions that would be necessary to reach the virus fivefold and threefold symmetry axes [63]. Therefore, our results could mean that structural constraints on the virus capsid limit replication under the transmission regime used, and selection of the A1 mutants helps to overcome those constraints. Some of the lineages evolved at 43°C showed the selection of substitution G1817A, also placed in the A1 protein. Although this is a synonymous substitution, we cannot discard that it has some beneficial effect either in stabilizing the viral RNA or in altering somehow its interaction with other molecules.

Adaptation to 43°C was associated to the selection of substitution A1088G (Asp342Gly) in the A2 protein, which is involved in the recognition of the conjugative pilus as receptor in F+ bacteria, the escort of the genomic RNA into the host cytoplasm, and the lysis of the bacteria to release the virus progeny. High temperatures might impair any of these processes, which would be restored, at least in part, in the mutant. It is interesting that during the evolution of bacteriophage phi6 in a novel temperature environment where heat shock imposed extreme virus mortality, a single amino acid substitution in the viral lysis protein was also selected [51].

The replicase substitutions selected at 43°C, U2776C, U3402C, and U3784C rendered the non synonymous changes Val141Ala, Ser350Pro, and Ile477Thr respectively. Ser350 is placed in the palm domain of the Qβ replicase, whereas Val141 is placed in the fingers domain, and Ile477 is in thumb domain [64], [65]. The three residues are not conserved among members of the virus family Leviviridae, which suggests that they do not form part of the active center of the Qβ replicase. Moreover, the related viruses MX1 and M11 contain proline instead of serine at position 350, the viruses NL95, GA, MS2, and FR contain alanine instead of valine at position 141, and threonine instead of isoleucine is found in the viruses NL95, SP, F1, and MX1 [64], [65]. The most probable explanation for the beneficial effects of these substitutions at high temperatures is that they increase the thermal stability of the replicase. The finding that some Qβ-related viruses posses in their wild type form the same amino acids as the high temperature Qβ adapted mutants suggests that what we have selected through experimental evolution in the lab was already selected in nature. Since we have shown that high temperature is the dominant force driving evolution in bacteriophage Qβ when it is exposed to fluctuations in this condition, the selection of thermostable replicases must not be an unusual fact in viruses exposed to transitory temperature increases. The determination of whether those viruses have higher tolerance to high temperatures would be very useful to check the validity of our interpretation.

Acknowledgments

We are very grateful to S Manrubia for useful discussions and critical reading of the manuscript.

Author Contributions

Conceived and designed the experiments: EL. Performed the experiments: MA KK LC. Analyzed the data: EL MA KK LC. Contributed to the writing of the manuscript: EL.

References

1. Levins R (1968) Evolution in changing environments. Princeton Univ. Press, Princeton, NJ.
2. Lynch M, Gabriel W (1987) Environmental tolerance. Am Nat 122:745–764.
3. Futuyma DJ, Moreno G (1988) The evolution of ecological specialization. Annu Rev Ecol Syst 19:207–233.
4. Via S (1990) Ecological genetics and host adaptation in herbivorous insects: the experimental study of evolution in natural and agricultural systems. Annu Rev Entomol 135:421–446.
5. Stearns SC (1992) The evolution of life histories, Oxford Univ Press, Oxford UK.
6. Hereford J (2009) A quantitative survey of local adaptation and fitness trade-offs. Am Nat 173:579–588.
7. Legros M, Koella JC (2010) Experimental evolution of specialization by a microsporidian parasite. BMC Evol Biol 10:159.
8. Gandon S (2004) Evolution of multihost parasites. Evolution 58:455–469.
9. Kawecki TJ, Ebert D (2004) Conceptual issues in local adaptation. Ecol Lett 7:1225–1241.
10. Magalhães S, Blanchet E, Egas M, Olivieri I (2009) Are adaptation costs necessary to build up a local adaptation pattern? BMC Evol Biol 9:182.
11. Poulin R (1998) Large-scale patterns of host use by parasites of freshwater fishes. Ecol Lett 1:118–128.
12. Krasnov BR, Poulin R, Shenbrot GI, Mouillot D, Khokhlova IS (2004) Ectoparasitic "jacks-of-all-trades": relationship between abundance and host specificity in fleas (Siphonaptera) parasitic on small mammals. Am Nat 164:506–516.
13. Palaima A (2007) The fitness cost of generalization: present limitations and future possible solutions. Bio J Linn Soc Lond 90:583–590.
14. Coffey LL, Vasilakis N, Brault AC, Powers AM, Tripet F, et al. (2008) Arbovirus evolution in vivo is constrained by host alternation. Proc Natl Acad Sci USA 105:6970–6975.
15. Hellgren O, Pérez-Tris J, Bensch S (2009) A jack-of-all-trades and still a master of some: prevalence and host range in avian malaria and related blood parasites. Ecology 90:2840–2849.
16. Kawecki TJ (1994) Accumulation of deleterious mutations and the evolutionary cost of being generalist. Am Nat 144:833–838.
17. Malpica JM, Sacristán S, Fraile A, García-Arenal F (2006) Association and host selectivity in multi-host pathogens. PLoS One 20;1:e41.
18. Straub CS, Ives AR, Gratton C (2011) Evidence for a trade-off between host-range breadth and host-use efficiency in aphid parasitoids. Am Nat 177:389–395.
19. Domingo E, Sheldon J, Perales C (2012) Viral quasispecies evolution. Microbiol Mol Biol Rev 76:159–216.
20. Novella IS, Hershey CL, Escarmis C, Domingo E, Holland JJ (1999) Lack of evolutionary stasis during alternating replication of an arbovirus in insect and mammalian cells. J Mol Biol 287:459–465.
21. Weaver SC, Brault AC, Kang W, Holland JJ (1999) Genetic and fitness changes accompanying adaptation of an arbovirus to vertebrate and invertebrate cells. J Virol 73:4316–4326.
22. Turner PE, Elena SF (2000) Cost of host radiation in an RNA virus. Genetics 156:1465–1470.
23. Cooper LA, Scott TW (2001) Differential evolution of eastern equine encephalitis virus populations in response to host cell type. Genetics 157:1403–1412.
24. Greene IP, Wang E, Deardorff ER, Milleron R, Domingo E, et al. (2005) Effect of alternating passage on adaptation of sindbis virus to vertebrate and invertebrate cells. J Virol 79:14253–14260.
25. Wallis CM, Stone AL, Sherman DJ, Damsteegt VD, Gildow FE, et al. (2007) Adaptation of plum pox virus to a herbaceous host (Pisum sativum) following serial passages. J Gen Virol 88:2839–2844
26. Agudelo-Romero P, de la Iglesia F, Elena SF (2008) The pleiotropic cost of host-specialization in Tobacco etch potyvirus. Infect Genet Evol 8:806–814.
27. Remold SK, Rambaut A, Turner PE (2008) Evolutionary genomics of host adaptation in vesicular stomatitis virus. Mol Biol Evol 25:1138–1147.
28. Vasilakis N, Deardorff ER, Kenney JL, Rossi SL, Hanley KA, et al. (2009) Mosquitoes put the brake on arbovirus evolution: experimental evolution reveals slower mutation accumulation in mosquito than vertebrate cells. PLoS Pathog 5(6):e1000467.
29. Coffey LL Vignuzzi M (2011) Host alternation of chikungunya virus increases fitness while restricting population diversity and adaptability to novel selective pressures. J Virol 85:1025–1035.
30. Deardorff ER, Fitzpatrick KA, Jerzak GV, Shi PY, Kramer LD, et al. (2011) West Nile virus experimental evolution in vivo and the trade-off hypothesis. PLoS Pathog 7(11):e1002335
31. Lalić J1, Cuevas JM, Elena SF (2011) Effect of host species on the distribution of mutational fitness effects for an RNA virus. PLoS Genet 7(11):e1002378.
32. Bedhomme S, Lafforgue G, Elena SF (2012) Multihost experimental evolution of a plant RNA virus reveals local adaptation and host-specific mutations. Mol Biol Evol 29:1481–1492.
33. Alto BW, Wasik BR, Morales NM, Turner PE (2013) Stochastic temperatures impede RNA virus adaptation. Evolution 67:969–979.
34. Lauring AS, Andino R (2010) Quasispecies theory and the behavior of RNA viruses. PLoS Pathog 6(7):e1001005.
35. Ruiz-Jarabo CM, Arias A, Baranowski E, Escarmís C, Domingo E (2000) Memory in viral quasispecies. J Virol 74:3543–3547.
36. Domingo E, Ruiz-Jarabo CM, Sierra S, Arias A, Pariente N, et al. (2002) Emergence and selection of RNA virus variants: memory and extinction. Virus Res 82:39–44.
37. Briones C, Domingo E (2008) Minority report: hidden memory genomes in HIV-1 quasispecies and possible clinical implications. AIDS Rev 10:93–109.
38. Cases-González C, Arribas M, Domingo E, Lázaro E (2008) Beneficial effects of population bottlenecks in an RNA virus evolving at increased error rate. J Mol Biol. 384:1120–1129.
39. Barrera I, Schuppli D, Sogo JM, Weber H (1993) Different mechanisms of recognition of bacteriophage Qβ plus and minus strand RNAs by Qβ replicase. J Mol Biol 232:512–521.
40. Arribas M, Cabanillas L, Lázaro E (2011) Identification of mutations conferring 5-azacytidine resistance in bacteriophage Qβ. Virology 417:343–352.
41. Cabanillas L, Arribas M, Lázaro E (2013) Evolution at increased error rate leads to the coexistence of multiple adaptive pathways in an RNA virus. BMC Evol Biol 13:11.
42. Whitlock MC (1996) The red queen beats the jack-of-all trades the limitations on the evolution of phenotypic plasticity and niche breadth. Am Nat 148:S65.
43. Domingo E (2003) Quasispecies and the development of new antiviral strategies. Prog Drug Res 60:133–158.
44. Domingo E, Menéndez-Arias L, Quiñones-Mateu ME, Holguín A, Gutiérrez-Rivas M, et al. (1997) Viral quasispecies and the problem of vaccine-escape and drug-resistant mutants. Prog Drug Res 48:99–128.
45. Sanjuán R, Lázaro E, Vignuzzi M (2012) Biomedical implications of viral mutation and evolution. Future Virology 7:391–402.
46. Holder KK, Bull JJ (2001) Profiles of adaptation in two similar viruses. Genetics 159:1393–1404.

47. Knies JL, Kingsolver JG, Burch CL (2009) Hotter is better and broader: thermal sensitivity of fitness in a population of bacteriophages. Am Nat 173:419–430.
48. Wang X, Minasov G, Shoichet BK (2002) Evolution of an antibiotic resistance enzyme constrained by stability and activity trade-offs. J Mol Biol 20:85–95.
49. Tokuriki N, Stricher F, Serrano L, Tawfik DS (2008) How protein stability and new functions trade off. PLoS Comput Biol 4(2):e1000002.
50. Lee KH, Miller CR, Nagel AC, Wichman HA, Joyce P, et al. (2011) First-step mutations for adaptation at elevated temperature increase capsid stability in a virus. PLoS One 6(9):e25640.
51. Dessau M, Goldhill D, McBride R, Turner PE, Modis Y (2012) Selective pressure causes an RNA virus to trade reproductive fitness for increased structural and thermal stability of a viral enzyme. PLoS Genet 8(11):e1003102.
52. DePristo MA, Weinreich DM, Hartl DL (2005) Missense meanderings in sequence space: a biophysical view of protein evolution. Nat Rev Genet 6:678–687.
53. Gerrish PJ, Lenski RE (1998) The fate of competing beneficial mutations in an asexual population. Genetica 102–103:127–144.
54. Sniegowski PD, Gerrish PJ (2010) Beneficial mutations and the dynamics of adaptation in asexual populations. Philos Trans R Soc Lond B Biol Sci 365:1255–1263.
55. Lang GI, Botstein D, Desai MM (2011) Genetic variation and the fate of beneficial mutations in asexual populations. Genetics 188:647–661.
56. Weinreich DM, Watson RA, Chao L (2005) Perspective: Sign epistasis and genetic constraint on evolutionary trajectories. Evolution 2005, 59:1165–1174.
57. Elena SF, Solé RV, Sardanyés J (2010) Simple genomes, complex interactions: epistasis in RNA virus. Chaos 20(2):026106.
58. Rokyta DR, Joyce P, Caudle SB, Miller C, Beisel CJ, et al. (2011) Epistasis between beneficial mutations and the phenotype-to-fitness Map for a ssDNA virus. PLoS Genet 7(6):e1002075.
59. Bull JJ, Badgett MR, Wichman HA, Huelsenbeck JP, Hillis DM, et al. (1997) Exceptional convergent evolution in a virus. Genetics 147:1497–1507.
60. Wichman HA, Badgett MR, Scott LA, Boulianne CM, Bull JJ (1999) Different trajectories of parallel evolution during viral adaptation. Science 285:422–424.
61. Bollback JP, Huelsenbeck JP (2009) Parallel genetic evolution within and between bacteriophage species of varying degrees of divergence. Genetics 181:225–234. 62.
62. García-Villada L, Drake JW (2013) Experimental selection reveals a trade-off between fecundity and lifespan in the coliphage Qβ. Open Biol 3(6):130043.
63. Rumnieks J, Tars K (2011) Crystal structure of the read-through domain from bacteriophage Qβ A1 protein. Protein Sci 20:1707–1712.
64. Kidmose RT, Vasiliev NN, Chetverin AB, Andersen GR, Knudsen CR (2010) Structure of the Qbeta replicase, an RNA-dependent RNA polymerase consisting of viral and host proteins. Proc Natl Acad Sci USA 107:10884–10889.
65. Takeshita D, Tomita K (2010) Assembly of Qbeta viral RNA polymerase with host translational elongation factors EF-Tu and -Ts. Proc Natl Acad Sci USA. 107:15733–15738.

8

Evolutionary Origin of GnIH and NPFF in Chordates: Insights from Novel Amphioxus RFamide Peptides

Tomohiro Osugi[1¤], Tomoki Okamura[1], You Lee Son[1], Makoto Ohkubo[1], Takayoshi Ubuka[1], Yasuhisa Henmi[2], Kazuyoshi Tsutsui[1]*

1 Laboratory of Integrative Brain Sciences, Department of Biology, Waseda University, and Center for Medical Life Science of Waseda University, Tokyo, Japan, **2** Aitsu Marine Station, Center for Marine Environmental Studies, Kumamoto University, Kumamoto, Japan

Abstract

Gonadotropin-inhibitory hormone (GnIH) is a newly identified hypothalamic neuropeptide that inhibits pituitary hormone secretion in vertebrates. GnIH has an LPXRFamide (X = L or Q) motif at the C-terminal in representative species of gnathostomes. On the other hand, neuropeptide FF (NPFF), a neuropeptide characterized as a pain-modulatory neuropeptide, in vertebrates has a PQRFamide motif similar to the C-terminal of GnIH, suggesting that GnIH and NPFF have diverged from a common ancestor. Because GnIH and NPFF belong to the RFamide peptide family in vertebrates, protochordate RFamide peptides may provide important insights into the evolutionary origin of GnIH and NPFF. In this study, we identified a novel gene encoding RFamide peptides and two genes of their putative receptors in the amphioxus *Branchiostoma japonicum*. Molecular phylogenetic analysis and synteny analysis indicated that these genes are closely related to the genes of GnIH and NPFF and their receptors of vertebrates. We further identified mature RFamide peptides and their receptors in protochordates. The identified amphioxus RFamide peptides inhibited forskolin induced cAMP signaling in the COS-7 cells with one of the identified amphioxus RFamide peptide receptors expressed. These results indicate that the identified protochordate RFamide peptide gene is a common ancestral form of GnIH and NPFF genes, suggesting that the origin of GnIH and NPFF may date back to the time of the emergence of early chordates. GnIH gene and NPFF gene may have diverged by whole-genome duplication in the course of vertebrate evolution.

Editor: Hubert Vaudry, University of Rouen, France, France

Funding: This work was supported in part by Grants-in-Aid for Scientific Research from the Ministry of Education, Science and Culture, Japan (18107002, 22132004 and 22227002 to K.T.). The funders had no role in study design, data collection and analysis, decision to publish, or preparation of the manuscript.

Competing Interests: The authors have declared that no competing interests exist.

* Email: k-tsutsui@waseda.jp

¤ Current address: Bioorganic Research Institute, Suntory Foundation for Life Sciences, Shimamoto, Mishima, Osaka, Japan

Introduction

Gonadotropin-inhibitory hormone (GnIH) is a newly identified hypothalamic neuropeptide that inhibits pituitary hormone secretion in vertebrates [1–4]. The discovery of GnIH has now changed drastically the classical understanding of the regulation of hypothalamic-pituitary-gonadal (HPG) axis in vertebrates [1–4]. In addition, a recent study demonstrated that hypothalamic GnIH inhibits socio-sexual behaviors by increasing neuroestrogen synthesis in male birds [5]. After the discovery of GnIH in birds, its orthologous peptides have been identified in a variety of vertebrates, such as primates and mammals [RFamide-related peptides (RFRP)] [6–9], amphibians [frog growth hormone-releasing peptide (fGRP) [10,11], *Rana* RFamide [12], and newt LPXRFamide peptide [13]], and teleosts [goldfish (gf)LPXRFamide peptide] [14]. All the identified GnIH and its orthologs possessed a C-terminal Leu-Pro-Xaa-Arg-Phe-NH_2 (Xaa = Leu or Gln) motif at their C-termini, and thus, they were designated structurally as LPXRFamide (X = L or Q) peptides (LPXRFa peptides). LPXRFa peptides constitute one of the largest groups in the RFamide peptide (RFa peptide) family in vertebrates [2–4]. Typically, the LPXRFa peptide precursors in vertebrates encode two to four LPXRFa peptides, such as RFRP-1 and RFRP-3 in primates and mammals [7–9]; GnIH, GnIH gene-related peptide (RP)-1 and GnIH-RP-2 in birds [15–18]; fGRP, fGRP-RP-1, fGRP-RP-2, and fGRP-RP-3 in amphibians [19]; and LPXRFa-1, LPXRFa-2, and LPXRFa-3 in fish [14,20,21] (see Table S1).

From a structural point of view, pain modulatory neuropeptides, such as neuropeptide FF (NPFF) and NPAF, share a C-terminal Pro-Gln-Arg-Phe-NH_2 motif (PQRFa peptides) [22]. Interestingly, the C-terminal motifs of GnIH and NPFF group peptides showed high sequence similarity, which are considered to be important for the interaction with their receptors (GPR147 and 74) [2–4,22]. The structure of their receptors also showed high sequence similarity. In addition, synteny analyses showed that GnIH gene and NPFF gene are both located near the *HOX* clusters [23,24]. Because the *HOX* clusters are considered to have duplicated from a common ancestral gene during whole-genome duplication events through vertebrate evolution [25], GnIH gene and NPFF gene may also have diverged through whole-genome duplication in the course of evolution in vertebrates [23,24]. In our recent studies,

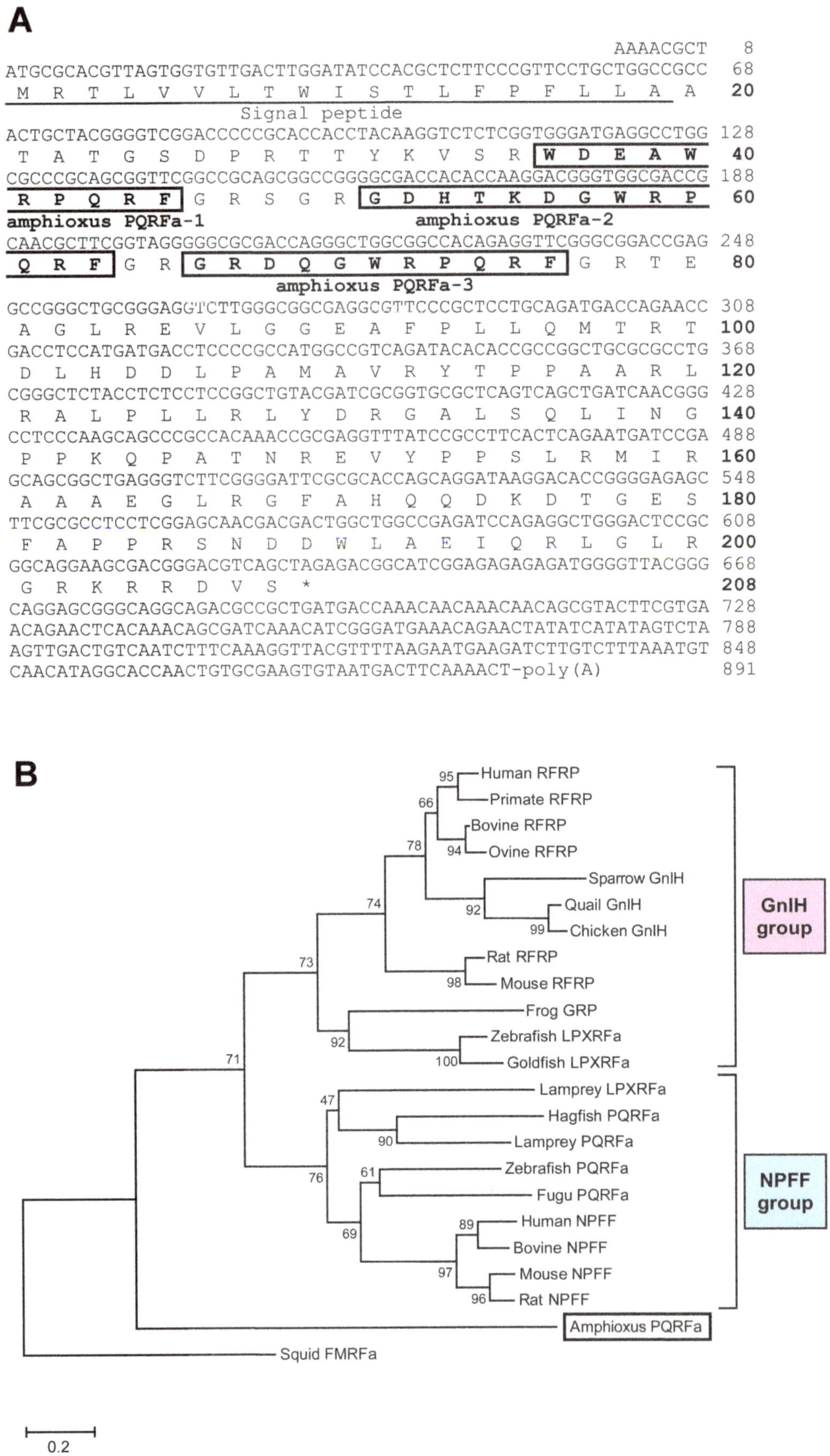
A
AAAACGCT 8
ATGCGCACGTTAGTGGTGTTGACTTGGATATCCACGCTCTTCCCGTTCCTGCTGGCCGCC 68
M R T L V V L T W I S T L F P F L L A A 20
Signal peptide
ACTGCTACGGGGTCGGACCCCCGCACCACCTACAAGGTCTCTCGGTGGGATGAGGCCTGG 128
T A T G S D P R T T Y K V S R W D E A W 40
CGCCCGCAGCGGTTCGGCCGCAGCGGCCGGGGCGACCACACCAAGGACGGGTGGCGACCG 188
R P Q R F G R S G R G D H T K D G W R P 60
amphioxus PQRFa-1
amphioxus PQRFa-2
CAACGCTTCGGTAGGGGGCGCGACCAGGGCTGGCGGCCACAGAGGTTCGGGCGGACCGAG 248
Q R F G R G R D Q G W R P Q R F G R T E 80
amphioxus PQRFa-3
GCCGGGCTGCGGGAGGTCTTGGGCGGCGAGGCGTTCCCGCTCCTGCAGATGACCAGAACC 308
A G L R E V L G G E A F P L L Q M T R T 100
GACCTCCATGATGACCTCCCCGCCATGGCCGTCAGATACACACCGCCGGCTGCGCGCCTG 368
D L H D D L P A M A V R Y T P P A A R L 120
CGGGCTCTACCTCTCCTCCGGCTGTACGATCGCGGTGCGCTCAGTCAGCTGATCAACGGG 428
R A L P L L R L Y D R G A L S Q L I N G 140
CCTCCCAAGCAGCCCGCCACAAACCGCGAGGTTTATCCGCCTTCACTCAGAATGATCCGA 488
P P K Q P A T N R E V Y P P S L R M I R 160
GCAGCGGCTGAGGGTCTTCGGGGATTCGCGCACCAGCAGGATAAGGACACCGGGGAGAGC 548
A A A E G L R G F A H Q Q D K D T G E S 180
TTCGCGCCTCCTCGGAGCAACGACGACTGGCTGGCCGAGATCCAGAGGCTGGGACTCCGC 608
F A P P R S N D D W L A E I Q R L G L R 200
GGCAGGAAGCGACGGGACGTCAGCTAGAGACGGCATCGGAGAGAGAGATGGGGTTACGGG 668
G R K R R D V S * 208
CAGGAGCGGGCAGGCAGACGCCGCTGATGACCAAACAACAAACAACAGCGTACTTCGTGA 728
ACAGAACTCACAAACAGCGATCAAACATCGGGATGAAACAGAACTATATCATATAGTCTA 788
AGTTGACTGTCAATCTTTCAAAGGTTACGTTTTAAGAATGAAGATCTTGTCTTTAAATGT 848
CAACATAGGCACCAACTGTGCGAAGTGTAATGACTTCAAAACT-poly(A) 891
B
95
Human RFRP
Primate RFRP
66
Bovine RFRP
78
94
Ovine RFRP
Sparrow GnIH
74
92
Quail GnIH
99
Chicken GnIH
Rat RFRP
73
98
Mouse RFRP
Frog GRP
92
Zebrafish LPXRFa
100
Goldfish LPXRFa
71
Lamprey LPXRFa
47
Hagfish PQRFa
90
Lamprey PQRFa
76
61
Zebrafish PQRFa
Fugu PQRFa
69
89
Human NPFF
Bovine NPFF
97
Mouse NPFF
96
Rat NPFF
Amphioxus PQRFa
Squid FMRFa
0.2
GnIH group
NPFF group

Figure 1. cDNA sequence encoding amphioxus PQRFa peptides and phylogenetic analysis. (*A*) Nucleotide sequence and deduced amino acid sequence of a cDNA encoding putative amphioxus PQRFa-1, amphioxus PQRFa-2, and amphioxus PQRFa-3. The sequences of these putative amphioxus peptides are shown in bold. The signal peptide (19 aa) is underlined. (*B*) Rooted phylogenetic tree of amphioxus PQRFa peptide precursor, identified or putative vertebrate GnIH precursors and NPFF precursors. Numbers on the branches indicate bootstrap percentage after 1000 replications in constructing the tree. The position of amphioxus PQRFa precursor is boxed. Lamprey LPXRFa precursor was included in the NPFF group due to the low amino acid homology to the precursors of GnIH group. Scale bar refers to a phylogenetic distance of 0.1 nucleotide substitutions per site.

the orthologs of GnIH and NPFF genes were identified from the brains of agnathans (lamprey and hagfish), the most ancient lineage of vertebrates [24,26,27]. Therefore, the origin of GnIH and NPFF genes may date back before the emergence of early vertebrates.

Amphioxus is considered to be the most basal chordates [28]. The recent genome sequencing analysis using *Branchiostoma floridae* revealed that a single *HOX* cluster is present in contrast to the multiple *HOX* clusters in vertebrates [29]. The single *HOX* cluster of amphioxus may reflect the primitive prevertebrate condition and this character suggests that the common ancestor of chordates

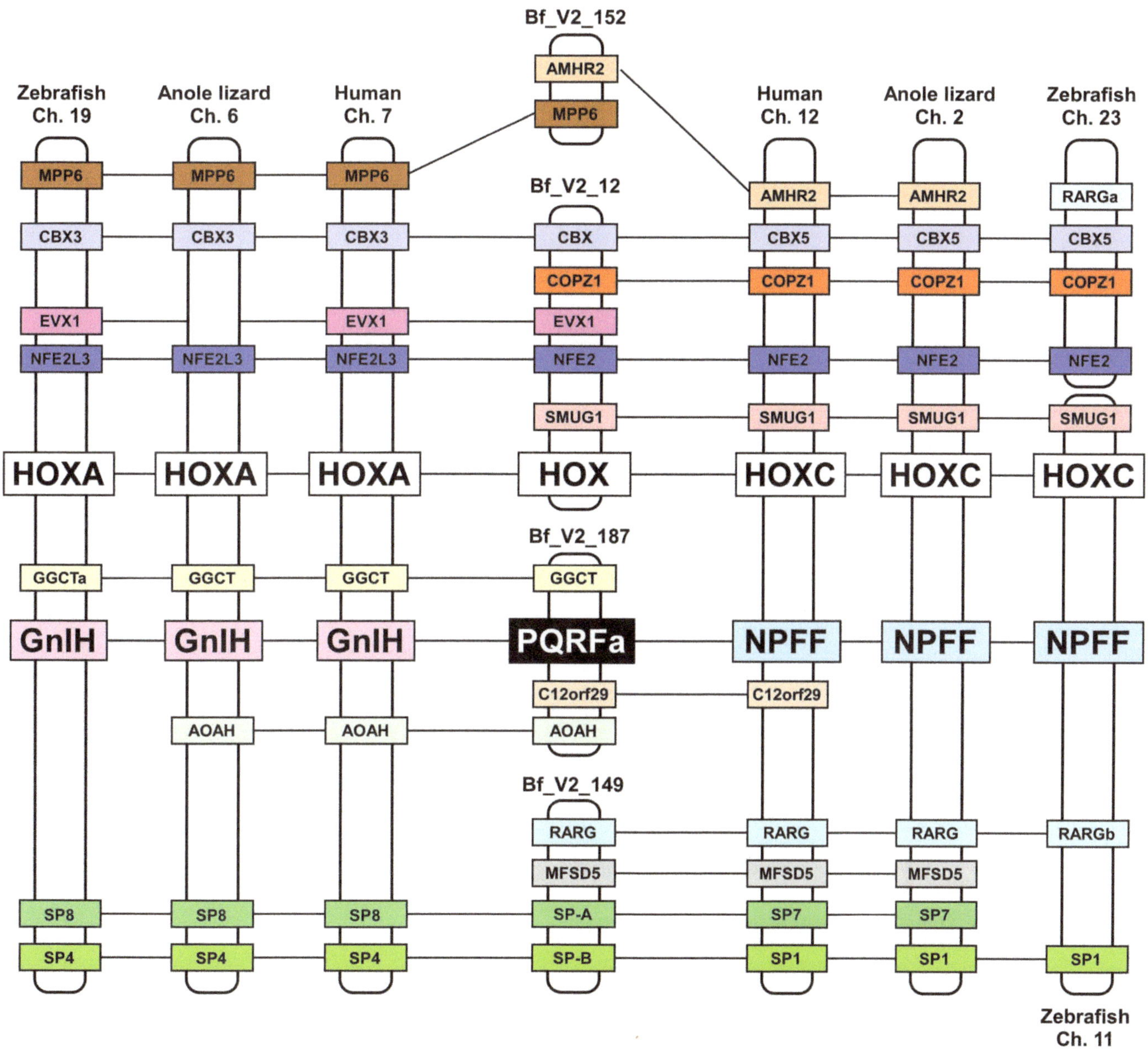

Figure 2. Synteny analysis of the amphioxus PQRFa peptide gene locus. The name of animals and chromosome numbers are shown on the top or bottom of each chromosome or scaffold region. Orthologous genes are linked by horizontal lines. The amphioxus PQRFa peptide gene, vertebrate GnIH genes or vertebrate NPFF genes are shown in black, pink or blue boxes. *HOX* clusters are shown in white boxes. The conserved synteny region exists around the loci of amphioxus PQRFa peptide gene, GnIH gene and NPFF gene.

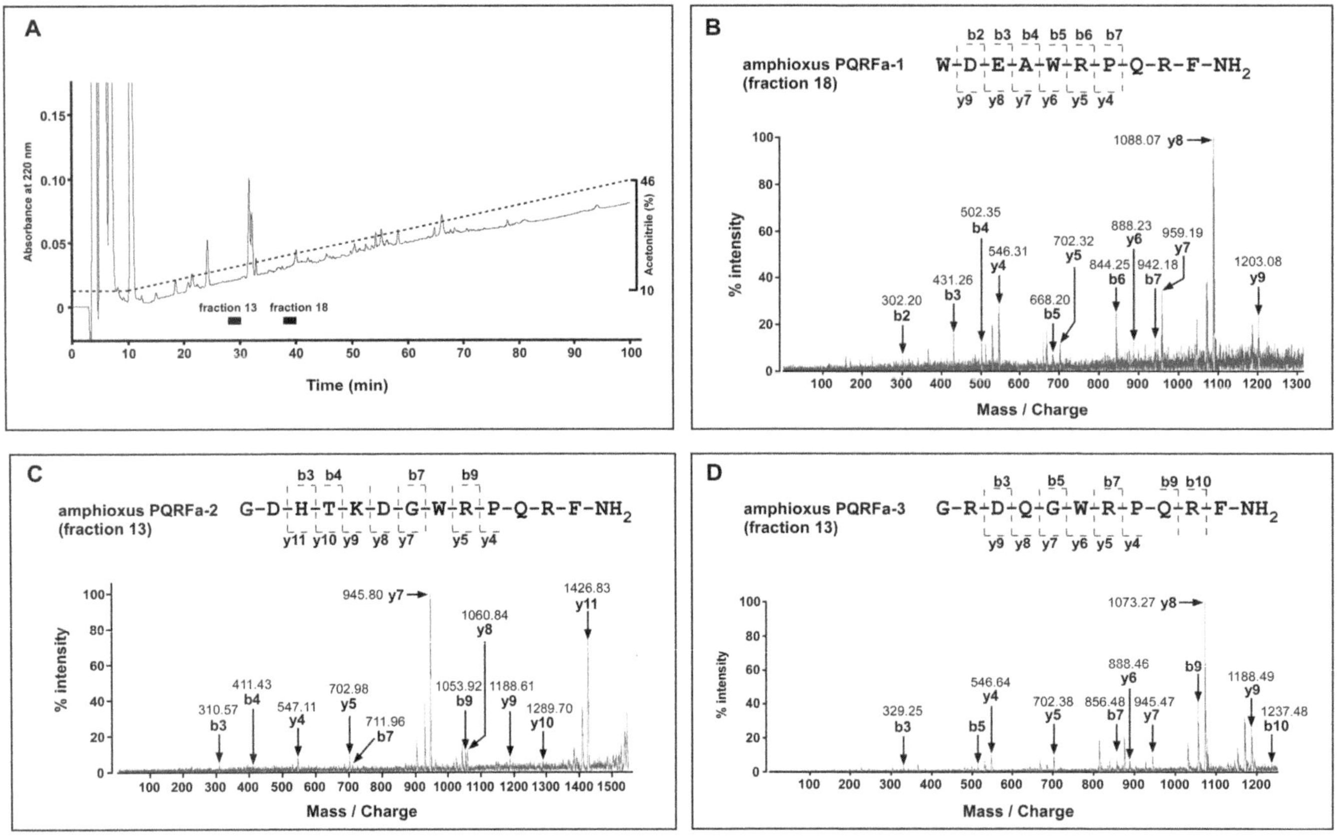

Figure 3. Purification and identification of endogenous amphioxus PQRFa peptides. (*A*) HPLC profile of the retained material obtained by immunoaffinity purification using the antiserum against PQRFa peptide. The immunoreactive substances were eluted in fraction 13 (17–18% acetonitrile) and fraction 18 (21–22% acetonitrile), indicated by the horizontal bars. (*B–D*) Detection of native amphioxus PQRFa peptides by MALDI-TOF MS. Fragmentation patterns of amphioxus PQRFa-1 (*B*), PQRFa-2 (*C*) and PQRFa-3 (*D*) by post-source decay analysis on MALDI-TOF MS. The spectrum shows typical mass values of predicted fragment ions. Observed N-terminal (b) and C-terminal (y) fragmentation ions are assigned in the sequence of each amphioxus PQRFa peptides.

may have not experienced the whole-genome duplication [30,31]. Based on these characters, the amphioxus is considered to be an excellent animal model to investigate the evolutionary origin of GnIH and NPFF genes in chordates. Therefore, in this study we sought to identify novel RFa peptides and their receptors in the amphioxus *Branchiostoma japonicum*.

Materials and Methods

Animals

Adult amphioxus (*Branchiostoma japonicum*) were collected in the southern half of the Ariake Sea (Shimabara Bay) in Kumamoto Prefecture, Japan in June 2011. The detailed method of the collection of amphioxus was described in Henmi and Yamaguchi [32]. The experimental protocols were approved by the committee of Waseda University and Kumamoto University, and performed according to the Guide for the Care and Use of Animals prepared by the Waseda University and Kumamoto University.

Bioinformatics Analyses

The Joint Genome Institute (JGI) genome database of the amphioxus *Branchiostoma floridae* (http://genome.jgi-psf.org/Brafl1/Brafl1.home.html) and Ensembl Genome Browser (http://www.ensembl.org/index.html) was used for the bioinformatics analyses. GENSCAN [33] (http://genes.mit.edu/GENSCAN.html) was used to obtain putative full length cDNA sequences of amphioxus PQRFa peptide gene or its receptor genes. The detail method is described in Materials and Methods S1.

Molecular Cloning

Based on the nucleotide sequences of the scaffold Bf_V2_187 for amphioxus PQRFa peptide precursor gene, scaffold BF_V2_167 for amphioxus PQRFa-R1 and scaffold BF_V2_95 for amphioxus PQRFa-R2 in the genome database of *Branchiostoma floridae*, we performed molecular cloning of the genes using the tissues of *Branchiostoma japonicum*. First-strand cDNA was synthesized with the oligo(dT)-anchor primer supplied in the 5′/3′ rapid amplification of cDNA ends (RACE) kit (Roche Diagnostics, Basel, Switzerland). The sequence of amphioxus PQRFa peptide precursor cDNA and PQRFa receptor cDNAs were determined as described previously [24,26,27]. All primers used in this study are summarized in Table S2. The detail method is described in Materials and Methods S1. The sequence reported in this paper has been deposited in the DDBJ, EMBL, and GenBank database (accession numbers AB863739, AB863740, AB863741 for cDNA sequences of amphioxus PQRFa peptide precursor, amphioxus PQRFa peptide receptor 1and 2).

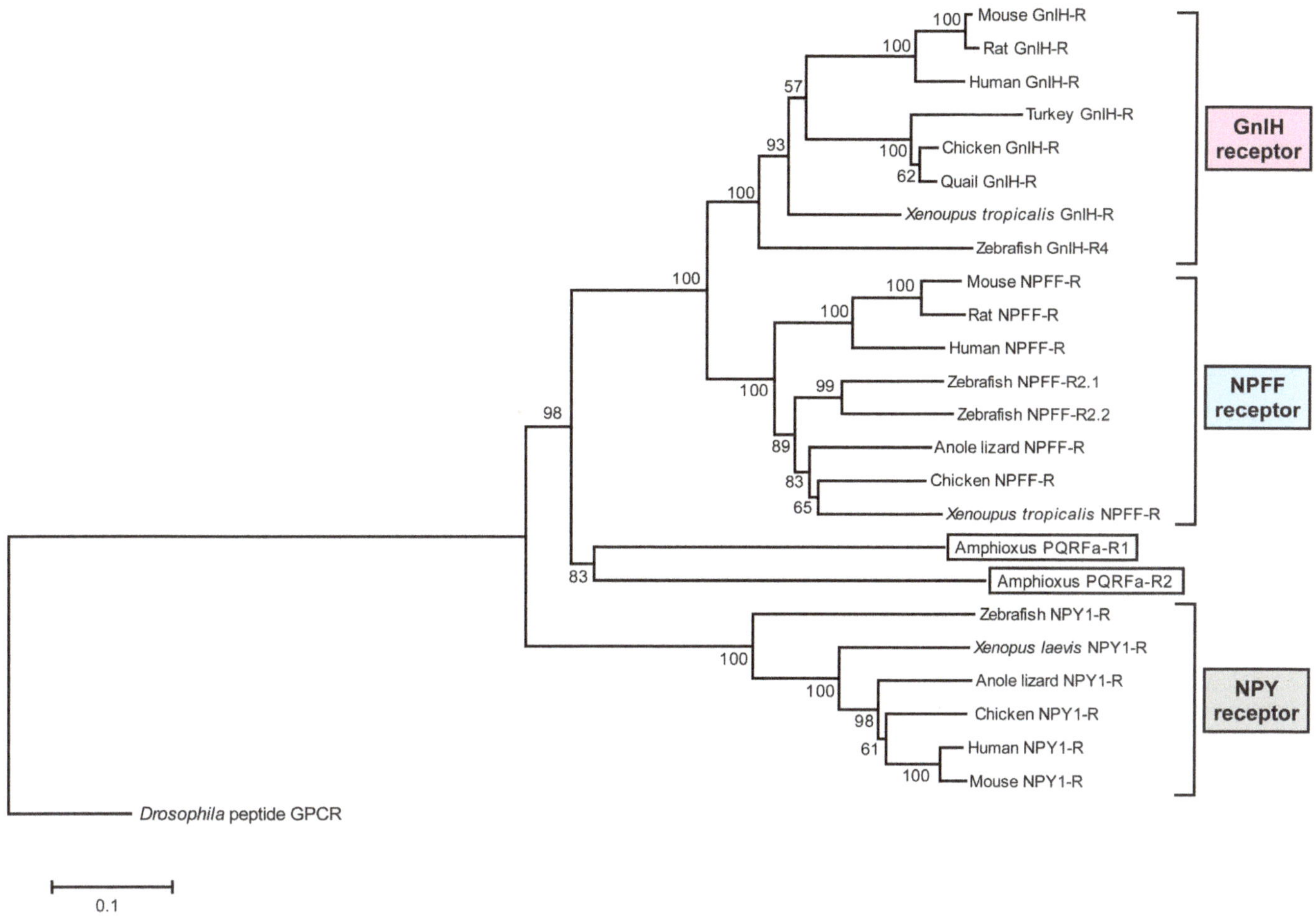

Figure 4. Phylogenetic tree of amphioxus PQRFa peptide receptors, vertebrate GnIH receptors, NPFF receptors and NPY receptors. Scale bar refers to a phylogenetic distance of 0.05 nucleotide substitutions per site. Numbers on the branches indicate bootstrap percentage after 1000 replications in constructing the tree.

Phylogenetic Analysis

Multiple sequence alignments and phylogenetic analyses of the amphioxus PQRFa peptide precursor and vertebrate GnIH and NPFF precursors were performed by using MEGA5 [34]. The phylogenetic tree was constructed by the neighbor-joining method. The data were obtained from 1000 bootstrap replicates to determine the confidence indices within the phylogenetic tree. The bootstrap values were shown as percentage after 1000 replications on the branches. The GenBank accession numbers of the sequences used in the phylogenetic analysis are shown in Tables S3 and S4. The alignments used for the construction of the phylogenetic trees are shown in Figures S1 and S2.

Peptide Extraction and Immunoaffinity Purification

The tissues of 250 adult amphioxus were used for the peptide extraction as described previously [24,26,27]. The affinity chromatography was performed as described previously [26,27]. The antiserum against PQRFa peptide [26] was conjugated to Protein A Sepharose 4B (Amersham Pharmacia Biotech, Uppsala, Sweden) as an affinity ligand. The tissue extract was applied to the column at 4°C, and the adsorbed materials were eluted with 0.3 M acetic acid containing 0.1% 2-mercaptoethanol. An aliquot of each fraction (1 ml) was analyzed by a dot immunoblot assay with the antiserum against PQRFa peptide according to our previous methods [24,26,27]. The detail method is described in Materials and Methods S1.

HPLC and Structure Determination

The immunoreactive materials obtained by immunoaffinity purification using the antiserum against PQRFa peptide [26] were subjected to a HPLC column (ODS-80TM; Tosoh, Tokyo, Japan) with a linear gradient of 10–50% acetonitrile containing 0.1% trifluoroacetic acid for 100 min at a flow rate of 0.5 ml/min, and the eluted fractions were collected every 2 min. All the fractions were analyzed by the MALDI-TOF MS (AXIMA-CFR plus; Shimadzu) to search the molecular mass values of the amphioxus PQRFa peptides that were predicted from the precursor protein sequence. Tandem mass spectra were acquired on the MALDI-TOF MS in post-source decay mode to confirm the amino acid sequences of amphioxus PQRFa peptides. The theoretical mass values of amphioxus PQRFa peptides and their fragment ions were calculated based on the peptide sequences with C-terminal amidation by Protein Prospector version 5.9.0 (http://prospector.ucsf.edu/prospector/mshome.htm).

Transient Transfection and Luciferase Assay

COS-7 cells were maintained in Dulbecco's modified Eagle's medium (DMEM; GIBCO/Invitrogen, Carlsbad, CA) supple-

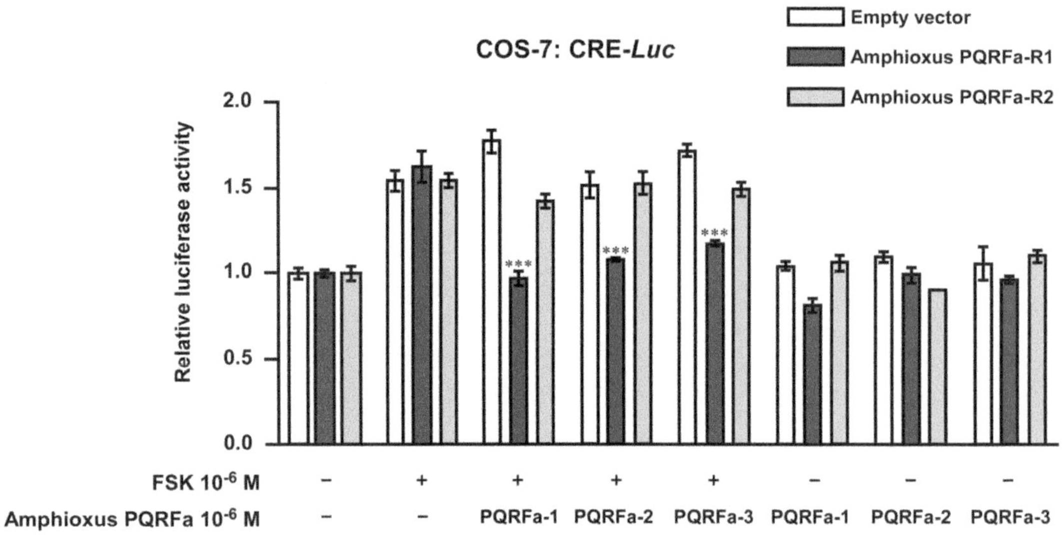

Figure 5. Effect of amphioxus PQRFa peptides on forskolin (FSK)-induced or basal CRE-luciferase activity. COS-7 cells were co-transfected with empty vector (white bars), amphioxus PQRFa-R1-expression vector (dark grey bars), or amphioxus PQRFa-R2- expression vector (light grey bars) along with the CRE-luciferase reporter. Cells were challenged by FSK alone or with amphioxus PQRFa peptides at the concentration of 10^{-6} M for 6 h. The effect of amphioxus PQRFa peptides on basal CRE-luciferase activity was also tested in COS-7 cells transfected with amphioxus PQRFa-R1 or R2.The relative luciferase activity was measured from cell lysates and expressed as fold activation over the respective basal levels. Each column and the vertical line represent the mean ±SEM (n = 5–8). ***, $P<0.001$ *vs.* challenged only FSK in each group, one-way ANOVA followed by Turkey's post-tests.

mented with high glucose (4.5 g/L) containing 10% fetal bovine serum (GIBCO/Invitrogen) and 1% penicillin/streptomycin antibiotics (GIBCO/Invitrogen) in a humidified 5% CO_2 atmosphere at 37°C. For luciferase assay, COS-7 cells were plated in 24-well plates and grown to 70–80% confluence for 24 h before transfection. Cells were then co-transfected with 200 ng of p3XFLAG-CMV-14 expression vector (Sigma-Aldrich Co., St Louis, MO) containing the amphioxus PQRFa-R1, PQRFa-R2, or empty vector, 200 ng of the pGL4.29[luc2P/CRE/Hygro] (firefly luciferase reporter construct; Promega, Madison, WI) and 5 ng of pRL-null (renilla luciferase reporter construct; Promega) using FuGENE HD Transfection Reagent (Promega) according to the manufacturer's instructions. Cells were starved overnight in serum-free DMEM and then challenged by forskolin (FSK) (Santa Cruz Biotechnology, Santa Cruz, CA) and amphioxus PQRFa peptides (PQRFa-1, -2 and -3) for 6 h. Cell extracts were prepared, and luciferase activity was measured using the dual-luciferase reporter assay system (Promega). The ratio of firefly luciferase activity to renilla luciferase activity was used as the results to coordinate the differences in transfection efficiency among samples. All assays were performed in duplicate and repeated three times. Statistical significance was assessed by Prism statistical software (GraphPad Software Inc., La Jolla, CA). The detail method is described in Materials and Methods S1.

Results

Molecular Characterization of Amphioxus PQRFa Peptide Precursor

We cloned an amphioxus PQRFa peptide precursor cDNA in *Branchiostoma japonicum* based on the nucleotide sequences obtained from the genome database of *Branchiostoma floridae* (Figure S3). Figure 1A shows that the deduced amphioxus PQRFa peptide precursor polypeptide encoded three putative amphioxus PQRFa peptides (amphioxus PQRFa-1, -2 and -3) that included PQRF sequence at their C-termini. The amphioxus PQRFa peptide precursor cDNA was composed of 891 nucleotides containing a short 5′ untranslated region (UTR) of 8 bp, a single open reading frame (ORF) of 624 bp, and a 3′ UTR of 259 bp with a poly(A) tail. The ORF region began with a start codon at position 9 and terminated with a TAG stop codon at position 633. We predicted that the amphioxus PQRFa transcript may be translated from Met^1 because a hydropathy plot analysis of the precursor showed that the most hydrophobic moiety, which is typical in a signal peptide region, followed Met^1. The cleavage site of the predicted signal peptide was the Ala^{19}–Ala^{20} bond, which is supported by the -3, -1 rule [35]. In this study, a phylogenetic tree was constructed by using neighbor joining method based on amino acid sequences of the amphioxus PQRFa peptide precursor, vertebrate GnIH precursors and vertebrate NPFF precursors. Squid FMRFa precursor was used as an out group. As shown in Figure 1B, the amphioxus PQRFa peptide precursor located before the split of GnIH group and NPFF group.

Synteny Analysis of Amphioxus PQRFa Peptide Gene Locus

Synteny analysis showed that the paralogous genes, such as *CBX3* and *CBX5* or *SP1* and *SP4*, or *HOXA* and *HOXC* are located in the conserved synteny region of GnIH and NPFF gene loci in vertebrates (Figure 2). Some orthologous genes seen around GnIH gene or NPFF gene loci are located near the amphioxus PQRFa peptide gene locus in the scaffold Bf_V2_187 (Figure 2). Although the amphioxus genome data are incomplete, many orthologous genes including *HOX* cluster are found in the scaffold Bf_V2_12, Bf_V2_149 or Bf_V2_152 (Figure 2). These data suggest that the synteny region was conserved among the loci around amphioxus PQRFa peptide gene, vertebrate GnIH genes and vertebrate NPFF genes.

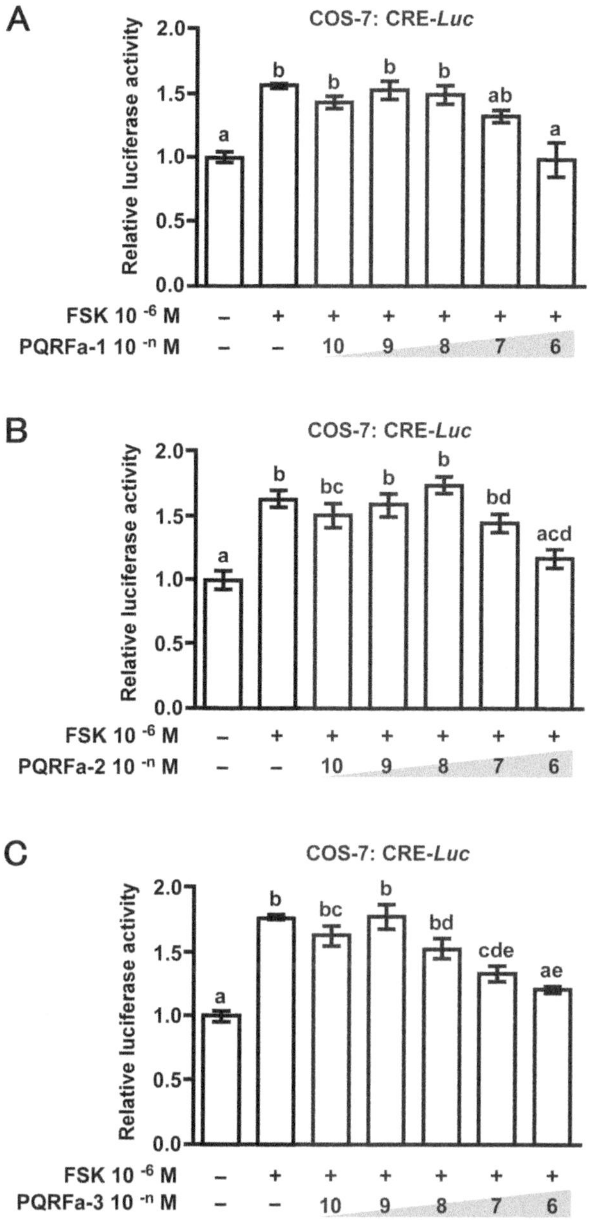

Figure 6. Dose-dependent inhibitory effects of amphioxus PQRFa peptides on forskolin (FSK)-induced CRE-luciferase activity. COS-7 cells were co-transfected with amphioxus PQRFa-R1-expression vector and the CRE-luciferase reporter. Cells were then challenged by FSK and amphioxus PQRFa-1 (*A*), PQRFa-2 (*B*) and PQRFa-3 (*C*) at the concentrations of 10^{-10}–10^{-6} M for 6 h. The relative luciferase activity was measured from cell lysates and expressed as fold activation over the respective basal levels. Each column and the vertical line represent the mean ±SEM (n = 4–6). Letters above the bars indicate significant differences among treatments (bars with the same letter are not significantly different; $P<0.05$; one-way ANOVA followed by Turkey's multiple comparisons test).

Isolation and Characterization of Endogenous Amphioxus PQRFa Peptides

The endogenous amphioxus PQRFa peptides were identified to test the biological activity of these peptides. As shown in Figure 3A, immunopurified material was subjected to reverse-phase HPLC purification with a linear gradient of 10–50% acetonitrile for 100 min. Each fraction was examined by matrix assisted laser desorption/ionization time-of-flight mass spectrometry (MALDI-TOF MS). Molecular ion peaks in the spectrum of the fraction corresponding to the elution times of 28–30 min was observed at 1401.72 m/z ($[M+H]^+$) and 1598.67 m/z ($[M+H]^+$) (Figure S2A). A molecular ion peak in the fraction corresponding to the elution times of 38–40 was observed at 1389.97 m/z ($[M+H]^+$) (Figure S4B). These values were close to the mass numbers of 1401.72 m/z $[(M+H)^+]$ (amphioxus PQRFa-3), 1598.79 m/z $[(M+H)^+]$ (amphioxus PQRFa-2), and 1389.68 m/z $[(M+H)^+]$ (amphioxus PQRFa-1) calculated for the deduced amidated peptide sequence, respectively. By tandem mass spectrometry, the typical fragment ion peaks of each peptide were observed (Figure 3 B–D). The fragment ion peaks of each peptide observed by tandem mass spectrometry were in agreement with the theoretical values. Therefore, the amino acid sequences of each peptide were determined as WDEAWRPQRF-NH_2 (amphioxus PQRFa-1), GDHTKDGWRPQRF-NH_2 (amphioxus PQRFa-2) and GRDQGWRPQRF-NH_2 (amphioxus PQRFa-3). The identified amphioxus PQRFa peptides were compared to vertebrate peptides of the GnIH group and the NPFF group that were previously identified from agnathans to humans (Table S1). The C-terminal six amino acids (WRPQRF) were conserved among amphioxus PQRFa peptides (Table S1). The C-terminal PQRFa motif was identical to several GnIH group peptides and all NPFF group peptides (Table S1).

Molecular Characterization of Amphioxus PQRFa Peptide Receptors

In this study, we cloned two putative amphioxus PQRFa peptide receptors named amphioxus PQRFa-R1 and PQRFa-R2, based on the nucleotide sequences of the scaffold BF_V2_167 and BF_V2_95 in the genome database. Analysis of these receptors for regional hydrophobicity revealed seven putative transmembrane domains (TMs; underlined in Figures S5 and S6). The phylogenetic analysis showed that the amphioxus receptors, amphioxus PQRFa-R1 and PQRFa-R2, are closer to GnIH and NPFF receptors compared to neuropeptide Y (NPY) receptors (Figure 4). NPY receptors are included in the phylogenetic tree because these receptors are evolutionarily close to GnIH and NPFF receptors [36].

Bioactivity of Endogenous Amphioxus PQRFa Peptides

Because GPR147 and GPR74 couple with $G_{\alpha i}$, it was hypothesized that activation of GnIH receptor inhibits adenylate cyclase (AC), thus reduces intracellular cAMP levels and PKA activity (AC/cAMP/PKA pathway) [3,37]. Indeed, mouse GnIH directly inhibited cAMP production induced by GnRH in mouse gonadotrope cell line (LβT2 cells) [38]. Therefore, we examined the effect of amphioxus PQRFa peptides on AC/cAMP/PKA pathway using a cAMP-response element (CRE) reporter system as described previously [38]. Forskolin (FSK), an AC activator, was used to investigate the inhibitory effect of amphioxus PQRFa peptides on AC/cAMP/PKA pathway through amphioxus receptors (Figure 5). As shown in Figure 5, FSK-induced CRE-luciferase activity was significantly inhibited by amphioxus PQRFa-1, PQRFa-2 and PQRFa-3 at the concentration of 10^{-6} M in COS-7 cells transfected with the amphioxus PQRFa-R1 (Figure 5, dark grey bars). In contrast, all amphioxus PQRFa peptides did not inhibit FSK-induced CRE-luciferase activity in COS-7 cells transfected with the amphioxus PQRFa-R2 (Figure 5, light grey bars). These results suggest that amphioxus PQRFa-R1 is the dominant receptor for amphioxus PQRFa peptides. On the other hand, all amphioxus PQRFa peptides did not modify basal

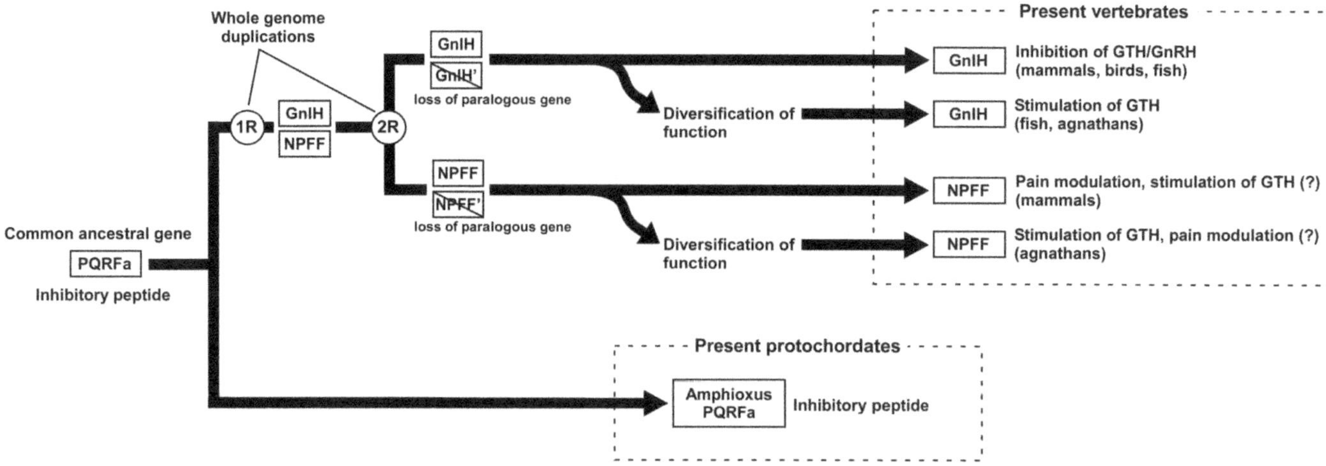

Figure 7. Proposed evolutionary history of GnIH and NPFF genes. GnIH and NPFF genes, originating from a common ancestral PQRFa gene, may have evolved through the whole-genome duplication. The amphioxus PQRFa peptides possess inhibitory action, suggesting that the common ancestor of GnIH and NPFF group peptides was an inhibitory peptide. After the 2 rounds of whole-genome duplications and subsequent loss of paralogous genes, the ancestral gene may have diverged into GnIH gene and NPFF gene. In the present vertebrates, GnIH group peptides inhibit GTH and GnRH in mammals, birds and fish, whereas they stimulate GTH in some fish and agnathans. NPFF group peptides modulate pain transmission in mammals and stimulate GTH in agnathans. It is possible that NPFF group peptides stimulate GTH in mammals and modulate pain transmission in agnathans.

CRE-luciferase activity in COS-7 cells even if they were transfected with the amphioxus PQRFa-R1 (Figure 5).

We then investigated the dose dependent effects of amphioxus PQRFa peptides on the inhibition of FSK-induced CRE-luciferase activity in COS-7 cells transfected with the amphioxus PQRFa-R1 (Figure 6). As shown in Figure 6, PQRFa-1 (A) and PQRFa-2 (B) inhibited FSK-induced CRE-luciferase activity at the concentration of 10^{-6} M. On the other hand, PORFa-3 (C) inhibited FSK-induced CRE-luciferase activity at 10^{-7} M and 10^{-6} M, suggesting that amphioxus PORFa-3 is the dominant ligand for the amphioxus PQRFa-R1.

Discussion

Based on previous studies [23,24], we hypothesized that GnIH gene and NPFF gene may have diverged from a common ancestral gene through the whole-genome duplication during vertebrate evolution. We have demonstrated that GnIH and NPFF genes are present in the agnathans, the most basal vertebrates [24,26,27]. To clarify the evolutionary origin of GnIH gene and NPFF gene, this study therefore investigated their existence in protochordates. We initially searched the genome database of the amphioxus *Branchiostoma floridae*, a most basal chordate, by using blast program to investigate whether amphioxus has similar gene(s) to GnIH gene or NPFF gene. As a result, we found a putative gene encoding three RFa peptides (amphioxus PQRFa peptides) that have the C-terminal PQRFa motif identical to that of vertebrate peptides of the GnIH group and the NPFF group. Subsequently, we cloned a cDNA of amphioxus PQRFa peptide precursor by a combination of 3′/5′ RACE in *Branchiostoma japonicum*. The deduced precursor polypeptide consisted of 208 amino acid residues, including the sequences of three putative amphioxus PQRFa peptides, i.e., amphioxus PQRFa-1, PQRFa-2 and PQRFa-3.

Phylogenetic analysis also showed that amphioxus PQRFa peptide precursor was located before the split of the GnIH group and NPFF group (see Figure 1B), suggesting that amphioxus PQRFa peptide precursor is closely related to the common ancestor of GnIH and NPFF precursors. Synteny analysis suggests that the conserved synteny region exists around the loci of amphioxus PQRFa peptide gene, GnIH gene and NPFF gene although the complete chromosome data are not available in the genome database of *Branchiostoma floridae*. Based on the phylogenetic analysis and synteny analysis, it is suggested that the characteristics of the common ancestral gene of GnIH and NPFF genes are conserved in the amphioxus PQRFa peptide gene. It is interesting that RFa peptides (Sm-npp-13) that possess the C-terminal PQRFa motif were reported by the analysis of EST database of the flatworm *Schistosoma mansoni* that belongs to the phylum Platyhelminthes [39]. The orthologous relationship between the genes of Sm-npp-13, GnIH and NPFF is obscure due to the lack of the synteny data. Further studies are needed to clarify the evolutionary relationship among amphioxus PQRFa peptides, GnIH, NPFF and flatworm PQRFa peptides. As mentioned above, GnIH gene and NPFF gene are located near the *HOXA* and *HOXC* clusters, respectively, suggesting that GnIH and NPFF genes may have duplicated through whole-genome duplications [23,24]. In primates, prolactin-releasing peptide (PrRP), one of the RFamide peptides, gene is located on the chromosome 2 where *HOXD* cluster is located. However, the location of PrRP gene is distant from *HOXD*, and PrRP does not possess a C-terminal PQRFamide motif that is conserved between GnIH and NPFF. Therefore, we consider that PrRP gene is not a paralogous gene of GnIH or NPFF gene. We could not find other RFamide peptide genes on the *HOXB* and *HOXD* loci. Accordingly, we consider that the paralogous genes of GnIH and NPFF were lost during evolution. The evolutionary relationship among all RFa peptides remains to be clarified.

To investigate the biological action of amphioxus PQRFa peptides, it is essential to identify their receptors. By using genome database search, we found putative amphioxus PQRFa peptide receptors (amphioxus PQRFa-R1 and PQRFa-R2) similar to vertebrate GnIH receptors and NPFF receptors (see Figure 4). Phylogenetic analysis showed that amphioxus PQRFa peptide receptors were closer to vertebrate GnIH and NPFF receptors compared to vertebrate NPY receptors. The phylogenetic positions of the putative amphioxus PQRFa peptide receptors suggest that these receptors are evolutionarily related to vertebrate GnIH and NPFF receptors.

It is important to identify endogenous amphioxus PQRFa peptides for the functional analyses of these ligands. The amphioxus PQRFa peptide sequences are flanked on both ends by the typical endoproteolytic sequences, i.e. KVSR, RSGR, or RFGR on the PQRFa peptide precursor polypeptide sequence (see Figure 1A), suggesting that mature peptides may be generated [40]. Therefore, we performed immunoaffinity purification to identify endogenous amphioxus PQRFa peptides. On the basis of mass spectrometry and post-source decay analysis, the primary structures of the isolated peptides were considered as follows: WDEAWRPQRFa (amphioxus PQRFa-1), GDHTKDGWRPQRFa (amphioxus PQRFa-2) and GRDQGWRPQRFa (amphioxus PQRFa-3). This is the first identification of RFa peptides in protochordates. The C-terminal four amino acids of amphioxus PQRFa peptides were identical to those of several GnIH group peptides and to all NPFF peptides in vertebrates, although the N-terminal sequences of these peptides were diverse (see Table S1).

We conducted *in vitro* assays to analyze whether the amphioxus PQRFa peptides have a biological activity. The synthetic amphioxus PQRFa peptides significantly inhibited FSK-induced CRE-luciferase activity in COS-7 cells expressing amphioxus PQRFa-R1. This result suggests that the amphioxus PQRFa-R1 couples with $G_{\alpha i}$ protein and thus amphioxus PQRFa peptides are considered to inhibit AC/cAMP/PKA pathway. Our previous study showed that GnIH directly inhibits cAMP production induced by GnRH in the mouse gonadotrope cell line, LβT2 [38]. Interestingly, NPFF also inhibited FSK-stimulated CRE-luciferase activity [41] or inhibited cAMP levels [42]. The present study suggests that amphioxus PQRFa peptides act as inhibitory peptides on the neuroendocrine system of reproduction and their mode of action is similar to those of vertebrate GnIH and NPFF group peptides. Therefore, the common ancestral peptide of GnIH and NPFF group peptides may have played an inhibitory role in reproduction of the ancestral protochordates. In mammals and birds, numerous studies showed that GnIH acts to inhibit GTH release and synthesis [2,3]. In fish, zebrafish GnIH (LPXRFa-3) inhibits luteinizing hormone (LH) release in goldfish [20]; whereas goldfish GnIHs (LPXRFa peptides) stimulate the expression of GTH mRNAs in sockeye salmon [43] and grass puffer [21]. Lamprey GnIH (LPXRFa peptide-2) also increased the expression of GTHβ mRNA in the pituitary [24]. Our recent studies further showed that the NPFFs (PQRFa peptides) stimulated the expression of GTHβ mRNA in the pituitary of hagfish [27] and increased the concentration of GnRH-II in the brain of lamprey [44]. Based on the present and previous studies, we hypothesized the evolutionary history of GnIH gene and NPFF gene (Figure 7). The inhibitory action of amphioxus PQRFa peptides suggests that the common ancestor of GnIH and NPFF might have been an inhibitory peptide (Figure 7). After the two rounds of whole-genome duplications and consequent loss of paralogous genes, GnIH and NPFF genes might have diverged from the common ancestral gene. Mammals, birds and some fish have retained the inhibitory role of GnIH in reproduction or NPFF in pain transmission, while some fish and lampreys have diversified the roles of GnIH and NPFF group peptides (Figure 7). The biological function of amphioxus PQRFa peptides is still unclear because GTHs have not been identified in amphioxus. The recent studies identified the genes of glycoprotein hormones (GPA2 and GPB5) in amphioxus that may be the ancestral form of vertebrate glycoprotein hormones including GTHs [45–47]. However, the function of these glycoprotein hormones is still unclear. Further studies on the identification of functional GTHs in amphioxus are needed to clarify whether amphioxus PQRFa peptides act as inhibitory peptides on GTH secretion in amphioxus as in mammals, birds and fish.

In conclusion, we identified novel RFa peptides in amphioxus as the first protochordate RFa peptides that have retained the characteristics of the common ancestor of vertebrate GnIH and NPFF group peptides, suggesting that GnIH and NPFF genes have originated from the ancestral PQRFa peptide gene of protochordates. The phylogenetic, synteny and functional analyses indicated that GnIH and NPFF genes have diverged from a common ancestral gene in the ancestral protochordates through whole-genome duplication event during vertebrate evolution (see Figure 7). The role of the common ancestral peptide of reproduction of GnIH and NPFF may have been an inhibitory peptide and the role has retained in mammals, birds and some fish during the course of vertebrate evolution.

Supporting Information

Figure S1 Amino acid alignments of the precursors of amphioxus PQRFa peptide, GnIH and NPFF. Amino acids identical to top sequence are indicated by dots. Gaps marked by hyphens were inserted to optimize homology.

Figure S2 Amino acid alignments of the receptors of amphioxus PQRFa peptide, GnIH and NPFF. Amino acids identical to top sequence are indicated by dots. Gaps marked by hyphens were inserted to optimize homology.

Figure S3 Nucleotide sequence and deduced amino acid sequence of a cDNA encoding amphioxus PQRFa peptides obtained from the genome database of the amphioxus *Branchiostoma floridae*. The signal peptide is underlined. The putative amphioxus PQRFa peptides are boxed.

Figure S4 Chromatogram of MALDI-TOF MS of native amphioxus PQRFa peptides. (*A*) Molecular ion peaks of 1401.72 m/z ($[M+H]^+$) (amphioxus PQRFa-3) and 1598.67 m/z ($[M+H]^+$) (amphioxus PQRFa-2) were observed in the fraction 13. (*B*) A molecular ion peak of 1389.97 m/z ($[M+H]^+$) (amphioxus PQRFa-1) was observed in the fraction 18.

Figure S5 Nucleotide sequence and deduced amino acid sequence of a cDNA encoding putative amphioxus PQRFa-R1 of *Branchiostoma japonicum*. The putative seven transmembrane domains are indicated by underline and TM.

Figure S6 Nucleotide sequence and deduced amino acid sequence of a cDNA encoding putative amphioxus PQRFa-R2 of *Branchiostoma japonicum*. The putative seven transmembrane domains are indicated by underline and TM.

Table S1 Comparison of the identified amphioxus PQRFa peptides with previously identified GnIH and NPFF vertebrates.

Table S2 Oligonucleotide sequences of primers used for cDNA cloning.

Table S3 Genbank accession numbers of the GnIH genes and NPFF genes used for the phylogenetic analysis.

Table S4 GenBank accession numbers of the GnIH receptor genes, NPFF receptor genes and NPY receptor 1 genes used for the phylogenetic analysis.

Materials and Methods S1 The detailed method for bioinformatics analysis, molecular cloning, DNA sequencing, peptide extraction and immunoaffinity purification, and transient transfection and luciferase assay are provided.

Author Contributions

Conceived and designed the experiments: T. Osugi YLS TU YH KT. Performed the experiments: T. Osugi T. Okamura YLS MO. Analyzed the data: T. Osugi T. Okamura YLS MO TU KT. Contributed reagents/materials/analysis tools: T. Osugi YLS MO YH KT. Wrote the paper: T. Osugi YLS TU KT.

References

1. Tsutsui K, Saigoh E, Ukena K, Teranishi H, Fujisawa Y, et al. (2000) A novel avian hypothalamic peptide inhibiting gonadotropin release. Biochem Biophys Res Commun 275: 661–667.
2. Tsutsui K (2009) Review: A new key neurohormone controlling reproduction, gonadotropin-inhibitory hormone (GnIH): Biosynthesis, mode of action and functional significance. Prog Neurobiol 88: 76–88.
3. Tsutsui K, Bentley GE, Bedecarrats G, Osugi T, Ubuka T, et al. (2010) Review: Gonadotropin-inhibitory hormone (GnIH) and its control of central and peripheral reproductive function. Front Neuroendocrinol 31: 284–295.
4. Ukena T, Tsutsui K (2005) Review: A new member of the hypothalamic RF-amide peptide family, LPXRF-amide peptides: Structure, localization, and function. Mass Spectrom Rev 24: 469–486.
5. Ubuka T, Haraguchi S, Tobari Y, Narihiro M, Ishikawa K, et al. (2014) Hypothalamic inhibition of socio-sexual behaviour by increasing neuroestrogen synthesis. Nat Commun 5: 3061.
6. Ukena K, Iwakoshi E, Minakata H, Tsutsui K (2002) A novel rat hypothalamic RFamide-related peptide identified by immunoaffinity chromatography and mass spectrometry. FEBS Lett 512: 255–258.
7. Ubuka T, Lai H, Kitani M, Suzuuchi A, Pham V, et al. (2009) Gonadotropin-inhibitory hormone identification, cDNA cloning, and distribution in rhesus macaque brain. J Comp Neurol 517: 841–855.
8. Ubuka T, Morgan K, Pawson AJ, Osugi T, Chowdhury VS, et al. (2009) Identification of human gonadotropin-inhibitory hormone homologs, RFRP-1 and RFRP-3, and the cognate receptor, GPR147 in the human hypothalamic pituitary axis. PLoS ONE 4: e8400.
9. Ubuka T, Inoue K, Fukuda Y, Mizuno T, Ukena K, et al. (2012) Identification, expression, and physiological functions of Siberian hamster gonadotropin-inhibitory hormone. Endocrinology 153: 373–385.
10. Koda A, Ukena K, Teranishi H, Ohta S, Yamamoto K, et al. (2002) A novel amphibian hypothalamic neuropeptide: isolation, localization, and biological activity. Endocrinology 143: 411–419.
11. Ukena K, Koda A, Yamamoto K, Kobayashi T, Iwakoshi-Ukena E, et al. (2003) Novel neuropeptides related to frog growth hormone-releasing peptide: isolation, sequence, and functional analysis. Endocrinology 144: 3879–3884.
12. Chartrel N, Dujardin C, Leprince J, Desrues L, Tonon MC, et al. (2002) Isolation, characterization, and distribution of a novel neuropeptide, *Rana* RFamide (R-RFa), in the brain of the European green frog *Rana esculenta*. J Comp Neurol 448: 111–127.
13. Chowdhury VS, Ubuka T, Osugi T, Shimura T, Tsutsui K (2011) Identification, localization and expression of LPXRFamide peptides, and melatonin-dependent induction of their precursor mRNA in the newt brain. J Endocrinol 209: 211–220.
14. Sawada K, Ukena K, Satake H, Iwakoshi E, Minakata H, et al. (2002) Novel fish hypothalamic neuropeptide. Cloning of a cDNA encoding the precursor polypeptide and identification and localization of the mature peptide. Eur J Biochem 269: 6000–6008.
15. Satake H, Hisada M, Kawada T, Minakata H, Ukena K, et al. (2001) Characterization of a cDNA encoding a novel avian hypothalamic neuropeptide exerting an inhibitory effect on gonadotropin release. Biochem J 354: 79–385.
16. Osugi T, Ukena K, Bentley GE, O'Brien S, Moore IT, et al. (2004) Gonadotropin-inhibitory hormone in Gambel's white-crowned sparrows: cDNA identification, transcript localization and functional effects in laboratory and field experiments. J Endocrinol 182: 33–42.
17. Tobari Y, Iijima N, Tsunekawa K, Osugi T, Okanoya K, et al. (2010) Identification of gonadotropin-inhibitory hormone in the zebra finch (*Taeniopygia guttata*): Peptide isolation, cDNA cloning and brain distribution. Peptides 31: 816–826.
18. Ubuka T, Kim S, Huang YC, Reid J, Jiang J, et al. (2008) Gonadotropin-inhibitory hormone neurons interact directly with gonadotropin-releasing hormone-I and -II neurons in European starling brain. Endocrinology 149: 268–278.
19. Sawada K, Ukena K, Kikuyama S, Tsutsui K (2002) Identification of a cDNA encoding a novel amphibian growth hormone-releasing peptide and localization of its transcript. J Endocrinol 174: 395–402.
20. Zhang Y, Li S, Liu Y, Lu D, Chen H, et al. (2010) Structural diversity of the GnIH/GnIH receptor system in teleost: its involvement in early development and the negative control of LH release. Peptides 31: 1034–1043.
21. Shahjahan M, Ikegami T, Osugi T, Ukena K, Doi H, et al. (2011) Synchronised expressions of LPXRFamide peptide and its receptor genes: seasonal, diurnal and circadian changes during spawning period in grass puffer. J Neuroendocrinol 23: 39–51.
22. Yang HY, Iadarola MJ (2006) Modulatory roles of the NPFF system in pain mechanisms at the spinal level. Peptides 27: 943–952.
23. Ikemoto T, Park MK (2005) Chicken RFamide-related peptide (GnIH) and two distinct receptor subtypes: identification, molecular characterization, and evolutionary considerations. J Reprod Dev 51: 359–377.
24. Osugi T, Daukss D, Gazda K, Ubuka T, Kosugi T, et al. (2012) Evolutionary origin of the structure and function of gonadotropin-inhibitory hormone: insights from lampreys. Endocrinology 153: 2362–2374.
25. Venkatesh B, Kirkness EF, Loh YH, Halpern AL, Lee AP, et al. (2007) Survey sequencing and comparative analysis of the elephant shark (*Callorhinchus milii*) genome. PLoS Biol 5: e101.
26. Osugi T, Ukena K, Sower SA, Kawauchi H, Tsutsui K (2006) Evolutionary origin and divergence of PQRFamide peptides and LPXRFamide peptides in the RFamide peptide family: Insights from novel lamprey RFamide peptides. FEBS J 273: 1731–1743.
27. Osugi T, Uchida K, Nozaki M, Tsutsui K (2011) Characterization of novel RFamide peptides in the central nervous system of the brown hagfish: Isolation, localization, and functional analysis. Endocrinology 152: 4252–4264.
28. Delsuc F, Brinkmann H, Chourrout D, Philippe H (2006) Tunicates and not cephalochordates are the closest living relatives of vertebrates. Nature 439: 965–968.
29. Holland LZ, Albalat R, Azumi K, Benito-Gutiérrez E, Blow MJ, et al. (2008) The amphioxus genome illuminates vertebrate origins and cephalochordate biology. Genome Res 18: 1100–1111.
30. Kuraku S, Meyer A (2006) The evolution and maintenance of Hox gene clusters in vertebrates and the teleost-specific genome duplication. Int J Dev Biol 53: 765–773.
31. Lemons D, McGinnis W (2006) Genomic evolution of Hox gene clusters. Science 313: 1918–1922.
32. Henmi Y, Yamaguchi T (2003) Biology of the amphioxus, Branchiostoma belcheri in the Ariake Sea, Japan I. Population structure and growth. Zool Sci 20: 897–906.
33. Burge C, Karlin S (1997) Prediction of complete gene structures in human genomic DNA. J Mol Biol 268: 78–94.
34. Tamura K, Peterson D, Peterson N, Stecher G, Nei M, et al. (2011) MEGA5: molecular evolutionary genetics analysis using maximum likelihood, evolutionary distance, and maximum parsimony methods. Mol Biol Evol 28: 2731–2739.
35. von Heijne G (1986) A new method for predicting signal sequence cleavage sites. Nucleic Acids Res 14: 4683–4690.
36. Larsson TA, Olsson F, Sundstrom G, Lundin LG, Brenner S, et al. (2008) Early vertebrate chromosome duplications and the evolution of the neuropeptide Y receptor gene regions. BMC Evol Biol 8: 184.
37. Bentley GE, Tsutsui K, Kriegsfeld LJ (2010) Recent studies of gonadotropin-inhibitory hormone (GnIH) in the mammalian hypothalamus, pituitary and gonads. Brain Res 1364: 62–71.
38. Son YL, Ubuka T, Millar RP, Kanasaki H, Tsutsui K (2012) Gonadotropin-inhibitory hormone inhibits GnRH-induced gonadotropin subunit gene transcriptions by inhibiting AC/cAMP/PKA-dependent ERK pathway in LβT2 cells. Endocrinology 153: 2332–2343.
39. McVeigh P, Mair GR, Atkinson L, Ladurner P, Zamanian M, et al. (2009) Discovery of multiple neuropeptide families in the phylum Platyhelminthes. Int J Parasitol 39: 1243–1452.
40. Seidah NG, Chrétien M (1999) Proprotein and prohormone convertases: a family of subtilases generating diverse bioactive polypeptides. Brain Res 848: 45–62.
41. Elshourbagy NA, Ames RS, Fitzgerald LR, Foley JJ, Chambers JK, et al. (2000) Receptor for the pain modulatory neuropeptides FF and AF is an orphan G protein-coupled receptor. J Biol Chem 275: 25965–25971.

42. Ankö ML, Panula P (2006) Regulation of endogenous human NPFF2 receptor by neuropeptide FF in SK-N-MC neuroblastoma cell line. J Neurochem 96: 573–584.
43. Amano M, Moriyama S, Iigo M, Kitamura S, Amiya N, et al. (2006) Novel fish hypothalamic neuropeptides stimulate the release of gonadotrophins and growth hormone from the pituitary of sockeye salmon. J Endocrinol 188: 417–423.
44. Daukss D, Gazda K, Kosugi T, Osugi T, Tsutsui K, et al. (2012) Effects of lamprey PQRFamide peptides on brain gonadotropin-releasing hormone concentrations and pituitary gonadotropin-β mRNA expression. Gen Comp Endocrinol 177: 215–219.
45. Dos Santos S, Bardet C, Bertrand S, Escriva H, Habert D, et al. (2009) Distinct expression patterns of glycoprotein hormone-α2 and -β5 in a basal chordate suggest independent developmental functions. Endocrinology 150: 3815–3822.
46. Tando Y, Kubokawa K (2009) A homolog of the vertebrate thyrostimulin glycoprotein hormone alpha subunit (GPA2) is expressed in amphioxus neurons. Zool Sci 26: 409–414.
47. Tando Y, Kubokawa K (2009) Expression of the gene for ancestral glycoprotein hormone beta subunit in the nerve cord of amphioxus. Gen Comp Endocrinol 162: 329–339.

9

Inferring Gene Family Histories in Yeast Identifies Lineage Specific Expansions

Ryan M. Ames[1]*, Daniel Money[2], Simon C. Lovell[1]

1 Computational and Evolutionary Biology, Faculty of Life Sciences, The University of Manchester, Manchester, United Kingdom, **2** Department of Ecology and Evolutionary Biology, University of Kansas, Lawrence, Kansas, United States of America

Abstract

The complement of genes found in the genome is a balance between gene gain and gene loss. Knowledge of the specific genes that are gained and lost over evolutionary time allows an understanding of the evolution of biological functions. Here we use new evolutionary models to infer gene family histories across complete yeast genomes; these models allow us to estimate the relative genome-wide rates of gene birth, death, innovation and extinction (loss of an entire family) for the first time. We show that the rates of gene family evolution vary both between gene families and between species. We are also able to identify those families that have experienced rapid lineage specific expansion/contraction and show that these families are enriched for specific functions. Moreover, we find that families with specific functions are repeatedly expanded in multiple species, suggesting the presence of common adaptations and that these family expansions/contractions are not random. Additionally, we identify potential specialisations, unique to specific species, in the functions of lineage specific expanded families. These results suggest that an important mechanism in the evolution of genome content is the presence of lineage-specific gene family changes.

Editor: Maria Anisimova, Swiss Federal Institute of Technology (ETH Zurich), Switzerland

Funding: This work is funded by BBSRC grant BB/I020489/1. The funders had no role in study design, data collection and analysis, decision to publish, or preparation of the manuscript.

Competing Interests: The authors have declared that no competing interests exist.

* E-mail: ryan.ames@manchester.ac.uk

Introduction

The creation of new genetic content in the form of new genes is a key component of genome evolution. New genes can arise through a variety of mechanisms including gene duplication, retroposition, horizontal gene transfer and *de novo* origination [1]; however, gene duplication has been recognised, since the 1970s, as the most prevalent source of new genes [2]. Indeed, in a comparison of the relative contributions of these gene creation mechanisms, gene duplication was shown to have produced ~80% of new genes in several *Drosophila* species [3].

Rates of gene gain by duplication have been shown to be high in a variety of species, and it has been suggested that in eukaryotes 50% of genes are expected to duplicate at least once every 35–350 million years [4]. In *Drosophila* the rate of gene gain has been estimated to be in the range of 5 to 11 genes every million years [3] and as high as 17 genes every million years [5]. However, current genome content is a balance between the rate of gene gain and the rate at which genes are lost. Indeed, the most common fate of duplicate genes is expected to be nonfunctionalisation [4], which may be followed by removal from the genome. That the size of the genome appears to be constant over time is probably due to the high rate of gene gain coupled with a high rate of gene loss. It is therefore important to consider the rates of both gene gain and gene loss independently in order to accurately understand genome evolution.

Reconstructing histories of gene families is currently an active area of research, and there have been several methods developed, along with some genome-wide studies of the evolution of gene families [6–11]. Inferences of gain and loss events along the human lineage of the mammalian phylogeny suggests that, between humans and chimpanzees, the complement of genes differs more than the sequences of orthologous nucleotides [12], leading to the argument that a "revolving door" of gain and loss leads to large differences between the genomes of humans and chimpanzees. Likewise, in comparisons of gene families in several species of *Drosophila* [5] large numbers of genes are both gained and lost with over 40% of families varying in size. Genome-wide studies of gene family evolution have also been conducted in species sets containing yeast [13–15]. In all species analysed a large turnover of genes is seen, and specific functions are associated with these changes.

Many different methods have been developed to study gene family evolution. Novel reconstruction algorithms [16,17] have been used to infer the histories of gene families. These algorithms a incorporate gain, loss and horizontal gene transfer and utilise phyletic profiles and a species tree. Parsimony [13] and weighted parsimony [9] have both been used to infer the evolution of gene families. Although these methods are quick and can be applied to genome-wide studies, they cannot account for multiple events (gain and subsequent loss) on a single branch.

As an alternative to parsimony methods that make use of phyletic profiles and a single species tree there are many methods that make use of gene trees produced from protein families. These methods aim to reconcile the gene trees with the single species tree and in the process of reconciliation infer the gain and loss events. Several software tools have been developed for gene/species tree reconciliation [18–22]. As with the parsimony methods these tools are quick, can be applied to whole genomes and have been shown to be accurate, although they may be affected by bias in some cases [23,24]. More recently, a range of reconciliation methods [6,8,15,25,26] have used probabilistic models or Bayesian methods to infer the gene trees and for gene/species tree reconciliation. These methods have shown to be both accurate and applicable to whole genome analyses.

The problem of inferring gene family histories has also been addressed with the development of likelihood-based methods that make use of probabilistic models [7,9,10,12,14]. These methods do not rely on individual gene trees but instead use a probabilistic model to infer the evolutionary histories of gene families based on a species phylogeny and phyletic profiles for each family. These methods have been shown to be accurate, applicable to genome-wide analyses and can infer multiple events on a single branch. As these methods are dependent on an underlying model, the biological realism of these models is important for their accuracy, and so, much research is focused on producing more biologically relevant models.

We have previously produced a method with several models of gene gain and loss, allowing for the inference of gene family histories, on a whole genome level, with variable duplication branch lengths and rates of evolution among families [9]. However, the model used in our previous study did not capture all the biological complexity of gene family evolution. We have considered innovation events such as *de novo* gene gain and horizontal gene transfer that may be much more widespread than originally thought, particularly in yeast [27]. However, ourselves, and many others have not considered the complete removal of a gene family by pseudogenisation. To gain a more complete understanding of these processes more biologically accurate models are needed.

Here, we apply a new evolutionary model, implemented in DupliPHY-ML [9], to infer the evolutionary histories of gene families from 9 yeast species, using data from the Génolevures project [28]. Our model, BDIE, allows the estimation of the relative rates of gene birth (new gene gain in existing families), death (gene loss by pseudogenisation), innovation (*de novo* gene gain or horizontal gene transfer creating a new family) and extinction (loss of a complete gene family) in yeast gene families on a whole genome scale. We are able to identify families that have experienced lineage specific expansions in the yeast species and link these to specific functions. We are also able to identify functions that appear to be repeatedly associated with expanding families in multiple lineages indicating that these functions may provide common adaptations in these yeast. In addition, we can also identify functions linked with expanding families in a single lineage suggesting that lineage specific gene family evolution plays a key role in specialisation.

Results

Rates of gene family evolution

We used gene family data from the Génolevures project [28]. Here, the families represent phylogenetic groups of genes related by homology and identified by the similarity between protein-coding genes from all the species. There are a total of 4,578 families (after removal of some families - see methods) with an average size of 9.72 genes and a largest family size, in any single taxa, of 54 members. We also looked at the average number of members in each taxa per family (Figure 1), which shows that in most families the each species contains <5 members. Lists of family membership can be found in File S1.

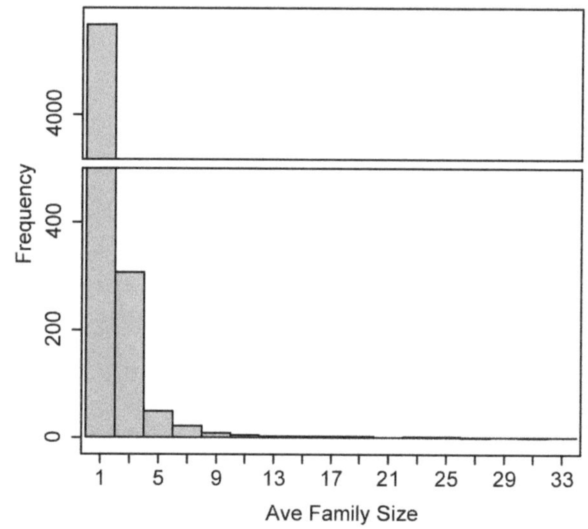

Figure 1. The distribution of average family sizes in the Génolevures data.

A key feature of DupliPHY-ML, is the ability to estimate the rates of gene loss (*d*), gene family extinction (*e*) and the innovation (*i*) of new gene families relative to the rate of gene gain (*b*). In this analysis the rate of gene gain (birth) refers to gene duplication in an existing gene family and gene loss (death) refers to the loss of a gene in a family with 2 or more members. Thus, extinction refers to the loss of a final member of a gene family and innovation refers to the gain of a new gene family by processes such as horizontal gene transfer or *de novo* gain. In our data set we find that $b=1.0$, $d=4.05$, $e=0.59$ and $i=0.05$. The estimates indicate that the rate of gene loss through pseudogenisation is much higher than the rate of gene duplication. This is a surprising result; although it is possible that genomes in at least some of these species are shrinking, it is perhaps more likely to indicate some artefacts from the data generation. We also find that the rate of extinction of a gene family is much lower than the rate of gene loss. The rate of new family formation is extremely low suggesting that in yeast, new gene gain by horizontal gene transfer or *de novo* gain is rare.

Using DupliPHY-ML to infer the ancestral history of gene families, estimates of ancestral family sizes are shown in File S2. DupliPHY-ML also allows us to infer how fast or slow individual families are evolving in terms of gains and losses over the phylogeny. For example, families with a fast evolutionary rate will have had many duplication and loss events in their inferred history. We see that the majority of families are evolving very slowly with only a minority of families displaying rapid evolution (Figure 2).

Data quality may affect the estimation of these rates. It is apparent that both the rates of gene loss and extinction are relatively high compared to the rate of gene gain. If some of the genome sequences for these species are not complete then this missing data may be construed as loss by DupliPHY-ML, which would act to erroneously increase the relative rates of loss and extinction. Indeed, the majority of gene families are very small (Figure 1) and regions with zero sequence coverage in a subset of species may erroneously appear to be extinction events. The effect of such missing sequences would lead to an increase in the extinction rate. However, we note that the rate of gene loss is much higher than that of extinction suggesting that even with

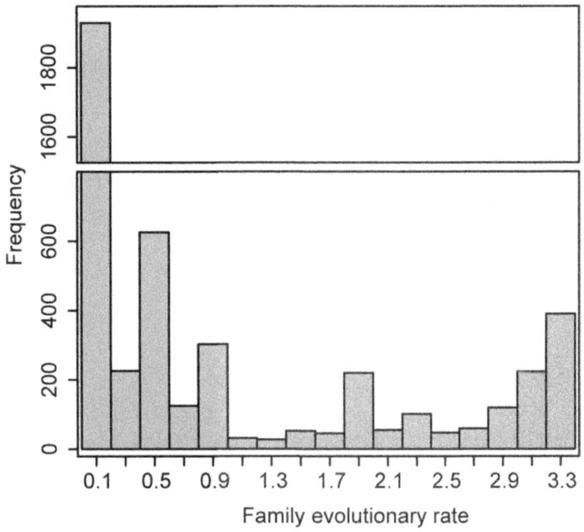

Figure 2. The rate of family evolution in the Génolevures data estimated by DupliPHY-ML as the mean posterior rate with gamma-distributed rates across families. Here a high estimated rate indicates a family that is evolving quickly and has undergone many duplication and loss events in its history. The average rate is 1.

these potential effect of missing data, extinction still seems rare compared to gene family contraction.

Validating the estimated rates of gene family evolution

As the estimated rates of gain and loss may be affected by both the model and the data used we attempted to validate our results. We see that on the Génolevures data our BDIE with gamma-distributed rates across families (+G) we estimate a very high rate of gene loss. In order to check that this is not a consequence of our model choice we repeated the analysis using our birth-death-innovation model [9]. The estimated rates for all the models can be found in Table S1. When we compare BDIE+G and BDI+G we see that the estimates of extinction are similar at 0.05 and 0.08 for the BDIE+G and BDI+G models respectively. The rate of death for the BDIE+G model is higher than that for the BDI+G model. This is likely due to the addition of the extinction parameter. As we think that extinction is a relatively rare event (only changes in size from 1 to 0 cause innovation and all other family size decreases are explained by death), and these rates are averaged over all families, by including extinction we remove rare events from this average, and as a consequence, the death rate increases.

We see similar patterns when we compare the the BDIE and BDI models. As with the gamma models the estimated rate of death is higher in the BDIE model for the reasons discussed above. The estimated rates of innovation are similar, 0.48 and 0.70 for BDIE and BDI respectively. Interestingly, in the non-gamma models there is a large increase in the estimates of the rate of innovation. We have seen that under the gamma model small families and families where one or more species has zero members tend to evolve faster. Our data also consists of many small families. Since the innovation parameter in the BDIE model describes going from 0 to 1 gene in a family, we expect that the innovation parameter will have the greatest effect on smaller families. This suggests that small families may need to be handled differently, which is not possible with the BDIE and BDI models. Further study of this result is warranted as it is suggestive of current models being misspecified.

In order to to test the effect of clustering parameters on the estimates of the rates of birth and death we generated 4 new gene families sets using Tribe-MCL [29] and a range of values to affect cluster granularity. The estimated rates for each clustering can be found in Table S2. As we increased the cluster granularity we identified more gene families and the average size of the gene families decreased. We find that with decreasing family size we see an increase in the estimated rate of death. We speculate that this may be due to the fact that as the death rate increases relative to the birth rate the stationary distribution of family size becomes more skewed towards smaller families and so better fits the distribution we see with larger granularity.

In order to quantify the effect of data quality on our results we generated a new set of gene families excluding *S. cerevisiae*, which has the highest quality sequence with the most annotated genes. When we repeat the DupliPHY-ML analysis we find that $b = 1.0$, $d = 2.89$, $e = 0.25$ and $i = 0.08$. The relative rates estimated without *S. cerevisiae* change from those estimated using all the species. Most notably, the rates of gene loss and extinction are reduced indicating that there is a higher rate of gene gain inferred in this analysis compared to our original analysis. Although the values of these estimates change, the overall pattern of these estimates remains the same.

We have used several methods of validating the estimated rates of gene family evolution. In each case by altering the gene family evolution model or the input data we see variation in the estimated rates. In many cases this variation can be explained. Importantly, in all cases the trends of the results remain the same; The rate of gene loss is higher than the rate of gene birth, the rate of extinction is lower than that of birth and innovation appears to be a rare event.

Gene family evolution over the phylogeny

DupliPHY-ML estimates the average number of gain and loss events per gene family on each branch. These events, represented as branch lengths, are useful for identifying lineages, including ancestral lineages, that have a high turnover of genes in gene families. It should be noted that model misspecification may lead to very short internal branches followed by longer branches at the tip of the tree. In the species phylogeny (Figure 3) we see a long internal branch after divergence from the outgroup *Y. lipolytica*. Notably, *Y. lipolytica* has a branch length of zero suggesting there are no lineage specific events on this branch. This is unlikely to be an accurate estimate, since *Y. lipolytica* would be expect to have at least a few branch-specific events. The long internal branch length from the outgroup indicates a large turnover of genes on this branch and recapitulates the larger genetic distance between the out groups and more closely related species in the tree. Other long internal branches may represent expansions or contractions of gene families that are common to multiple species within the phylogeny. There are short branch lengths leading to *S. kluyveri* and *K. thermotolerans* (Figure 3), suggesting that these species have a very similar evolutionary history and have not had much lineage-specific evolution of gene families.

Terminal branches with long lengths represent lineage specific turnover of genes in gene families. Species with long branches may represent many gains of genes in a family and/or many losses. *D. hansenii* and *E. gossypii* both have longer terminal branches than the other species, indicating a large turnover of genes in these species. Since gains, and potentially losses, of genes in families may indicate some adaption to a new environment or evolution after speciation. In order to identify signals of environmental adaptation

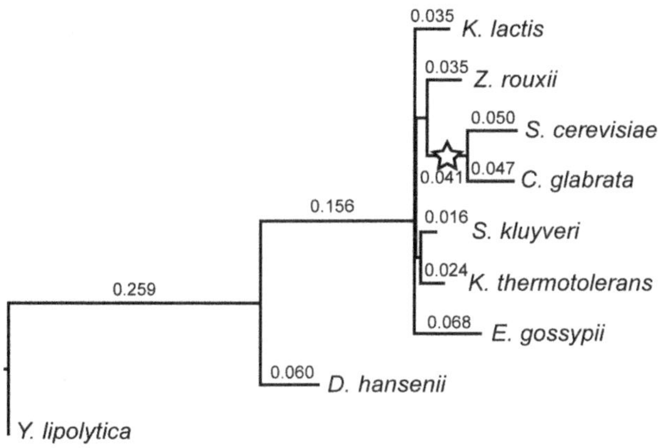

Figure 3. The phylogeny for the Génolevures data. Branch lengths represent the average number of gain and loss events per gene family as inferred by DupliPHY-ML. The branch on which the whole-genome duplication occurred is indicated with a star.

we first need to identify significant lineage specific expansions and contractions of gene families.

Lineage specific expansions of gene families

We used the net change in family size on each terminal branch compared with the net change on all other branches to identify those families with significant lineage specific expansion. We specifically chose the net change in order to identify those families that have undergone expansion and those genes have been retained, rather than identifying families with just a high turnover of genes. We have identified a large number of lineage specific expanding gene families (Table 1). In *S. cerevisiae* and *D. hansenii* we can identify more than 50 such families. The ranges of expanding family sizes also varies greatly within and between species. In all species with identified expanding families we observe families with only 2 members identified as expanding. In these cases it may be that all other species lack the family or have single member families (singletons) and the expansion to 2 members is lineage specific and significant by our definition.

The observed patterns of evolution of gene families, including both lineage-specific expansions and slowly-evolving families, may be due to either selection or neutral evolution. If the process of expansion is neutral we expect the newly introduced genes to nonfunctionalise and later be removed. If selection is acting to increase family size in specific families we would expect that these families would be associated with specific functions that may aid adaptation, and, in particular, may be relatable to known environmental adaptation routes in yeast. Additionally, in the presence of selection we expect that some of these functions may be common to multiple species. In order to determine whether selection is operating, we determined the functional classes of genes where families are changing rapidly.

In order to identify the functions relating to the lineage specific expanding gene families we used the Gene Ontology (GO) [30]. First, we identified enriched GO slim terms for each of the expanding families (File S3). We then visualised the GO slim terms on a Voronoi treemap and highlighted the enriched terms. These visualisations allow us to detect similarities and differences in types of gene families that are expanding in these species.

The species analysed show a wide variance in enrichment of GO terms (Figure 4). Most species show few enriched GO terms in expanding families. *S. cerevisiae* shows the most enrichment with some terms enriched in multiple expanding gene families. Enriched *S. cerevisiae* terms include those related to transport of amino acids and carbohydrates as well as enzyme activity such as isomerase, helicase and hydrolase activity. Transport of carbohydrates and amino acids are commonly enriched GO terms and are also enriched in the expanding families of *D. hansenii* and *K. thermotolerans*. Finally, we can identify expanding families that might be indicative of adaptation to a specific environment given the family's functional annotation; *Z. rouxii* is the only species to show enrichment for terms relating to the response to chemical stimuli and response to oxidative stress.

The observation that many of the expanding families have similar functional annotation suggests that many of the expansions we observe are driven by selection rather than random turnover of genes, or systematic gaps in sequence data. The common identification of nutrient uptake terms enriched in expanding gene families suggests that the expansion of gene in nutrient transporter families may be a common response to new environments or selection pressures. In this scenario following speciation which may have been precluded by a change in environmental conditions, selection for more efficient or varied uptake of nutrients from the environment may result in the observed expansions. In contrast, the identification of families showing unique functional annotation, such as the response to oxidative stress in *Z. rouxii*, suggests that we are also able to detect species specific adaptations.

Interestingly, we can identify rapidly expanding families of unknown function. As not all genes from these species can be

Table 1. Lineage specific gene family expansions in the Génolevures data.

Taxa	Num Genes	Num families	Ave size	Min size	Max size
S. kluyveri	5321	19	5.21	2	24
K. lactis	5076	19	4.15	2	12
K. thermotolerans	5092	15	5.93	2	22
Y. lipolytica	6448	0	0.00	0	0
C. glabrata	5203	13	5.46	2	13
Z. rouxii	4991	20	6.15	2	21
S. cerevisiae	6663	53	6.86	2	33
E. gossypii	4715	12	3.50	2	6
D. hansenii	6272	51	7.45	3	54

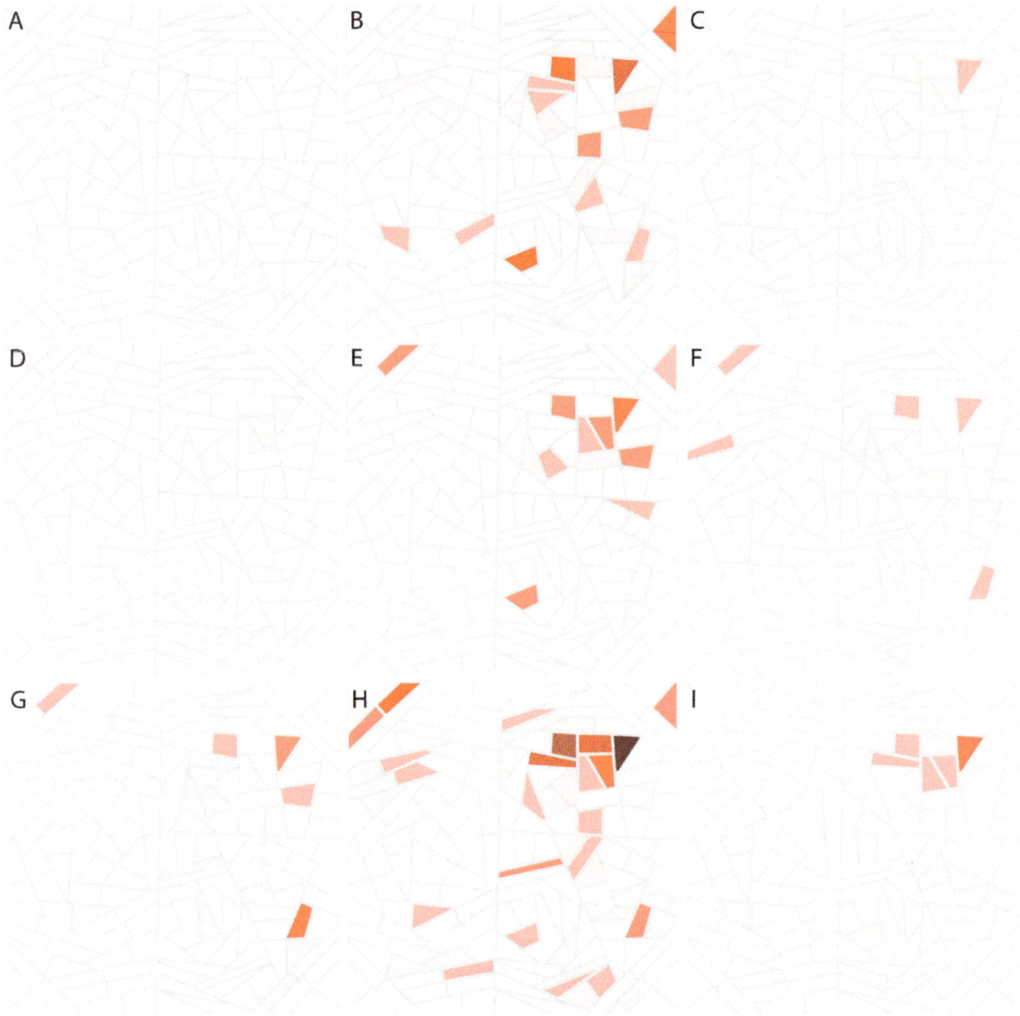

Figure 4. Functional enrichment of lineage specific expanding gene families for GO slim terms. Each cell represents a GO slim term from the biological process, molecular function or cellular component ontology. The positioning of the cells is determined by a term overlap metric so that more similar terms are positioned close together. Cells are shaded if one or more expanding gene family shows enrichment for that term. The intensity of the shading represents the number of gene families showing enrichment for that term. The most intense colour shows > 10 families are enriched. Each panel represents a species from the Génolevures data set: *Y. lipolytica* (A), *D. hansenii* (B), *E. gossypii* (C), *K. lactis* (D), *S. kluyveri* (E), *K. thermotolerans* (F), *Z. rouxii* (G), *S. cerevisiae* (H) and *C. glabrata* (I). Lists of enriched terms in expanding families are available in File S3

annotated with *S. cerevisiae* identifiers, our data contain many genes with no known functions. By annotating these genes with an "unknown" function, we can test to see whether genes with this unknown function are enriched in any rapidly expanding families. Indeed, we find enrichment for unknown functions in all species except *S. cerevisiae* and *Y. lipolytica*, which has no rapidly expanding families. The presence of these expanding families suggests that there are species specific adaptations using novel functions. We also identify two families enriched for the unknown function that are rapidly expanding in multiple species. Family GL3C0410 is enriched for the unknown function in *C. glabrata*, *D. hansenii* and *S. kluyveri*. Family GL3R0575 is also enriched for the unknown function in *D. hansenii* and *K. lactis*. The same family expanding in multiple species may indicate that the function represented by these families is important to multiple species.

Lineage specific contractions of gene families

In addition to looking at families showing rapid expansion we also looked at families showing significant contraction along a terminal branch. Contraction of gene families may signify the removal of non-adaptive functions or a streamlining of the genome and may result from specific adaptations. We find fewer families experiencing significant contraction than families experiencing expansion. *E. gossypii* has the largest number of contracting families with 28. Interestingly, *D. hansenii*, which has more than 50 expanding families only contains a single significantly contracting family suggesting the *D hansenii* genome has recently undergone an increase in size.

As before we used the GO to identify the functions of the contracting gene families. We find several terms common to multiple species again suggesting that species specific losses may reflect common adaptations. *S. cerevisiae*, *E. gossypii*, *S. kluyveri* and *K. thermotolerans* all show contractions of gene families related to the golgi apparatus. Furthermore, *C. glabrata*, *E. gossypii*, *S. kluyveri*, *K. lactis* and *Z. rouxii* all show enrichment for transmembrane transport terms in contracting families. Full lists of the GO terms associated with contracting families can be found in File S4.

Whole-genome duplication

Two of the species in our data set (*S. cerevisiae* and *C. glabrata*) have a whole-genome duplication (WGD) in their evolutionary past [31,32]; the branch on which this event occurred is indicated by a star on Figure 3. Subsequent to this duplication, a large number of duplicate genes were lost [32,33]. In common with all previously-published methods, our BDIE model does not incorporate any parameters that represent whole genome duplication. The presence of whole-genome duplication therefore has the potential to give rise to biases in the results, especially since the subset of genes retained after the WGD is enriched for specific functions [33,34].

There are three possible patterns of gene loss that may have occurred subsequent to WGD. First, a gene duplicated by WGD may be lost in both daughter species; in this case neither the duplication nor loss would be present in our input data. Second, a gene duplicated by WGD may be retained in both daughters. DupliPHY should assign this duplication to the WGD branch, leading to a large number of gains, and so long branch length in Figure 3. Third, a duplicated gene may be differentially lost in one daughter species and retained in the other. In this case the only signal for the duplication is the presence of a gene in one species. Here the duplication event may be assigned to the terminal branch rather than the ancestral WGD branch and then inferring a loss in the other species.

The first case is invisible to all methods of ancestral reconstruction that are based on analysis of extant genomes, including ours. However, we see that on the WGD branch of the phylogeny 4,161 gene families do not appear to change size. Since, by definition, all gene families must change size during a WGD, this figure gives an indication of how many events are invisible to our method and similar methods. We see some evidence of case two (retention in both species). There are 285 gene gains (in 263 families) on the WGD branch, and only 172 losses (in 152 families). This pattern is the reverse of that seen throughout the tree as a whole, where gene loss is four times more likely than gene gain.

In order to investigate the possible causes of bias that arise from case 3 (differential retention being miss-assigned), we determined how many of the gains on the *S. cerevisiae* and *C. glabrata* lineages have arisen via WGD. We used the assignment of duplications to either the WGD set (data from Byrne and Wolfe [35]) or small-scale duplication set (SSD, data from Hakes *et al* [36]). We find that, for *S. cerevisiae* 64 of the lineage specific gene gains arise from the WGD event, 203 from SSD events, with 97 gains unknown. For *C. glabrata* only 2 genes arise from WGD, 23 from SSD and 46 genes arise from unknown origins. The rapidly expanding families that contain WGDs for both *S. cerevisiae* and *C. glabrata* can be found in File S5. As the method of Hakes *et al* is conservative in assigning SSDs and much research has focussed on identifying WGDs, we believe that the majority of the "unknown" gains are likely to be SSDs. Although it is still possible that some of these genes may be unannotated WGDs or may have arisen from innovation events such as horizontal gene transfer between species. Since there are many more WGDs than SSDs (551 and 272 gene pairs, respectively), the observed pattern is unlikely to be an artefact arising from differing number of duplications in the input set. We conclude, therefore, that although the whole-genome duplication must have affected the pattern of observed duplications, particularly for the *S. cerevisiae* and *C. glabrata* lineage-specific duplications, the effect is unlikely to change our conclusions substantially.

In addition we also repeated the inference of duplicate gene family histories excluding any gene families in the Génolevures data that contained a known WGD from Byrne and Wolfe [35]. This data set contained 4,064 families with an average size of 8.88 genes. Here we find that $b=1.0$, $d=5.09$, $e=1.01$ and $i=0.05$. Overall the relative rates show the same pattern as the original analysis. The rate of loss is high compared to duplication and innovation is rare. Notably, the relative rates of both gene loss and gene family extinction have increased compared to the original analysis. This suggests that most genes retained from the WGD were inferred as gains in our analysis and the removal of these families has increased the estimation of the rate of loss and extinction relative to the rate of gene gain.

Discussion

Here we have used the software DupliPHY-ML to infer the evolutionary history of gene families in a range of yeast species and present evidence that suggests lineage specific gene family expansions lead to species-specific functional adaptations. We have implemented a new model with parameters for birth, death, innovation and extinction.

We note that in our model the birth parameter represents the retention of a gene after duplication rather than simply the occurrence of a duplication event. The observed birth rate is therefore a combination of the underlying rate of duplication, and the various factors that affect fixation rate, including selection. Although we may define the biological event represented by birth differently from previous work the model is similar to that of many other birth-death models used to study the evolution of gene families [6,8,14]. We make this distinction because it may be expected that the underlying duplication rate would correlate with gene family size, since large families contain more genes that are candidates for duplication. However, there is clear evidence for selection to remove or silence recently duplicated genes [4]. The combination of these factors means that the size of families are not well described by a simple model where birth rate increases linearly with family size [37,38]. There is, however, a weak correlation between the rate of change and family size; this conclusion holds for a range of species, including yeast. It is therefore likely that our model could be improved by the addition of a suitably weak correlation parameter.

DupliPHY-ML estimates the rate of gene loss, gene family extinction and gene family innovation relative to the rate of gene gain. The rates we estimate from the data show that the rate of gene loss is much higher than the rate of gene gain. Pseudogenisation followed by gene loss is thought to be the most common fate of duplicate genes [39]. The rate of innovation is much lower than that of the rate of gene gain by duplication within an existing family. Indeed, *de novo* gain of genes seems to be very rare, with only a handful of reported cases in yeast [40,41], *Drosophila* [3] and human [42]. We identified innovation events in our data by looking for examples of gene family gain within our phylogeny *i.e.* going from 0 to 1 or more members along a branch. The gene *BSC4* has been reported as a *de novo* gene in *S. cerevisiae* and has been associated with DNA repair during the stationary phase in *S. cerevisiae* when shifted to a nutrient-poor environment [41]. Here we find it is a member of a singleton family, and therefore, as arising from an innovation event. Hall *et al.* [40] identified two genes, *BDS1* and *URA1*, in *S. cerevisiae* that originated by horizontal gene transfer. In our data we also identify *BDS1*, an aryl- and alkyl-sulfatase gene required for the use of specific sulfates as sulphur sources, as arising from an innovation event. Interestingly, this gene appears to be present in both *S. cerevisiae* and *K. thermoltolerans*, suggesting that this gene may have been introduced to these genomes by separate horizontal transfer events. *URA1*, a dihydroorotate dehydrogenase involved in the in the *de novo* biosynthesis of pyrimidines, is not identified in our data as originating from an innovation event as all species studied here contain a gene with similarity to *URA1*, meaning the gene family was present at the root of our phylogeny.

Recently, Carvunis *et al* [27] described a process of gene formation from non-genic sequences through the formation of proto-genes that later evolve into genes. The finding of a significant number of proto-genes ($\sim$ 1900) in the yeast genome suggests that the rate of innovation identified in this study may be well below the true level of *de novo* gene gain. Interestingly, we find evidence of 1, 292 innovation events across all 8 species used in this study. We point out that although are estimates of the rate of innovation are much smaller than that of gene duplication we still find a large number of cases of innovation in our data. Additionally, it is possible that the exclusion of singleton families from this analysis has led to the low estimate of the rate of innovation of genes.

As with innovation, we find that gene family extinction is rare compared to the loss of a single gene in a family with more than one member. In mammals the rate of extinctions is lower than both expansions and contractions [12]. It has been suggested that gene family extinction may occur when there are shifts in nutrient availability, meaning specialised families are no longer necessary for growth [43].

We note that the estimated rates of death and extinction appear to be artificially high. Although the pairwise comparisons of the rates of birth/innovation and death/extinction make biological sense there are alternative explanations for the high estimates of death and extinction. Firstly, there are differences in the set of genes identified in these species that may have arisen because of low sequencing coverage or errors in gene identification. We have demonstrated that the estimated rates are not overly affected by the presence of one well sequenced species by removing *S. cerevisiae*. When *S. cerevisiae* is removed from the analysis the relative rates estimated by DupliPHY-ML are affected. In particular the estimated rate of loss is reduced from 4.05 to 2.89. These estimates of loss may be artificially high because repeated lineage specific losses of ancestral genes may inflate the estimated rate of loss. Alternatively, the genomes may be shrinking from some large ancestral genome at, or before, the last common ancestor of the set of species used in our analysis. Finally, large differences in the number of identified genes (Table 1) across all species may explain the high estimates of loss. CAFE 3 [11] attempts to correct for missing data but these types of corrections could not be used for this study as we have no accurate estimate of the amount of data missing from these genomes.

We have repeated much of our analysis to validate our estimates of the rates of birth, death, innovation and extinction by removing families, species, running different models and using data generated with different clustering methods. In some cases we find that the estimates of these parameters vary depending on the data or the model used. This indicates that the model used in this study still does not capture the complete biological processes of gene family evolution and suggests that further research of these models is warranted. Importantly, for all of the validation the trends in the estimates of these parameters remains the same and so, we can draw biological conclusions from our results.

Extended internal or terminal branches may represent areas of the phylogeny that have undergone accelerated gene family evolution in response to some selection pressure. In the species phylogeny *D. hansenii* and *E. gossypii* show the longest terminal branch lengths. These long branches may well be representative of adaptation to their respective environments. Indeed, these yeasts appear to occupy very specialised environments with *D. hansenii* being a cryotolerant, halotolerant marine yeast often found in cheese [44] and *E. gossypii* being a cotton pathogen [45]. Alternatively, long branch lengths may represent some stochastic turnover of genes in areas of the genome that experience high rates of gene gain or loss, such as subtelomeric regions [46,47].

The DupliPHY-ML method allows the determination of rates of gene gain and loss in a branch-specific manner. Families with low rates of changes may be tightly constrained by selection, with changes in the numbers of members having a deleterious effect. By contrast, the fast evolving families may represent those that are expanding or contracting in response to selection from the environment. Indeed, Demuth *et al.* [12] identified a set of rapidly evolving gene families in a mammalian phylogeny and showed that these families were associated with the same biological functions as quickly evolving genes and regulatory regions. Similarly, gene families that are evolving quickly in yeast may have important biological functions if there is selection pressure from the environment to increase the membership in these families. Alternatively, these families may be experiencing stochastic turnover of genes. In order to distinguish between evolution by adaptive or neutral processes we first needed to identify gene families that show significant expansion in a lineage. Families showing significant expansion on the terminal branches of the phylogeny represent lineage specific examples and may provide insight into the genetic mechanisms that lead to species specific adaptations.

We find that the ASP family is rapidly expanding; the ASP genes are asparaginases involved in the catabolism of alternative nitrogen sources [48,49]. This family is shown to be expanding on the lineage leading to *S. cerevisiae* which is consistent with previous analyses [50]. Other *S. cerevisiae* lineage specific expansions show an expansion of two of the major subtelomeric gene families. There are expansions in the seripauperin multigene (*PAU*) family and the *DUP240* integral membrane protein family. The *DUP240* family is a collection of nonessential genes that have been linked with membrane trafficking processes [51]. Subtelomeric regions have been shown to contain many duplicated genes that may be a consequence of higher rates of recombination in these regions [46,47]. Interestingly, in *S. cerevisiae* some genes gained in these families have relocated to internal sites on chromosomes [50] suggesting that duplication events that increase gene family membership may be coupled with relocation events.

The use of GO [30] gives an overview of the types of genes in expanding gene families. There is a wide range of variation in functional annotation, through enrichment of GO terms, in the expanding families. Several species show very little enrichment whereas *S. cerevisiae* shows wide ranging enrichment. The species in these data are sufficiently diverged that it is possible to detect differences in the types of families that are expanding in these species (Figure 4). We can identify unique functions associated with families expanding in a single species that might indicate lineage specific adaptation. A single expanding family in the species *Z* shows enrichment for response to chemical stimuli and response to oxidative stress. This family is comprised of 5 genes, 3 of which have high similarity to *S. cerevisiae CTA1* and 2 with similarity to *CTT1*, suggesting lineage specific duplication of these genes. *CTA1* is a catalase A involved in hydrogen peroxide detoxification and is important during the oxidative stress response [52]. *CTT1* is a catalase T and is also involved in hydrogen peroxide detoxification and is known to be induced under oxidative stress conditions [52,53]. In *C. albicans* it has been shown that trehalose mobilisation is important for tolerance of hydrogen peroxide [54]. *Z. rouxii* is well known to be tolerant to salt and osmotic stress [55,56] and it has been suggested that the high expression of trehalose synthesis genes under non-stress conditions [57] may be important in a range of stress responses [58]. It may be the case that *Z. rouxii's* naturally high rate of trehalose synthesis makes the species tolerant of a wide range of stresses and the expansion of a family associated with hydrogen peroxide resistance is a complementary adaptation to a specific environmental stress. Additionally, the presence of enrichment for an "unknown" GO term in expanding gene families suggests that there are also species specific adaptations using novel functions.

It is also possible to identify common functions associated with families expanding in multiple species. As these expansions are lineage specific and thus, independent, any common functional enrichment of these families may represent repeated selection for a specific adaptation. Many species contain lineage specific expanding families that are associated with carbohydrate transport and metabolism as well as amino acid transport. *S. cerevisiae* is well adapted to the uptake of sugars and has a range of transporters for different sugars [59,60]. It seems that there has been expansion in the sugar transporter family in *S. cerevisiae* through a combination of whole genome and tandem duplication possibly to facilitate the evolution of aerobic fermentation [61,62]. Indeed, we also see a large expansion of a sugar transporter family in *S. cerevisiae. D. hansenii* also shows expansion of families containing carbohydrate transporters and has been shown to be able to transport and utilise a range of sugars including hexoses and pentoses [63]. The prevalence of sugar transporter family expansion across several species of yeast may reflect a general adaptation strategy to environments containing an array of sugars with different genes amplified to fine-tune a yeast's metabolism to a specific sugar. Indeed, *K. lactis* which is found mostly in dairy products has acquired the ability to utilise lactose. We find that few *HXT* genes have been retained [62] and no expanding families show enrichment for sugar transporters.

Interestingly, we can also identify gene families that are contracting and the functions represented by these families. As with expanding families it is possible to detect different families with the same functional annotation contracting in multiple species. These functions tend to be associated with general transmembrane transporter activity and exocytosis (File S4). It has been argued that loss of genes may be adaptive [64] by the removal of non-adaptive functions and streamlining of the genome. It is tempting in this case to speculate that the losses shown here, coupled with gains of specific transporters, represent a restructuring of the uptake capabilities of the yeast in response to a new environment. Thus contractions as well as expansions may play a role in species specific adaptations.

The evolutionary histories of gene families can provide important insights into the past and present adaptations of species. We have inferred the evolutionary histories of all gene families across a variety of species, using the most sophisticated published models. We find that families evolve at a variety of rates and that these rates vary on different branches of the phylogeny. This variability suggests that rates are likely to be dictated by a range of selection pressures that act on a particular species. Additionally, we have identified families that show significant expansions on the terminal branches of the phylogeny. These families may be experiencing high rates of expansion because of neutral mechanisms or alternatively, may be under selection for increased membership. Indeed, we have identified several families showing rapid lineage specific expansion that are located in subtelomeric regions of the genome that are known to undergo regular recombination events. We also see that functional annotations associated with these expanding families suggest that there is selection for expansion of stress response and sugar transporter families. These expansions contribute to species differences and their individual functional specialisations.

Methods and Materials

Genomic and gene familiy data

Annotated genome sequences for 9 species of hemiascomycete yeast were downloaded from the Génolevures project [28]. These species were *Eremothecium gossypii, Candidia glabrata, Debaryomyces hansenii var. hansenii, Kluyveromyces lactis var. lactis, kluyveromyces thermotolerans, Saccharomyces cerevisiae, Saccharomyces kluyveri, Yarrowia lipolytica* and *Zygosaccharomyces rouxii.* These specific species were selected because they have previously been used to generate a set of gene families using a consistent and comprehensive method [65]. For all species we used the predicted ORFs and protein sequences identified by the sequencing project. Genes were annotated using BLAST [66] and the *S. cerevisiae* annotated genome as a reference. An E-value cutoff of 1×10^{-4} was used to call annotations in the BLAST results.

We used gene families from the Génolevures data as previously identified [65]. Briefly, the authors aligned the proteomes of the yeast species using BLAST [66] and the Smith-Waterman algorithm. Following this Tribe-MCL [29] was used to cluster the dat set using a range of inflation coefficients to produce clusters at different granularities. Finally, the authors use consensus clustering and an election algorithm to compare clusters and categorise them as robust, consensus, multiple choice or unique. In order to use the most reliable predictions of gene families we used those classified as "robust" or "consensus" families. Additionally, we removed families where any single taxa contains more than 75 members. We did this to make the analysis computationally tractable but it might also be the case that extremely large families are under specific selection pressures that may not be well defined by our current models. We note that this removes some notoriously complicated families such as polyproteins and repeat domains as well as single gene families which represent unique genes in the tree.

The specific clustering parameters used to generate the Génolevures data and our selection of "robust" and "consensus" families, may have introduced some bias into the identified families. In order to assess any bias in the data performed our own clustering on the raw Génolevures data. Here we used the protein sequences available for each species from the Génolevures website to perform an all against all BLAST search. The resulting network of BLAST hits was then clustered using Tribe-MCL [29] with a range of inflation values that control cluster granularity. Tribe-MCL was run with 4 inflation values of 1.4, 2, 4 and 6. For each of the resulting sets of gene families we ran DupliPHY-ML with the BDIE+G model (described below). We note that for computational tractability we still removed families where any single taxa contains more than 75 members and removed any singleton families. This analysis allows us to determine the effect of clustering on our results.

Phylogenetic tree

A phyogenetic species tree is necessary to infer the evolutionary histories of gene families. DupliPHY-ML does not use gene trees to infer events but instead will infer gain and loss events on the species phylogeny. Here, we used a subset of the cladogram presented in [67] to get the phylogenetic tree structure. Branch lengths were estimated for this structure using Baseml of the PAML package [68]. We used MUSCLE [69] to align the coding sequences of 2324 common genes between all species (based on

BLAST annotation described above), excluding any genes that appeared to have more than one copy in any species, and these alignments were concatenated. The list of common genes used in this step can be found in File S6. Baseml was run using the general time reversible model with no molecular clock to infer branch lengths. These branch lengths were used to test for significantly expanding or contracting gene families after the inference of gene family histories.

Inferring the evolutionary history of gene families

Given the identified gene families and phylogenetic trees for these data sets we then used DupliPHY-ML [9] to infer the evolutionary history of each gene family. These histories were inferred under the Birth-Death, Innovation and Extinction (BDIE) model with gamma-distributed rates across families. This model is similar to the Birth-Death-Innovation (BDI) model described previously [9], with the addition of an extinction parameter. DupliPHY-ML estimates each of the parameters from the available gene family data. Here the rate of gene birth corresponds to the rate of gene gain by duplication in existing families. The rate of gene death simulates the loss of a duplicate gene in a family with more than 1 member. Innovation represents the gain of a novel gene family through *de novo* gene gain or horizontal gene transfer and extinction represents the loss of a gene family from the tree. Once a gene family has been lost, it may only arise again through an innovation event. The BDIE model has an instantaneous rate matrix, **Q**, defined by equation 1.

$$Q_{i,j} = \begin{cases} b \text{ if } j-i=1 \text{ and } i\neq 0 \text{ (birth)} \\ d \text{ if } i-j=1 \text{ and } j\neq 0 \text{ (death)} \\ h \text{ if } i=0 \text{ and } j=1 \text{ (innovation)} \\ k \text{ if } j = 0 \text{ and } i=1 \text{ (extinction)} \\ 0 \text{ if} |i-j|>1 \text{ (maximum one event)} \end{cases} \quad (1)$$

We also supplied DupliPHY-ML with a set of unobservable families. Unobservable families represent potential phyletic patterns that can occur in these species but have been purposely removed from the analysis. In this case these are families that contain only a single gene. DupliPHY-ML corrects for these cases using the method of Felsenstein [70].

In order to validate the BDIE model we repeated the analysis of the Génolevures data using the previously described BDI model [9]. We also ran these models without gamma-distributed rates across families. All models were run on the same dataset and with all other run options kept constant.

To identify lineage specific gene family expansions we compared the change in number of members of an inferred gene family on an internal node to the observed gene family size the tips of the tree. The change in the number of members was normalised by the terminal branch length to control for greater divergence time allowing for higher turnover. If this change was positive and greater than the mean expansion or contraction on all other branches in the tree we used a Wilcoxon Rank Sum test to check for a significant expansion. We controlled for false discovery rate using the method of Benjamini and Hochberg [71]. All tests with $P<0.05$ after false discovery rate correction were taken as the set of rapidly expanding families. We note that by simply calculating the change in family membership from an ancestral node to a child node is likely to underestimate the true number of changes along a branch as some genes may be gained and lost on that branch. This simplification means that the calculations of the amount of change along a branch are likely to be an underestimate and so our results may be conservative. It also means that we focus on those gene gains that have been retained in the extant species.

We also identified lineage specific gene family contractions. Much like for expanding families we looked for the change in number of family members on each branch of the tree normalised by the branch length. If the change was negative (i.e. loss) and the number of losses was greater than the mean number of events on all other branches we used a Wilcoxon Rank Sum test to check for a significant expansion. We controlled for false discovery rate using the method of Benjamini and Hochberg [71]. All tests with $P<0.05$ after false discovery rate correction were taken as the set of rapidly contracting families.

Controlling for genome annotation and whole genome duplicates

The presence of genomes with better quality sequence and annotation may affect our analysis by artificially altering the estimated rates of gain and loss. In order to determine the effects of variable genome quality on our analysis we removed *S. cerevisiae* and repeated the analysis. To create a new set of gene families that exclude *S. cerevisiae* we first performed an all against all BLAST [66] search of all remaining species. We then used Tribe-MCL [29] with an inflation parameter of 1.2 to cluster the BLAST similarity network and generate a new set of gene families. Finally, we repeated the inference of gene family history using DupliPHY-ML as described above.

We also note that the new model of genome evolution (and all other existing models) doesn't have a parameter for large scale duplication events. As the phylogeny used in this study is known to contain a whole genome duplication event [32] we controlled for the effect of whole duplication on our estimates of gain and loss by removing all families that contained a known whole genome duplicate. After the removal of these families we repeated the DupliPHY-ML analysis as described above.

Functional annotation, enrichment and visualisation of gene families

In order to assign function to the expanding gene families we used the slim version of the Gene Ontology (GO) [30]. As GO terms are associated with genes via *S. cerevisiae* identifiers we annotated all genomes with an *S. cerevisiae* identifier (see above). We note that some genes that have no GO term annotation and as a consequence some expanding gene families or some members of these families have no functional annotation. There are also cases of genes in some species having no associated gene identifier from *S. cerevisiae*, in these cases we have assigned these genes a pseudo "unknown" GO term. This allows us to identify families that are enriched for an unknown function and may perform a novel function not performed by *S. cerevisiae*. We then used the GO slim term associations available at the *Saccharomyces* Genome Database (SGD) to assign GO terms to genes.

To test for enrichment of a GO term in a gene family we used Fisher's exact test. Here, the sample was all annotated genes in a gene family and the background population was all the annotated genes in the yeast genome. The sample successes therefore were all genes in the gene family annotated with the specific GO term and the population successes were all genes in the genome annotated with the specific GO term. All P-values were false discovery rate corrected using the method described by Benjamini and Hochberg [71] with a significance cutoff of $P<0.05$. This analysis was applied to both the rapidly expanding and contracting gene families.

In order to visually compare functional enrichment, the Term Overlap (TO) metric of the GLASS software (available at http://www.bioinformatics.ic.ac.uk/glass/) was used to determine the pairwise distances between GO slim terms. A tree-structure was generated from these pairwise distances using the neighbour-joining algorithm implemented in Quicktree [72]. The tree structure was then represented in two dimensions using Voronoi Treemaps [73,74], implemented with GLASS [75]. In this visualisation each cell represents a GO slim term, whose location within the panel is determined by the TO distance to all other terms. A cell is coloured if one of more expanding gene families are enriched for that GO term, with the intensity of the colour indicating the number of gene families enriched for that term.

Supporting Information

File S1 This file lists all of the Génolevures families including the identified present in each family for each species.

File S2 The ancestral reconstructions for each family analysed with DupliPHY-ML. The first line of the file is a newick string representation of the yeast phylogeny used in the study. The rest of the file is a tab delimited list of gene family ID and the number of members present at each node of the tree.

File S3 The functional enrichment of GO slim terms for rapidly expanding families in each species. The file lists the enriched terms, the ontology within GO slim and the raw and corrected P-values.

File S4 The functional enrichment of GO slim terms for rapidly contracting families in each species. The file lists the enriched terms, the ontology within GO slim and the raw and corrected P-values.

File S5 Expanding families in *S. cerevisiae* and *C. glabrata* that contain whole genome duplicates.

File S6 The list of 2324 common genes used to infer the branch lengths on the yeast phylogeny.

Table S1 Comparison of parameter estimates of 4 DupliPHY-ML models run on the Génolevures data.

Table S2 The effects of gene family data on the estimation of rates under the BDIE+G model.

Author Contributions

Conceived and designed the experiments: RMA. Performed the experiments: RMA. Analyzed the data: RMA. Contributed reagents/materials/analysis tools: RMA DM. Wrote the paper: RMA SCL.

References

1. Long M, Betrán E, Thornton K, Wang W (2003) The origin of new genes: glimpses from the young and old. Nature Reviews Genetics 4: 865–875.
2. Ohno S (1970) Evolution by Gene Duplication. Springer, New York.
3. Zhou Q, Zhang G, Zhang Y, Xu S, Zhao R, et al. (2008) On the origin of new genes in Drosophila. Genome Research 18: 1446–1455.
4. Lynch M, Conery J (2000) The evolutionary fate and consequences of duplicate genes. Science 290: 1151–1155.
5. Hahn M, Han M, Han S (2007) Gene family evolution across 12 Drosophila genomes. PLoS Genetics 3: e197–e209.
6. Akerborg O, Sennblad B, Arvestad L, Lagergren J (2009) Simultaneous Bayesian gene tree reconstruction and reconciliation analysis. Proceedings of the National Academy of Sciences of the United States of America 106: 5714–5719.
7. Csuros M (2010) Count: evolutionary analysis of phylogenetic profiles with parsimony and likelihood. Bioinformatics 26: 1910–1912.
8. Rasmussen MD, Kellis M (2011) A bayesian approach for fast and accurate gene tree reconstruction. Molecular Biology and Evolution 28: 273–290.
9. Ames R, Money D, Ghatge V, Whelan S, Lovell S (2012) Determining the evolutionary history of gene families. Bioinformatics 28: 48–55.
10. Librado P, Vieira F, Rozas J (2012) Badirate: estimating family turnover rates by likelihood-based methods. Bioinformatics 28: 279–281.
11. Han MV, Thomas GW, Lugo-Martinez J, Hahn MW (2013) Estimating gene gain and loss rates in the presence of error in genome assembly and annotation using cafe 3. Molecular Biology and Evolution: In Press.
12. Demuth J, Bie T, Stajich J, Cristianini N, Hahn M (2006) The evolution of mammalian gene families. PLoS ONE 1: e85.
13. Krylov DM, Wolf YI, Rogozin IB, Koonin EV (2003) Gene loss, protein sequence divergence, gene dispensability, expression level, and interactivity are correlated in eukaryotic evolution. Genome Research 13: 2229–2235.
14. Hahn M, De Bie T, Stajich J, Nguyen C, Cristianini N (2005) Estimating the tempo and mode of gene family evolution from comparative genomic data. Genome Research 15: 1153–1160.
15. Wapinski I, Pfeffer A, Friedman N, Regev A (2007) Natural history and evolutionary principles of gene duplication in fungi. Nature 449: 54–61.
16. Kunin V, Ouzounis C (2003). GeneTRACE-reconstruction of gene content of ancestral species.
17. Mirkin B, Fenner T, Galperin M, Koonin E (2003) Algorithms for computing parsimonious evolutionary scenarios for genome evolution, the last universal common ancestor and dominance of horizontal gene transfer in the evolution of prokaryotes. BMC Evolutionary Biology 3: 2.
18. Page R (1998) GeneTree: comparing gene and species phylogenies using reconciled trees. Bioinformatics 14: 819–820.
19. Chen K, Durand D, Farach-Colton M (2000) NOTUNG: A Program for Dating Gene Duplications and Optimizing Gene Family Trees. Journal of Computational Biology 7: 429–447.
20. Durand D, Halldórsson B, min Vernot B (2006) A Hybrid Micro-Macroevolutionary Approach to Gene Tree Reconstruction. Research in Computational Molecular Biology: 9th Annual International Conference, RECOMB 2005, Cambridge, MA, USA, May 14–18, 2005: Proceedings.
21. Vernot B, Stolzer M, Goldman A, Durand D (2008) Reconciliation with non-binary species trees. Journal of Computational Biology 15: 981–1006.
22. Wehe A, Bansal MS, Burleigh JG, Eulenstein O (2008) Duptree: a program for large-scale phylogenetic analyses using gene tree parsimony. Bioinformatics 24: 1540–1541.
23. Górecki P (2004) Reconciliation problems for duplication, loss and horizontal gene transfer. Proceedings of the eighth annual international conference on Resaerch in computational molecular biology: 316–325.
24. HahnM(2007) Bias in phylogenetic tree reconciliation methods: implications for vertebrate genome evolution. Genome Biology 8: R141–R149.
25. Arvestad L, Berglund AC, Lagergren J, Sennblad B (2003) Bayesian gene/species tree reconciliation and orthology analysis using mcmc. Bioinformatics 19: i7–i15.
26. Larget BR, Kotha SK, Dewey CN, Ané C (2010) Bucky: gene tree/species tree reconciliation with bayesian concordance analysis. Bioinformatics 26: 2910–2911.
27. Carvunis AR, Rolland T, Wapinski I, Calderwood MA, Yildirim MA, et al. (2012) Proto-genes and de novo gene birth. Nature: In Press.
28. Sherman D, Martin T, Nikolski M, Cayla C, Souciet J, et al. (2009) Genolevures: protein families and synteny among complete hemiascomycetous yeast proteomes and genomes. Nucleic Acids Research 37: D550–D554.
29. Enright A, Van Dongen S, Ouzounis C (2002) An efficient algorithm for large-scale detection of protein families. Nucleic acids research 30: 1575–1584.
30. Ashburner M, Ball C, Blake J, Botstein D, Butler H, et al. (2000) Gene ontology: tool for the unification of biology. The Gene Ontology Consortium. Nature Genetics 25: 25–29.
31. Wolfe K, Shields D (1997) Molecular evidence for an ancient duplication of the entire yeast genome. Nature 387: 708–713.
32. Kellis M, Birren B, Lander E (2004) Proof and evolutionary analysis of ancient genome duplication in the yeast *Saccharomyces cerevisiae*. Nature 428: 617–624.
33. Scannell D, Byrne K, Gordon J, Wong S, Wolfe K (2006) Multiple rounds of speciation associated with reciprocal gene loss in polyploid yeasts. Nature 440: 341–345.
34. Conant G, Wolfe K (2008) Turning a hobby into a job: how duplicated genes find new functions. Nature Reviews Genetics 9: 938–950.
35. Byrne K, Wolfe K (2005) The Yeast Gene Order Browser: combining curated homology and syntenic context reveals gene fate in polyploid species. Genome research 15: 1456–1461.

36. Hakes L, Pinney J, Lovell S, Oliver S, Robertson D (2007) All duplicates are not equal: the difference between small-scale and genome duplication. Genome Biology 8: R209–R222.
37. Karev G, Wolf Y, Rzhetsky A, Berezovskaya F, Koonin E (2002) Birth and death of protein domains: A simple model of evolution explains power law behavior. BMC Evolutionary Biology 2: 18–34.
38. Novozhilov AS, Karev GP, Koonin EV (2006) Biological applications of the theory of birth-anddeath processes. Briefings in bioinformatics 7: 70–85.
39. Lynch M, Force A (2000) The probability of duplicate gene preservation by subfunctionalization. Genetics 154: 459–473.
40. Hall C, Brachat S, Dietrich FS (2005) Contribution of horizontal gene transfer to the evolution of saccharomyces cerevisiae. Eukaryotic cell 4: 1102–1115.
41. Cai J, Zhao R, Jiang H, Wang W (2008) De novo origination of a new protein-coding gene in Saccharomyces cerevisiae. Genetics 179: 487–496.
42. Knowles D, McLysaght A (2009) Recent de novo origin of human protein-coding genes. Genome Research 19: 1752–1759.
43. Demuth J, Hahn M (2009) The life and death of gene families. BioEssays 31: 29–39.
44. Fleet G (1990) Yeasts in dairy products. Journal of Applied Microbiology 68: 199–211.
45. Ashby S, Nowell W (1926) The fungi of stigmatomycosis. Annals of Botany: 69–84.
46. Ames RM, Rash BM, Hentges KE, Robertson DL, Delneri D, et al. (2010) Gene duplication and environmental adaptation within yeast populations. Genome Biology and Evolution 2: 591–601.
47. Brown C, Murray A, Verstrepen K (2010) Rapid expansion and functional divergence of subtelomeric gene families in yeasts. Current Biology 20: 895–903.
48. Dunlop P, Meyer G, Ban D, Roon R (1978) Characterization of two forms of asparaginase in *Saccharomyces cerevisiae*. Journal of Biological Chemistry 253: 1297–1304.
49. Sinclair K, Warner J, Bonthron D (1994) The asp1 gene of Saccharomyces cerevisiae, encoding the intracellular isozyme of l-asparaginase. Gene 144: 37–43.
50. Gordon J, Byrne K, Wolfe K (2009) Additions, losses, and rearrangements on the evolutionary route from a reconstructed ancestor to the modern *Saccharomyces cerevisiae* genome. PLoS Genetics 5: e1000485–e1000499.
51. Despons L, Wirth B, Louis V, Potier S, Souciet J (2006) An evolutionary scenario for one of the largest yeast gene families. Trends in Genetics 22: 10–15.
52. Cohen G, Fessl F, Traczyk A, Rytka J, Ruis H (1985) Isolation of the catalase a gene of *Sacchu-romyces cerevisiae* by complementation of the cta1 mutation. Molecular and General Genetics 200: 74–79.
53. Lee J, Godon C, Lagniel G, Spector D, Garin J, et al. (1999) Yap1 and skn7 control two specialized oxidative stress response regulons in yeast. Journal of Biological Chemistry 274: 16040–16046.
54. Alvarez-Peral F, Zaragoza O, Pedreno Y, Argüelles J (2002) Protective role of trehalose during severe oxidative stress caused by hydrogen peroxide and the adaptive oxidative stress response in *Candida albicans*. Microbiology 148: 2599–2606.
55. Onishi H (1960) Studies on osmophilic yeasts. Bulletin of the agricultural chemical society of japan 24: 126–130.
56. D'Amore T, Crumplen R, Stewart G (1991) The involvement of trehalose in yeast stress tolerance. Journal of Industrial Microbiology & Biotechnology 7: 191–195.
57. Kwon H, Yeo E, Hahn S, Bae S, Kim D, et al. (2003) Cloning and characterization of genes encoding trehalose-6-phosphate synthase (tps1) and trehalose-6-phosphate phosphatase (tps2) from *Zygosaccharomyces rouxii*. FEMS Yeast Research 3: 433–440.
58. Smits G, Brul S (2005) Stress tolerance in fungi—to kill a spoilage yeast. Current Opinion in Biotechnology 16: 225–230.
59. Lagunas R (1993) Sugar transport in Saccharomyces cerevisiae. FEMS Microbiology Letters 104: 229–242.
60. Ozcan S, Johnston M (1999) Function and Regulation of Yeast Hexose Transporters. Microbiology and Molecular Biology Reviews 63: 554.
61. Brown C (1998). Multiple duplications of yeast hexose transport genes in response to selection in a glucose-limited environment.
62. Lin Z, Li W (2011) Expansion of hexose transporter genes was associated with the evolution of aerobic fermentation in yeasts. Molecular Biology and Evolution 28: 131–142.
63. Nobre A, Lucas C, Leão C (1999) Transport and utilization of hexoses and pentoses in the halotolerant yeast *Debaryomyces hansenii*. Applied and environmental microbiology 65: 3594–3598.
64. Olson M (1999) When less is more: gene loss as an engine of evolutionary change. American journal of human genetics 64: 18–23.
65. Nikolski M, Sherman D (2007) Family relationships: should consensus reign?—consensus clustering for protein families. Bioinformatics 23: e71–e76.
66. Altschul S, Gish W, Miller W, Myers E, Lipman D (1990) Basic local alignment search tool. Journal of Molecular Biology 215: 403–410.
67. Dujon B (2010) Yeast evolutionary genomics. Nature Reviews Genetics 11: 512–524.
68. Yang Z (2007) PAML 4: phylogenetic analysis by maximum likelihood. Molecular Biology and Evolution 24: 1586–1591.
69. Edgar R (2004) MUSCLE: multiple sequence alignment with high accuracy and high throughput. Nucleic Acids Research 32: 1792–1797.
70. Felsenstein J (1992) Phylogenies from restriction sites: A maximum-likelihood approach. Evolution 46: 159–173.
71. Benjamini Y, Hochberg Y (1995) Controlling the false discovery rate: a practical and powerful approach to multiple testing. Journal of the Royal Statistical Society Series B (Methodological) 57: 289–300.
72. Howe K, Bateman A, Durbin R (2002) Quicktree: building huge neighbour-joining trees of protein sequences. Bioinformatics 18: 1546–1547.
73. Balzer M, Deussen O (2005) Exploring relations within software systems using treemap enhanced hierarchical graphs. In: Visualizing Software for Understanding and Analysis, 2005. VISSOFT 2005. 3rd IEEE International Workshop on. IEEE, pp. 1–6.
74. Balzer M, Deussen O, Lewerentz C (2005) Voronoi treemaps for the visualization of software metrics. In: Proceedings of the 2005 ACM symposium on Software visualization. ACM, pp. 165–172.
75. Ames RM, MacPherson JI, Pinney JW, Lovell SC, Robertson DL (2013) Modular biological function is most effectively captured by combining molecular interaction data types. PLoS ONE 8: e62670.

10

Patterns of Variation at *Ustilago maydis* Virulence Clusters 2A and 19A Largely Reflect the Demographic History of Its Populations

Ronny Kellner[1,2]*, Christian Hanschke[2], Dominik Begerow[2]

1 MPRG Fungal Biodiversity, Max Planck Institute for Terrestrial Microbiology, Marburg, Germany, 2 Department of Geobotany, Ruhr-Universität Bochum, Bochum, Germany

Abstract

The maintenance of an intimate interaction between plant-biotrophic fungi and their hosts over evolutionary times involves strong selection and adaptative evolution of virulence-related genes. The highly specialised maize pathogen *Ustilago maydis* is assigned with a high evolutionary capability to overcome host resistances due to its high rates of sexual recombination, large population sizes and long distance dispersal. Unlike most studied fungus-plant interactions, the *U. maydis* – *Zea mays* pathosystem lacks a typical gene-for-gene interaction. It exerts a large set of secreted fungal virulence factors that are mostly organised in gene clusters. Their contribution to virulence has been experimentally demonstrated but their genetic diversity within *U. maydis* remains poorly understood. Here, we report on the intraspecific diversity of 34 potential virulence factor genes of *U. maydis*. We analysed their sequence polymorphisms in 17 isolates of *U. maydis* from Europe, North and Latin America. We focused on gene cluster 2A, associated with virulence attenuation, cluster 19A that is crucial for virulence, and the cluster-independent effector gene *pep1*. Although higher compared to four house-keeping genes, the overall levels of intraspecific genetic variation of virulence clusters 2A and 19A, and *pep1* are remarkably low and commensurate to the levels of 14 studied non-virulence genes. In addition, each gene is present in all studied isolates and synteny in cluster 2A is conserved. Furthermore, 7 out of 34 virulence genes contain either no polymorphisms or only synonymous substitutions among all isolates. However, genetic variation of clusters 2A and 19A each resolve the large scale population structure of *U. maydis* indicating subpopulations with decreased gene flow. Hence, the genetic diversity of these virulence-related genes largely reflect the demographic history of *U. maydis* populations.

Editor: Gustavo Henrique Goldman, Universidade de Sao Paulo, Brazil

Funding: This work was supported by the Deutsche Forschungsgemeinschaft (DFG,BE 2201/8-1) and the Ruhr-University Bochum. The funders had no role in study design, data collection and analysis, decision to publish, or preparation of the manuscript.

Competing Interests: The authors have declared that no competing interests exist.

* E-mail: ronny.kellner@mpi-marburg.mpg.de

Introduction

Fungal plant parasites exhibit a large repertoire of biotrophic interactions. Domestication and practices of modern agro-ecosystems can enforce iterative bottlenecks and expansions on host and pathogen populations such that the genetic structure of extant pathogen populations more strongly reflect demographic than adaptive processes [1,2]. The fungal pathogen *Ustilago maydis* establishes an intimate biotrophic interaction with its domesticated host plant *Zea mays* (maize) [3,4]. As shown by population studies, agro-ecosystem-dependent bottlenecks and radiations dramatically imprinted the genetic structure of *U. maydis* such that the extant intraspecific variation reflects mainly evolution since the time of early domestication [5]. Studies based on amplified fragment length polymorphism fingerprinting and simple sequence repeat analyses demonstrated low gene flow among major populations of *U. maydis* from North and South America [5,6]. This indicated the majority of genetic variation within *U. maydis* accumulated in the last ~10,500 years. During domestication maize was successively transformed from its progenitor teosinte into more than 2,000 known genetic varieties, which include modern maize inbred lines [7,8]. Domestication traits of maize underwent a reduction of diversity, and population size dramatically increased, especially in the last 200 years [9–11]. However, a considerably high level (~70%) of genetic variation was retained in current maize lines. [11]. Studies on modern maize varieties demonstrated that quantitative resistance against *U. maydis* involves several quantitative trait loci (QTL), which each contribute to the frequency and severity of infection [12].

Ustilago maydis originated as a pathogen on teosinte and parasitises all maize varieties along with their wild progenitor in agricultural and wild populations. Owing to the lack of a gene-for-gene interaction many gene products quantitatively contribute to the virulence of *U. maydis*. As shown by ultrastructural studies, the interaction in this pathosystem is based on a characteristic vesicular system [13,14] that highlights the important role of fungal secreted effector proteins. Effectors secreted into the apoplast by *U. maydis* are supposed to cross the plant cell membrane [3,15] and to interfere with host defenses, similar to other fungal parasites [16]. A large portion of the transcriptome of *U. maydis* is predicted to entail secreted proteins, approximately 426 out of 6786 genes. These include, for example, the "maize induced genes" (*mig*) [17]. However, this number may be even higher when modes of non-conventional secretion are considered [4,18]. Remarkably, many secreted protein-encoding genes of *U. maydis* are organised in gene clusters that are co-regulated and induced during pathogenesis. Moreover, deletion strains that lacked individual gene clusters exhibited altered virulence on the maize variety 'Early Golden Bantam' [3,4].

Several virulence gene clusters of *U. maydis* have homologs in the related smut fungus *Sporisorium reilianum*. This provides the opportunity to assess evolutionary patterns in homologous genome regions as well as homologous genes. Comparative genomics revealed that *U. maydis* and *S. reilianum* have largely syntenic genomes and high sequence identities of homologous genes [19]. However, virulence gene clusters and secreted protein-encoding genes are enriched in genomic regions of very low sequence identity [19]. Hitherto, little is known about the role of these so-called divergence clusters but the accumulation of potential virulence genes in rapidly evolving genomic regions suggests that they promote adaptive processes in the molecular "arms race" of host resistance and fungal virulence. In particular, the impact of demographic processes on virulence-related genes of *U. maydis* within these regions is poorly understood.

To gain first insights into the intraspecific genetic diversity of virulence in *U. maydis*, we examined a selection of virulence-related genes from single isolates across larger geographic distances. We analysed genetic polymorphisms in 34 potential virulence factor genes of *U. maydis* in 17 isolates from Europe, North America and Latin America. We focused on gene cluster 2A, which is associated with virulence attenuation; cluster 19A, which is crucial for virulence, and the cluster independent effector gene, *pep1* [4,20]. These were compared to 14 putative non-virulence genes and 4 house-keeping genes. Low levels of genetic diversity were revealed in the putative virulence factor genes across large geographic distances. In addition, coalescence analyses of virulence clusters 2A and 19A indicated a subpopulation structure with limited gene flow between isolates of geographically distinct regions. This was indicative of strong impact from demographic processes on these virulence gene clusters.

Results

Gene content and single nucleotide polymorphisms

In order to study polymorphisms of virulence-associated genes we completely or partially sequenced 48 genes from up to 17 isolates of *U. maydis* that represented geographical variation from Europe, North America and Latin America (Tables 1, S1, S2). Of these 48 genes, 34 encoded putatively secreted proteins and 14 genes encoded putatively non-secreted proteins. In particular, the entire virulence cluster 2A was sequenced spanning 20,725 basepairs (bp) from the left cluster-flanking gene *um01233* to the right cluster-flanking gene *um01242* [16]. We partially sequenced 37 genes from virulence cluster 19A and its flanking regions. Of these 37 genes, 25 encoded putatively secreted proteins of cluster 19A and 12 genes encoded putatively non-secreted proteins. Gene-specific primer pairs were designed for amplification of a 267 to 645 bp region of the respective gene (Table S3). The cluster 19A dataset of each strain comprised 15,613 bp, which represents 62% and 18% of the sequences of genes encoding secreted and non-secreted proteins, respectively. The entire virulence gene *pep1* (*um01987*) was sequenced using primers directed against non-coding cis- and trans-flanking regions of *pep1*. For comparison, we sequenced parts of the house-keeping genes *ef1-α* (*um00924*), *gapdh* (*um02491*), *rpb1* (*um03863*) and ITS containing *5.8S* rDNA, encoding elongation factor 1-alpha, glyceraldehyde 3-phosphate dehydrogenase, RNA polymerase II subunit 1, and the internal

Table 1. Summary of virulence and house-keeping genes sequence data.

Parameters	*pep1*	Cluster 2A secreted	Cluster 19A secreted	Cluster 19A non-secreted	ITS1/2 + *5.8S*	*gapdh*	*ef1-α*	*rpb1*
Δ phenotype	non-pathogenic	hypervirulent	reduced virulence	NA	lethal	NA	NA	lethal
N	17	13	10	10	15	16	17	17
sequenced length in bp	537 (537)	20725 (10026)	10624 (15291)	5423 (25617)	703	864 (1433)	969 (1380)	624 (5325)
haplotypes	5	12	8	7	1	5	3	1
π (JC) x 10^{-3}	3.96	2.53 (4.16)	2.72	2.54	0	1.87	0.68	0
SNP/kb	11.17	7.18 (11.10)	6.97	4.61	0	5.78	3.10	0

Δ phenotype, knockout phenotype (*pep1*: [20]; cluster 2A and cluster 19A: [4]); N, number of isolates analysed; π, nucleotide diversity based on the average number of nucleotide differences between two random sequences; brackets enclose sequence lengths and values of complete coding sequences; NA, not available.

transcribed spacer containing *5.8S* rDNA, respectively (Table 1). Details of strains, amplicons and primers are given in Tables S1, S2 and S3. Sequences have been submitted to GenBank with the accession numbers JF313619-JF313649 and HQ002861-HQ003216 listed in Table S4.

The applied sequencing approach revealed that each gene was present in all investigated strains (Tables S1, S2). Furthermore, sequences of the entire cluster 2A revealed a conserved gene content and gene order in all strains. In total, we obtained 282 polymorphic sites in 41 out of 52 genes. In particular, 102, 89 and 91 single nucleotide polymorphisms (SNPs) represent non-synonymous, synonymous and unique polymorphisms, respectively (Table 1). Each observed SNP consists of two alleles, and each polymorphic codon is based on one SNP except for one polymorphic amino acid in the gene *um05301* of the European strain RK124, which is based on two SNPs.

Nucleotide diversity of house-keeping genes and genes encoding secreted and non-secreted proteins

To survey genetic variation within *U. maydis* we calculated the nucleotide diversity π for all genes and non-coding regions. The overall mean nucleotide diversity π of all genes of this study was 0.0032. Student's t-tests of π value datasets from cluster-associated genes including *pep1*, house-keeping genes and non-coding genes revealed significant differences in nucleotide diversity ($p<0.001$). Thus, house-keeping genes exhibited significantly lower π values (10-fold) and non-coding regions revealed significantly higher π values (approximately 2-fold) compared to the virulence cluster-associated genes (Figure 1A). In particular, 5.8S, non-coding ITS rDNA and *rpb1* exhibited no sequence polymorphisms and are identical to the *U. maydis* reference genome (Table 1; MUMDB). Exon sequences of *ef1-α* and *gapdh* revealed π-values of 0.00068 and 0.00028, respectively (Table 1). However, there was no significant overall difference in nucleotide diversity π between genes encoding secreted and non-secreted proteins (Figure 1A). Thus, the overall nucleotide diversity in *U. maydis* is low with significant differences between house-keeping genes and cluster-associated genes but no significant difference between secreted and non-secreted protein-encoding genes.

Tests for natural selection

We tested whether single virulence-related genes and non-virulence genes underwent positive selection as the nucleotide diversity π values of the investigated genes of *U. maydis* varied between 0.000 to 0.007 (Table S2). To this end, we first calculated Tajima's D, Fu & Li's D* and Fu & Li's F* summary statistics for all genes and determined their probabilities (Table S2) [21,22]. This revealed eight genes that showed signatures of non-neutral evolution with a p-value below 0.05 for at least one of the three applied summary statistics. Of these, one gene encoded the putatively secreted effector gene *um01236* of cluster 2A, six genes encoded putatively secreted proteins of cluster 19A and one gene encoded the putatively non-secreted cluster 19A-flanking gene *um10560*. In particular, *um01236*, *um05290*, *um12302*, *um05303* and *um10560* revealed positive summary statistics that indicated a significant excess of intermediate frequency mutations. Furthermore, *um05301*, *um10556* and *um05310* revealed negative values that indicated an excess of rare mutation. To further evaluate the summary statistics and to analyse signatures of selection at the amino acid level we calculated the ratio of non-synonymous polymorphisms per non-synonymous site (Pa) to the number of synonymous polymorphisms per synonymous site (Ps). From 52 analysed genes only 27 exhibited both, non-synonymous and synonymous polymorphisms, and could be used for comparison. Of these, 20 encoded secreted proteins, 6 encoded non-secreted proteins and 1 gene encoded the house-keeping gene *ef1-α* (Figure 2, Table S5). To identify genes under natural selection, we applied a Z-test [23] that compared the rates of Pa (proportion of non-synonymous polymorphisms) and Ps (proportion of synonymous polymorphisms). Due to the low levels of genetic variation the Z-tests revealed no significant evidence for non-neutral or neutral evolution (Figure 2, Table S6). In summary, nucleotide-based tests for selection identified signatures of non-neutral evolution for eight genes of virulence clusters 2A and 19A. However, they were not confirmed by codon-based Pa/Ps statistics.

Phylogeographic structure

Population genetic analyses of *U. maydis* demonstrated genetic differentiation between geographically distinct populations and suggested the presence of genetic isolation between major populations during recent evolutionary time [5,6]. To evaluate the genetic divergence in our dataset we first conducted separate Maximum Likelihood (ML) analyses of cluster 2A and 19A alignments. ML recovered two, strongly-supported clades, one that contained all Latin American strains (LA) and a second with all North American and European strains (NA-E; Figure S1A, B). RAxML does not account for intragenic recombination, so separate coalescence analyses were performed for both alignments. These analyses recovered the same clades from the ML analyses (Figure 3A, B). The dataset of cluster 19A favoured a third group of all European strains (E; Figure 3B), which was supported with a weak *posteriori* probability for the dataset of cluster 2A (Figure 3A). Analyses of sub-datasets of genes encoding either secreted or non-secreted proteins revealed the same clades with lower support values (data not shown). Based on these results we concluded the presence of subpopulations in terms of decreased gene flow between individuals from LA, NA and E.

Intraspecific diversity of cluster 19A gene sequences

Within cluster 19A, we identified genes containing no polymorphisms (*um05312*, *um05318*, *um05319* and *um10561*), genes containing only synonymous mutations (*um10705*, *um05293*, *um05308*, *um05314*, *um10557*, *um10558* and *um05322*), genes containing only non-synonymous mutations (*um10552*, *um05295*, *um05305*, *um05306*, *um05309*, *um05311* and *um05317*) as well as genes containing both, synonymous and non-synonymous mutations (*um05290*, *um05294*, *um12302*, *um10553*, *um10554*, *um05299*, *um05300*, *um05301*, *um05302*, *um05303*, *um10555*, *um10556*, *um05310*, *um10559*, *um10560*; Figure 4A). In both Latin American strains (RK122 and RK217), one nucleotide polymorphism in *um10553* introduced a stop codon at position 173 of the encoded protein. In addition, *um05311* of RK217 encodes a stop codon at amino acid position 167 that should lead to a truncated version of the protein (Figure 4A). However, as sequence information of cluster 19A is clearly limited due to short partial sequences of single genes the power of the applied summary statistics is likely reduced.

Based on nucleotide diversity π, the genetic variation seemed to be heterogeneously distributed along cluster 19A (Figure 4A). Therefore we split the cluster in two distinctive regions, which we named 19A-L (ranging from *um05294* to *um10555*) and 19A-R (ranging from *um05305* to *um05319*). Student's t-tests revealed significant differences in nucleotide diversity between 19A-L and 19A-R (Figure 1B, $p<0.01$). Thus, 19A-L carries most of the cluster variation whereas 19A-R predominantly contains genes with low variation, genes that exhibit only synonymous polymorphisms and genes without polymorphisms (Figure 4A). In addition,

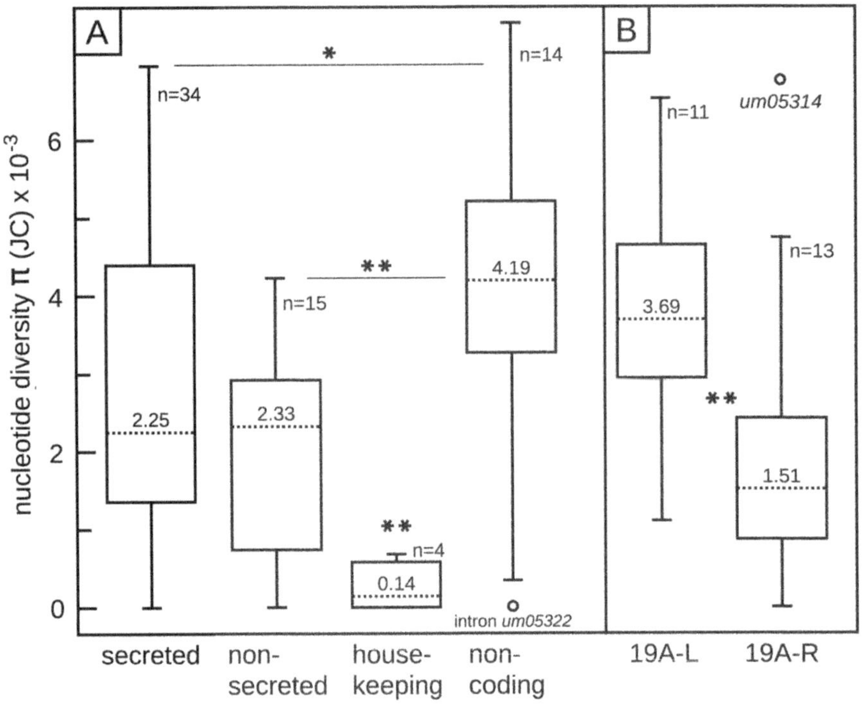

Figure 1. Boxplot comparisons of nucleotide diversity π datasets. (A) Nucleotide diversity boxplot diagrams of putatively secreted and non-secreted protein-coding as well as house-keeping genes (*ef1-α, rpb1, gapdh* and ITS including *5.8S* rDNA) and non-coding regions. (B) Nucleotide diversity boxplot diagrams of sub-datasets of cluster 19A. 19A-L: left part of cluster 19A genes encoding secreted proteins (*um05294 – um10555*); 19A-R right part of cluster 19A genes encoding secreted proteins (*um05305 – um05319*); Dotted line: median; n: number of sequences per dataset. *: $p \leq 0.05$; **: $p \leq 0.01$ in Student's t-test.

sequences of 19A-L together with cluster 19A left flanking gene sequences are enriched in fixed differences compared to 19A-R and cluster 19A right flanking gene sequences (Figure 4B). However, this bilateral distribution of polymorphisms is not reflected by Tajima's D, Fu & Li's D* and Fu & Li's F* summary statistics. They revealed a homogeneous distribution of seven selectively non-neutral loci including the cluster-flanking genes *um10560* and *um05290* as well as the cluster 19A genes *um12302*, *um05301*, *um05303*, *um10556* and *um05310* (Figure 4C).

We tested for linkage disequilibrium (LD) between all polymorphic sites of cluster 19A as the evolutionary rate of locally linked loci can hitchhike with favorable or non-favorable mutations of neighboring loci due to recombinational linkage. The LD plot revealed a patchy distribution of LD between both close and distant sites of cluster 19A, which rejected the hypothesis of local LD in cluster 19A (Figure S2). In summary, cluster 19A displays a significant, heterogenic pattern of sequence variation that has been created by recombination.

Intraspecific diversity along cluster 2A

In contrast to cluster 19A the genes of cluster 2A revealed no heterogenic pattern of variation as shown by a sliding window analysis (window length: 200, step size: 50; Figure 5A). However, there are some regions spanning up to 650 bp that lack polymorphisms, e.g. encoding the C terminus as well as the 3′ non-coding region of *um01240* and the C terminus of *um01242* (Figure 5A).

To test for natural selection we estimated Tajima's D, Fu & Li's D* and Fu & Li's F* summary statistics in sliding window analyses of cluster 2A sequence alignments (window size 200 bp, step size 50 bp). The null hypothesis is significantly rejected in five regions. Thereof, a 5′ non-coding region of *um01239* is supported by all three methods, the 5′non-coding regions of *um01234* and *um01236* as well as a C-terminal region of *um01236* are supported by Tajima's D and Fu & Li's F*, whereas a 3′ non-coding region of *um01238* is solely supported by Tajima's D (Figure 5A). However, codon-based tests for natural selection did not support non-neutral evolution for any gene of cluster 2A. Because coalescent analyses suggested a subpopulation structure for *U. maydis* (Figure 3) we examined whether the non-neutrally evolving regions of cluster 2A contain polymorphisms that are fixed between the two isolates from North America (NA), the five isolates from Europe (E) and the six isolates from Latin America (LA). Out of 166 polymorphic sites, 39 sites were polymorphic in a single strain. Isolates from NA, E and LA had polymorphic sites at 44, 82 and 46 respectively. In total 50 polymorphic sites (Figure 5B) constitute fixed differences that were non-polymorphic within the subpopulation but polymorphic between the subpopulations NA, E and/or LA. This includes 12 non-synonymous polymorphisms, 7 synonymous polymorphisms and 31 polymorphisms in non-coding regions. Of these, 24 fixed differences coincide with regions that revealed signatures of non-neutral selection (Figure 5C). In addition, eight fixed differences were present in a single gene (*um01236*) comprising all of its non-synonymous polymorphisms.

Hence, coding regions and non-coding regions of cluster 2A in *U. maydis* revealed signatures of non-neutral evolution. In addition, these regions are enriched with polymorphisms that seem to be fixed in geographically isolated subpopulations.

Discussion

In molecular "arms races" host resistance and fungal virulence underlie strong selection with recurrent adaptive evolution of

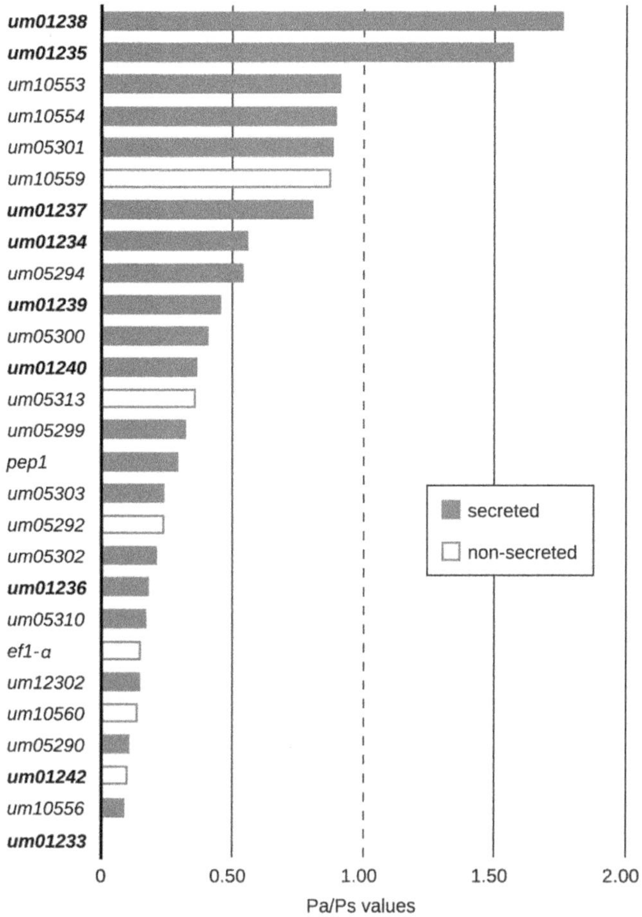

Figure 2. Ratios of Pa and Ps. Listed are all genes that contain non-synonymous and synonymous mutations. Grey bars indicate putatively secreted protein-coding genes and white bars indicate putatively non-secreted protein-encoding genes. Bold um gene numbers represent genes of cluster 2A. Light um gene numbers represent genes of cluster 19A.

resistance and virulence-related genes. Consequently, virulence genes frequently show signatures of positive selection [24,25]. However, in agro-ecosystems such genetic signatures are often obliterated due to demographic fluctuations [2]. In the present study we examined the genetic diversity of two virulence gene clusters, 2A and 19A, of the maize pathogen *U. maydis*. This demonstrated an overall low intraspecific genetic variation of these virulence-related genes across large geographic distances and across subpopulation barriers.

The nucleotide diversities π of all investigated *U. maydis* genes were remarkably low compared to the diversity levels within the fungal hemi-biotrophic ascomycete *Zymoseptoria tritici* [26]. The comparison of three genomes of *Z. tritici* documented mean π values above 0.01 for single chromosomes. In contrast, the virulence clusters 2A and 19A, the cluster-flanking genes, the house-keeping genes as well as the non-coding regions of 17 isolates of *U. maydis* revealed π values below 0.003, which indicated a loss of genetic diversity in *U. maydis*. The overall sequence divergence is too low to rely on tests for natural selection of genes or single sites [27,28]. Furthermore, sequence alignments of the investigated virulence genes of *U. maydis* with their putative orthologs from *S. reilianum* and *U. hordei* are highly unreliable, which renders interspecific evolutionary analyses impossible [19,29]. Hence, future grass smut analyses of fast evolving genes are in need of genome information from isolates with altered host specificities, e.g. *U. maydis* isolates from teosinte, the wild progenitor of maize, or more closely related species of *U. maydis*, e.g. *U. bouriquetii* or *U. vetiveriae* [30,31].

In contrast to increased levels of divergence on the interspecies level between *U. maydis* and the related maize pathogen *Sporisorium reilianum* [8], our comparison within *U. maydis* revealed no significant difference between virulence-related and non-virulence genes (Figure 1). These low levels of genetic variation demonstrated that virulence clusters 2A and 19A diverged before the adaptation to modern maize. In addition, they reflect the young age of recent *U. maydis* populations [5]. If a maximum age of 10,000 years for populations of *U. maydis* is assumed substantial levels of intraspecific genetic variation were retained at numerous virulence and non-virulence loci, which likely reflects its large population sizes. In addition, we noted that the genetic diversity of virulence clusters 2A and 19A both reflect the large scale sub-population structure of *U. maydis* (Figure 4). Hence, the investigated virulence genes largely reflect the demographic history of these populations and signatures of natural selection are, if present at all, strongly imprinted. The discrepancy between the non-significant codon-based tests for natural selection and the summary statistics, e.g. Tajima's D confirm this finding as the latter tests are known to be sensitive for demographic impacts. However, only three genes of cluster 19A and two cluster 19A-flanking genes seem to be subject to purifying selection as they revealed no polymorphisms among individuals of distant geographic regions (Figure 5). For the other genes Pa/Ps tests for selection are non-significant due to low sampling depth. In conclusion, the demographic history of *U. maydis* requires deeper sampling to determine methods of natural selection.

It remains unclear whether virulence cluster genes of *U. maydis* were directly involved in the adaptation to domesticated maize. For example, the non-virulent *pep1* deletion mutant of *U. maydis* can be fully rescued by the *pep1* ortholog of the barley smut *U. hordei*, which questions a host-specific function in *U. maydis* [20]. Similarly, gene products of clusters 2A and 19A could act as general rather than host-specific virulence factors. In consequence, they would carry out conserved functions that are subject to purifying selection, hence, leading to less genetic variation. The recent functional characterization of cluster 19A provided evidence for its role in quantitative virulence (Brefort *et al.* 2014, in press in PLoS Pathogens). Although significantly impaired in virulence cluster 19A, deletion mutants were able to complete their life cycle with the production of few spores. Cluster 19A sub-deletion experiments identified 4 individual genes and one gene family that independently contributed to virulence or anthocyanin production. Interestingly, deletion of the left part of cluster 19A led to a dramatic reduction of spore production and the loss of anthocyanin production (Brefort *et al.* 2014, in press in PLoS Pathogens) whereas the right part of cluster 19A was only weekly attenuated in virulence. Our study revealed that the left part including *um05294* to *um19554* (*tin1-1* to *tin1-5*) and *um05302* (*tin2*) contains significantly more polymorphisms compared to the right part of cluster 19A (Figures 1B, 4). This indicates that this region might undergo diversification in *U. maydis*. However, a correlation of these virulence phenotypes and our patterns of intraspecific genetic variation is problematic because single genotypes of *U. maydis* can show dramatic differences in virulence on different maize varieties (Kellner and Begerow unpublished data). Additionally, the maize variety Early Golden Bantam, which is mainly used in functional analyses, likely does not represent the host

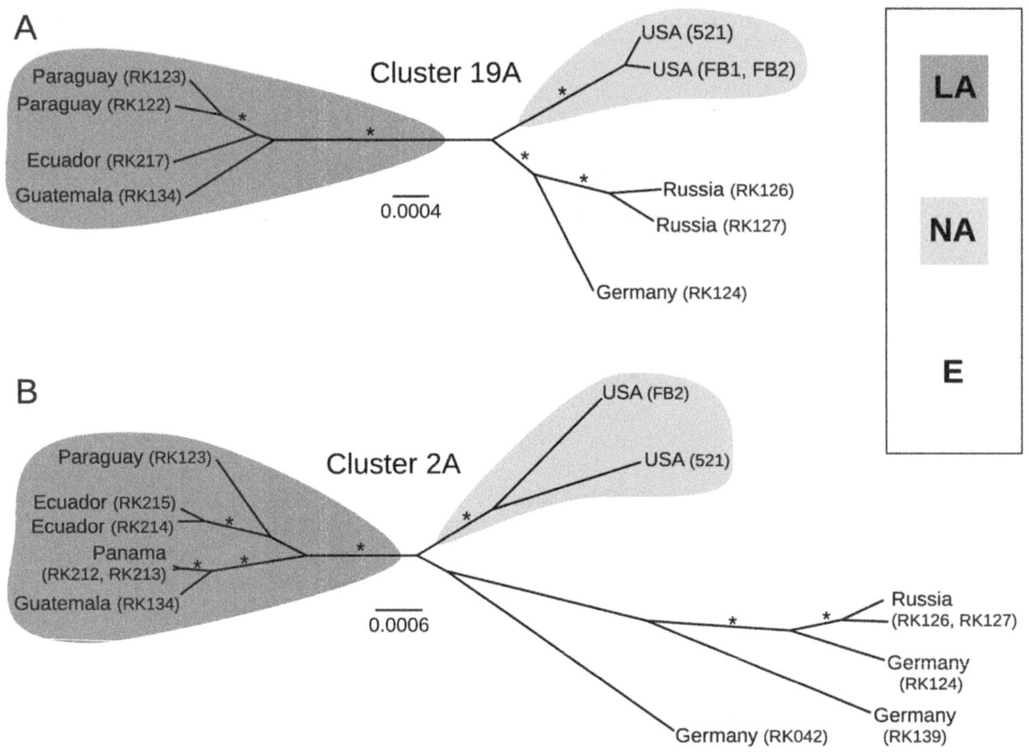

Figure 3. Coalescence genealogies of cluster 19A and cluster 2A sequence datasets. Coalescence-based genealogies (Beast v1.6.1) of datasets comprising (A) partial sequences of cluster 19A and (B) entire cluster 2A. Asterisks above branches correspond to a posteriori probabilities higher than 0.98. Branch lengths correspond to substitutions per site.

genotype that the ancestors of investigated isolates of *U. maydis* evolved on.

Domestication selects for specific phenotypes and involves the fixation of specific alleles. In crops these selective sweeps largely acted on regulatory elements of domestication trait-related genes involving altered transcriptional activity [32]. Similarly, the 5′non-coding regions of *um01234*, *um01236* and *um01239* of cluster 2A revealed weak evidence for non-neutral evolution. In case of non-neutral evolution these regions could play a role in transcriptional or post-transcriptional control indicating the relevance of altered gene expression or altered mRNA stability for the evolution of virulence in *U. maydis* (Figure 5). Similar to RNA-binding proteins of *U. maydis*, e.g. Khd4 and Rrm4 that regulate their targets by binding to specific sequences in 5′ and 3′ untranslated regions (UTRs) [33], SNPs within UTRs of virulence genes could alter gene expression and stability. A linkage of altered gene expression and adaptive evolution is proclaimed by several studies. In *Saccharomyces cerevisiae* experimental evolution altered gene expression levels of several hundred genes within 250 asexual generations and resulted in genotypes that were better adapted to glucose limitation and produced less ethanol [34]. In a recent study, the transcription factor Efg1p of *Candida albicans* revealed a linkage of expression levels and immune status of the host [35]. Hence, the intraspecific sequence variation in 5' non-coding regions of *U. maydis* virulence cluster genes could reflect recent adaptive processes in gene regulation although the linkage of sequence polymorphism and altered gene expression remains to be experimentally verified and evaluated.

In summary, the overall low levels of genetic variation of virulence clusters 2A and 19A across large geographic distances and subpopulations reflect the demographic history of *U. maydis* populations. Future comparative genomic studies of *U. maydis* are well advised to consider more in depth sampling compared to other fungal parasite model species.

Materials and Methods

Fungal isolates, primer design and sequencing

Isolates of *Ustilago maydis* used in this study were isolated from herbarium material originating from Europe, North America and South America. In total, 18 isolates representing 11 different localities were isolated and grown as described in [31] (Table S1). The isolate collection contained two isolates from teosinte (*Zea mays ssp. parviglumis*), the wild progenitor of maize. Genomic DNA of haploid *U. maydis* cultures was isolated by the method of [36]. Primers for *pep1*, cluster 2A and cluster 19A were designed based on the genome sequence of *U. maydis* (MUMDB). Primer properties were evaluated with OligoCalc [37,38] or Clonemanager v9.0 (Sci-Ed Software, Cary). ITS rDNA containing *5.8S* was amplified using the primers ITS1 and ITS4 [39]. *rpb1* was amplified using the primers RoK157 and RoK158 [31]. *ef1-α* was amplified using primers 983F and 2018R, and sequenced with primers 1567R and 2018R [40]. *gapdh* was amplified with primers G3PD-581F and G3PD-2020R, and sequenced with these external primers and the internal primer G3PD-1501R [41]. Detailed primer descriptions are given in Table S3.

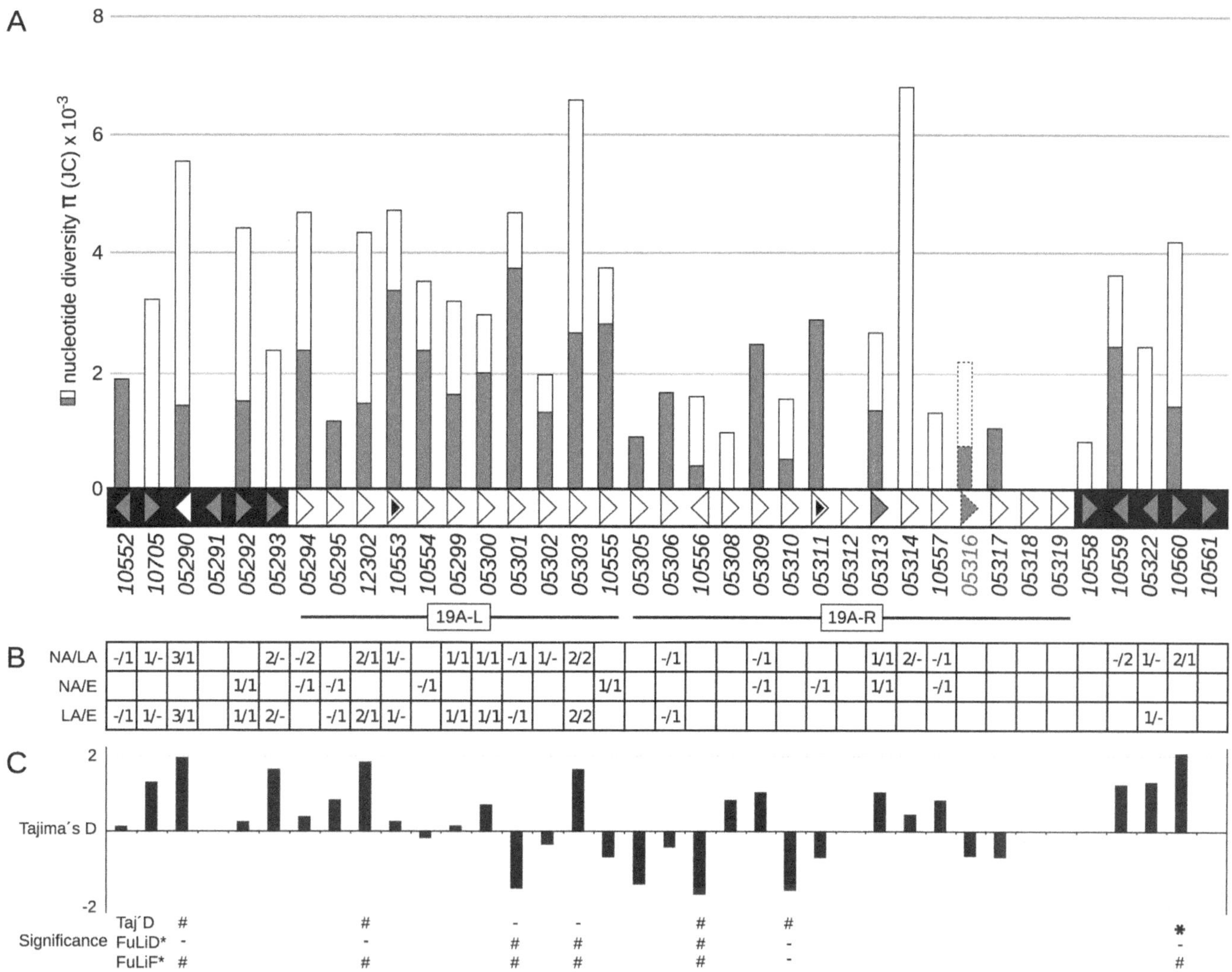

Figure 4. Cluster 19A nucleotide diversity π, fixed polymorphisms and neutrality tests. (A) Genes are illustrated in the genomic gene order. Arrowheads below the graph indicate their reading direction. White and grey arrows depict secreted and non-secreted protein-coding genes, respectively. Black sections are cluster-flanking regions. Black triangles highlight genes that revealed truncated alleles. Numbers correspond to *U. maydis* gene numbers in (MUMDB). Two-section columns height corresponds to nucleotide diversity π with Jukes-Cantor (JC) correction (scaled on left axis) and its sections reflect the ratio of non-synonymous (grey) and synonymous (white) substitutions. The pseudogene *um05316* is dotted. (B) Number of fixed polymorphisms between subpopulations of North America (NA), Europe (E) and Latin America (LA). First and second numbers refer to non-synonymous and synonymous substitutions, respectively. (C) Tajima's D statistics for gene datasets of cluster 19A. P-values of Tajima's D (Taj'D) and corresponding Fu & Li's D* (FuLiD) and F* (FuLiF) statistics are highlighted with # ($p<0.05$) and * ($p<0.01$).

For DNA amplification ≤5 kb, Phusion High-Fidelity Polymerase (Finnzymes, Espoo) and for amplification >5 kb, KOD Xtreme Polymerase (Merck Biosciences, Nottingham) were used following manufacturer's instructions. PCR products were purified directly or through gel purification using my-Budget Double Pure Kit (Bio-Budget, Krefeld). The complete gene *pep1* (537 bp) was sequenced from 16 isolates. The entire cluster 2A was amplified from twelve isolates spanning 20,725 bp from the left flanking gene *um01233* to the right flanking gene *um01242*. Sequencing of cluster 2A fragments was done by LGC Genomics (Berlin) applying 454 and shot gun sequencing. 26 genes of cluster 19A and 11 of its flanking genes were partially sequenced in nine isolates. PCR fragments were sequenced on an ABI 3130XL Genetic Analyser (Applied Biosystems) from the sequencing service of the Biochemistry Department at the Ruhr-Universität Bochum. Nucleotide sequences of *pep1*, *gapdh*, *ef1-α*, cluster 2A and cluster 19A genes have been deposited in GenBank under the accession numbers listed in Table S4.

Sequence alignment, gene identification and domain prediction

Obtained sequences were quality checked and edited by hand if needed using Sequencher 5.0 (Gene Codes Corporation, Ann Arbor). For phylogenetic and coalescence analyses, sequences were aligned with MAFFT 6.707 applying default settings and a maximum number of 1,000 iterations. Subsequently, leading and trailing gaps were manually removed from the alignment. Genes, exons, introns and intergenic regions within respective regions were identified by comparison with sequences of the *U. maydis* genome using the genome browser (MUMDB). In order to evaluate correlations of gene function and substitution patterns we adopted information from MUMDB and additively predicted

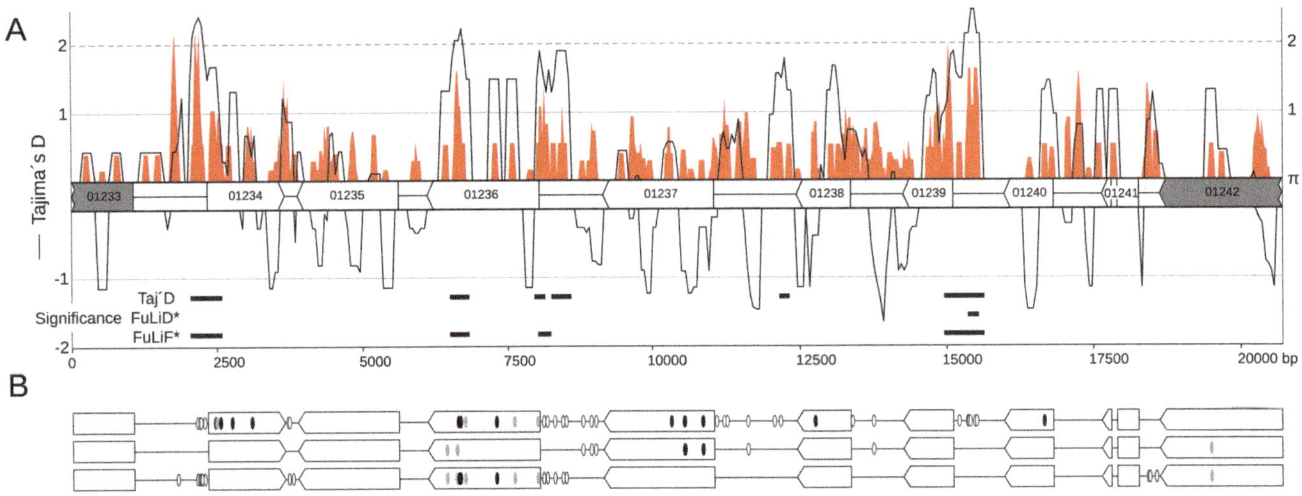

Figure 5. Cluster 2A nucleotide diversity π, fixed polymorphisms and neutrality tests. (A) Genes and non-coding regions are illustrated in the genomic gene order. Arrows indicate their reading direction. Numbers correspond to *U. maydis* gene numbers in [51]. Grey arrows depict genes that encode non-secreted proteins and white arrows depict genes that encode secreted proteins. The red graph illustrates sequence diversity π (x 10^{-3}) of a sliding window analysis (window length: 200; step size: 50) scaled on the right axis. The black curve illustrates a sliding window analysis (window length: 200; step size: 50) of Tajima's D statistics scaled on the left axis. Bars below the graph indicate segments with p-values <0.05 for Tajima's D (Taj'D) and corresponding Fu & Li's D* (FuLiD) and F* (FuLiF) statistics. (B) Fixed differences between datasets of subpopulations North America (NA), Europe (E) and Latin America (LA). Ovals highlight positions of fixed polymorphisms in non-coding (white) and coding regions comprising non-synonymous (black) and synonymous (grey) substitutions.

functional domains applying SMART blast searches [42,43]. For comparison with *S. reilianum* we used additional information from the genome browser [44]. Information on synteny conservation and genetic homology between *U. maydis* and *S. reilianum* were obtained from MUMDB.

Phylogenetic and coalescence analyses

Maximum Likelihood (ML) analyses were performed with RAxML 7.0.4 [45,46]. RAxML 7.0.4 conducted 1000 bootstrap replicates using a rapid bootstrap algorithm applying GTRMIX approximation [47]. In the subsequent ML search for the best scoring ML tree starting from each 5th bootstrap tree the more accurate GTRCAT approximation was applied. Bootstrap support values were mapped on the most likely tree that was visualised and edited in FigTree v1.3.1. For the analysis of cluster 19A, sequences were concatenated. For each partition, RAxML estimated and optimised individual α-shape parameters, GTR-rates and empirical base frequencies.

The Bayesian Markov Chain Monte Carlo (MCMC) method implemented in Beast v1.6.1 was used to estimate coalescence-based genealogies of cluster 19A and 2A datasets [48]. Coalescent-based gene genealogy of a combined analysis of all partial sequences of cluster 19A and its flanking genes was conducted applying the GTR+I+G model. For each partition Beast estimated individual base frequencies, α-shape parameters and GTR-rates. The entire cluster 2A was analysed applying the same settings as used for cluster 19A to each gene and each non-coding locus individually. To consider different demographic histories of *U. maydis* we, furthermore, applied and compared three different tree prior settings (constant size, exponential growth and expansion growth) in the coalescence analysis.

DNA polymorphism and analyses of linkage disequilibrium

Identification of single nucleotide polymorphisms (SNP), synonymous and non-synonymous mutations as well as further statistics of DNA polymorphism, nucleotide diversity π, summary statistics (Tajima's D, Fu & Li's D* and Fu & Li's F* [21,22]), ratios of Pa and Ps were conducted using DnaSP v5.10.01 [49]. For datasets of cluster 2A sliding window analyses (window length: 200, step size: 50) were performed. To analyse the differences in nucleotide diversity between datasets, we created boxplot diagrams [50] and tested for significant differences between datasets applying student's t-tests in R (www-R-project.org). In order to calculate mean π values of cluster 2A and 19A, respective loci were concatenated. Pairwise linkage disequilibrium was analysed at all polymorphic sites between all nucleotide variants. Sites containing alignment gaps, or polymorphic sites segregating for three or four nucleotides, were excluded from the analysis. Polymorphic sites with statistical significance for linkage disequilibrium ($p<0.05$ in two tailed Fisher's exact test and chi-square test) were plotted along the sequence using standard packages in R (www-R-project.org).

Supporting Information

Figure S1 Phylogeny of cluster 19A and cluster 2A. Maximum Likelihood phylogeny (RAxML 7.0.4) of sequence alignments of cluster 19A and cluster 2A from U. maydis. From cluster 19A partial coding sequences of 37 genes and from cluster 2A complete coding sequences of 8 genes were analysed. Bootstrap values of 1000 replicates are given next to branches (asterisks indicate values ≥93). Branch lengths are scaled with respect to the expected number of nucleotide substitutions per site.

Figure S2 Linkage disequilibrium dot plot cluster 19A. Test for linkage disequilibrium between nucleotide variants at all different polymorphic sites along cluster 19A. Sequences of cluster 19A are stringed together in genomic order. Sites containing alignment gaps, or polymorphic sites segregating for three or four nucleotides, are excluded from the analysis. Plotted are polymorphic sites with statistical significance for linkage disequilibrium (p<

0.05 in two tailed Fisher's exact test and chi-square test). 19A-L and 19A-R represent the sub-clusters of Figure 4.

Table S1 *U. maydis* strain collection and amplicon summary. Zmm: *Zea mays spp. mays*; Zmp: *Zea mays ssp. parviglumis*; Pop: amount of populations sampled; p.com.: personal communication; nd: not determined.

Table S2 Summary of sequence analyses. N: number of analysed *U. maydis* strains; haplotypes: number of haplotypes in the dataset; length: sequence length of complete genes in bp; sequenced: number of sequenced basepairs; SNPs: single nucleotide polymorphisms; nonsyn: non-synonymous mutations; syn: synonymous mutations; π (JC): nucleotide diversity calculated with DnaSP v4.0 applying the Jukes Cantor correction; TajD: Tajima's D summary statistics and corresponding probability value (SigTajD); FuLiD*: Fu & Li's D* summary statistics and corresponding probability value (SigLiD*); FuLiF*: Fu & Li's F* summary statistics and corresponding probability value (SigF*); #: p-value <0.05, *: p-value <0.01; NA: not available; NS: not significant.

Table S3 Primers used in this study.

Table S4 Accession numbers.

Table S5 Rates of Pa and Ps. Rates of Pa and Ps were calculated according to the number of non-synonymous and synonymous sites and non-synonymous and synonymous mutations per locus. stdv: standard deviation.

Table S6 Z-test statistics on rates of Pa and Ps. Codon-based test for neutral evolution averaged over all sequence pairs. The probability of rejecting the null hypothesis of strict-neutrality (Pa = Ps), positive selection (Pa > Ps) and purifying selection (Pa < Ps) is shown. Values of p less than 0.05 are considered significant at the 5% level. The test statistic (Pa - Ps) is shown in the Stat column. Pa and Ps are the numbers of synonymous and non-synonymous polymorphisms per site, respectively. The variance of the difference was computed using the bootstrap method (1,000 replicates). Analyses were conducted using the Nei-Gojobori method [52]. The analysis involved 13 and 10 coding sequences of cluster 2A and 19A, respectively. All positions with less than 95% site coverage were eliminated. Total amount of positions as amino acids in the final dataset is given in column "length in aa". Evolutionary analyses were conducted in MEGA5 [53].

Acknowledgments

We are thankful to Alistair McTaggart, Eva Stukenbrock, Evelyn Vollmeister and Kerstin Schipper for helpful suggestions and critical reading of the manuscript. We also thank Regine Kahmann at the Max Planck Institute Marburg for facility supply and fruitful discussions. Strain supply by Regine Kahmann, José Paulo Sampaío and Jan Schirawski is gratefully acknowledged.

Author Contributions

Conceived and designed the experiments: DB RK. Performed the experiments: RK CH. Analyzed the data: RK CH. Contributed reagents/materials/analysis tools: DB. Wrote the paper: RK.

References

1. Kohn LM (2005) Mechanisms of fungal speciation. Annu Rev Phytopathol 43: 279–308.
2. Stukenbrock EH, McDonald BA (2008) The Origins of Plant Pathogens in Agro-Ecosystems. Ann Rev Phytopathol 46: 75–100.
3. Brefort T, Doehlemann G, Mendoza-Mendoza A, Reissmann S, Djamei A, et al. (2009) *Ustilago maydis* as a Pathogen. Annu Rev Phytopathol 47: 423–445.
4. Kämper J, Kahmann R, Bölker M, Ma L-JJ, Brefort T, et al. (2006) Insights from the genome of the biotrophic fungal plant pathogen *Ustilago maydis*. Nature 444: 97–101.
5. Munkacsi AB, Stoxen S, May G (2008) *Ustilago maydis* populations tracked maize through domestication and cultivation in the Americas. Proc Biol Sci 275: 1037–1046.
6. Barnes CW, Szabo LJ, May G, Groth J V (2004) Inbreeding levels of two *Ustilago maydis* populations. Mycologia 96: 1236–1244.
7. Liu K, Goodman M, Muse S, Smith JS, Buckler E, et al. (2003) Genetic structure and diversity among maize inbred lines as inferred from DNA microsatellites. Genetics 165: 2117–2128.
8. Van Heerwaarden J, Doebley J, Briggs WH, Glaubitz JC, Goodman MM, et al. (2010) Genetic signals of origin, spread, and introgression in a large sample of maize landraces. Proc Natl Acad Sci U S A 108: 1088–1092.
9. Wright SI, Bi IV, Schroeder SG, Yamasaki M, Doebley JF, et al. (2005) The effects of artificial selection on the maize genome. Science (80-) 308: 1310–1314.
10. Tian F, Stevens NM, Buckler ES 4th (2009) Tracking footprints of maize domestication and evidence for a massive selective sweep on chromosome 10. Proc Natl Acad Sci U S A 106 Suppl: 9979–9986.
11. Hufford MB, Xu X, van Heerwaarden J, Pyhäjärvi T, Chia J-M, et al. (2012) Comparative population genomics of maize domestication and improvement. Nat Genet 44: 808–811.
12. Baumgarten AM, Suresh J, May G, Phillips RL (2007) Mapping QTLs contributing to *Ustilago maydis* resistance in specific plant tissues of maize. Theor Appl Genet 114: 1229–1238.
13. Bauer R, Mendgen K, Oberwinkler F (1995) Cellular interaction of the smut fungus *Ustacystis waldsteiniae*. Can J Bot 73: 867–883.
14. Bauer R, Oberwinkler F, Vánky K (1997) Ultrastructural markers and systematics in smut fungi and allied taxa. Can J Bot 75: 1273–1314.
15. Bauer R, Mendgen K, Oberwinkler F (2008) Cellular interaction of the smut fungus. 73: 867–883.
16. Catanzariti AM, Dodds PN, Ellis JG (2007) Avirulence proteins from haustoria-forming pathogens. FEMS Microbiol Lett 269: 181–188.
17. Basse CW, Stumpferl S, Kahmann R (2000) Characterization of a *Ustilago maydis* gene specifically induced during the biotrophic phase: evidence for negative as well as positive regulation. Mol Cell Biol 20: 329–339.
18. Stock J, Sarkari P, Kreibich S, Brefort T, Feldbrügge M, et al. (2012) Applying unconventional secretion of the endochitinase Cts1 to export heterologous proteins in *Ustilago maydis*. J Biotechnol 161(2): 80–91.
19. Schirawski J, Mannhaupt G, Munch K, Brefort T, Schipper K, et al. (2010) Pathogenicity determinants in smut fungi revealed by genome comparison. Science (80-) 330: 1546–1548.
20. Doehlemann G, van der Linde K, Assmann D, Schwammbach D, Hof A, et al. (2009) Pep1, a secreted effector protein of *Ustilago maydis*, is required for successful invasion of plant cells. PLoS Pathog 5: e1000290.
21. Tajima F (1989) Statistical method for testing the neutral mutation hypothesis by DNA polymorphism. Genetics 123(3): 585–595.
22. Fu YX, Li WH (1993) Statistical Tests for Neutrality of Mutations. Genetics 133: 693–709.
23. Nei Kumar (2000) Molecular Evolution and Phylogenetics. Oxford: Oxford University Press.
24. Stergiopoulos I, de Wit PJ (2009) Fungal effector proteins. Annu Rev Phytopathol 47: 233–263.
25. Stukenbrock EH, McDonald BA (2009) Population Genetics of Fungal and Oomycete Effectors Involved in Gene-for-Gene Interactions. Mol Plant Microbe Interact 22: 371–380.
26. Stukenbrock EH, Bataillon T, Dutheil JY, Hansen TT, Li R, et al. (2011) The making of a new pathogen: insights from comparative population genomics of the domesticated wheat pathogen *Mycosphaerella graminicola* and its wild sister species. Genome Res 21: 2157–2166.
27. Yang Z (1997) Likelihood Ratio Tests for Detecting Positive Selection and Application to Primate Lysozyme Evolution: 568–573.
28. Yang Z (2007) PAML 4: Phylogenetic Analysis by Maximum Likelihood. Mol Biol Evol 24: 1586–1591.
29. Laurie JD, Ali S, Linning R, Mannhaupt G, Wong P, et al. (2012) Genome comparison of barley and maize smut fungi reveals targeted loss of RNA silencing components and species-specific presence of transposable elements. Plant Cell 24: 1733–1745.

30. Stoll M, Begerow D, Oberwinkler F (2005) Molecular phylogeny of Ustilago, *Sporisorium*, and related taxa based on combined analyses of rDNA sequences. Mycol Res 109: 342–356.
31. Kellner R, Vollmeister E, Feldbrügge M, Begerow D (2011) Interspecific sex in grass smuts and the genetic diversity of their pheromone-receptor system. PLoS Genet 7: e1002436.
32. Meyer RS, Purugganan MD (2013) Evolution of crop species: genetics of domestication and diversification. Nat Rev Genet 14: 840–852.
33. Vollmeister E, Feldbrügge M (2010) Posttranscriptional control of growth and development in *Ustilago maydis*. Curr Opin Microbiol 13: 693–699.
34. Ferea T, Botstein D, O.Brown P, Rosenzweig R (1999) Systematic changes in gene expression patterns following adaptive evolution in yeast. PNAS 96: 9721–9726.
35. Pierce J V, Kumamoto CA (2012) Variation in Candida albicans EFG1 Expression Enables Host-Dependent Changes in Colonizing Fungal Populations. MBio 3. doi:10.1128/mBio.00117-12.Updated.
36. Hoffman CS, Winston F (1987) A ten-minute DNA preparation from yeast efficiently releases autonomous plasmids for transformation of *Escherichia coli*. Gene 57: 267–272.
37. Kibbe WA (2007) OligoCalc: an online oligonucleotide properties calculator. Nucleic Acids Res 35: W43–46 doi:10.1093/nar/gkm234.
38. Oligo-Calc: Oliginucleotide Properties Calculator. http://basic.northwestern.edu/biotools/OligoCalc.html.
39. White TJ, Bruns TD, Lee S, Taylor DR (1990) Amplification and direct sequencing of fungal ribosomal RNA genes for phylogenetic. In: Innis MA, Gelfand DH, Sninsky J, White TJ, editors. PCR Protocols: a Guide to Methods and Amplification. San Diego: Academic Press. pp. 315–322.
40. Rehner SA, Buckley E (2005) A *Beauveria* phylogeny inferred from nuclear ITS and EF1-alpha sequences: evidence for cryptic diversification and links to *Cordyceps* teleomorphs. Mycologia 97: 84–98.
41. Munkacsi AB, Stoxen S, May G (2007) Domestication of maize, sorghum, and sugarcane did not drive the divergence of their smut pathogens. Evolution (N Y) 61: 388–403.
42. Schultz J, Milpetz F, Bork P, Ponting CP (1998) SMART, a simple modular architecture research tool: identification of signaling domains. Proc Natl Acad Sci U S A 95: 5857–5864.
43. Letunic I, Doerks T, Bork P (2009) SMART 6: recent updates and new developments. Nucleic Acids Res 37: D229–32.
44. MSRDB (n.d.) MIPS *Sporisorium Reilianum* DataBase. http://mips.helmholtz-muenchen.de/genre/proj/sporisorium/.
45. Felsenstein J (1981) Evolutionary trees from DNA sequences: a maximum likelihood approach. J Mol Evol 17: 368–376.
46. Stamatakis A, Ludwig T, Meier H (2005) RAxML-III: a fast program for maximum likelihood-based inference of large phylogenetic trees. Bioinformatics 21: 456–463.
47. Stamatakis A, Hoover P, Rougemont J (2008) A rapid bootstrap algorithm for the RAxML Web servers. Syst Biol 57: 758–771.
48. Drummond AJ, Rambaut A (2007) BEAST: Bayesian evolutionary analysis by sampling trees. BMC Evol Biol 7: 214.
49. Librado P, Rozas J (2009) DnaSP v5: A software for comprehensive analysis of DNA poly- morphism data: 5–6.
50. Kirkman TW (1996) Statistics to use. http://www.physics.csbsju.edu/stats/.
51. MUMDB (n.d.) MIPS *Ustilago Maydis* DataBase. Available: http://mips.helmholtz-muenchen.de/genre/proj/ustilago.
52. Nei M, Gojobori T (1986) Simple methods for estimating the numbers of synonymous and nonsynonymous nucleotide substitutions. Mol Biol Evol 3: 418–426.
53. Tamura K, Peterson D, Peterson N, Stecher G, Nei M, et al. (2011) MEGA5: molecular evolutionary genetics analysis using maximum likelihood, evolutionary distance, and maximum parsimony methods. Mol Biol Evol 28: 2731–2739.

11

Expression and Molecular Evolution of Two *DREB1* Genes in Black Poplar (*Populus nigra*)

Yanguang Chu, Qinjun Huang, Bingyu Zhang, Changjun Ding, Xiaohua Su*

State Key Laboratory of Tree Genetics and Breeding, Research Institute of Forestry, Chinese Academy of Forestry, Key Laboratory of Tree Breeding and Cultivation, State Forestry Administration, Beijing, China

Abstract

Environmental stresses such as low temperature, drought, and high salinity significantly affect plant growth and yield. As selective forces, these adverse factors play essential roles in shaping phenotypic variation in plant populations. Black poplar (*Populus nigra*) is an economically and ecologically important forest tree species with widely distributed populations and is thus suitable for experiments detecting evolutionary footprints left by stress. Here, we performed expression and evolutionary analysis of two duplicated DREB A1-subgroup (DREB1) genes, *PnDREB68* and *PnDREB69*, encoding transcription factors that are involved in stress responses. The two genes showed partially overlapping but distinct expression patterns in response to stresses. These genes were strongly and rapidly induced by cold stress in leaves, stems, and roots. In leaf tissue, dehydration stress induced the expression of *PnDREB68* but not *PnDREB69*. *PnDREB69* displayed more rapid responses and longer expression durations than *PnDREB68* under salt and ABA stress, respectively. Based on single nucleotide polymorphism (SNP) analysis, we found significant population genetic differentiation, with a greater F_{ST} value (0.09189) for *PnDREB69* than for *PnDREB68* (0.07743). Nucleotide diversity analysis revealed a two-fold higher π_T for *PnDREB68* than for *PnDREB69* (0.00563 vs. 0.00243), reflecting strong purifying selection acting on the former. The results suggest that positive selection acted on *PnDREB69*, as evidenced by neutral testing using Tajima's *D* statistic. The distinct selective forces to which each of the genes was subjected may be associated with expression divergence. Linkage disequilibrium (LD) was low for the sequenced region, with a higher level for *PnDREB68* than for *PnDREB69*. Additionally, analysis of the relationship among carbon isotope ratios, SNP classes and gene expression, together with motif and domain analysis, suggested that 14 polymorphisms within the two genes may be candidates for an association study of important traits such as water use efficiency/drought tolerance in black poplar.

Editor: Manoj Prasad, National Institute of Plant Genome Research, India

Funding: This work was financially supported by the National Natural Science Foundation of China (Grant No. 31000310), the National High-tech R&D Program of China (Grant No. 2011AA100201), and the National Key Technology R&D Program for the Eleventh Five-Year-Plan of China (Grant No. 2006BAD01A15). The funders had no role in study design, data collection and analysis, decision to publish, or preparation of the manuscript.

Competing Interests: The authors have declared that no competing interests exist.

* E-mail: suxh@caf.ac.cn

Introduction

The growth and development of plants are frequently challenged by environmental stresses such as drought, high salinity, and temperature change [1], resulting in a significant reduction in productivity in both crops and forest trees. These adverse factors represent major selective forces that act on plant phenotypes [2]. Plant species with widely distributed natural populations constantly exhibit phenotypic differentiation in many traits due to adaptation to their specific environments. This intraspecific variation is thought to be largely shaped by the forces of natural selection [3,4]. Thus, understanding the genetic basis underlying adaptive phenotypic variation is an important aim for both evolutionary genetics and molecular breeding in plants [5].

Stress tolerance is a quantitative trait that is often modulated by a group of transcription factor genes. CBF/DREB (C-repeat binding factor/dehydration-responsive element binding) is a gene family belonging to the AP2/ERF (APETALA2/EREBP) superfamily in plants, which encodes a large number of transcription factors [6]. CBF/DREB genes specifically bind to the C-repeat (CRT)/dehydration-responsive element (DRE) motif and regulate the expression of various stress-responsive genes that harbor these elements [7]. The DREB subfamily is composed of six subgroups of genes (A1–A6) [8]. Among these genes, in addition to containing a highly conserved AP2/ERF DNA-binding domain that is ubiquitous to all DREB genes, most A1-subgroup (DREB1) proteins possess conserved sequences including the nuclear localization signal (NLS) sequence PKRAGRTKFRETRHP, the DSAW motif, and the LWSY motif [9]. DREB1 genes, namely *CBF1–CBF4*, *DDF1*, and *DDF2*, were first identified in the model plant *Arabidopsis thaliana* [10–13]. Genes belonging to the DREB1 subgroup were subsequently identified and functionally characterized in many annual herbaceous plants, such as rice (*Oryza sativa*), wheat (*Triticum aestivum*), barley (*Hordeum vulgare*), maize (*Zea mays*), and *Atriplex hortensis* [7]. Regarding perennial woody plants, studies of *DREB1* genes have been only documented in a limited number of species, including *Populus* spp [14,15], *Eucalyptus* spp [16,17], and peach (*Prunus persica*) [18,19].

The majority of studies of DREB genes have focused on the molecular functions of individual gene; an understanding of the mechanisms of functional divergence among DREB genes, and studies aimed at detecting the evolutionary footprints on these genes, are currently lacking. A previous study has proposed that the functional divergence of three duplicated members of the DREB1 gene family (CBF1, CBF2, and CBF3) may have resulted from specific sequence polymorphisms that were produced by contrasting evolutionary forces in *Arabidopsis* [20]. However, to date, the evolutionary history and molecular genetic analysis of the functional divergence of *DREB1* genes have not been reported for any long-lived woody tree species.

Poplars are employed as a model woody species in plant biology studies, particularly studies involving adaptation, secondary growth, and the natural variation of traits [21]. *Populus nigra* L. (black poplar) belongs to section *Aigeiros* within the genus *Populus* (family Salicaceae) [22]. *P. nigra* is predominantly outbreeding, with a wide distribution across Europe, the Mediterranean coast of Northern Africa, and Western and Central Asia [23]. *P. nigra* has long been used to obtain interspecific hybrids (e.g., *Populus.* × *euramericana*) with favorable traits for commercialized planting by crossing this plant with related species, such as *Populus deltoides*. Natural populations of *P. nigra* also play important roles in the balance of riparian ecosystems by restoring riverbanks and controlling flooding [24]. Populations of *P. nigra* are distributed across a wide range of areas, from arid regions in Italy to relatively humid regions in Germany. This ecological feature makes *P. nigra* a suitable species to investigate the genetic factors that contribute to the natural variation of adaptation-related traits in plants such as drought tolerance. In this study, we combined expression analysis and molecular evolution estimation, including nucleotide diversity, neutrality testing, and linkage disequilibrium, using polymorphism data from two duplicated *DREB1* genes, *PnDREB68* and *PnDREB69*, to infer their diverged expression patterns in response to stresses and the selection forces that may have contributed to this divergence. We also examine the potential implications of our results for association studies of drought tolerance/water use efficiency traits.

Materials and Methods

Ethics Statement

No specific permissions were required for the described field studies. The Jade Spring Hill Nursery Garden in Beijing is not privately-owned.

Plant Materials

One-year-old *P. nigra* (clone 'N21') trees were grown in a greenhouse under controlled conditions at the Chinese Academy of Forestry (CAF) for gene expression analysis. To examine molecular evolution based on single nucleotide polymorphisms and phenotypic analysis, a total of 65 unrelated individuals representing the major distribution of *P. nigra* were planted in a field trial at the Jade Spring Hill Nursery Garden in Beijing. These individuals were initially sourced from natural populations across Europe and Asia, including four geographic regions (WE, Western Europe; CE, Central Europe; SE, Southern Europe; and CA, Central Asia) that cover 15 countries. Detailed information about each sample tree is presented in the Table S1. Young, healthy leaf tissues were sampled from the 65 individuals, quickly frozen in liquid nitrogen, and stored at −70°C until use.

Cloning of cDNAs and Amplification of their Genomic DNA

A genome-wide analysis of the AP2/ERF superfamily of poplar (*Populus trichocarpa*) revealed six gene members of the DREB1 subgroup [25]. Within this subgroup, two *DREB1* genes, *DREB68* (XM_002299529) and *DREB69* (XM_002298031), are located on chromosome 1 and appear to be tandem duplicates arranged in a tail-to-tail manner. The two genes were not involved in the whole-genome duplication (WGD) event and the intergenic region between them is relatively short; thus, they are suitable for the analysis of expression divergence and molecular evolution. The counterparts in *P. nigra* were therefore named *PnDREB68* and *PnDREB69*. Total RNA was extracted from leaf tissue using an Ambion Plant RNA Isolation Aid (Life Technologies, Carlsbad, CA, USA) according to the manufacturer's instructions. Gene-specific primers were designed for cDNA cloning via RT-PCR based on the sequences of *DREB68* and *DREB69* transcripts and the publicly available sequences of the two loci from JGI (http://genome.jgi-psf.org/Poptr1_1/Poptr1_1.home.html; http://www.phytozome.net/search.php?method = Org_Ptrichocarpa) [26]. Genomic DNA for each individual tree was extracted from leaf tissue using a DNeasy Plant Mini Kit (Qiagen, Hilden, Germany) following the manufacturer's instructions. Primer pairs for amplifying genomic DNA in *P. nigra* were designed according to the genome assembly for *P. trichocarpa* described above. The web-based program Primer 3 [27] was used to design primers for RT-PCR and genomic DNA amplifications. Expected PCR products were purified and subcloned into the pGEM-T vector (Promega, Fitchburg, WI, USA). To amplify genomic DNA, reactions were carried out in a volume of 50 μl containing 50 ng of template DNA, 0.2 mM of each dNTP, 0.5 μM of each primer, 1× PCR buffer plus $MgCl_2$, and 0.5 U LA-Taq polymerase (TaKaRa, Dalian, China). All PCR reactions were run on a PE 9700 Thermal Cycler (Applied Biosystems, Foster City, CA, USA). Detailed information about the primers and PCR programs can be found in Table S2. Using a BigDye Terminator 3.1 Kit (Applied Biosystems), recombinant plasmids from RT-PCR were sequenced, and PCR products produced from genomic DNA templates were purified and subjected to direct sequencing. Sequencing reactions were conducted using an ABI 3730XL Automated DNA Sequencer (Applied Biosystems). Each amplicon was sequenced on both strands to generate a consensus sequence and to minimize PCR and sequencing errors.

Cis-acting elements within the predicted promoter sequences were deduced by searches in the PlantCARE and PLACE databases [28–30]. The deduced amino acid sequences of *PnDREB68* and *PnDREB69* were aligned using Clustal X [31] to perform a comparison analysis of gene structure.

Stress Treatments, qRT-PCR, and Microarray Analysis

Cuttings of *P. nigra* were cultivated in plastic pots containing an experiment-specific soil mixture in a greenhouse. When the plants were approximately 20 cm tall, they were exposed to cultivation solution containing 300 mM NaCl, 20% polyethylene glycol (PEG6000), or 100 mM abscisic acid (ABA) and grown at a low temperature (4°C) for 0 (normal growth conditions without stress), 2, 8, 24, or 48 h. Each treatment consisted of five replicates. Following the treatments, the leaves, stems, and roots were collected, frozen immediately in liquid nitrogen, and stored until use. Total RNA was extracted from the collected tissues using an Ambion Plant RNA Isolation Aid (Applied Biosystems) according to the manufacturer's instructions. The cDNA was synthesized using a PrimeScript RT Reagent Kit (TaKaRa). Then, qRT-PCR was performed in an ABI 7500 FAST Sequence Detector (Applied Biosystems) with SYBR Green Real-time PCR Master Mix (TaKaRa). Gene-specific primers were designed to amplify 120–130 bp fragments of *PnDREB68* and *PnDREB69*, and parallel PCRs were carried out using a gene-specific primer pair for poplar *ACTIN1* (GenBank Accession XM_002298674) as a reference gene. Primer sequences for the real-time PCR assay of *PnDREB68*,

PnDREB69, and *ACTIN1* are listed in Table S2. Four technical replicates for each PCR reaction were performed per RNA sample.

To perform comparisons of the orthologs within the genus *Populus*, expression profiles of *DREB68* and *DREB69* from *P. trichocarpa* in a range of samples were analyzed using the publicly available Affymetrix microarray data for poplar from the PopGenExpression server (http://www.barutoronto.ca) [32]. The dataset can also be downloaded from the Gene Expression Omnibus (http://ncbi.nlm.hih.gov/geo) under the accession number GSE13990. Probe sets for *P. trichocarpa DREB68* and *DREB69* gene models were retrieved using the Probe Match tool in the NetAffx Analysis Center (http://www.affymetrix.com/analysis/index.affx).

Phylogenetic Analysis

To infer the general evolutionary history of plant *DREB1* genes, amino acid sequences of representative members of the A1 group of DREB family (DREB1) proteins from five species (*P. trichocarpa* [25], *Arabidopsis thaliana*, *Oryza Sativa*, *Sorghum bicolor*, and *Vitis vinifera*) were randomly selected and retrieved from the whole-genome databases in Phytozome v9.1 at www.phytozome.net (Dataset S1). Thus, the sequences included in this analysis did not specifically consider genes created by WGD or tandem duplication. These sequences, along with those of *PnDREB68* and *PnDREB69*, were subjected to phylogenetic analysis using MEGA 5 [33]. Sequences from members of the A2 group of DREB family (DREB2) proteins in moss (*Physcomitrella patens*) were also downloaded from Phytozome and used as an outgroup (Dataset S1). The neighbor-joining method was used to build phylogenetic trees with the p-distance amino acid substitution model. Phylogeny testing was performed using 1,000 bootstrap replicates.

Nucleotide Diversity and Population Differentiation

Genomic DNA regions containing *PnDREB68* and *PnDREB69* were sequenced in 65 unrelated individuals. Sets of alleles for each locus were aligned using Clustal W [31]. Single nucleotide polymorphisms (SNPs) were identified and counted by visually inspecting each sequence alignment for each locus. The corresponding chromatogram files were checked to confirm each polymorphic site and its genotype class. Nucleotide diversity was estimated using DnaSP v5.10 [34] (http://www.ub.es/dnasp/). The levels of nucleotide diversity of the candidate genes were estimated as θw and π, which are based on the number of segregating sites [35] and the average number of nucleotide differences per site between sequences [36], respectively. Insertions and deletions (indels) were excluded from these analyses. Population genetic differentiation was estimated as the fixation index, F_{ST} [37,38] between pairs of populations (geographic regions in this study) using DnaSP. The significance of population differentiation was evaluated using the S_{nn}-test statistic of Hudson [39].

Tests of Neutrality

To test for departures from the standard Wright-Fisher model of neutral evolution, statistics including Tajima's D [40] and Fu and Li's D [41] were calculated for both whole-gene and coding regions, synonymous (syn) sites, nonsynonymous (nonsyn) sites, and silent (synonymous and noncoding) sites using DnaSP. Fu and Li's D is believed to be more sensitive than Tajima's D for detecting evolutionary events such as population expansion and positive selection [41]. Neutral testing was performed using SNP data obtained from each geographic region or whole population. The significance of these statistics was tested by 1,000 replicates of coalescent simulation based on Pi [42].

Linkage Disequilibrium

Pairwise linkage disequilibrium across the *PnDREB68* and *PnDREB69* loci was estimated as the squared correlation of allele frequencies, r^2 [43]. The value between pairs of SNPs (minor allele frequency >0.10) was measured using TASSEL software [44], with one-tailed Fisher's exact tests and Bonferroni corrections applied for each locus. The statistical significance of each value was determined using Fisher's exact tests.

Measurement of Carbon Isotope Ratio

Carbon isotope ratio ($\delta^{13}C$, ‰), or carbon isotope composition, was proposed as an indicator of water use efficiency, which can reflect drought tolerance in plants [45]. Leaf tissues were collected from the 65 clones in September. This time point was chosen because during September, bud set was observed in all trees, which indicates the end of the growing season. The leaves were dried in an oven at 70°C and ground to a fine powder. The samples then were sent to the MASS Laboratory of Institute of Atom at the Chinese Academy of Agricultural Sciences, and the carbon isotope ratio of each sample ($^{13}C/^{12}C_{sample}$) was measured using a MAT-251 Isotope Ratio Mass Spectrometer (Finnigan, Waltham, MA, USA). These ratios were expressed relative to the V-PDB standard (R_{PDB}), where $\delta^{13}C$ (‰) $= (^{13}C/^{12}C_{sample} - {}^{13}C/^{12}C_{PDB})/(^{13}C/^{12}C_{PDB}) \times 1{,}000$. Carbon isotope discrimination values were not shown, as the carbon signature of the trial field was not identified. For each treatment, the ninth leaf from each of three trees was collected and mixed for subsequent analysis. Thus, a total of nine leaves per individual were sampled and three replicated values were thereby obtained. Differences in carbon isotope ratios among 65 individuals, as well as among individuals for each SNP, were evaluated by one-way analysis of variance (ANOVA, $\alpha = 0.05$). Statistical analyses were performed using Data Processing System (DPS) 3.01 software [46].

Results

Characterization of cDNA and Genomic DNA Sequences at the *PnDREB68* and *PnDREB69* Loci

In *P. trichocarpa*, the two *DREB1* genes examined are annotated in NCBI as *DREB68* (XM_002299529) and *DREB69* (XM_002298031), and the corresponding gene models in the genome database are annotated as Potri.001G110800 and Potri.001G110700, respectively. Using primers designed with these sequences, the cDNAs with complete coding sequences (CDSs) for the two genes were isolated from *P. nigra* using RT-PCR (GenBank accession numbers KJ000090 and KJ000091). These two genes were designated as *PnDREB68* and *PnDREB69* according to the nomenclature for the *P. trichocarpa* orthologs in NCBI. *PnDREB68* and *PnDREB69* have 753-bp and 693-bp open reading frames encoding polypeptides of 251 and 231 amino acid residues, respectively. We also isolated 111 bp and 93 bp of the 5′UTR and 183 bp and 234 bp of the 3′UTR for *PnDREB68* and *PnDREB69*, respectively. These two genes share 57.32% nucleotide sequence and 54.75% amino acid sequence identities. A comparison of *PnDREB68* and *PnDREB69* with homologs in *P. trichocarpa* indicated homologous DNA sequence identities of 60.94% and 97.09%, and amino acid sequence homologies of 77.99% and 98.27%, against *DREB68* and *DREB69*, respectively.

We obtained 5,348 bp of the genomic DNA sequence across the *PnDREB68* and *PnDREB69* loci through four rounds of PCR, each of which produced a fraction of the entire sequenced region

(Figure 1A). This genomic region was re-sequenced in 65 unrelated individuals (GenBank accession numbers KJ019894–KJ020023). The structure of the amplified region is illustrated in Figure 1A, showing a tail-to-tail orientation for transcription of the two genes, which is consistent with that in *P. trichocarpa*. Within this region, *PnDREB69* is intronless, with a 696-bp CDS and a 996-bp 5′ upstream sequence, while *PnDREB68* possesses one intron (75 bp) that separated the CDS into two fragments (456 bp and 300 bp, respectively), and a 1,395 bp 5′ upstream sequence. Regarding the 5′-upstream sequences, we assigned the region upstream of the partial 5′UTR (obtained from cloned cDNA) as the promoter for *PnDREB69* (903 bp) and *PnDREB68* (1284 bp), as the transcription start sites (TSS) for the two genes were not experimentally determined. A 1,430-bp intergenic sequence between the translation stop codons of the genes was also amplified (Figure 1A). Two overlapping PCR primers (RP2 and FP3) (Table S2) covered a 45-bp sequence in the intergenic region (Figure 1A). We separated the intergenic sequence in the middle of the region comprising the two primers and assigned the upstream and downstream sequences to the *PnDREB69* and *PnDREB68* loci, respectively, to perform molecular genetic analyses for each gene. The cDNA sequences were further confirmed by analyses of their genomic DNA sequences.

A 903-bp promoter sequence of *PnDREB69* and *PnDREB68* were compared to infer whether differences in regulatory elements could contribute to the different expression profiles (see below) between the two genes. Various *cis*-acting elements were detected, in which eleven were found to be shared by *PnDREB69* and *PnDREB68* with differences in copy number in most cases (Table S3). Moreover, heat stress-responsive (HSE), drought-inducible (MBS), and auxin-responsive elements were detected in the promoter region of *PnDREB68* but not in *PnDREB69*, while a salicylic acid-responsive (TCA) element was only observed in the *PnDREB69* promoter. Alignment of amino acid sequences revealed conserved DREB1 signature sequences shared by both proteins, including a nuclear localization signal (NLS), an AP2 DNA-binding domain, and a DSAW motif (Figure 1B). A LWSY motif was also identified in PnDREB68, but it was modified to LWSS in PnDREB69. A putative C-terminus activation domain similar to *Arabidopsis* AtDREB1B/CBF1 could also be observed for PnDREB69 and PnDREB68 (Figure 1B).

Expression Patterns of *PnDREB68* and *PnDREB69* under Normal Growth Conditions and under Abiotic Stress

The expression patterns of the duplicated *DREB68* and *DREB69* genes were determined by quantitative RT-PCR under normal growth conditions, abiotic stress conditions, and ABA treatment. The expression of *PnDREB68* and *PnDREB69* was rapidly induced by cold stress in leaf and stem tissues (more than two-fold higher than control levels at the 2 h time point), and a slow response to cold stress was observed in roots (with higher relative abundance than the control beginning at the 8 h time point) (Figure 2A, 2B. and 2C). In most cases, *PnDREB69* had higher expression levels than *PnDREB68* when the plants were exposed to cold stress. When salt stress (NaCl) was applied, *PnDREB69* and *PnDREB68* displayed elevated expression levels in stems at the 8 and 48 h time point, respectively (Figure 2B), but the expression levels were reduced in roots (Figure 2C). Interestingly, in leaf tissue, dehydration (PEG6000) stress significantly induced the expression of *PnDREB68* at the early stage of

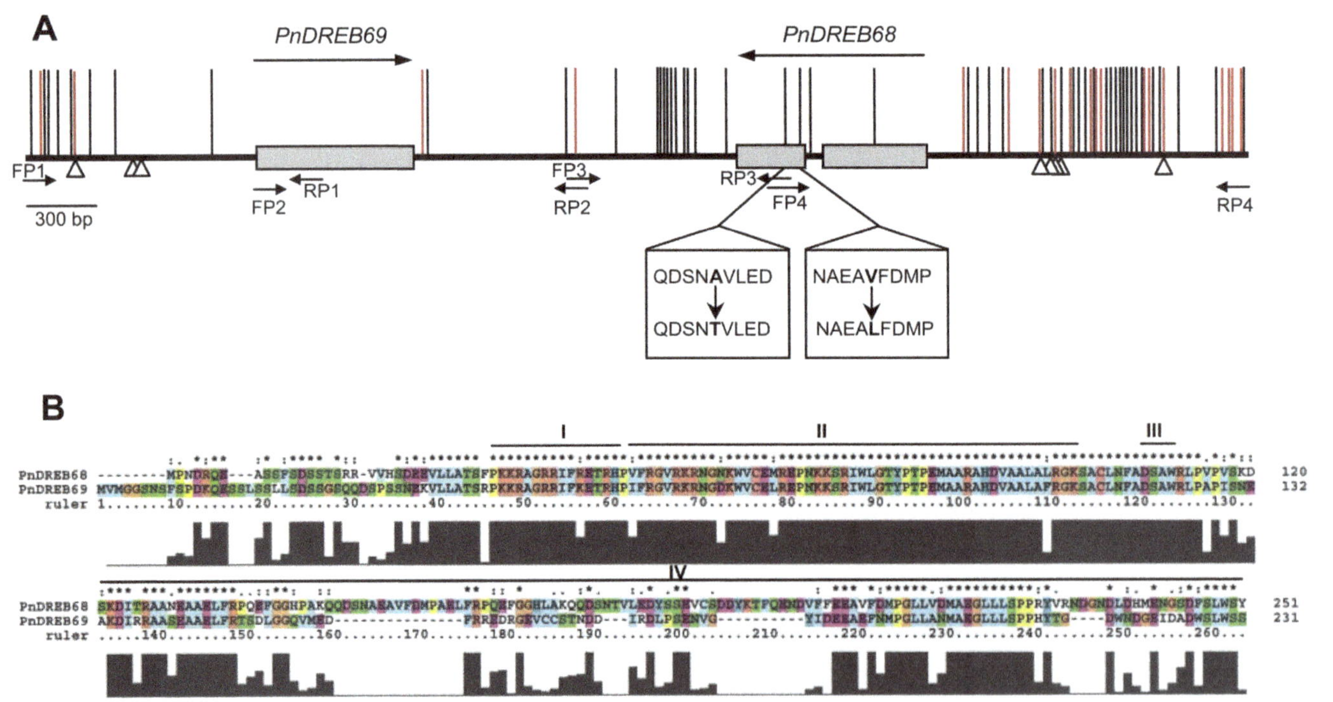

Figure 1. Genomic organization, polymorphism distribution (A), and comparison of protein sequences (B) of *PnDREB69* and *PnDREB68*. (A) Noncoding regions, including the promoter, intron, and intergenic sequence are represented by lines, and coding regions are indicated by boxes. The arrows represent the orientation of transcription (above the gene structure) or the locations of PCR primers (below the gene structure). The vertical lines above the gene structure indicate common SNPs (frequency >0.10) (black) and indels (red). Triangles below the gene structure indicate *cis*-acting elements with changed sequences caused by SNPs or indels. Two nonsynonymous SNPs that caused changed amino acid codons are illustrated below the coding regions. (B) Alignment of PnDREB69 and PnDREB68 proteins. Nuclear localization signal (NLS, I), AP2 domain (II), DSAW motif (III), and putative activation domain are indicated by lines.

treatment (2 and 8 h) but not the expression of *PnDREB69* (Figure 2A), whereas in stem tissue, both genes were slightly induced (with different relative abundances) by dehydration (Figure 2B). ABA treatment induced the expression of *PnDREB69* at all time points in leaves and stems, while *PnDREB68* showed elevated expression only at a single time point (24 h in leaves and 48 h in stems) (Figure 2A,2B). Very few transcripts were detected in root tissues under dehydration or ABA treatment conditions (Figure 2C), which suggested that the expression levels for both genes were reduced. The microarray analysis of orthologs from *P. trichocarpa* showed that the expression levels of *DREB68* were higher than those of *DREB69* across most samples, with the exception of xylem tissue. In mature leaf and seedlings grown in continuous darkness, the two genes exhibited inverse trend of transcript accumulations (Figure S1).

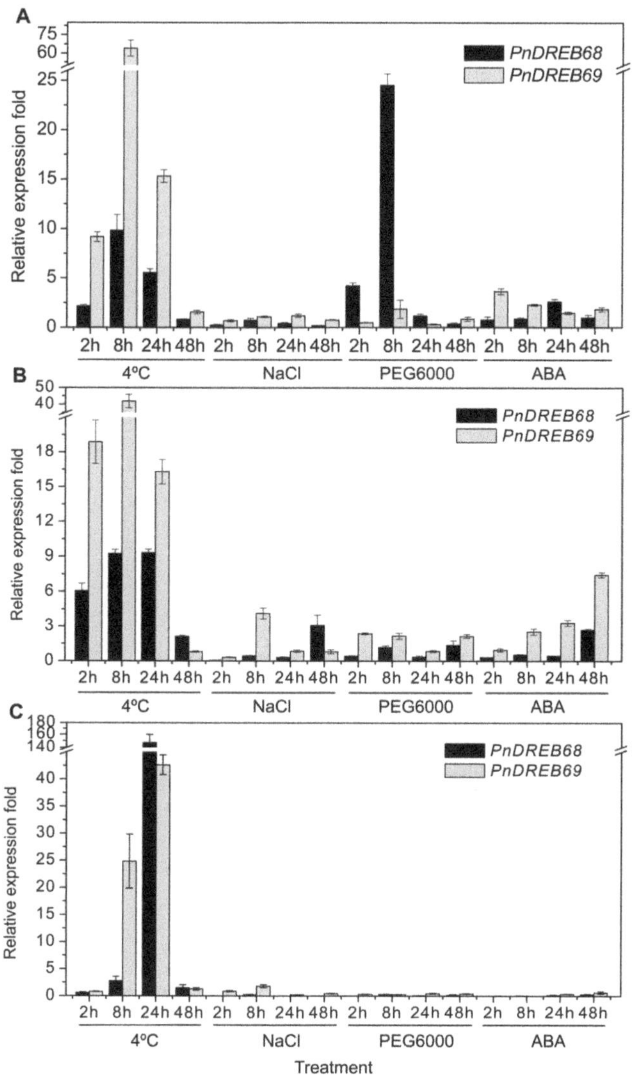

Figure 2. Quantitative RT-PCR analysis of expression patterns of *PnDREB69* and *PnDREB68* in response to abiotic stresses and ABA treatment in (A) leaf, (B) stem, and (C) root tissues. The expression levels for each gene after treatment were normalized to the expression level under normal growth conditions, which was set at 1.0. The relative expression levels of the genes were calculated using the standard $2^{-\Delta\Delta Ct}$ method.

Phylogenetic Relationships and Motif Composition

To investigate the evolutionary relatedness of duplicated *PnDREB68*/*PnDREB69* with the A1 group of DREB (*DREB1*) genes in other plant species, a NJ phylogenetic tree was constructed based on 26 sequences from *P. nigra* and seven additional land plant species. The rooted phylogenetic tree (based on DREB1 proteins) shows that *PnDREB69* and *PnDREB68* have the closest relationships with the homologs in *P. trichocarpa* (*DREB69* and *DREB68*) and *Populus tomentosa* (PtCBF2; Figure 3A). The phylogenetic tree forms two well-defined branches. One branch contains only sequences from dicots, while the other branch contains sequences from monocots, suggesting that the diversification of the *DREB1* genes occurred before the divergence of dicots and monocots.

Twenty distinct motifs were detected in the 26 represented DREB1 proteins using the MEME program (Figure 3B). Details of the 20 motifs were listed in Table S4. Common motif composition occurred in most of the closely related members, which suggested similar functional roles of the motifs among these proteins. Motifs 1, 2, and 3 were most commonly present in all tested proteins. Motif 8 was only found in dicots, while motif 19 was uniquely present in monocots. Additionally, motif 10 appeared to be specific to perennial woody plants (poplar and grape). The basal plant species moss had the lowest number of motifs [47], in which motif 17 was not found in other seed plant species.

Nucleotide Diversity and Neutrality Tests

Based on the alignment analysis of the 5,348-bp genomic DNA sequences from 64 unrelated individuals, fifty common SNPs (minor allele frequency >0.10) were detected, including three from coding regions and 47 from noncoding regions. Among these SNPs, 11 were identified within *PnDREB69*, while 39 (including three located in CDS) were detected from *PnDREB68* (Figure 1A). Two nonsynonymous SNPs were found in the coding region of *PnDREB68*, causing amino acid changes in the putative activation domain from Val and Ala to Leu and Thr, respectively (Figure 1A). A total of 18 indels were revealed, in which 14, 2, and 2 were located in the *PnDREB68* promoter, *PnDREB69* promoter and intergenic region, respectively (Figure 1A). Of these, eight polymorphic sites (four SNPs and four indels) occurred in several *cis*-acting elements, including the AT1-motif, Box 4, G-box, and TATA-box (Table 1).

Within the entire sequenced region, an analysis of the SNP data found that the promoters and intergenic region showed the highest level of nucleotide diversity relative to coding regions and intron (Figure 1A, Figure 4A). When the two loci were considered separately, a higher average nucleotide diversity was observed for *PnDREB68*, as evidenced by two-fold greater values of θ_W (0.00777) and π_T (0.00563), than those of *PnDREB69* ($\theta_W = 0.00389$, $\pi_T = 0.00243$) (Figure 4A, Table 2). The overall nucleotide diversity of samples from each geographic region of *PnDREB68* was also found to be higher. Except for the CA samples, larger values of π_{sil}, π_{syn}, and π_{nonsyn} for *PnDREB68* from the three remaining geographic regions (WE, CE, and SE) were observed, indicating higher levels of nucleotide diversity at these sites. A greater value for the π_{nonsyn}/π_{syn} ratio was also inferred for *PnDREB68* relative to that of *PnDREB69* (0.54984 vs. 0.31333). However, compared with *PnDREB68*, samples from CA at *PnDREB69* displayed much higher but similar levels of nucleotide diversity at the syn sites (0.00998 vs. 0.00183) and nonsyn sites (0.00195 vs. 0.00192), respectively. Tests of neutrality of the two genes produced generally negative values for Tajima's *D* and Fu and Li's *D*, except for CA at *PnDREB69* and SE at *PnDREB68* for Tajima's *D* and CA at *PnDREB68* for Fu and Li's *D*. Within the

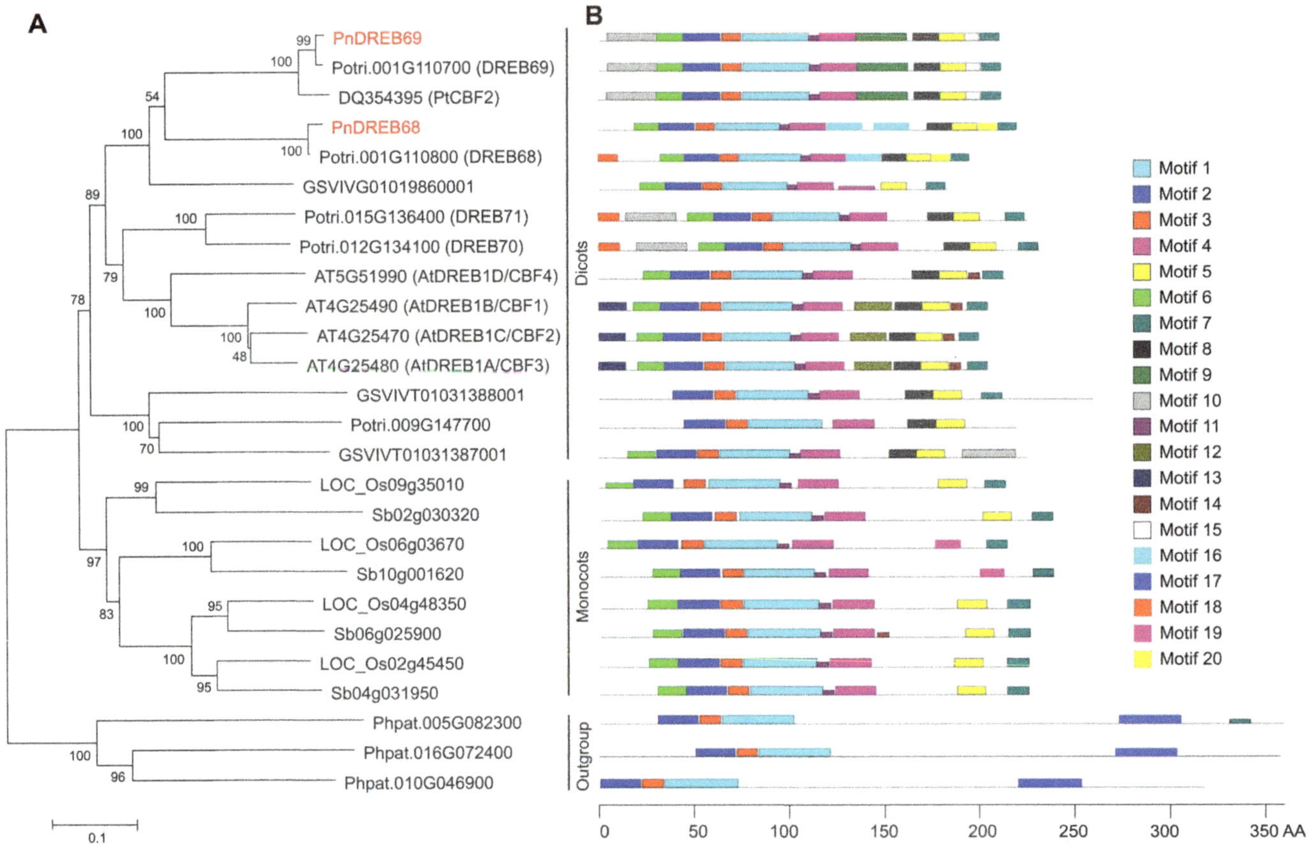

Figure 3. Phylogenetic relationships (A) and conserved motif analysis (B) of DREB1 members from poplar and other plant species. (A) Three DREB2 proteins from *Physcomitrella patens* (Phpat) were used as outgroups. The sequence of PtCBF2 from *P. tomentosa* was obtained from GenBank of the National Center for Biotechnology Information (http://www.ncbi.nlm.hih.gov/). Protein sequences of *P. trichocarpa* (Potri), *Arabidopsis thaliana* (AT), *Vitis vinifera* (GSVIV), *Oryza sativa* (LOC_Os), *Sorghum bicolor* (Sb), and *P. patens* were downloaded from Phytozome v9.1 (http://www.phytozome.net/). (B) Motifs are illustrated as boxes with different colors.

PnDREB69 locus, significantly negative Tajima's D values were estimated for the coding region ($D = -1.85645$, $P < 0.05$) and nonsyn sites ($D = -1.96094$, $P < 0.05$), suggesting that positive selection acted on these sites.

Population Differentiation

We estimated the genetic differentiation within pairs of *P. nigra* populations using the fixation index F_{ST}. Both genes showed significant genetic differentiation among populations, with a S_{nn} value of 0.36564 ($P = 0.0150$) and 0.38653 ($P = 0.0030$) for *PnDREB68* and *PnDREB69*, respectively. No significant pairwise estimate of F_{ST} was detected between populations. The results show that low to moderate values of F_{ST} were detected for six different population pairs (Western-Southern, Western-Central, Western-Central Asia, Southern-Central, Southern-Central Asia, and Central-Central Asia), ranging from −0.01400 and 0.01953

Table 1. Polymorphisms located in regulatory elements in the promoter regions of *PnDREB69* and *PnDREB68*.

Locus	Polymorphism	Position[a]	Frequency (%)	Regulatory element	Function of element
PnDREB69	T/C	−863	35.38	TATA-box	Core promoter element around −30 of transcription start
	A/G	−581	20.00	TATA-box	Core promoter element around −30 of transcription start
	1-base indel	−563	1.54	TATA-box	Core promoter element around −30 of transcription start
PnDREB68	1-base indel	−867	18.46	TATA-box	Core promoter element around −30 of transcription start
	C/A	−818	13.85	AT1-motif	Part of a light-responsive module
	1-base indel	−810	3.08	TATA-box	Core promoter element around −30 of transcription start
	A/T	−803	16.92	Box 4	Part of a conserved DNA module involved in light responsiveness
	16-base indel	−462	12.31	G-box	*cis*-Acting regulatory element involved in light responsiveness

[a]number of base pairs from the translation start site, as the transcription start sites of *PnDREB69* and *PnDREB68* were not experimentally determined.

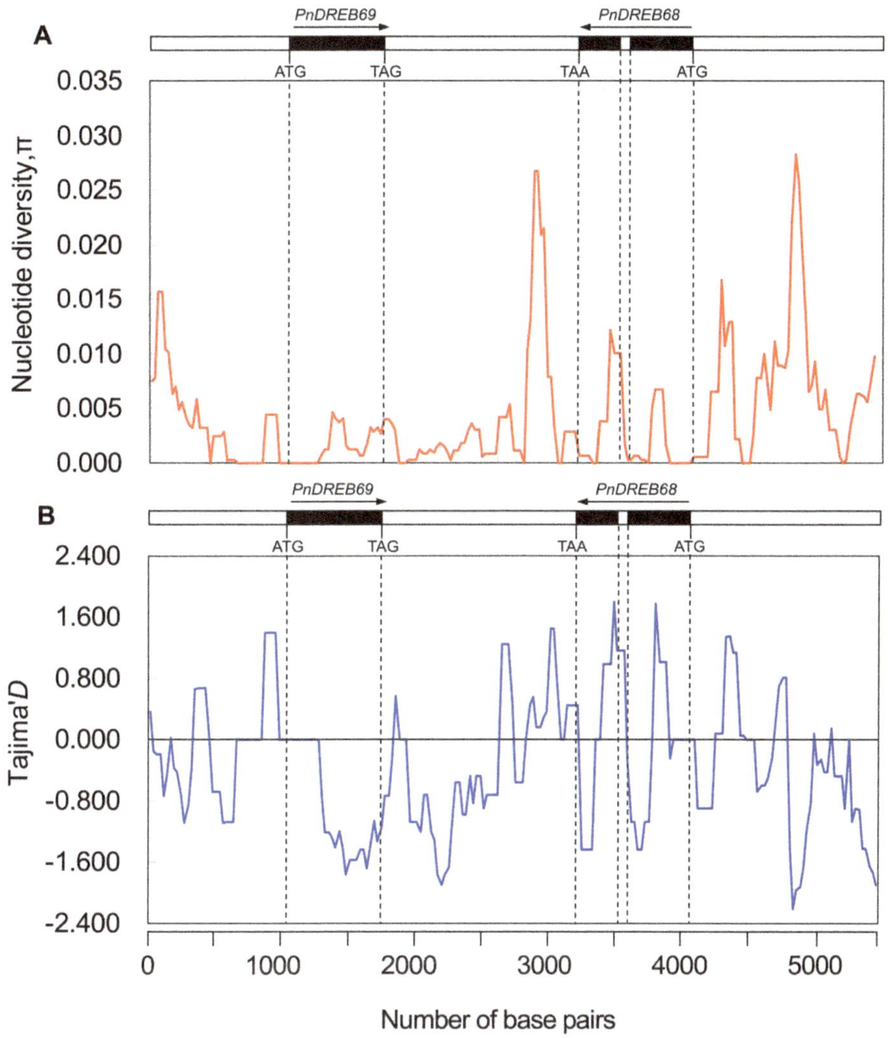

Figure 4. Nucleotide diversity (π) (A) and Tajima's *D* (B) patterns along the *PnDREB69* and *PnDREB68* loci. The sliding window approach was applied with a step size of 25 bp and a window size of 100 bp. The genomic organization of the sequenced region is shown above each plot. Dark boxes indicate coding regions, while white boxes indicate intron, intergenic sequences, or promoters. Arrows indicate transcriptional orientation.

to 0.17093 and 0.13471 for *PnDREB68* and *PnDREB69*, respectively. Higher F_{ST} values were detected when the samples from Central Asia was compared with each of the three geographic regions from Europe (Table 3). A comparison of Central Europe and Central Asia populations yielded the highest value of F_{ST} (0.17093) for *PnDREB68*, and a Southern Europe-Central Asia comparison yielded the highest value of F_{ST} (0.13471) for *PnDREB69*. A comparison between genes revealed a higher level of genetic differentiation in *PnDREB69* (F_{ST} value 0.09189) than in *PnDREB68* (F_{ST} value 0.07743).

Linkage Disequilibrium

The extent of linkage disequilibrium (LD) across the *PnDREB69* and *PnDREB68* loci, including the coding regions, intron, intergenic region, and promoters, were estimated using r^2 as the parameter. A total of 50 common SNPs (frequency >0.10) were included in this analysis, yielding 1,225 pairs of sites for the r^2 calculations. The average level of LD was low, with an r^2 value of 0.140251 for the entire region sequenced. In addition, r^2 values of 0.10641 and 0.21352 were estimated for the *PnDREB69* and *PnDREB68* loci, respectively. The architecture of LD is illustrated in Figure 5. The LD values varied across different regions, with higher values for 5′UTR (average $r^2 = 0.3031$) and intergenic regions and lower values for exons (average $r^2 = 0.0647$) and introns (average $r^2 = 0.0854$). Four (from SNP site 3864 to 4378) and five (from SNP site 4424 to 4688) SNPs in high LD with each other were found to form two distinct haplotype blocks (Figure 5). Within each haplotype block, LD among the SNPs was high ($r^2 >$ 0.8279).

Identification of Candidate SNPs Affecting the Carbon Isotope Ratio in *P. nigra*

The expression of *PnDREB68* was strongly induced by dehydration (PEG6000) stress in leaves, which was not observed for *PnDREB69*. PEG6000 treatment is frequently used to mimic drought stress [48]. Therefore, we further measured an important indicator for plant drought tolerance/water use efficiency (WUE), the carbon isotope ratio ($\delta^{13}C$), to investigate the natural variation of this phenotype and its potential relationship with SNP polymorphisms. The carbon isotope ratio showed abundant

Table 2. Nucleotide diversity and neutrality tests in *PnDREB69* and *PnDREB68* from *P. nigra*.

Locus	Geographic regions	Polymorphism									
		N	*S*	θ_W	π_T	π_{sil}	π_{syn}	π_{nonsyn}	π_{nonsyn}/π_{syn}	Tajima's *D*	Fu and Li's *D*
PnDREB69	WE	15	24	0.00309	0.00230	0.00398	0.00171	0.00096	0.56140	−1.19260	−1.69321
	CE	20	17	0.00201	0.00160	0.00190	0.00000	0.00056	0.00000	−0.76189	−0.90690
	SE	18	22	0.00268	0.00226	0.00286	0.00135	0.00021	0.15556	−0.60492	−0.99244
	CA	12	22	0.00305	0.00336	0.00377	0.00998	0.00195	0.19539	0.44303	0.38302
	Total	65	44	0.00389	0.00243	0.00286	0.00300	0.00094	0.31333	−1.24244	−1.40772
PnDREB68	WE	15	70	0.00737	0.00528	0.00620	0.00357	0.00164	0.45938	−1.23222	−1.67901
	CE	20	59	0.00571	0.00523	0.00617	0.00310	0.00149	0.48065	−0.34058	−0.14845
	SE	18	57	0.00569	0.00616	0.00733	0.00299	0.00153	0.51171	0.26964	0.15085
	CA	12	39	0.00439	0.00417	0.00473	0.00183	0.00192	0.49180	−0.23144	0.01017
	Total	65	107	0.00777	0.00563	0.00661	0.00311	0.00171	0.54984	−0.95162	−1.96457

WE, Western Europe; SE, Southern Europe; CE, Central Europe; and CA, Central Asia.

phenotypic variation; the values ranged from −25.99 to −30.40‰ (mean value of −28.49‰). ANOVA indicated a significant difference ($P<0.05$) in $\delta^{13}C$ values among the 65 individuals tested. This result suggests that the carbon isotope ratio trait bears significant natural variation in *P. nigra* populations and thus is suitable for population genetic analysis. We recorded the genotype (homozygous or heterozygous for each SNP) of each individual tree in our experimental panel (65 individuals), and the carbon isotope ratio of each individual was assigned to each genotyped SNP. For each SNP assayed, phenotypic data from trees with the same genotype were pooled and compared with each other. Significant differences in carbon isotope ratios were found among individuals for four SNPs (M645, M779, M374, and M519) (Figure 6A). M645 and M519 showed three genotype classes whereas M779 and M374 displayed only two genotype classes. All of these SNPs are located in promoter region of *PnDREB68*. No correlation was found between carbon isotope ratios and the SNPs located in *PnDREB69*.

To further examine whether *PnDREB68* expression in the different genotype classes of SNPs is correlated with carbon isotope ratio, expression levels of the four SNPs, M645, M779, M374, and M519, were compared using qRT-PCR. In this analysis, leaf tissues from 24 (8 for each of the three genotype classes of M645, M779 or M519, Figure 6A) or 16 (8 for each of the two genotype classes of M374, Figure 6A) individuals treated with PEG6000 for eight hours were used. These individuals were randomly selected from the 65 trees used for SNP identification and $\delta^{13}C$ measurement (Table S1). Among these SNPs examined, significant differences ($P<0.01$, ANOVA) in the expression levels among the genotype classes were found only in M519 (Figure 6B). Individuals of AC group showed the highest *PnDREB68* transcript levels, and this gene was expressed at the lowest level in the AA group. These data were negatively correlated with the $\delta^{13}C$ values observed for the genotype classes of M519 (Figure 6A). In these genotype classes, lower (AC group) and greater (AA group) $\delta^{13}C$ corresponded with higher and lower expression of *PnDREB68*, respectively.

Discussion

Gene duplication is believed to be a source of evolutionary forces leading to functional diversification [49,50]. The consequences of retaining duplicated genes can include the following: (1) retaining the function of its ancestors; (2) loss of gene function (pseudogenization); (3) retaining part of the function of its ancestors (subfunctionalization), and (4) acquiring a new function (neofunctionalization) [51]. Tandem duplication refers to the occurrence of two sequences, one following the other, in the same chromosomal segment. The most remarkable difference in expression patterns revealed in this study is that the expression of *PnDREB68* in leaves was dramatically triggered at the early stage of PEG6000 treatment, whereas no significant increase in expression of *PnDREB69* was observed under the same stress conditions. Expression of *PnDREB69* in leaves and stems was more sensitive to ABA than that of *PnDREB68*, as evidenced by higher expression of *PnDREB69* relative to that of *PnDREB68* throughout the duration of treatment. In leaf tissues, *PnDREB68* may trigger a series of dehydration/drought responses by regulating the expression of downstream genes, while the ABA signal may be essential for the expression and function of *PnDREB69* in both leaves and stems. This phenomenon may suggest functional divergence between the tandemly duplicated *PnDREB68* and *PnDREB69* genes involving in subfunctionalization and/or neofunctionalization, which need to be experimentally validated.

Table 3. Genetic population differentiation of *P. nigra* from four geographic regions.

Geographic regions	*PnDREB68* F_{ST}	*PnDREB69* F_{ST}
WE-SE	−0.00275	0.07663
WE-CE	−0.01400	0.03739
WE-CA	0.13004	0.12584
SE-CE	−0.00952	0.01953
SE-CA	0.16574	0.13471
CE-CA	0.17093	0.10741
Total	0.07743*	0.09189**

WE, Western Europe; SE, Southern Europe; CE, Central Europe; and CA, Central Asia. *$P<0.05$, **$P<0.01$.

The differences in the type and copy number of regulatory elements between the two genes suggested that these differences could be partly responsible for the differential expression of *PnDREB69* and *PnDREB68*. Specifically, the *PnDREB68* promoter possesses a drought-inducible (MBS) element that is absent in *PnDREB69*, which was consistent with the observation that the expression of *PnDREB68* was strongly induced by drought stress (PEG6000) in leaf tissues (Figure 2A). Moreover, the finding of sequence differences in the LWSY(S) motif and the putative activation domain also support the potential functional differences between *PnDREB69* and *PnDREB68*. Expression divergence of orthologs from the close relative, *P. trichocarpa*, was also revealed by microarray analysis, as reflected by distinct expression patterns across a range of tissues, organs and treatments.

The expression of *CBF* genes (DREB1 members) has been well-documented in other plant species. In *Populus balsamifera* subsp. *trichocarpa*, four *CBF* genes (*CBF1–CBF4*) are cold-inducible, with various levels of induction observed among genes, depending on the tissue (leaf or stem) in which they are expressed [14]. Divergent expression was also observed among three *CBF* genes in wild tomatoes. These genes are all cold-induced, with various transcript accumulation levels observed among accessions, and they are drought-responsive in *Solanum peruvianum*, although *CBF2* is not drought-responsive in *Solanum chilense* [52]. Genome-wide analysis of the expression pattern of *DREB1* genes has also been reported in *A. thaliana* and rice. The Arabidopsis DREB1 family is composed of six gene members. Among these, *CBF1–CBF3* are tandem duplicated and are involved in the cold-response pathway [53], while *CBF4* regulates drought adaptation [11], and *DDF1–DDF2* function in the salt response and in development [12,13]. In rice (*Oryza sativa* L.), six out of ten *DREB1* genes are expressed in response to low temperatures (4°C or 12°C), and two of these genes are also induced by drought or salt stress [54]. Previous studies on other plant species and our data from *P. nigra* show that the response to cold stress is a common phenomenon in plant *DREB1* genes, whereas other expression behaviors such as responses to drought, salt, and ABA are less common.

The phylogenic relationships and conserved motif analysis revealed diversification and a general evolutionary history of plant DREB1 proteins. Two distinct branches in the phylogenetic tree indicated that *DREB1* genes have common ancestors before the split of dicots and monocots (Figure 3A). MEME analysis revealed an increasing number of motifs from moss to seed plants, and several motifs are limited to specific species (dicots, monocots or woody plants) (Figure 3B). These data suggests a possible divergence of functions among different clades of plants in which these motifs are involved. Motif 17 in moss is absent in other seed plants (Figure 3B), which is probably because of a motif-loss event during the evolutionary history of DREB1 proteins.

Evaluations of the level of genetic differentiation can provide information about the extent of population structure, which is crucial for further genetic studies such as association mapping. Significant genetic differentiation in the natural populations was found for both loci (Table 3), where *PnDREB69* displayed a higher fixation index (F_{ST}) value than *PnDREB68*, indicating a greater level of population differentiation in the former. The degree of genetic differentiation observed in this study was somewhat lower than that previously reported for *P. nigra* (overall $F_{ST}=0.268$ based on AFLP, and 0.081 based on microsatellite analysis) [24], while still confirming the existence of considerable population subdivision in this species. The estimated level of population differentiation in *P. nigra* is comparable to estimates from other close relative poplar species including *Populus tremula* ($F_{ST}=0.117$ for five loci) [55], *P. balsamifera* (F_{ST} range, 0.018–0.256 for three loci) [56], and *P. tomentosa* (F_{ST} range, 0.3552–0.12686 for two loci) [57].

Investigating intraspecific variation of nuclear genes (mainly inferred from the level of nucleotide diversity) in natural populations is essential for understanding the evolutionary forces that act on plant genomes. In this study, we evaluated the nucleotide diversity of two tandemly duplicated *DREB1* genes. We detected two-fold higher total nucleotide diversity (π_T) for *PnDREB68* than for *PnDREB69* (Table 2), indicating that more relaxed evolutionary constraints occur in *PnDREB68* than in *PnDREB69*. This diversity originated from a much higher value of nucleotide diversity of silent sites (π_{sil}) for *PnDREB68* than for *PnDREB69*. The higher level of nucleotide diversity of *PnDREB68* may be a reflection of strong purifying selection acting on this gene. The levels of nucleotide diversity for the two *DREB1* genes are lower than previous estimates for nine loci in *P. nigra* [58]. A comparison within the genus *Populus* revealed significantly lower nucleotide diversity in *PnDREB* loci than in two sucrose synthase genes (*PtSUS1*/*PtSUS2*, $\pi_T=0.00924/0.01093$) in *P. tomentosa* [57]. For each gene, the nucleotide diversity of *PnDREB69* is similar to that of *P. balsamifera* ($\pi_T=0.0025$) [56] but lower than that of glycosyl-phosphatidylinositol-anchored protein gene (*PtCOBL4*) in *P. tomentosa* [59] and *P. tremula* ($\pi_T=0.0042$) [60]. However, *PnDREB68* showed comparable levels of diversity relative to *PtCOBL4* and to that in *P. tremula* but was two-fold higher than that in *P. balsamifera*. This interspecific variation in nucleotide diversity within the same genus can be explained by several factors, including differences in sampling procedures, the genomic regions studied, demographic histories, genetic backgrounds, and muta-

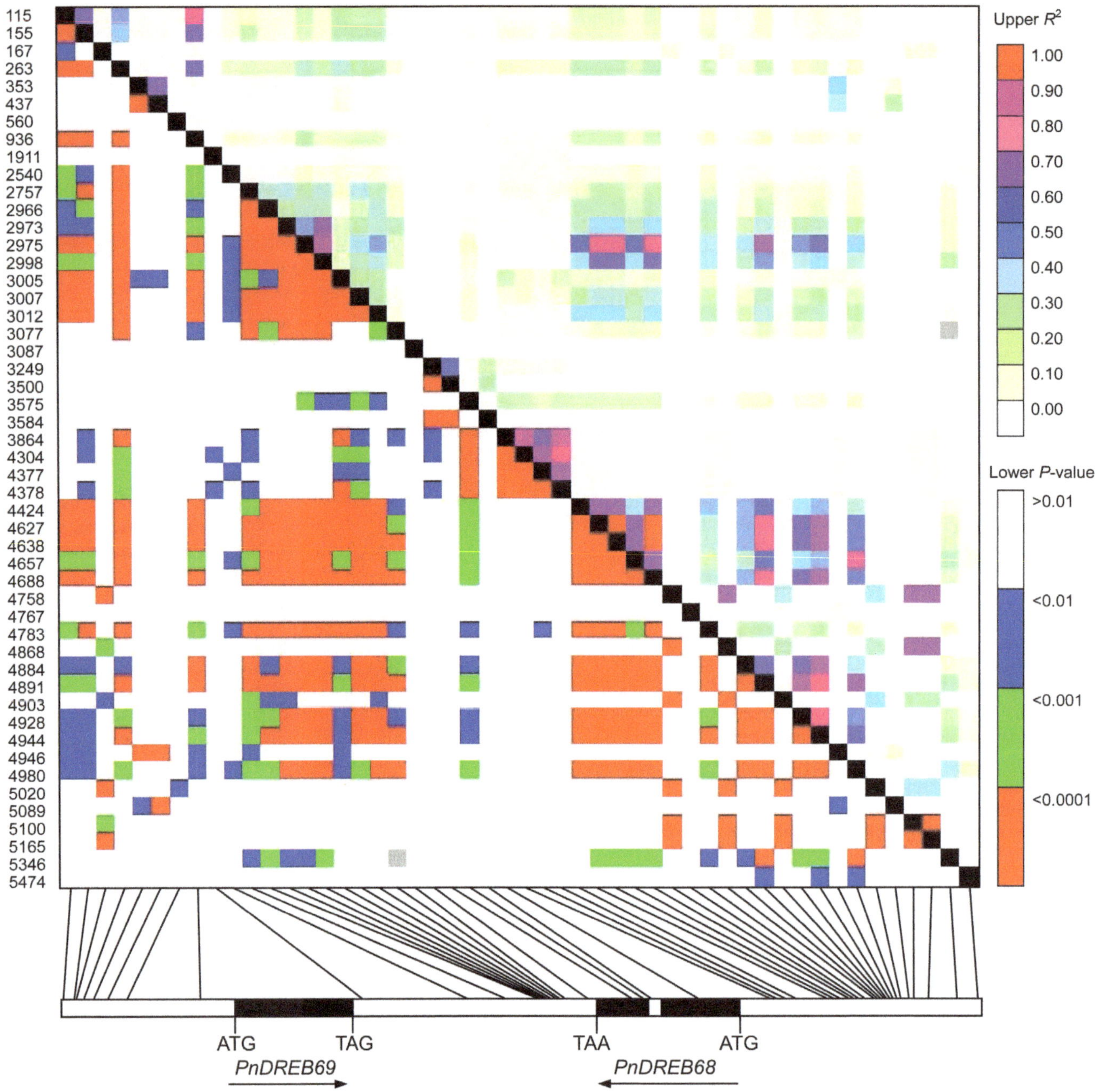

Figure 5. Pairwise linkage disequilibrium (LD) in sequenced region across the *PnDREB69* and *PnDREB68* loci. LD was estimated using the squared allelic correlation coefficient (r^2) between each pair of SNPs. Boxes below represent the genomic organization of the sequenced region as described in Figure 3. The physical position of each common SNP is labeled on the left and is linked to the corresponding location on the schematic diagram of the genomic region. Each grid in the upper-right matrix represents the level of LD reflected by r^2 for comparison of each SNP pair.

tion rates. Variation in nucleotide diversity was also demonstrated in other *DREB1* genes, including seven *CBF* genes in rye (*Secale cereale* L.; π ranging from 0.0015 to 0.0145) [61] and *Arabidopsis thaliana*, in which the π value of promoters for *CBF* genes ranges from 0.00109 to 0.02749, while the nucleotide diversity of transcript units ranges from 0.00264 to 0.00685 [20]. An analysis of orthologs of three *CBF* genes in wild tomatoes indicated that the average nucleotide diversity ranges from 0.021 to 0.079 in *S. chilense* and from 0.016 to 0.106 in *S. peruvianum* [52].

Tajima'*D* and Fu and Li's *D* tests are frequently used estimates that determine whether a gene has departed from neutral evolutionary expectations. The generally negative values for Tajima's *D* indicate an excess of rare variants in the two genes, which may have resulted from a recent population bottleneck (hitchhiking event) or a process such as background selection. Negative and not significant values for Tajima's *D* and Fu and Li's *D* were estimated for the entire region, coding sequence, synonymous sites, and nonsynonymous sites of *PnDREB68*, which is consistent with the results of nucleotide diversity analysis, suggesting that purifying selection was the main force driving the evolution of *PnDREB68*. By contrast, a much lower level of nucleotide diversity was detected for *PnDREB69*, with a smaller

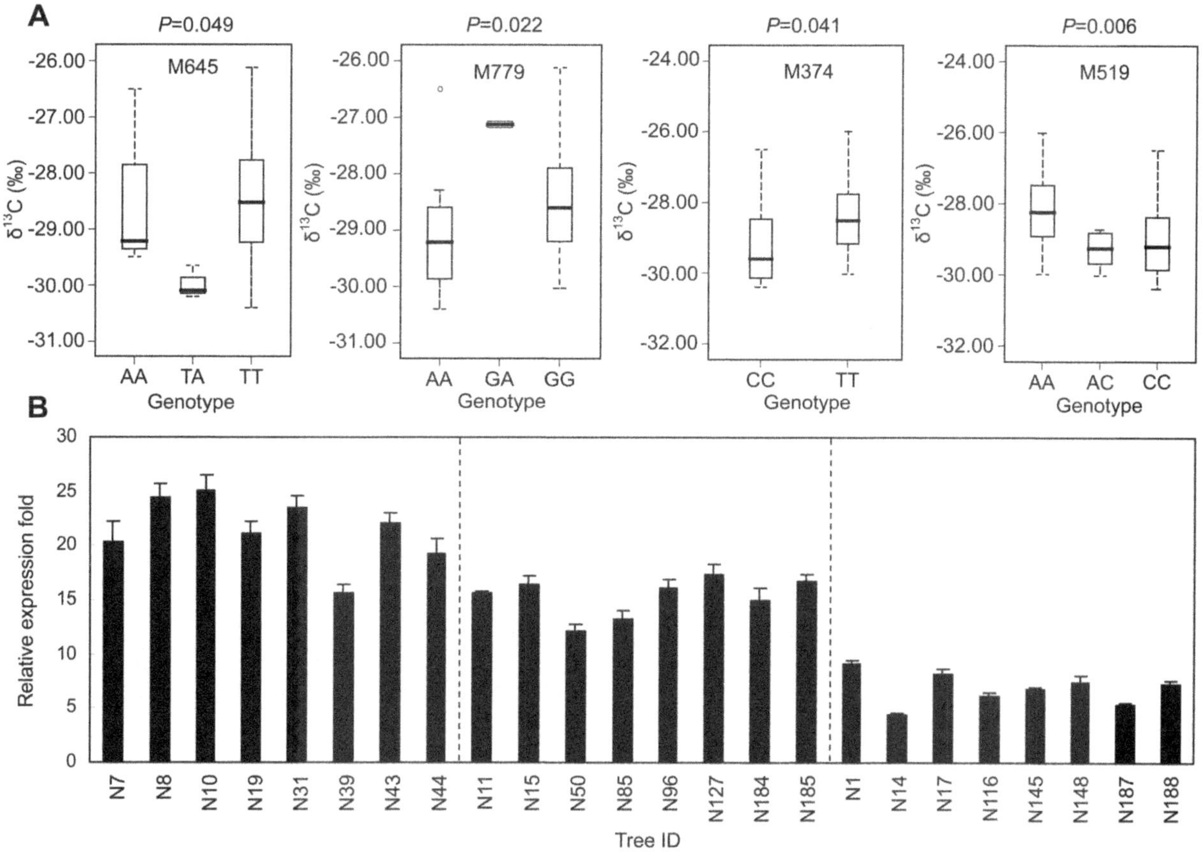

Figure 6. Effects of four SNPs (box plots) of *PnDREB68* on carbon isotope ratios ($\delta^{13}C$) (A) and *PnDREB68* expression of three genotype classes of M519 (B) in *P. nigra*. (A) Values of $\delta^{13}C$ were categorized by genotype class for each SNP. $N = 65$ for each SNP examined. (B) For qRT-PCR analysis, 24 representative individuals (8 for each of the three genotype classes) were selected from 65 trees used for SNP identification and $\delta^{13}C$ measurement. N7–N44 represent the AC group, N11–N185 represent the CC group, and N1–N188 represent the AA group. Expression levels among different genotype classes of M645, M779, or M374, were not significantly different (data not shown).

nonsynonymous/synonymous nucleotide substitutions (π_{nonsyn}/π_{syn}) ratio than that observed for *PnDREB68*. Furthermore, a significant negative value for Tajima's *D* for this locus was inferred for the coding region and for nonsynonymous sites of *PnDREB69*, supporting the hypothesis that this gene has undergone positive selection.

The response to dehydration stress in leaf tissue in *PnDREB68* was absent from the expression patterns observed for *PnDREB69*, which may be associated with positive selection acting on *PnDREB69*. The estimates of LD level for *PnDREB68* or *PnDREB69*, and the entire sequenced region spanning the two loci, are consistent with our previously calculated LD values for nine genes in black poplar, revealing low LD levels in this species [58]. Compared with other species in the genus *Populus*, the strength of LD of black poplar is slightly higher than that of *P. tremula* (squared allelic correlation coefficient, $r^2<0.1$; ~200 bp for 77 genes) [60] and the gibberellin 20-oxidase gene (*PtGA20Ox*) in *P. tomentosa* ($r^2 = 0.1$ within a 500-bp region) [62] but much lower than that of *P. balsamifera* ($r^2 = 0.52$ for 372 gene fragments) [63]. This interspecific variation in LD levels may be partially attributed to the different gene regions considered, and it also indicates differences in recombination history among species within the genus *Populus*. Despite its high nucleotide diversity, *PnDREB68* exhibited more extended LD than *PnDREB69*, indicating that more favorable mutations were fixed during the evolutionary history of *PnDREB68* than that of *PnDREB69*. This distinct spectrum of LD extension also explains the expression divergence of the two duplicated *DREB1* genes.

DREB family genes have recently been incorporated into association studies of stress-tolerance-related traits in crop species. Two and one nonsynonymous SNPs in *ScCbf15* and *ScCbf12*, respectively, were found to be associated with frost tolerance in rye (*Secale cereale* L.) [64]. Lata et al. reported an association between a synonymous SNP in the *DREB2* subgroup gene *SiDREB2* and dehydration tolerance in foxtail millet (*Setaria italica* L.) [65]. However, molecular evolution and/or association studies of *DREB* family genes have not previously been documented in any woody tree species. In this study, preliminary analysis of phenotypic data and SNP genotypes revealed four noncoding SNPs that may have affected the carbon isotope ratio in the genetic materials studied. QRT-PCR analysis revealed significant differences in expression levels among the three genotype classes of M519, supporting the contribution of this SNP to the natural variation of carbon isotope ratio of black poplar. Based on the fact that plants can discriminate against ^{13}C during photosynthesis and the isotopic ratio of ^{13}C to ^{12}C ($\delta^{13}C$) varies in C_3 plants, the important parameter related to drought tolerance, instantaneous water use efficiency (WUE), is found to be positively related to $\delta^{13}C$ [66,67]. This theory has been demonstrated in many plant species [68]. Our results suggest the potential importance of regulatory polymorphisms (located in gene promoters) in contributing to phenotypic variation [69]. Moreover, eight polymorphisms led to nucleotide changes in regulatory elements in *PnDREB69* and *PnDREB68*, including the AT1-motif, Box4, G-box, and TATA-box, which are important for light-regulated gene expression. As *cis*-acting elements in promoters are crucial for gene expression and changes in their sequences could influence gene regulation, further studies need to be conducted on these polymorphisms to address their potential effects on gene expression and phenotypic variation. Two nonsynonymous SNPs with unknown functions were also found in the putative activation domain of *PnDREB68*. An activation domain is essential for the activity of a protein, and this has been demonstrated in *Arabidopsis CBF1*, whose C-terminal 98 amino acids activation domain functions in *trans*-activation. The 14 polymorphisms described above, therefore, are candidates for further association studies of important traits in *P. nigra*, such as water use efficiency/drought tolerance.

In summary, we presented expression and molecular evolution analysis of two duplicated *DREB1* genes in black poplar, *P. nigra*. We detected expression divergence; the most remarkable difference was that the expression of *PnDREB68* varied in response to dehydration stress, which was not observed in *PnDREB69*. This result may suggest that functional divergence have occurred between the two genes. *PnDREB68* showed much higher nucleotide diversity than *PnDREB69*, suggesting that strong purifying selection has acted on the former gene. Evidence of positive selection on *PnDREB69* was inferred by neutrality tests, which revealed a significantly negative value of Tajima's *D* for its coding region. The different evolutionary forces acting on *PnDREB69* and *PnDREB68* may explain the expression divergence observed between these genes. Additionally, the two genes also displayed different degrees of population genetic differentiation and linkage disequilibrium. Finally, a set of polymorphisms in *PnDREB68* and *PnDREB69* may be candidates for association analysis of important traits, such as drought tolerance.

Supporting Information

Figure S1 Expression of *P. trichocarpa DREB69* and *DREB68* genes across a range of tissues, organs, and treatments. The patterns of relative transcript accumulation of the two genes were determined by microarray analysis. *Red* indicates higher levels and *blue* indicates lower levels of transcript accumulation. Each column represents the average of biological triplicates. ML, mature leaf; YL, young leaf; RT, root; CL, seedlings grown in continuous light; DL, seedlings grown in continuous darkness and then transferred to light for 3 h; CD, seedlings grown in continuous darkness; XY, differentiating xylem; FC, female catkins; and MC, male catkins.

Table S1 Geographical origin of the 65 individuals used in this study.

Table S2 PCR primers used for RT-PCR, qRT-PCR, and genomic DNA amplifications.

Table S3 Putative *cis*-acting elements in the promoter regions of *PnDREB69* and *PnDREB68*.

Table S4 Motif sequences of DREB1 proteins used in phylogenic analysis.

Dataset S1 Amino acid sequences of DREB1 and outgroup proteins used in phylogenic analysis.

Author Contributions

Conceived and designed the experiments: XS. Performed the experiments: YC QH BZ CD. Analyzed the data: YC. Wrote the paper: YC.

References

1. Debnath M, Pandey M, Bisen PS (2011) An omics approach to understand the plant abiotic stress. OMICS 15: 739–762.
2. Bohnert HJ, Nelson DE, Jensen RG (1995) Adaptations to environmental stresses. Plant Cell 7: 1099–1111.
3. Linhart YB, Grant MC (1996) Evolutionary significance of local genetic differentiation in plants. Annu Rev Ecol Syst 27 237–277.
4. MacColl AD (2011) The ecological causes of evolution. Trends Ecol Evol 26: 514–522.
5. Orr HA, Coyne JA (1992) The genetics of adaptation – a reassessment. Am Nat 140: 725–742.
6. Stockinger EJ, Gilmour SJ, Thomashow MF (1997) *Arabidopsis thaliana CBF1* encodes an AP2 domain-containing transcriptional activator that binds to the C-repeat/DRE, a *cis*-acting DNA regulatory element that stimulates transcription in response to low temperature and water deficit. Proc Natl Acad Sci USA 94: 1035–1040.
7. Lata C, Prasad M (2011) Role of DREBs in regulation of abiotic stress responses in plants. J Exp Bot 62: 4731–4748.
8. Sakuma Y, Liu Q, Dubouzet JG, Abe H, Shinozaki K, et al. (2002) DNA-binding specificity of the ERF/AP2 domain of *Arabidopsis* DREBs, transcription factors involved in dehydration- and cold-inducible gene expression. Biochem Biophys Res Commun 290: 998–1009.
9. Cong L, Chai TY, Zhang YX (2008) Characterization of the novel gene *BjDREB1B* encoding a DRE-binding transcription factor from *Brassica juncea* L. Biochem Biophys Res Commun 371: 702–706.
10. Gilmour SJ, Zarka DG, Stockinger EJ, Salazar MP, Houghton JM, et al. (1998) Low temperature regulation of the *Arabidopsis* CBF family of AP2 transcriptional activators as an early step in cold-induced COR gene expression. Plant J 16: 433–442.
11. Haake V, Cook D, Riechmann JL, Pineda O, Thomashow MF, et al. (2002) Transcription factor CBF4 is a regulator of drought adaptation in *Arabidopsis*. Plant Physiol 130: 639–648.
12. Magome H, Yamaguchi S, Hanada A, Kamiya Y, Oda K (2004) dwarf and delayed-flowering 1, a novel *Arabidopsis* mutant deficient in gibberellin biosynthesis because of overexpression of a putative AP2 transcription factor. Plant J 37: 720–729.
13. Magome H, Yamaguchi S, Hanada A, Kamiya Y, Oda K (2008) DDF1 transcriptional activator upregulates expression of a Gibberellin deactivating gene, *GA2ox7*, under high-salinity stress in *Arabidopsis*. Plant J 56: 613–626.
14. Benedict C, Skinner JS, Meng R, Chang Y, Bhalerao R, et al. (2006) The CBF1-dependent low temperature signalling pathway, regulon and increase in freeze tolerance are conserved in *Populus* spp. Plant Cell Environ 29: 1259–1272.
15. Maestrini P, Cavallini A, Rizzo M, Giordani T, Bernardi R, et al. (2009) Isolation and expression analysis of low temperature-induced genes in white poplar (*Populus alba*). J Plant Physiol 166: 1544–1556.
16. Navarro M, Marque G, Ayax C, Keller G, Borges JP, et al. (2009) Complementary regulation of four Eucalyptus *CBF* genes under various cold conditions. J Exp Bot 60: 2713–2724.
17. Navarro M, Ayax C, Martinez Y, Laur J, El Kayal W, et al. (2011) Two *EguCBF1* genes overexpressed in *Eucalyptus* display a different impact on stress tolerance and plant development. Plant Biotechnol J 9: 50–63.
18. Wisniewski M, Norelli J, Bassett C, Artlip T, Macarisin D (2011) Ectopic expression of a novel peach (*Prunus persica*) CBF transcription factor in apple (*Malus × domestica*) results in short-day induced dormancy and increased cold hardiness. Planta 233: 971–983.
19. Artlip TS, Wisniewski ME, Bassett CL, Norelli JL (2013) CBF gene expression in peach leaf and bark tissues is gated by a circadian clock. Tree Physiol 33: 866–877.
20. Lin YH, Hwang SY, Hsu PY, Chiang YC, Huang CL, et al. (2008) Molecular population genetics and gene expression analysis of duplicated *CBF* genes of *Arabidopsis thaliana*. BMC Plant Biol 8: 111.
21. Jansson S, Douglas CJ (2007) *Populus*: a model system for plant biology. Annu Rev Plant Biol 58: 435–458.
22. Eckenwalder JE (1996) Systematic and evolution of *Populus*. In: Stettler RF, Bradshaw HD Jr, Heilman PE, Hinckley TM (eds) Biology of *Populus* and its implications for management and conservation. NRC Research Press, National Council of Canada, Ottawa, 7–32.

23. Vanden BA (2003) EUFORGEN Technical Guidelines for genetic conservation and use for European black poplar (*Populus nigra* L.). International Plant Genetic Resources Institute, Rome, 1–6.
24. Smulders MJM, Cottrell JE, Lefèvre F, van der Schoot J, Arens P, Vosman B, et al. (2008) Structure of the genetic diversity in black poplar (*Populus nigra* L.) populations across European river systems: Consequences for conservation and restoration. Forest Ecol Manag 255: 1388–1399.
25. Zhuang J, Cai B, Peng RH, Zhu B, Jin XF, et al. (2008) Genome-wide analysis of the AP2/ERF gene family in *Populus trichocarpa*. Biochem Biophys Res Commun 371: 468–474.
26. Tuskan GA, Difazio S, Jansson S, Bohlmann J, Grigoriev I, et al. (2006) The genome of black cottonwood, *Populus trichocarpa* (Torr. & Gray). Science 313: 1596–1604.
27. Rozen S, Skaletsky HJ (2000) Primer 3 on the WWW for general users and for biologist programmers. In: Krawetz S, Misener S (eds) Bioinformatics: Methods and Protocols (Methods in Molecular Biology). Humana Press, Totowa, NJ, 365–386.
28. Prestridge DS (1991) SIGNAL SCAN: a computer program that scans DNA sequences for eukaryotic transcriptional elements. Comput Appl Biosci 7: 203–206.
29. Higo K, Ugawa Y, Iwamoto M, Korenaga T (1999) Plant *cis*-acting regulatory DNA elements (PLACE) database: 1999. Nucleic Acids Res 27: 297–300.
30. Lescot M, Déhais P, Thijs G, Marchal K, Moreau Y, et al. (2002) PlantCARE, a database of plant *cis*-acting regulatory elements and a portal to tools for in silico analysis of promoter sequences. Nucleic Acids Res 30: 325–327.
31. Larkin MA, Blackshields G, Brown NP, Chenna R, McGettigan PA, et al. (2007) Clustal W and Clustal X version 2.0. Bioinformatics 23: 2947–2948.
32. Wilkins O, Nahal H, Foong J, Provart NJ, Campbell MM (2009) Expansion and diversification of the *Populus* R2R3-MYB family of transcription factors. Plant Physiol 149: 981–993.
33. Tamura K, Peterson D, Peterson N, Stecher G, Nei M, et al. (2011) MEGA5: Molecular evolutionary genetics analysis using maximum likelihood, evolutionary distance, and maximum parsimony methods. Mol Biol Evol 28: 2731–2739.
34. Librado P, Rozas J (2009) DnaSP v5: a software for comprehensive analysis of DNA polymorphism data. Bioinformatics 25: 1451–1452.
35. Watterson GA (1975) On the number of segregating sites in genetical models without recombination. Theor Popul Biol 7: 256–276.
36. Nei M, Li WH (1979) Mathematical model for studying genetic variation in terms of restriction endonucleases. Proc Natl Acad Sci USA 76: 5269–5273.
37. Hudson RR, Boos DD, Kaplan NL (1992) A statistical test for detecting geographic subdivision. Mol Biol Evol 9: 138–151.
38. Charlesworth B (1998) Measures of divergence between populations and the effect of forces that reduce variability. Mol Biol Evol 15: 538–543.
39. Hudson RR (2000) Genetics. A new statistic for detecting genetic differentiation 155: 2011–2014.
40. Tajima F (1989) Statistical method for testing the neutral mutation hypothesis by DNA polymorphism. Genetics 123: 585–595.
41. Fu YX, Li WH (1993) Statistical tests of neutrality of mutations. Genetics 133: 693–709.
42. Nei M (1987) Molecular evolutionary genetics. Columbia University Press, New York, NY.
43. Remington DL, Thornsberry JM, Matsuoka Y, Wilson LM, Whitt SR, et al. (2001) Structure of linkage disequilibrium and phenotypic associations in the maize genome. Proc Natl Acad Sci USA 98: 11479–11484.
44. Bradbury PJ, Zhang Z, Kroon DE, Casstevens TM, Ramdoss Y, et al. (2007) TASSEL: Software for association mapping of complex traits in diverse samples. Bioinformatics 23: 2633–2635.
45. Martin B, Thorstenson YR (1988) Stable carbon isotope composition ($\delta^{13}C$), water use efficiency, and biomass productivity of *Lycopersicon esculentum*, *Lycopersicon pennellii*, and the F_1 Hybrid. Plant Physiol 88: 213–217.
46. Tang QY, Feng MG (2002) DPS data processing system for practical statistics. Beijing: Science Press.
47. Bailey TL, Williams N, Misleh C, Li WW (2006) MEME: discovering and analyzing DNA and protein sequence motifs. Nucleic Acids Res 34: W369–373.
48. Chazen O, Hartung W, Neumann PM (2006) The different effects of PEG 6000 and NaCl on leaf development are associated with differential inhibition of root water transport. Plant Cell Environ 18: 727–735.
49. Ohno S (1970) Evolution by gene duplication. Springer-Verlag, Heidelberg, Germany.
50. Lynch M, Conery JS (2000) The evolutionary fate and consequences of duplicate genes. Science 290: 1151–1155.
51. Lan T, Yang ZL, Yang X, Liu YJ, Wang XR, et al. (2009) Extensive functional diversification of the *Populus* glutathione S-transferase supergene family. Plant Cell 21: 3749–3766.
52. Mboup M, Fischer I, Lainer H, Stephan W (2012) Trans-species polymorphism and allele-specific expression in the *CBF* gene family of wild tomatoes. Mol Biol Evol 29: 3641–3652.
53. Chinnusamy V, Zhu JH, Zhu JK (2007) Cold stress regulation of gene expression in plants. Trends Plant Sci, 12: 10.
54. Mao D, Chen C (2012) Colinearity and similar expression pattern of rice *DREB1s* reveal their functional conservation in the cold-responsive pathway. PLoS ONE 7: e47275.
55. Ingvarsson PK (2005) Nucleotide polymorphism and linkage disequilibrium within and among natural populations of European aspen (*Populus tremula* L., Salicaceae). Genetics 169: 945–953.
56. Breen AL, Glenn E, Yeager A, Olson MS (2009) Nucleotide diversity among natural populations of a North American poplar (*Populus balsamifera*, Salicaceae). New Phytol 182: 763–773.
57. Zhang DQ, Xu BH, Yang XH, Zhang ZY, Li BL (2011) The sucrose synthase gene family in *Populus*: structure, expression, and evolution. Tree Genet Genomes 7: 443–456.
58. Chu Y, Su X, Huang Q, Zhang X (2009) Patterns of DNA sequence variation at candidate gene loci in black poplar (*Populus nigra* L.) as revealed by single nucleotide polymorphisms. Genetica 137: 141–150.
59. Zhang DQ, Yang XH, Zhang ZY, Li BL (2010) Expression and nucleotide diversity of the poplar *COBL* gene. Tree Genet Genomes 6: 331–344.
60. Ingvarsson PK (2008) Multilocus patterns of nucleotide polymorphism and the demographic history of *Populus tremula*. Genetics 180: 329–340.
61. Li Y, Haseneyer G, Schön CC, Ankerst D, Korzun V, et al. (2011) High levels of nucleotide diversity and fast decline of linkage disequilibrium in rye (*Secale cereale* L.) genes involved in frost response. BMC Plant Biol 11: 6.
62. Tian JX, Du QZ, Chang MQ, hang DQ (2012) Allelic variation in *PtGA20Ox* associates with growth and wood properties in *Populus* spp. PLoS ONE 7: e53116.
63. Olson MS, Robertson AL, Takebayashi N, Silim S, Schroeder WR, et al. (2010) Nucleotide diversity and linkage disequilibrium in balsam poplar (*Populus balsamifera*). New Phytol 186: 526–536.
64. Li Y, Böck A, Haseneyer G, Korzun V, Wilde P, et al. (2011) Association analysis of frost tolerance in rye using candidate genes and phenotypic data from controlled, semi-controlled, and field phenotyping platforms. BMC Plant Biol 11: 146.
65. Lata C, Bhutty S, Bahadur RP, Majee M, Prasad M (2011) Association of an SNP in a novel DREB2-like gene *SiDREB2* with stress tolerance in foxtail millet [*Setaria italica* (L.)]. J Exp Bot 62: 3387–3401.
66. Farquhar GD, Richards RA (1984) Isotopic composition of plant carbon correlates with water-use efficiency of wheat genotypes. Aust J Plant Physiol 11: 539–552.
67. Farquhar GD, Ehleringer JR, Hubick KT (1989) Carbon isotope discrimination and photosynthesis. Annu Rev Plant Physiol Plant Mol Biol 40: 503–537.
68. Condon AG, Richards RA, Rebetzke GJ, Farquhar GD (2004) Breeding for high water-use efficiency. J Exp Bot 55: 2447–2460.
69. Maurano MT, Humbert R, Rynes E, Thurman RE, Haugen E, et al. (2012) Systematic localization of common disease-associated variation in regulatory DNA. Science 337: 1190–1195.

Niche Construction on Environmental Gradients: The Formation of Fitness Valley and Stratified Genotypic Distributions

Xiaozhuo Han[1]*, Cang Hui[2,3]

1 School of Applied Mathematics, Guangdong University of Technology, Tianhe District, Guangzhou, China, 2 Centre for Invasion Biology, Department of Mathematical Sciences, University of Stellenbosch, Matieland, South Africa, 3 Mathematical and Physical Biosciences, African Institute for Mathematical Sciences, Muizenberg, South Africa

Abstract

The process of niche construction can alter the trajectory of natural selection through organism-environment feedback. As such, the mechanism and impact of niche construction can be better investigated along environmental gradients. Here we investigate how the process of niche construction affects the distribution of genotypes and fitness landscape along an environmental gradient under three selection regimes, namely heterozygote superiority, genetic loci which dictates niche construction ability being either selectively neutral or non-neutral. Using a spatially explicit cellular automaton, we show that niche construction can stratify genetic diversity by forming band-like distributions consisting of different genotypic compositions and promote reproduction isolation by forming a divide with reduced average fitness along the gradients, termed a fitness valley. The band structure and the presence of a fitness valley depend on heterogeneous environments, resource-dependent fitness and the selection acting on the gene loci affecting the niche-constructing ability. Our work adds to the growing body of evidence on criticizing species distribution models which assume that the environment alone can determine species distributions. Based on the results, we argue that conservation planning should target preserving or restoring environmental gradients.

Editor: Boris Alexander Vinatzer, Virginia Tech, United States of America

Funding: This work was supported by the National Natural Science Foundation of China (31100308). C.H. also received support from the National Research Foundation (Grants 76912 and 81825) and the University of Stellenbosch. The funders had no role in study design, data collection and analysis, decision to publish, or preparation of the manuscript.

Competing Interests: The authors have declared that no competing interests exist.

* E-mail: hanxzh03@163.com

Introduction

Organisms are not passively selected by their ambient environment but coevolve with their environment through either direct niche construction (also called ecosystem engineering; e.g. digging burrows and spreading webs) or indirect life-history activities (e.g. photosynthesis in plants that can fix atmospheric carbon into the soil and thus dramatically alter the soil profile) [1–6]. The process of niche construction can form a positive or amplified feedback between the demand and supply of limiting resources and can, thus, potentially affect the evolutionary trajectory by modifying the selective pressure of certain traits, especially those that are responsible for the niche construction [7–8]. Indeed, the concept of niche construction emphasizes the importance of organism-environment coupling during adaptive evolution and has been suggested to be capable of promoting stable polymorphism [9], altering competition outcomes [10–11], fixing deleterious alleles [12–15], and posing evolutionary momentum of directional selection [16–17]. However, studies on the effect of niche construction on the spatial distribution of species and their genetic structures are lacking, especially regarding how the process of niche construction can affect the distribution of different genotypes and how it can potentially promote the formation of range boundaries which further promote diversification and polymorphism.

The fitness of an organism is a rather context-based term, depending on whether its niche requirement matches the characteristics of inhabited environment. A highly fit genotype in one environment does not warrant a high fitness in other environments. Consequently, species often exhibit large variation of morphological traits and life-history strategies at regional scales (e.g. [18–20]). Such variation is maintained by both genetic and environmental factors and often shows a systematic shift along environmental gradients (e.g. along rainfall or altitudinal gradients; [21–24]). To this end, studies along environmental gradients become particularly appealing because they provide an ideal experiment for examining how the coupling of genetic structures and environmental factors interact to affect fitness along the environmental gradient. This further helps to resolve the long-standing debate on the role of genetic variation in species' adaptation to environmental heterogeneity [25]. It is thus important to further assess how the process of niche construction affects the genetic structure and fitness landscape along environmental gradients.

We here investigate how the process of niche construction interacts with resources along an environmental gradient and how it further affects genetic structures and polymorphism patterns.

Using a spatially explicit model of population genetics on lattices, we demonstrate the pattern of genetic diversity and population fitness along the environmental gradient. Specifically, two questions on how population genetics and fitness landscape change along environmental gradients are addressed. First, species distribution models (also known as ecological niche modelling) have been widely used in conservation management to project the potential distribution of a focal species or genotype based on detected relationships between species observed distribution and environmental characteristics (e.g. [26–29]). An underlying assumption of such models is that environment dictates species distribution. That is, population density and genotypic frequency change gradually in response to the change of resources along environmental gradients, without clear boundaries. In other words, clear boundaries of species distributions should reflect sudden change in environmental resources or dispersal barriers. Although this top-down effect of environmental filtering on species distribution can be strong at regional scales [30–32], it does not explain the common observation of clear distributional boundaries of species and genotypes at local scales even if the environmental gradient is subtle or continuously changing. We here demonstrate how the process of niche construction affects the formation of clear distributional boundaries along a linear environmental gradient.

Second, to promote diversification and polymorphism in adaptive evolution, species need to possess certain reproduction barriers after disruptive selection to prohibit remixing [33–35]. This reproduction barrier can be set up either via assortative behaviors (e.g. [23], [36]) or a physical barrier (e.g. mountains and rivers as dispersal barriers; [37]). Can the process of niche construction along a linear environmental gradient promote the formation of a reproduction barrier? We here examine the potential of a fitness valley (reproductive barrier) driven by niche construction in the fitness landscape along a linear environmental gradient that separates species distributions and stratifies species genetic structures along the environmental gradient. This fitness valley could restrict potential gene flows and function as a reproduction barrier during allopatric diversification.

Model

To examine the spatial distributions of different genotypes that differentially affect local nutrient content via the process of niche construction along a linear environmental gradient, we built a spatially-explicit individual-based cellular automaton (CA) on 200×200 lattices, where each cell contains a random-mating diploid individual with two dialelic loci **E** and **A** [15]. We assume that the frequency of allele E at generation t (P_{Et}) affects the individual's capacity of niche construction [13] and that the niche construction can affect the within-cell environmental resource positively or negatively by either producing or consuming the resource. Specifically, in each generation, the amount of resource (R) in a specific cell is governed by three processes (independent depletion, renewal and niche construction):

$$R_t=\lambda_1 R_{t-1}(1-\gamma P_{Et})+\lambda_2 P_{Et}+\lambda_3 \quad (1)$$

where λ_1 and λ_3 are coefficients of independent resource depletion and renewal; λ_2 and γ are coefficients of positive and negative niche construction. We assume $0\leq\lambda_1, \lambda_2, \lambda_3, \gamma<1$ and $\lambda_1+\lambda_2+\lambda_3\leq1$, where the latter keeps the amount of resource within [0,1] interval. If there is no niche construction (i.e. $\lambda_2=0$ and $\gamma=0$), the resource will converge to a stable level ($R=\lambda_3/(1-\lambda_1)$). In the following, we ignore negative niche construction (i.e. $\gamma=0$).

Following Laland et al. [13], we assume that both genotypes at loci **E** and **A** contribute to a two-locus fitness,

$$\omega=f_E \cdot f_A+\varepsilon \cdot f(R) \quad (2)$$

where f_E represents the fitness contribution from locus **E** ($=\alpha_1$ for genotype *EE*, 1 for *Ee* and β_1 for *ee*); f_A represents the fitness contribution from locus **A** ($=\alpha_2$ for genotype *AA*, 1 for *Aa* and β_2 for *aa*); the coefficient ε ($-1<\varepsilon<1$) determines the strength of the resource-dependent component ($f(R)$) relative to the fixed-fitness component ($f_E \cdot f_A$).We consider three selection regimes: heterozygote advantage ($\alpha_i<1$ and $\beta_i<1$), selection only at the ***A*** locus ($\alpha_1=\beta_1=1$), and selection only at the ***E*** locus ($\alpha_2=\beta_2=1$). As above, the resource level was affected by the frequency of allele E; in return, the resource level then affects the individual fitness (Eq. (2)) by the additive term ($\varepsilon \cdot f(R)$) and interferes the fitness at locus **A** ($f(R)=R$ for *AA*, $\sqrt{R(1-R)}$ for *Aa* and $1-R$ for *aa*) (Laland et al., 1999).

In the individual-based CA model, we introduced a linear environmental gradient along the vertical direction of the lattices (increasing from bottom to top). Specifically, we assume that the coefficient of independent resource renewal (λ_3) is not a constant but a linear function of the vertical coordinates (y) of the cell, $\lambda_3=ky$, where k can be considered an indicator of the gradient of the environmental resource. To ensure the resource ranges from 0 to 1 along y axis we let $k=0.003$ in the following analysis.

We chose periodic boundaries for the left and right edges to diminish the boundary effect and reflective boundaries for the top and bottom edges. Each cell of the lattices was initially randomly assigned one of the nine genotypes. During each time step, the individual in a focal cell chose to mate with the individual having the highest fitness in the four nearest neighboring cells, and then the individual was replaced one of its offspring randomly chosen according to the following fitness-dependent probabilityP_i:

$$P_i=\frac{\omega_i}{\sum_{i\in\Omega}\omega_i} \quad (3)$$

where Ω is the set of all possible genotypes that the parent can produce; ω_i is the fitness of the i th genotype. The resource level (R) of this cell was then updated according to Eq.(1).

We used the Shannon H index to describe the genetic diversity for individuals on each row in the lattices [38]–[39],

$$H=-\sum_{n=1}^{S}P_n\ln P_n \quad (4)$$

where S represents the number of genotypes and P_n the proportion of genotype n. Shannon's H ranges from 0 to $\ln S$ and increases either when there is a high number of genotypes or when genotype frequencies are even. The minimum genetic diversity occurs when only one genotype exists ($H = 0$), whilst the maximum genetic diversity occurs when the frequencies of all nine genotypes are equal (i.e. $P_n=1/S$ and $H_{\max} = 2.198$).

Results

When there was no environmental gradient, no clear patterns emerged (Fig.1A), contrasting to the clear patterns of triple-band (Fig.1B & C) and double-band distributions of genotypes (Fig.1D) along the environmental gradient. Genotypic diversity, as depicted by the Shannon H index, also showed a step-wise form (Fig.2), corresponding to the triple- and double-band distributions. With the increase of positive niche construction intensity (λ_2), this step-wise form further shifted towards the direction of lower resources (Fig.2A), and the genotypic distribution was transferred from a double-band to a triple-band pattern (Fig.2B). Note that, even without niche construction ($\lambda_2=0$), the environmental gradient can still stratify the genotypic distributions.

The step-wise form of genotypic diversity suggests that the genotypic diversity reached its peak at an intermediate resource level along the environmental gradient. This is because the transition of genetic composition happens at the intermediate resource level, with one or three genotypes being completely replaced by another three genotypes towards the other end of the gradient (Fig.3A, C & E). When further examining the average fitness of the individuals on each row, we found that the fitness landscape along the environmental gradient formed a valley at the intermediate resource level where the transition of genetic composition occurred (Fig.3B & F). When the selection was not acting on the niche construction locus, the fitness valley was inconspicuous (Fig.3D), even though the transition of genetic composition still occurred (Fig.3C).

Both the intensity of selection (α_i and β_i) and the coefficientε can affect the genotypic distribution. If individual fitness is independent from the resource level ($\varepsilon=0$), no band-like distribution of genotypes will emerge, regardless of the selection regimes. When the species experiences the selection regime of heterozygote superiority (i.e. $\alpha_i<1$ and $\beta_i<1$, Fig.4), the effects of α_1 and β_1 on the genetic diversity H are rather similar (Figure 4A & B), showing the triple-band distribution that is independent from the selection intensity on locus $\boldsymbol{E}$. In contrast, with the increase of selection intensity on locus $\boldsymbol{A}$, the distribution of genotype changed from a double-band pattern ($0.07<\alpha_2$, $\beta_2<0.8$) to a triple-band pattern ($0.8<\alpha_2$, $\beta_2<1$) (Fig.4C & D). In addition, the double-band distribution of genotypes will emerge only if $\varepsilon<0.14$, and a triple-band distribution will emerge only if $\varepsilon>0.14$ (Fig.4E). That is, the more affected the individual fitness is by the resource-dependent component relative to the fixed-fitness component, the more likely that the stratified distributions of genotypes will occur.

When the selection acts on locus $\boldsymbol{A}$ ($\alpha_1=\beta_1=1, \alpha_2, \beta_2 \neq 1$, Fig.5), there is no stratified genetic diversity if $\alpha_2, \beta_2>1.2$; otherwise, the distribution of genotype would be hyper sensitive to the change of coefficient $\alpha_2, \beta_2$and hence forms double- or triple-band patterns (Fig.5A & B). The genetic diversity varying with the coefficient $\varepsilon$is similar to the selection regime of heterozygote superiority (Fig.5C, comparing with Fig.4E), showing a triple-band pattern when $\varepsilon>0.13$ and a double-band pattern for other values of ε.

When the selection acts on locus $\boldsymbol{E}$ ($\alpha_2=\beta_2=1$, $\alpha_1, \beta_1 \neq 1$, Fig.6), there are three genotypes in each band when $\alpha_1<1$ (Fig.6A), with the transition between bands occurring instantly (referring to Fig.3E). Only one genotype (eeAA) exists in the lower half of the environmental gradient if $\alpha_1>1$, (Fig.6A). The sensitivity of genotype distributions to coefficient β_1 is similar to the sensitivity to coefficient α_1 and is thus not shown. As the alleles on locus $\boldsymbol{E}$ affect both the resource level through niche construction and the fitness, the genetic diversity H becomes insensitive to the relative contribution (ε) of the resource-dependent component to the overall fitness, except for the extreme case when $\varepsilon=0$ (Fig.6B). Each band consists of three genotypes when $\varepsilon \neq 0$ and switches to another three genotypes instantly at the middle point of the environmental gradient (Fig.6B).

Discussion

The concept of niche construction emphasizes the change that organisms bring about in their selective environments and is considered an evolutionary process rather than an evolutionary product [40]. The effects of niche construction can, arguably, persist over geological time, modulating macro-evolutionary patterns and species diversity [4], forming the momentum of evolution towards specific directions [12], [16]. Niche construction can accelerate the formation of steady polymorphism especially under deteriorating habitats and thus impede the negative impact of harmful environments [15]. For instance, the adaptive feedback between plants and their soil environment could account for why plants partially regulate soil nutrient content and thus possess evolutionary advantage during ecological succession and species packing [41–42]. Most of these works are based on context-based

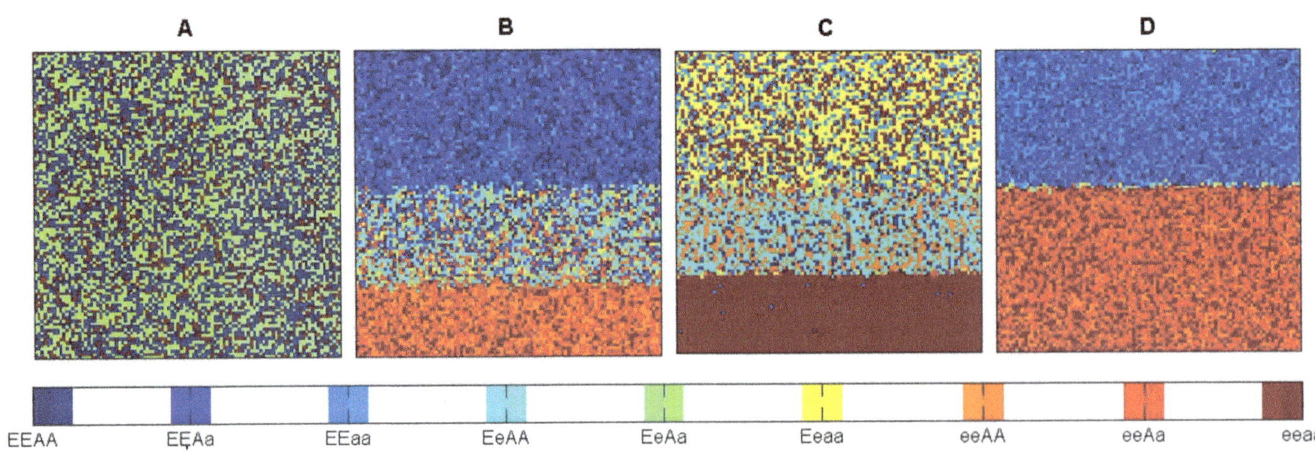

Figure 1. The distribution of genotypes on environmental gradients under three selection regimes. No environmental gradient in (A) where $\lambda_3=0.3$; linear gradient in (B), (C) and (D), with $\lambda_3=ky$, where $k=0.003$ along y axis that having 100 coordinated points. Heterozygote superiority is assumed in (A) and (B), with $\alpha_1=\alpha_2=0.99$, $\beta_1=\beta_2=0.9$; (C) selection only acts on locus A, with $\alpha_1=\beta_1=1$, $\alpha_2=0.7$, $\beta_2=0.9$); (D) selection only acts on locus E, with $\alpha_1=0.7$, $\beta_1=0.9$, $\alpha_2=\beta_2=1$. Other parameters are set in the cellular automaton: $\lambda_1=0.64$, $\lambda_2=0.05$, $\varepsilon=0.3$.

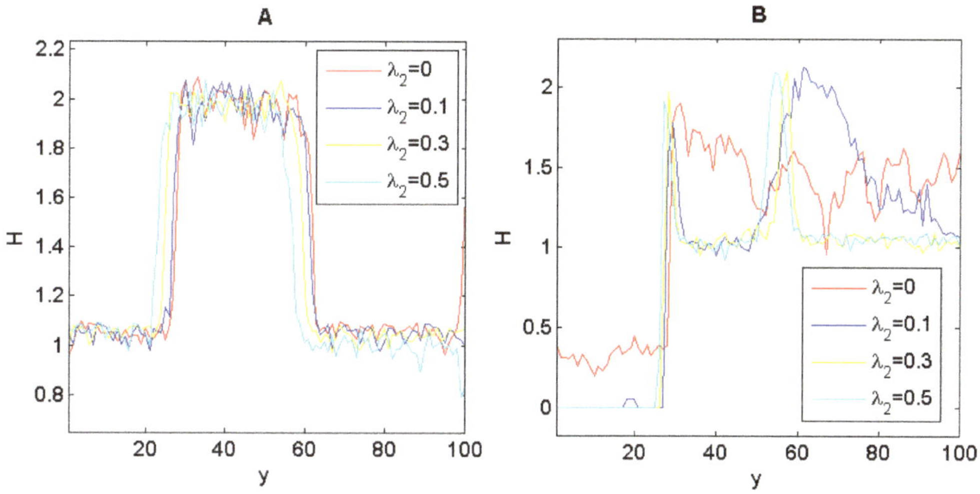

Figure 2. Genotypic diversity (*H*) as a function of the linear environmental gradient. Different lines show the change of genotypic diversity under different intensities of positive niche construction (λ_2). Plot (A) and (B) are for heterozygote superiority and selection on locus *A*, respectively. Other parameters are same as in Fig.1B & C.

genetics and feedbacks between species' niche constructing traits and environments which often cause novel evolutionary trajectories [2], [43]. Our work here adds to the growing body of evidence criticizing species distribution models which assume that the environment alone can determine species distributions.

Not only should we consider the role of genetic diversity in species' adaptation to environmental heterogeneity [25] but also its role in affecting the ability and outcomes of niche construction on heterogeneous environments. Firstly, environmental gradients are necessary for the formation of stratified genotypic diversity (Fig.1). That is, environmental heterogeneity is a prerequisite for maintaining stable genetic variation. Secondly, band-like patterns of genotypic distribution only formed when the fitness is resource or context dependent. The stronger the resource-dependent component is relative to the fixed-fitness component in the fitness, the more bands will likely occur (Fig.4E & 5C). The strength of niche construction alone, through affecting the resource-dependent component in the fitness, can change the evolutional direction and genetic structures. Thirdly, the fitness landscape forms a valley at which the transition of genetic composition happens along the environmental gradient. This fitness valley will disappear when the selection is not acting on the gene loci affecting the niche-constructing ability (Fig.3B, 3D &3F).

Focusing on niche construction could provide new insights into biological conservation. Meffe and Carroll [44] emphasized the necessity for conservation biologists to take an evolutionary perspective, while niche construction is an underappreciated evolutionary process in shaping local environments and ecosystems [1], [3]. Traditional conservation planning works only on available genetic resources which are rapidly changing in this era of Anthropocene [45], often facing an increasingly bleak future [3]. When niche-constructing organisms cause physical changes in abiotic environments, these changes could become evolutionarily significant to other species due to modified selection pressure [1]. Jones et al. [46] envisaged defining the utility of ecosystem engineering in conservation, especially when those key engineering or niche-constructing species can be pre-identified (see also [1]). Here, our results emphasize the need to preserve environmental gradients for essential or limiting resources.

Our results raise questions on the effectiveness of using species distribution models (SDMs) for predicting species' potential range in novel environment, and forecasting range shift due to environmental changes. Four assumptions are typical of a SDM [47–48]: (i) species current distribution is at equilibrium, (ii) the fitted relationship between species known occurrence and habitat characteristics is an adequate representation of the realized niche, (iii) this relationship of the realized niche does not change across space and time (known as the niche conservatism; [49]), and (iv) species can access all niches via dispersal. Limited dispersal capacity and time window prohibit species to access remote and isolated niches, questioning the first and last assumptions [50–51], and the hybrid model which implements a dynamic process of spreading has been proposed as a remedy [28], [52–55]. Our results here further raise questions on assumptions (ii) and (iii), as the realized niche is often mediated by context-specific biotic interactions, with no sufficient evidence supporting a constant niche [50], [56–57]. In particular, the crux of niche construction is the coevolution of organisms and their environments, often causing a positive feedback known as the ecological inheritance. The observed distribution of a species is resultant from a long-term coupling of the species and its habitat, often extending beyond the bounds of its niche; that is, species can persist in areas where it cannot invade. As such, when projected in novel environments, the species potential distribution is likely inflated [1], [58]. Some introduced species can eventually establish and become invasive through the reinforcing feedback facilitated by humans, resulting in a regime shift in the recipient ecosystem [59]. Our results further showed that the positive feedback of niche construction can form sharp boundaries and a rapid transition of genetic composition along smooth environmental gradients, whereas slight changes in the slope of the environmental gradient can lead to drastic changes in species distribution. Therefore, not only can the environment affect species distribution, but also the environmental gradient. To this end, environmental homogenization and fragmentation will likely affect species distribution strongly; this has been ignored in current SDMs.

Individual-based models (IBMs) are a power tool to examine complex dynamic behavior and emerging patterns [60], and have increasingly been used in ecological studies [48]. Our model

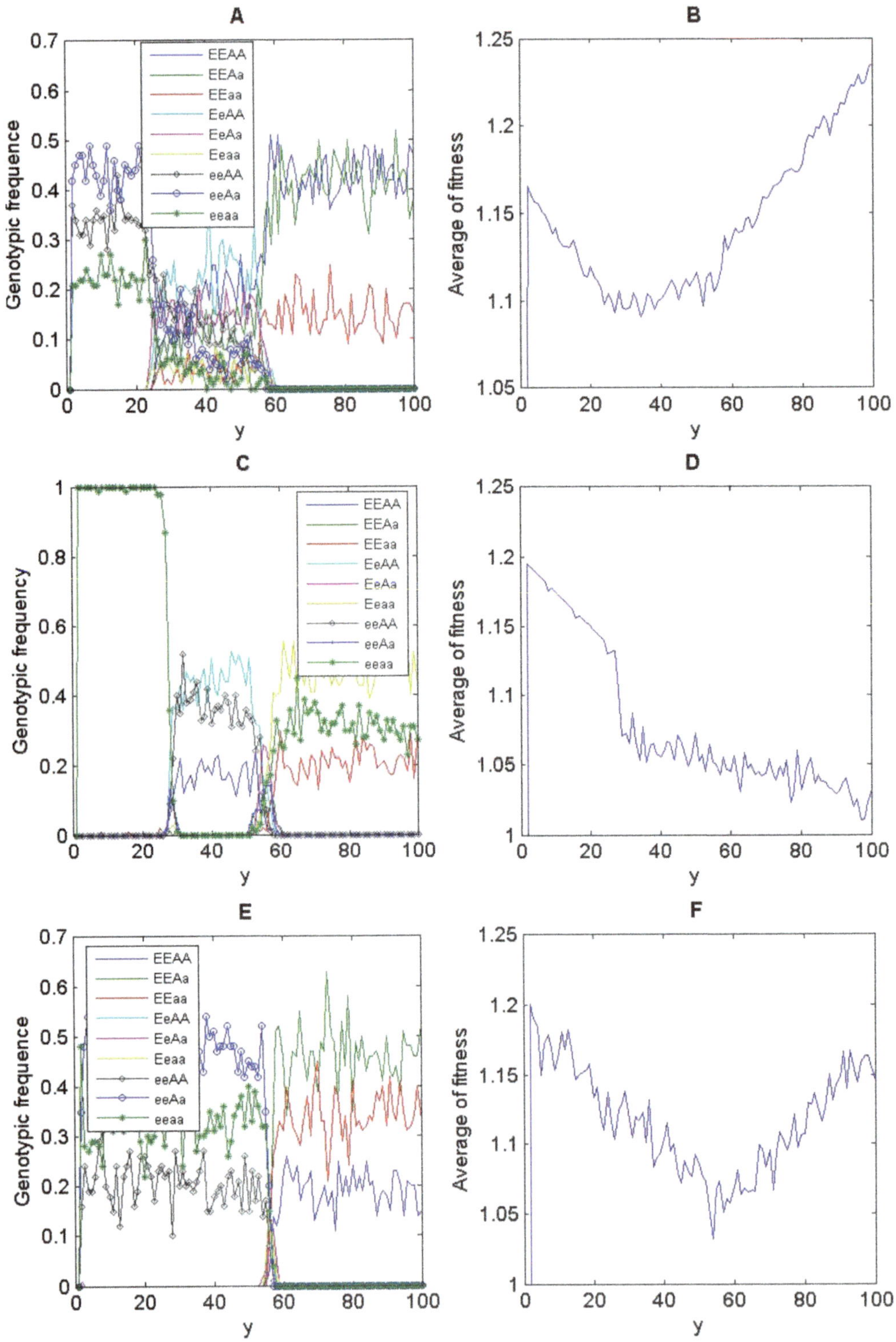

Figure 3. Genotypic frequency and average fitness as a function of the linear environmental gradient. Heterozygote superiority is assumed in (A) and (B); selection on locus *A* in (C) and (D); selection on locus *E* in (E) and (F). Parameters are the same as in Fig.1B, C & D.

assumes that the individual of a focal cell selects the fittest neighbors to mate. As the fitness is determined by altered environment from past activities of niche constructing genotypes, we are essentially dealing with the process of ecological inheritance in a standard evolutionary model [1]. Our model has four limitations and can be expanded in future work. First, we simulated the spatial interaction using only the Von Neumann neighborhood, meaning that the gene flow only happens locally, therefore ignoring the potential effect of long-distance dispersal [24]. Along an environmental gradient, only the fittest has the opportunity for reproduction, therefore an individual that disperses away from the current optimal habitat will likely land

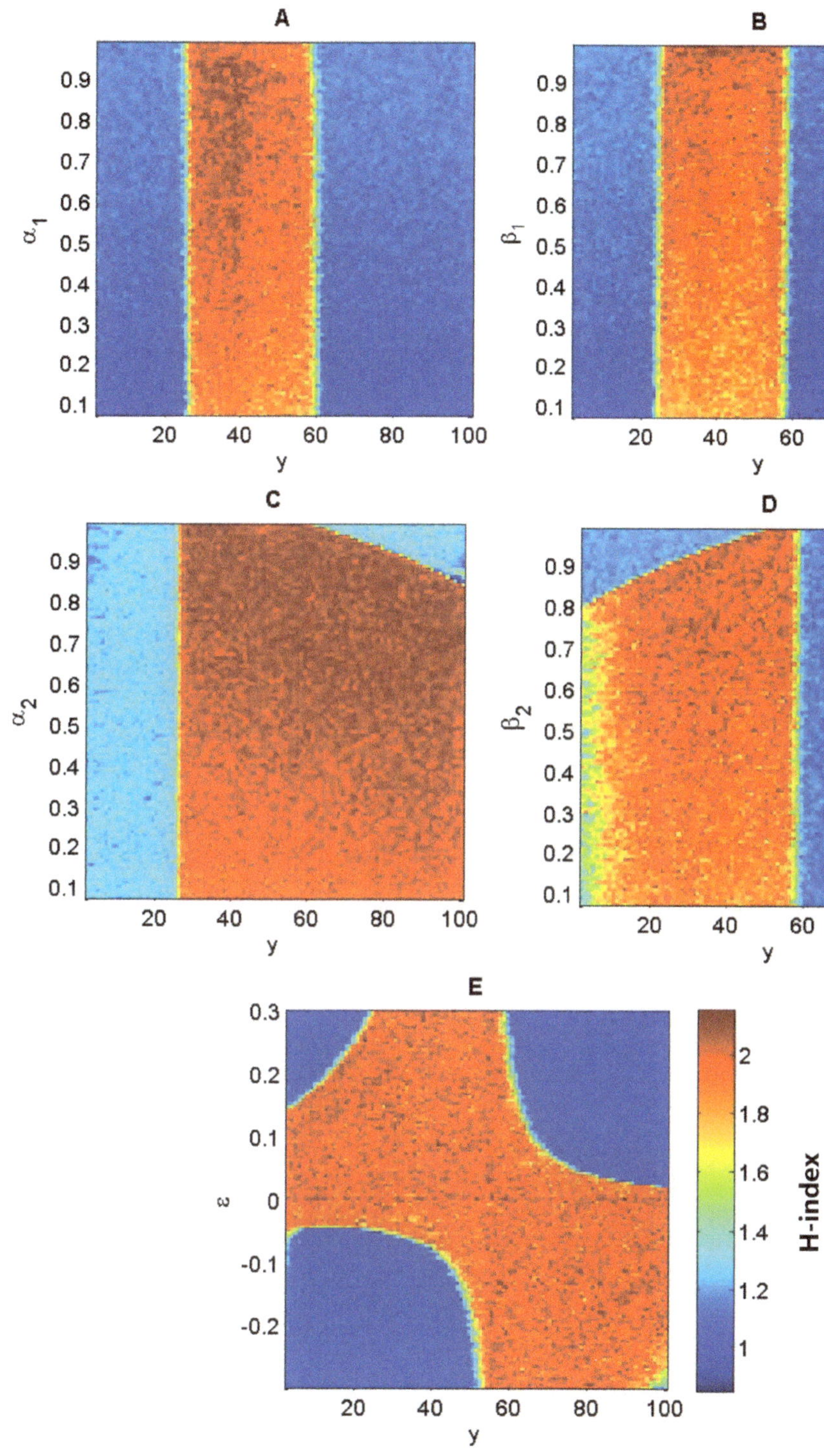

Figure 4. Genotypic diversity (*H*) as a function of the linear environmental gradient and fitness parameters (α, β and ε) under heterozygote superiority ($\alpha_i < 1$ and $\beta_i < 1$). Other parameters are the same as in Fig.1B.

itself in a suboptimal environment, and thus long-distance dispersal is selected against in our model. How long-distance dispersal affects the coupling of organisms and their environment is yet to be explored. Second, our model depicts the scenario of frequency-dependent selection in a zero-sum community [12]. Of course, as fitness can surely affect population demographics, considering a density-dependent selection could be more realistic and beget richer evolutionary dynamics [9-10], [61]. Third, our model only considers the feedback of a species' niche construction on its own fitness and distribution. A more realistic scenario could involve multiple species that affect each other through niche construction [10], [18]. Finally, the current model can be further expanded by allowing a time lag between the activity of niche construction and its impact on fitness. Such a time lag is often due

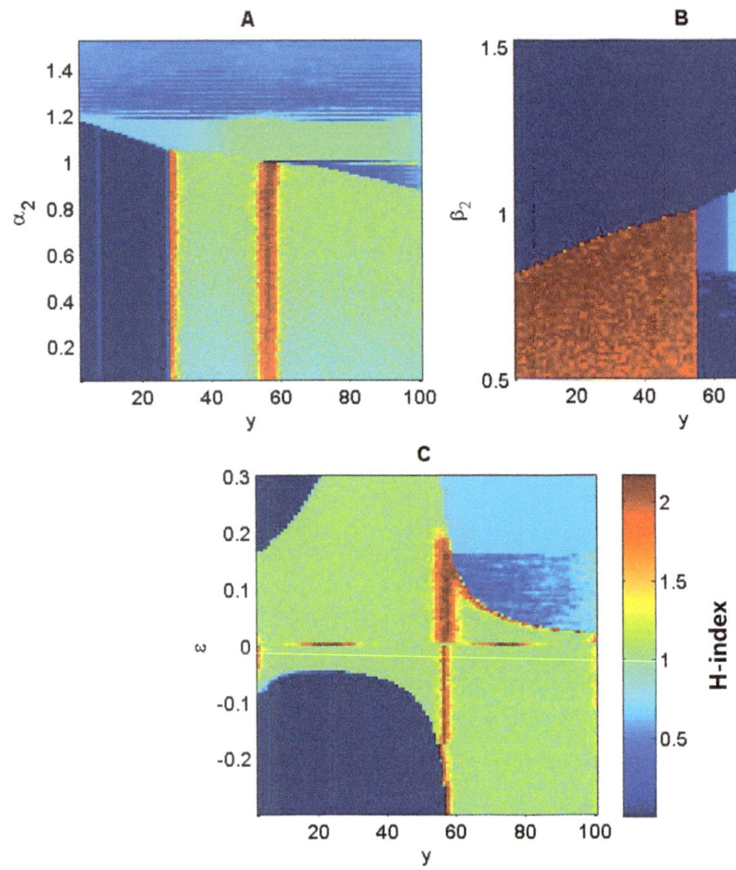

Figure 5. Genotypic diversity (*H*) as a function of the linear environmental gradient and fitness parameters (α, β and ε) when selection acts on locus *A* (i.e. $\alpha_1 = \beta_1 = 1$). Parameters: $\beta_2 = 0.9$ in (A); $\alpha_2 = 0.99$ in (B), with other parameters $\lambda_1 = 0.64$, $\lambda_2 = 0.05$, $\varepsilon = 0.3$; $\alpha_2 = 0.99$ and $\beta_2 = 0.9$ in (C).

to that sufficient resource change is needed for relevant genotypes to be selected, which can only be achieved through gradual and slow accumulation, forming evolutionary momentum and inertia [12], and complex spatial patterns (e.g. phase-lock oscillation; [62]).

Speciation most often happens via allopatric divergence where new species arise from reproduction isolation after separation by dispersal barriers. Doebeli & Dieckmann [63] offered a new theoretical perspective on the importance of environmental gradients to diversification through fostering frequency-dependent selection. The fitness valley identified here divides the distributions of high-fitness populations into two, with each consisting of unique genotypes. That is, intrinsically sympatric processes of organism-environment feedback can generate sharp geographical boundaries that separate genetically unique populations by the fitness valley. This fitness valley will further restrict potential gene flows

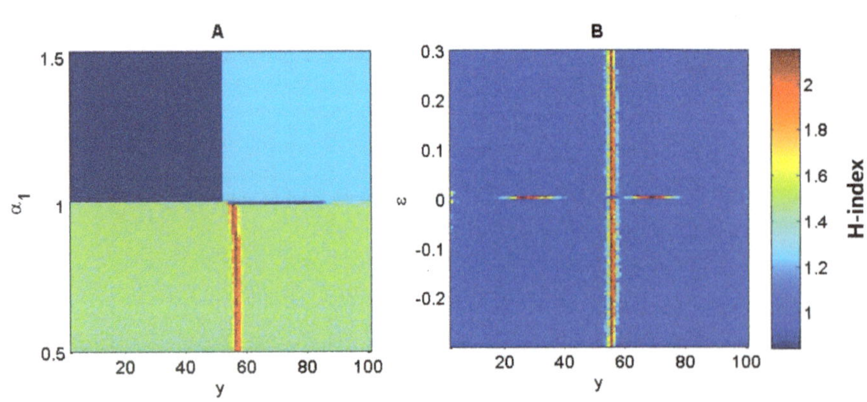

Figure 6. Genotypic diversity (*H*) as a function of the linear environmental gradient and fitness parameters (α and ε) when the selection acts on locus *E* (i.e. $\alpha_2 = \beta_2 = 1$). Parameters: $\beta_1 = 0.9$ and $\varepsilon = 0.3$in (A); $\alpha_1 = 0.99$ and $\beta_1 = 0.9$ in (B); others $\lambda_1 = 0.64$, $\lambda_2 = 0.05$.

between separated populations and thus promote allopatric speciation and diversification. The spatially local process of niche construction along a linear environmental gradient is capable of forming a reproduction barrier which facilitates speciation through genotype-environment feedback [15], [17]. Putting evolutionary processes into a spatially heterogeneous context with the organism and its environment co-affecting and co-adapting to each other could finally help to understand how future species will survive and adapt in the era of Anthropocene [64].

Acknowledgments

We are grateful to Beverley Laniewski for their constructive comments and kind help with the English of this manuscript. We also thank the help comments received from the anonymous reviews and the editors.

Author Contributions

Conceived and designed the experiments: XH CH. Performed the experiments: XH. Analyzed the data: XH. Contributed reagents/ materials/analysis tools: XH. Wrote the paper: CH XH. Conceptualized the idea of this work: XH CH. Wrote the initial simulation code in analysis: XH. Obtained results from the simulation: XH CH.

References

1. Odling-Smee FJ, Laland KN, Feldman MW (2003) Niche Construction: The Neglected Process in Evolution. Princeton: Princeton University Press.
2. Donohue K (2005) Niche construction through phonological plasticity: life history dynamics and ecological consequences. New Phytol 166: 83–92.
3. Boogert NJ, Paterson DM, Laland KN (2006) The implications of niche construction and ecosystem engineering for conservation biology. Bioscience 56: 570–8.
4. Erwin DH (2008) Macroevolution of ecosystem engineering, niche construction and diversity. Trends Ecol Evol 23: 304–10.
5. Wcislo WT, Tierney SM (2009) Behavioural environments and niche construction: the evolution of dim-light foraging in bees. Biol Rev 84: 19–37.
6. Dassonville N, Guillaumaud N, Piola F, Meerts P, Poly F (2011) Niche construction by the invasive Asian knotweeds (species complex *Fallopia*): impact on activity, abundance and community structure of denitrifiers and nitrifiers. Biol Invasions 13: 1115–33.
7. Odling-Smee FJ (1988) Niche constructing phenotypes. In: Plotkin HC, editor. The Role of Behaviour in Evolution. MIT: Cambridge. pp. 73–132.
8. Odling-Smee FJ, Laland KN, Feldman MW (1996) Niche construction. Am Nat 147: 641–8.
9. Hui C, Yue D (2005) Niche construction and polymorphism maintenance in metapopulation. Ecol Res 20: 115–9.
10. Hui C, Li Z, Yue D (2004) Metapopulation dynamics and distribution, and environmental heterogeneity induced by niche construction. Ecol Model 177: 107–18.
11. Zhang F, TaoY, Hui C (2012) Organism-induced habitat restoration leads to bi-stability in metapopulations. Math Biosci 240: 260–6.
12. Laland KN, Odling-Smee FJ, Feldman MW (1996) The evolutionary consequences of niche construction: A theoretical investigation using two-locus theory. J Evol Biol 9: 293–316.
13. Laland KN, Odling-Smee FJ, Feldman MW (1999) Evolutionary consequences of niche construction and their implications for ecology. Proc Natl Acad Sci USA 96: 10242–47.
14. Han X, Li Z, Hui C, Zhang F (2006) Polymorphism maintenance in a spatially structured population: a two-locus genetic model of niche construction. Ecol Model 192: 160–74.
15. Han X, Hui C, Zhang Y (2007) Formation and maintenance of spatial polymorphism induced by niche construction. ICNC, Vol III: 705–14.
16. Taylor T (2004) Niche construction and the evolution of complexity. In: Proceedings of Artificial Life IX. pp. 375–80.
17. Silver M, Di Paolo E (2006) Spatial effects favour the evolution of niche construction. Theor Popul Biol 70: 387–400.
18. Kylafis G, Loreau M (2011) Niche construction in the light of niche theory. Ecol Lett 14: 82–90.
19. Berthouly-Salazar C, van Rensburg BJ, le Roux JJ, van Vuuren BJ, Hui C (2012) Spatial sorting drives morphological variation in the invasive bird, *Acridotheres tristis*. PLoS ONE 7: e38145.
20. Hui C, Roura-Pascual N, Brotons L, Robinson RA, Evans KL (2012) Flexible dispersal strategies in native and non-native ranges: environmental quality and the 'good-stay, bad-disperse' rule. Ecography 35: 1024–32.
21. Azevedo RB, French V, Partridge L (1996) Thermal evolution of egg size in *Drosophila melanogaster*. Evolution 50: 2338–45.
22. Johnsto TA, Leggett WC (2002) Maternal and environmental gradient in the egg size of an iteroparous fish. Ecology 83: 1777–91.
23. Zhang F, Hui C, Pauw A (2013) Adaptive divergence in Darwin's race: how coevolution can generate trait diversity in a pollination system. Evolution 67: 548–60.
24. Berthouly-Salazar C, Hui C, Blackburn TM, Gaboriaud C, van Rensburg BJ, et al. (2013) Long-distance dispersal maximizes evolutionary potential during rapid geographic range expansion. Mol Ecol 22: 5793–804.
25. Schmidt PS, Serrao EA, Pearson GA, Riginds C, Riginos C, et al. (2008) Ecological genetics in the north Atlantic: environmental gradient and adaptation at specific loci. Ecology 89: 91–107.
26. Guisan A, Thuiller W (2005) Predicting species distribution: offering more than simple habitat models. Ecol Lett 8: 993–1009.
27. Elith J, Leathwick JR (2009) Species distribution models: ecological explanation and prediction across space and time. Ann Rev Ecol Evol Syst 40: 677–97.
28. Roura-Pascual N, Bas JM, Thuiller W, Hui C, Krug RM, et al. (2009) From introduction to equilibrium: reconstructing the invasive pathways of the Argentine ant in a Mediterranean region. Global Change Biol 15: 2101–15.
29. Roura-Pascual N, Hui C, Ikeda T, Leday G, Richardson DM, et al. (2011) The relative roles of climate suitability and anthropogenic influence in determining the pattern of spread in a global invader. Proc Natl Acad Sci USA 108: 220–5.
30. Cornell HV, Lawton JH (1992) Species interactions, local and regional processes, and limits to the richness of ecological communities - a theoretical perspective. J Anim Ecol 61: 1–12.
31. Whittaker RJ, Willis KJ, Field R (2001) Scale and species richness: towards a general, hierarchical theory of species diversity. J Biogeogr 28: 453–70.
32. Hui C, Richardson DM, Pyšek P, Le Roux JJ, Kučera T, et al. (2013) Increasing functional modularity with residence time in the co-distribution of native and introduced vascular plants. Nature Communications 4: 2454. DOI: 10.1038/ncomms3454
33. Coyne JA, Orr HA (2004) Speciation. Sunderland: Sinauer Associates.
34. Hui C, McGeoch MA (2006) Evolution of body size, range size and food composition in a predator-prey metapopulation. Ecol Complex 3: 148–59.
35. Mallet J (2008) Hybridization, ecological races, and the nature of species: empirical evidence for the ease of speciation. Phil Trans R Soc B 363: 2971–86.
36. Bridle JR, Ritchie MG (2001) Assortative mating and the genic view of speciation. J Evol Biol 14: 878–9.
37. Hewitt G (2000) The genetic legacy of the Quaternary ice ages. Nature 405: 907–13.
38. Lewontin RC (1972) The Apportionment of Human Diversity. Evol Biol 6: 391–8.
39. Izsák J, Papp L (2000) A link between ecological diversity indices and measures of biodiversity. Ecol Model 130: 151–6.
40. Laland KN, Sterelny K (2006) Seven reasons (not) to neglect niche construction. Evolution 60: 1751–62.
41. Kylafis G, Loreau M (2008) Ecological and evolutionary consequences of niche construction for its agent. Ecol Lett 11: 1072–81.
42. Corenblit D, Steiger J, Gurnell AM, Naiman RJ (2009) Plants intertwine fluvial landform dynamics with ecological succession and natural selection: a niche construction perspective for riparian systems. Global Ecol Biogeogr 18: 507–20.
43. Donohue K (2009) Completing the cycle: maternal effects as the missing link in plant life histories. Philos Trans R Soc Lond B Biol Sci 364: 1059–74.
44. Meffe GK, Carroll CR (1997) What is conservation biology? In: Meffe GK, Carroll CR, editors. Principles of Conservation Biology. Sunderland: Sinauer Associates. pp. 3–27.
45. Myers N (1997) Global Biodiversity II: Losses and Threats. In: Meffe GK, Carroll CR editors. Principles of Conservation Biology. Sunderland: Sinauer Associates. pp. 123–58.
46. Jones CG, Lawton JH, Shachak M (1997) Positive and negative effects of organisms as physical ecosystem engineers. Ecology 78: 1946–57.
47. Franklin J (1995) Predictive vegetation mapping: geographic modelling of biospatial patterns in relation to environmental gradients. Prog Phys Geogr 19: 474–99.
48. Grimm V, Railsback SF (2005) Individual-based modeling and ecology. Princeton: Princeton University Press.
49. Wiens JJ, Graham CH (2005) Niche conservatism: integrating evolution, ecology, and conservation biology. Annu Rev Ecol Evol Syst 36: 519–39.
50. Sinclair SJ, White MD, Newell GR (2010) How useful are species distribution models formanaging biodiversity under future climates? Ecol Soc 15: 8–21.
51. Barve N, Barve V, Jimenez-Valverde A, Lira-Noriega A, Maher SP, et al. (2011) The crucial role of the accessible area in ecological niche modeling and species distribution modeling. Ecol Model 222: 1810–9.
52. Davies SJ, Clusella-Trullas S, Hui C, McGeoch MA (2013) Farm dams facilitate amphibian invasion: Extra-limital range expansion of the painted reed frog in South Africa. Austral Ecol 38: 851–63.
53. Horak J, Hui C, Roura-Pascual N, Romportl D (2013) Changing roles of propagule, climate, and land use during extralimital colonization of a rose chafer beetle. Naturwissenschaften, 100: 327–36.
54. Caplat P, Hui C, Maxwell BD, Peltzer DA (2014) Cross-scale managment strategies for optimal control of tree invading from source plantations. Biol Invasions 16: 677–90.

55. Donaldson JE, Hui C, Richardson DM, Robertson MP, Webber BL, et al. (2014) Invasion trajectory of alien trees: the role of introduction pathway and planting history. Global Change Biol 20: 1527–37.
56. Schröder B, Seppelt R (2006) Analysis of pattern-process-interactions based on landscape models - overview, general concepts, and methodological issues. Ecol Model 199: 505–16.
57. Pearman PB, Guisan A, Broennimann O, Randin CF (2007) Niche dynamics in space and time. Trends Ecol Evol 23: 149–58.
58. Guisan A, Petitpierre B, Broennimann O, Daehler C, Kueffer C (2014) Unifying niche shift studies: insights from biological invasions. Trends Ecol Evol. In press.
59. Gaertner M, Biggs R, Te Beest M, Hui C, Molofsky J, et al. (2014) Invasive plants as drivers of regime shifts: Identifying high priority invaders that alter feedback relationships. Divers Distrib. In press. DOI:10.1111/ddi.12182.
60. Bullock J (1994) Correspondence: individual-based models. Trends Ecol Evol 9: 299.
61. Yue D, Hui C, Li Z (2004) Niche construction for desert plants in individual and population scales: theoretical analysis and evidences from saksaul (Haloxylon ammodendron) forests. Isr J Plant Sci 52: 235–44.
62. Han X, Hui C, Zhang Y (2009) Effects of time-lagged niche construction on metapopulation dynamics and environmental heterogeneity. Appl Math Comput 215: 449–58.
63. Doebeli M, Dieckmann U (2003) Speciation along environmental gradients. Nature 421: 259–64.
64. Fogarty L, Rendell L, Laland KN (2012) The importance of space in models of social learning, cultural evolution and niche construction. Adv Complex Syst 15: 1–2.

13

Molecular Evolution of Candidate Genes for Crop-Related Traits in Sunflower (*Helianthus annuus* L.)

Jennifer R. Mandel[1¤], Edward V. McAssey[1], Savithri Nambeesan[1], Elena Garcia-Navarro[2], John M. Burke[1*]

1 Department of Plant Biology, University of Georgia, Athens, Georgia, United States of America, 2 Instituto de Agricultura Sostenible, Consejo Superior de Investigaciones Científicas, Córdoba, Spain

Abstract

Evolutionary analyses aimed at detecting the molecular signature of selection during crop domestication and/or improvement can be used to identify genes or genomic regions of likely agronomic importance. Here, we describe the DNA sequence-based characterization of a pool of candidate genes for crop-related traits in sunflower. These genes, which were identified based on homology to genes of known effect in other study systems, were initially sequenced from a panel of improved lines. All genes that exhibited a paucity of sequence diversity, consistent with the possible effects of selection during the evolution of cultivated sunflower, were then sequenced from a panel of wild sunflower accessions an outgroup. These data enabled formal tests for the effects of selection in shaping sequence diversity at these loci. When selection was detected, we further sequenced these genes from a panel of primitive landraces, thereby allowing us to investigate the likely timing of selection (i.e., domestication vs. improvement). We ultimately identified seven genes that exhibited the signature of positive selection during either domestication or improvement. Genetic mapping of a subset of these genes revealed co-localization between candidates for genes involved in the determination of flowering time, seed germination, plant growth/development, and branching and QTL that were previously identified for these traits in cultivated × wild sunflower mapping populations.

Editor: Tom Gilbert, Natural History Museum of Denmark, Denmark

Funding: This work was supported by funding from the USDA National Institute of Food and Agriculture (2008-35300-19263) and the NSF Plant Genome Research Program (DBI-0820451). The funders had no role in study design, data collection and analysis, decision to publish, or preparation of the manuscript.

Competing Interests: The authors have declared that no competing interests exist.

* E-mail: jmburke@uga.edu

¤ Current address: Department of Biological Sciences, University of Memphis, Memphis, Tennessee, United States of America

Introduction

Strong directional selection is thought to be responsible for the dramatic phenotypic differences between domesticated lineages and their wild progenitors [1,2]. Genetic map-based approaches, including both QTL and association analyses, have been used to identify numerous genomic regions and, in some cases, genes underlying these phenotypic transitions (e.g., *fw2.2*, [3]; *fw3.2* [4,5] *tb1*, [6]; *sh4*, [7]). An alternative to mapping studies is the use of molecular population genetic methods to identify genes or genomic regions that may have experienced past selection. These efforts typically involve statistical tests to determine if the observed pattern of genetic diversity in a particular gene or genomic region can be explained by the standard neutral model (e.g., [8–13]). Rejection of the null hypothesis of neutrality provides evidence of past selection.

Overall, crop lineages are expected to exhibit a genome-wide loss of genetic diversity relative to their wild progenitors due to the occurrence of population bottlenecks during domestication and/or improvement [2,14]. But because selection influences genetic diversity in a locus-specific manner, genes targeted by positive selection will exhibit a greater than expected loss of diversity as compared to the genome-wide, neutral expectation [12,13,15]. Importantly, this provides a means for identifying genes, or at least genomic regions, that are likely to be of agronomic importance even though they may be recalcitrant to map-based analyses due to a lack of segregating variation within the crop lineage. Here, we describe molecular evolutionary analyses aimed at identifying genes that were targeted by selection during the evolution of cultivated sunflower (*Helianthus annuus* L.) from amongst a pool of candidates identified based on homology to genes of known effect from other study systems.

Cultivated sunflower was domesticated from wild sunflower (both *H. annuus*) approximately 4,000 years ago by Native Americans as a source of edible seeds and as well as for non-food purposes (e.g., as a source of dyes for textiles; [16]). More recently, sunflower has been the subject of intensive breeding as it has been transformed into a globally-important oilseed crop [17]. Wild sunflower exhibits seed dormancy, variable flowering time, extensive and variable branching, and it also produces relatively small seeds that are dispersed upon maturity (i.e., the mature heads "shatter"). In contrast, cultivated sunflower typically exhibits a loss of seed dormancy, more rapid/consistent flowering, strong apical dominance (i.e., a loss of branching), and considerably larger seeds that are retained in the head until harvest. Previous studies have primarily employed map-based approaches to identify genomic regions involved in the evolution of cultivated sunflower (e.g., [18–21]), though evolutionary analyses have also shown great promise [22–24].

In the present study, we mined the literature for genes from other species that are known to influence traits related to the evolution of cultivated sunflower. These included genes with known effects on floral development and flowering time, seed/fruit development, germination, plant growth/development, and branching. We then identified homologs of these genes in sunflower and sequenced them from panels of wild, primitive, and improved sunflowers. These data allowed us to test for evidence of positive selection during the domestication and/or improvement of sunflower. When possible, we also genetically mapped genes showing evidence of selection and compared their positions to those of QTL that had previously been mapped in cultivated x wild sunflower mapping populations. We found strong evidence for positive selection in a number of these genes as well as evidence of QTL co-localization in several cases. As such, these genes are excellent candidates for future functional studies aimed at understanding the molecular mechanisms underlying the evolution of cultivated sunflower.

Materials And Methods

Gene identification and primer design

To identify candidates for genes underlying traits related to sunflower domestication and improvement, we searched the literature for genes influencing relevant aspects of floral development and flowering time, seed/fruit development, seed germination, plant growth/development, and branching. We then performed BLAST searches of these genes against sunflower expressed sequence tags (ESTs) from the Compositae Genome Project EST Database (http://compgenomics.ucdavis.edu/) as well as the reference transcriptome assembly from Bachlava et al. [25]. Reciprocal best BLAST hits with an E-value of less than 10E-10 were identified as putative sunflower homologs and retained for further analysis (Table S1). Primers specific to a portion of each sunflower unigene identified via BLAST were then designed using either PrimerPlus 3.0 (http://www.bioinformatics.nl/) or PrimerQuest (http://www.idtdna.com/). To help avoid designing primers across splice sites, we used a tblastx-based intron finding Perl script (http://www.citrusgenome.ucr.edu/usa/ucr/Files.php) with *Arabidopsis* genome sequence information (v. 10) available from TAIR (http://www.arabidopsis.org/) to identify putative intron positions in sunflower unigenes. The resulting primer sequences can be found in Table S2. We also included a set of 11 presumptively neutral control genes that were previously identified by Chapman et al. [22] (Table S2).

Plant materials and DNA sequencing

The focus of this study was a collection of 28 *H. annuus* individuals including 8 wild sunflowers, 6 primitive domesticates (i.e., Native American landraces which represent an intermediate stage between wild sunflower and modern cultivars), and 14 improved lines (Table 1). Achenes for these individuals were obtained, with permissions, from the USDA North Central Regional Plant Introduction Station (NCRPIS) and French National Institute for Agricultural Research (INRA). The improved lines included the parents of a well-characterized sunflower recombinant inbred line (RIL) mapping population (RHA 280 and RHA 801) as well as the "Core 12" from Mandel et al. [26], which includes 12 inbred lines that capture ca. 50% of the allelic diversity present within the cultivated sunflower gene pool. We also included two outgroups: *H. argophyllus*, which is sister to *H. annuus*, and *H. petiolaris*, another closely related species that is sister to the *H. annuus*/*H. argophyllus* clade. A single individual from each of the 28 genotypes was grown to the seedling stage, and total DNA was extracted from each using a CTAB extraction protocol [27] and quantified using Picogreen (Applied Biosystems). The quantity/quality of DNA was also evaluated using a Nanodrop 1000 spectrophotometer. For candidate gene sequencing, we employed a tiered approach, as follows: (1) we first sequenced all candidate genes in the improved panel; (2) following the recommendations of Yamasaki et al. [28] and Chapman et al. [29], those candidates that had no or very low nucleotide diversity ($\pi<0.01$) in the improved panel were retained for sequencing in the wild panel and the outgroups, thereby enabling tests for selection; and (3) when a gene showed evidence for selection in the improved panel, it was also sequenced in the panel of primitive accessions. This last piece of data allowed us to further infer the likely timing of selection (i.e., domestication vs. improvement by investigating these genes in a panel containing primitive sunflower varieties vs. recently improved sunflower varieties).

PCR was performed in a total volume of 20 µL containing 5 ng of template DNA, 30 mM Tricine pH 8.4-KOH, 50 mM KCl, 2 mM $MgCl_2$, 125 µM of each dNTP, 0.2 µM reverse primer, 0.2 µM forward primer and 2 units of *Taq* polymerase. The PCR conditions involved a 'touchdown' protocol, as follows: 3 min at 95C; 10 cycles of 30 s at 94C, 30 s at 65C and 45 s at 72C, annealing temperature decreasing to 55C by 1C per cycle, followed by 30 cycles of 30 s at 94C, 30 s at 55C, 45 s at 72C, followed by 20 min at 72C. PCR products were checked for single-banded amplification via electrophoresis on 1% agarose gels. Amplification conditions were modified for loci that exhibited weak or non-specific amplification (i.e., faint or multiple bands, respectively) by either decreasing (i.e., starting at 60C and descending to 50C) or increasing (i.e., starting at 70C and descending to 60C) the annealing temperature.

PCR products were treated with 4 units Exonuclease I and 0.8 units Shrimp Alkaline Phosphatase (USB) at 37C for 45 min followed by enzyme denaturation at 80C for 15 min to prepare for sequencing using BigDye v3.1 (Applied Biosystems). The sequencing reactions were cleaned using Sephadex (Amersham) before being run on an ABI 3730xl (Applied Biosystems). In cases where direct sequencing results were unclear due to unresolvable heterozygous bases, indels, or short repeats, PCR products were TA-cloned into pGEM-T vectors (Promega), transformed into competent *Escherichia coli* (JM109; Promega), and screened for the presence of an insert. At least five positive colonies per individual were then sequenced as above except that vector primers [T7 and SP6] were used. Sequences have been deposited in the National Center for Biotechnology Information (NCBI) as BioProject PRJNA248055.

Sequence analyses and tests for selection

For all genes, sequences were aligned using Sequencher version 4.10 (GeneCodes), and FASTA files were generated for each. These FASTA files were then imported into DnaSP version 4.50.2 [30] for analysis. Where possible, individuals exhibiting heterozygous bases were resolved into haplotypes using the PHASE algorithm in DnaSP (or they were cloned and re-sequenced; see above). We then used DnaSP to compute the number of synonymous segregating sites (S), synonymous nucleotide diversity (π), Watterson's [31] estimate of synonymous diversity (θ), number of segregating indels, and the synonymous genetic distance from the outgroup for each gene. The distance (D), from the outgroup was determined by calculating the number of synonymous segregating sites in all pairwise comparisons among sequenced individuals within each panel and the outgroup and then averaging to obtain D (the authors of the program recommend using one sequence from each species; we used all individuals within a panel in order to provide a more robust value for D). We also compared levels of genetic diversity amongst the three panels:

Table 1. List of sunflower accessions analyzed in this study.

Panel	Name	USDA Accession Number
Wild	Ames 14400	PI 649851
	Ann-1114	PI 613727
	Ames 1473	PI 413027
	Ames 1455	PI 413011
	Ames 1516	PI 413067
	Ames 23238	PI 649853
	Ames 23940	PI 649854
	Ann-646	PI 435552
Primitive	Havasupai	PI 369358
	Hidatsa	PI 600721
	Hopi	PI 432504
	Mandan	PI 600717
	Maiz Negro	PI 650761
	Seneca	PI 369360
Improved	Mammoth	PI 476853
	HA 234	PI 599778
	HA 316	NSL 208764
	HA 404	PI 597368
	HA 821	PI 599984
	RHA 280	PI 552943
	RHA 328	NSL 202284
	RHA 358	PI 531071
	VIR 847	PI 386230
	RHA 408	PI 603989
	RHA 426	PI 617099
	RHA 801	PI 599768
	SF 33*	---
	SF 230*	---

*Accessions from the French National Institute for Agricultural Research.

wild, primitive (for those genes with preliminary evidence of selection), and improved. For these comparisons, S and the angular transformation of π and θ values were analyzed. For all genes that were sequenced in all three panels, two-factor ANOVAs (panel and locus) were performed using JMP version 9 (SAS Institute, Cary, NC) and posterior Tukey-Kramer tests were used to test for significant differences amongst means.

For the selection analyses, we used the maximum likelihood (ML) version [32] of the Hudson-Kreitman-Aguade (HKA) test [33], which allows for an explicit test of selection at individual loci in a multilocus framework. The neutral theory of molecular evolution [34] predicts that the amount of within-species polymorphism should be correlated with levels of between-species divergence. The ML-HKA test evaluates this prediction in a locus-specific fashion, thereby allowing for the identification of individual genes showing evidence of selection. It does this by comparing the fit of two models to the observed data. In one model, all loci (a set of neutral controls +a locus of interest) are assumed to be evolving neutrally. In the second model, the locus of interest is deemed to be under selection. Significance is then evaluated by comparing twice the difference in the likelihoods of the two models against a chi-square distribution with one degree of freedom [35].

We first confirmed that the collection of neutral genes utilized herein were not themselves under selection. To do this, we performed a "round-robin" test of all 11 putative neutral genes against each other to confirm selective neutrality (i.e., that selection has not influenced their pattern of nucleotide polymorphism). For each gene, this entailed using the ML-HKA test to compare two models, one model in which all 11 genes were assumed to be neutral, and one in which 10 genes are neutral and one is selected. This process was repeated with five different seed numbers for each of the 11 putatively neutral genes in each of the panels and none exhibited evidence of selection.

Following confirmation that our control loci were indeed behaving in a neutral fashion, we performed the ML-HKA test for each of the candidate genes of interest. This initially involved comparing the levels of polymorphism and divergence in the wild panel and the improved panel. As noted above, when a gene showed significant evidence of positive selection in the improved lines, we also sequenced it in the primitive panel and tested for selection at that stage. All ML-HKA tests were performed with five different seed numbers and a chain length of 100,000 as recommended by the authors. The resulting maximum likelihoods were averaged across the five replicates and used to perform a likelihood-ratio test.

Genetic mapping of candidate genes

Candidate genes that demonstrated evidence of positive selection during domestication and/or improvement were screened for polymorphism in eight arbitrarily chosen individuals of one of two recombinant inbred line (RIL) mapping populations via Sanger sequencing: an improved x wild sunflower cross (HA89 x ANN1238; [18,36]) for domestication candidates and an improved x primitive sunflower cross (NMS373 x Hopi; [37]) for improvement candidates. When a mappable polymorphism was identified, the locus was amplified from a larger set of 96 RILs from the appropriate cross and scored as either a length variant or via PCR-restriction fragment length polymorphism (PCR-RFLP). Loci were added to previously published linkage maps using either

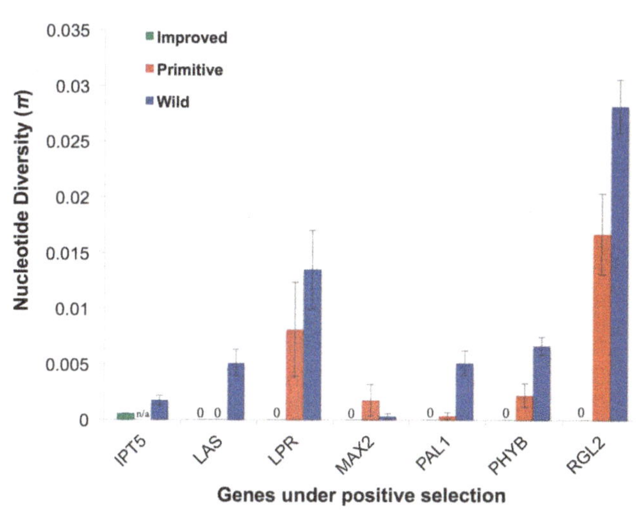

Figure 1. Nucleotide diversity in wild, primitive, and improved sunflower for the genes shown to be under positive selection during the evolution of cultivated sunflower. Error bars indicate standard deviations. Note that we were unable to obtain sequence information for the *IPT5* gene in the primitive panel.

Table 2. Population genetic diversity statistics for candidate and neutral/control genes.

Type	Panel	N	S (SE)	Sig.¹	π (SE)	Sig.¹	θ (SE)	Sig.¹
Candidate	W	24	6.25 (1.48)	X	0.022 (0.005)	X	0.024 (0.005)	X
	P	6	3.77 (1.68)	XY	0.012 (0.008)	XY	0.013 (0.006)	XY
	I	24	0.91 (0.33)	Y	0.0028 (0.001)	Y	0.0029 (0.001)	Y
Neutral	W	11	7.18 (1.55)	A	0.039 (0.006)	A	0.043 (0.008)	A
	P	11	3.55 (0.93)	B	0.027 (0.005)	B	0.025 (0.005)	B
	I	11	3.27 (1.10)	B	0.020 (0.005)	B	0.014 (0.004)	C

Panel: W = wild, P = primitive, I = improved; N = number of genes sequenced; S = number of segregating synonymous sites; π = nucleotide diversity for synonymous sites; θ = Waterson's theta for synonymous sites; Sig. = results of posterior Tukey-Kramer test with different letters representing significantly different values. ANOVA results: Neutral genes: θ ($F_{2,20}$=8.58, p<0.0001), π ($F_{2,20}$=8.73, p<0.0001), S ($F_{2,20}$=4.89, p=0.0009). Candidate genes: π ($F_{2,10}$=5.26, p=0.001), θ ($F_{2,10}$=5.19, p=0.01), S, ($F_{2,10}$=3.56, p=0.03).

¹ANOVA for candidate genes performed on the six candidate genes that were sequenced in all three panels.

the data and methods of Baack et al. [36] for the HA89 x ANN1238 RILs or Bowers et al. [37] for the NMS373 x Hopi RILs. If no polymorphism in the amplified region was readily identified, we used BLAST similarity of the selected candidate genes to the consensus genetic map of Bowers et al. [37] or to an existing sequence-based map [38] to locate the genomic position of the locus. When possible, we also compared the genetic map positions of genes under selection to those of previously mapped QTL for the phenotypes of interest. This was done by projecting QTL from several previous studies [18,19,39,40] onto the sunflower consensus map [37] based on shared markers. For the QTL mapping populations of Burke et al. [18] and Wills et al. [19], we used the sunflower CMap database (http://sunflower.uga.edu/cmap/) to identify markers that flanked the QTL of interest (i.e, those located near candidate genes for the same trait) and were also present on the sunflower consensus map [37]. We then used the genetic positions from the consensus map to display the QTL relative to the positions of the candidate genes (see below; note: when available, we used the 2-LOD interval for marking QTL regions as presented in the studies). The studies of Dechaine et al. [39] and Brunick [40] were not included in the CMap database, so the shared flanking markers were identified from the original source documents and projected as above.

Results And Discussion

As expected, we observed an overall, progressive loss of genetic diversity in the primitive and improved sunflower lines as compared to their wild progenitor (Table 2). Of the 76 candidate genes that we sequenced, 24 exhibited little or no nucleotide variation in a panel of 14 improved sunflower lines (π<0.0202, though most of these 24 were much lower than this value; see Table 3 and Table S3 for additional statistics), consistent with the possibility that they experienced positive selection during domestication and/or improvement. We then sequenced these 24 genes in a geographically diverse panel of eight wild sunflower individuals as well as outgroups and used the resulting data to test for selection. Seven of these genes exhibited significant departures from neutrality in the ML-HKA tests (Table 3; Figure 1), thereby providing strong evidence of past selection. Note also that some of the candidate genes harbored low genetic diversity in the wild panel as compared to the neutral genes. In fact, two candidate genes for branching (*IPT5* and *MAX2*) were shown to be under selection in the wild. After sequencing these genes in the primitive landraces, we found that *LATERAL SUPPRESSOR* (*LAS*) showed evidence for selection in both the primitive and improved lines. Five other genes, including *LOW PHOSPHATE ROOT* (*LPR*), *MORE AXILLARY GROWTH 2* (*MAX2*), *PHENYLALANINE AMMONIA-LYASE 1* (*PAL1*), *PHYTOCHROME B* (*PHYB*), and *RGA-LIKE* 2 (*RGL2*), showed evidence of selection in the improved lines only. We thus conclude that *LAS* likely experienced selection during the initial phase of sunflower domestication, whereas selection on the remainder was likely restricted to the subsequent period of improvement. Note that a reliable PCR product was not obtained for *ISOPENTENYL-TRANSFERASE 5* in the primitive lines. We were thus unable to investigate the timing of selection for this gene.

One of the selected genes, *PHYB*, is a photoperiod response gene that is thought to play a role in the transition from vegetative to reproductive growth (e.g., [41,42]). This gene also has a possible role in seed germination [43] (see discussion of germination and seed dormancy in sunflower below). The control of flowering is an important agricultural trait, and the evolution of flowering time is known to have played a critical role in the success of many crop

Table 3. Population genetic diversity statistics for the invariant genes tested for evidence of positive selection.

Gene	Trait	Panel	L	*l*	S	π	θ	Sig.
ACTIN-RELATED PROTEIN 7	Pleiotropic effects on	W	549	48.67	2	0.01761	0.01585	ns
	plant development	P	-	-	-	-	-	-
		I	549	48.67	0	0	0	ns
BLIND	Branching	W	383	12.17	0	0	0	ns
		P	-	-	-	-	-	-
		I	383	12.17	0	0	0	ns
CHALCONE SYNTHASE	Plant growth; regulation	W	310	40.81	0	0	0	ns
	of auxin	P	-	-	-	-	-	-
		I	310	40.67	0	0	0	ns
CONSTANS	Flowering	W	483	114.17	9	0.03213	0.02787	ns
		P	-	-	-	-	-	-
		I	483	114.11	4	0.00518	0.00919	ns
CYTOKININ OXIDASE/	Plant growth	W	375	88.67	3	0.01015	0.0136	ns
DEHYDROGENASE 1		P	-	-	-	-	-	-
		I	375	88.67	0	0	0	ns
ELONGATION FACTOR	Plant growth	W	380	37.98	2	0.00943	0.01587	ns
BINDING SITE 1B		P	-	-	-	-	-	-
		I	380	38	0	0	0	ns
FASCIATED	Fruit development	W	435	43.17	0	0.02978	0.0268	ns
		P	-	-	-	-	-	-
		I	435	43.17	0	0	0	ns
GLABRA 2	Seed development	W	639	88.31	7	0.028370.02493	0.02493	ns
		P	-	-	-	-	-	-
		I	639	89	4	0.0124	0.01204	ns
GLABRA2-EXPRESSION	Cell fate/division affecting	W	625	90.04	0	0	0	ns
MODULATOR	plant development	P	-	-	-	-	-	-
		I	625	90.58	0	0	0	ns
HEADING DATE 6	Flowering	W	737	38.67	0	0	0	ns
		P	-	-	-	-	-	-
		I	737	38.67	0	0	0	ns
INOSITOL POLYPHOSPHATE	Branching	W	367	86.11	11	0.02669	0.0385	ns
6-/3-KINASE 2B		P	-	-	-	-	-	-
		I	367	85.98	2	0.00391	0.00598	ns
ISOPENTENYLTRANSFERASE 5	**Plant growth**	**W**	**486**	**116.52**	**1**	**0.00107**	**0.00108**	*
		P	**N/A**	**N/A**	**N/A**	**N/A**	**N/A**	**N/A**

Table 3. Cont.

Gene	Trait	Panel	L	*l*	S	π	θ	Sig.
		I	486	116.50	0	0	0	**
LATERAL SUPPRESSOR	Branching	W	532	103.67	7	0.02244	0.02035	ns
		P	532	103.67	0	0	0	*
		I	532	103.67	0	0	0	**
LOW PHOSPHATE ROOT	Root growth/development	W	413	97.75	10	0.02774	0.03391	ns
		P	413	98.06	11	0.01868	0.03715	ns
		I	413	98	0	0	0	***
MORE AXILLARY GROWTH 2	Branching	W	468	86.5	1	0.00165	0.00364	*
		P	468	86.44	4	0.00774	0.01532	ns
		I	468	86.5	0	0	0	**
MORE AXILLARY GROWTH 4	Branching	W	104	0	-	-	-	-
		P	-	-	-	-	-	-
		I	104	0	-	-	-	-
METHYL-CPG BINDING 9	Branching	W	227	52.25	11	0.05391	0.06345	ns
		P	-	-	-	-	-	-
		I	225	52.37	4	0.0202	0.0202	ns
PHENYLALANINE AMMONIA-LYASE 1	Plant growth	W	473	105.54	12	0.02178	0.03427	ns
		P	473	105.5	1	0.00158	0.00314	ns
		I	473	105.5	0	0	0	*
PHYTOCHROME A	Flowering	W	447	46.5	2	0.02085	0.01424	ns
		P	-	-	-	-	-	-
		I	447	46.5	0	0	0	ns
PHYTOCHROME B	Flowering	W	683	166.83	13	0.02648	0.02639	ns
		P	683	166.83	5	0.00908	0.00992	ns
		I	683	166.83	0	0	0	**
PHYTOCHROME E	Flowering	W	457	103.48	14	0.0379	0.04254	ns
		P	-	-	-	-	-	-
		I	457	103.42	5	0.01029	0.01267	ns
PIN-FORMED 1	Root/shoot development	W	411	101.92	11	0.02743	0.03253	ns
		P	-	-	-	-	-	-
		I	411	101.93	2	0.0063	0.00504	ns
PHOTOPERIOD -H1	Flowering	W	246	50.31	3	0.017560.01797	0.01797	ns
		P	-	-	-	-	-	-
		I	246	50.33	1	0.00841	0.00511	ns
RGA-LIKE 2	Germination	W	377	89.5	31	0.11417	0.12297	ns

Table 3. Cont.

Gene	Trait	Panel	L	l	S	π	θ	Sig.
		P	**377**	**89.06**	**13**	**0.07069**	**0.04834**	**ns**
		I	377	88.83	0	0	0	***

Panel, W = wild, P = primitive, I = improved; L = alignment length in basepairs; l = number of synonymous sites; S = number of segregating synonymous sites; π = nucleotide diversity for synonymous sites; θ = Waterson's theta for synonymous sites; Sig. = ML-HKA significance: ns = not significant, $P<0.001$ = ***, $P<0.01$ = **, $P<0.05$ = *. Bold genes are those that showed significant evidence of selection. Note: we were unable to successfully sequence the *IPT5* gene in P.

species, including sunflower [17,23,44,45]. Wild sunflower exhibits extensive variation in flowering time [46,47], whereas the primitive sunflowers typically flower later in the season [48], and modern varieties have been selected for relatively early flowering [44,49] making it possible to produce sunflower across a broader range of environments [17].

Interestingly, Blackman et al. [23] included *PHYB* in their analysis of the role of flowering time genes in the evolution of cultivated sunflower and found marginally significant evidence for selection during improvement ($P = 0.07$). Though *PHYB* was found to have identical predicted protein sequences in the parents of an improved x wild sunflower mapping population that exhibits extensive variation in flowering time, this gene was consistently expressed at higher levels in the cultivar parent. We also found that *PHYB* co-localized with a previously identified QTL in an improved x wild sunflower mapping study ([39]; Figure 2). These results suggest that post-domestication selection may have targeted a *cis*-regulatory element that influences *PHYB* expression and that the diversity loss within *PHYB* itself is a byproduct of this selection.

Like flowering time, plant architecture changed dramatically during the evolution of cultivated sunflower. Initially, selection for increased apical dominance is thought to have resulted in a complete loss of branching [48,50,51]. In the mid-20th century, however, apical branching was re-introduced in a subset of cultivated sunflower lines as part of a transition to hybrid breeding and a concomitant desire to produce male lines with indeterminate flowering [52]. Consequently, branching is polymorphic within the cultivated sunflower gene pool. This re-introduction of branching is, however, due to the effects of a recessive allele at a single locus that maps to the upper third of linkage group (LG) 10 [53]. As such, other branching-related genes that were targeted by selection during sunflower domestication or improvement would still be expected to harbor low diversity. In fact, homologs of three genes known to influence branching in other species exhibited evidence of positive selection in sunflower, including one during domestication (*LAS*), one during improvement (*MAX2*), and one with unknown timing (*IPT5*) (Figure 1). *LAS* is a transcription factor and a positive regulator of bud, or branch, initiation [54–56], *MAX2* is an F-box protein that is thought to influence lateral shoot growth [57], and *IPT5* is known to be involved in cytokinin biosynthesis [58]. The genomic locations of *LAS* and *IPT5* could not be determined in this study, but *MAX2* (which also showed evidence of selection in wild sunflower) co-localized with a previously identified QTL for branching on LG 17.

Of the genes selected for analysis due to their potential role in other aspects of plant growth and development, two (*LPR* and *PAL1*) showed evidence of positive selection during improvement. *LPR* is a multicopper oxidase affecting root growth/development in *Arabidopsis* [59], whereas *PAL1* is a component of the phenylpropanoid pathway having broad effects on plant growth/development [60]. These genes co-localized with previously known QTL for numerous plant growth traits in sunflower including inflorescence size, plant architecture, leaf shape, and seed size [18,19,36] (Figure 2), though a better understanding of the likely phenotypic effects of variation at these genes awaits further study. Finally, *RGL2*, which is a DELLA protein that represses germination in *Arabidopsis* [61], exhibited evidence of selection during improvement. Wild sunflower exhibits strong seed dormancy whereas the primitive and improved varieties have little or no dormancy [40]. Interestingly, *RGL2* co-localized with a QTL for seed dormancy in an improved x wild sunflower cross [40] (Figure 2). These findings make this genomic region, and the *RGL2* gene in particular, a promising target for functional studies involving seed dormancy and germination.

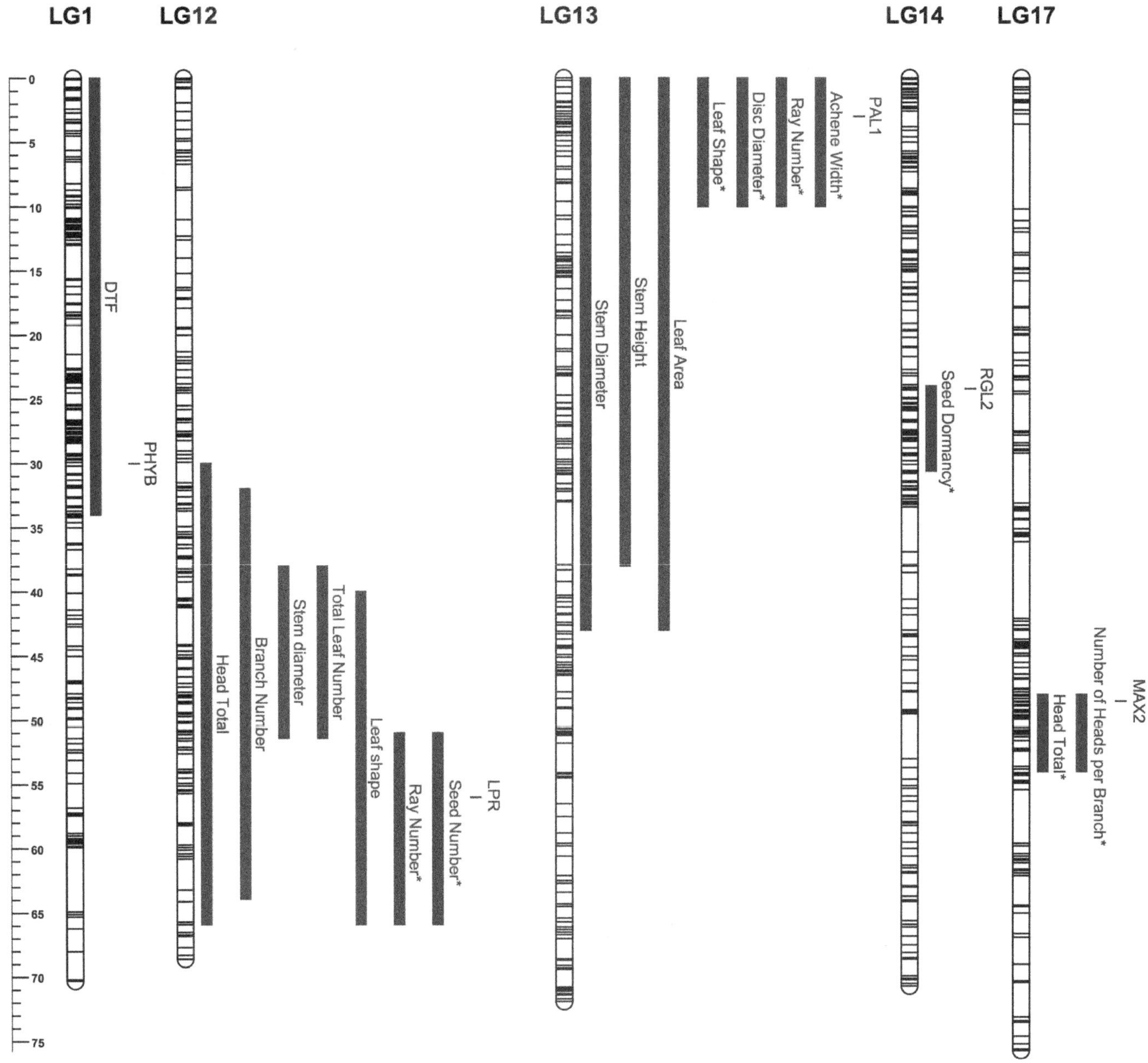

Figure 2. Co-localization of candidates for genes under selection and QTL for flowering time, branching, germination., and a number of plant growth and development traits that were previously identified in an improved x wild (shown in blue) or in both an improved x wild and a primitive x wild (shown in purple) sunflower QTL mapping population. QTL presented as 1-LOD are marker with an asterisk. All QTL and candidate gene positions are presented in the context of the sunflower consensus map (Bowers et al. 2012).

Of course, it is always possible that genes bearing the signature of selection such as those identified above were not themselves targeted by selection. Rather, these genes may simply be linked to the actual targets of selection (i.e., genetic hitchhiking). Though early studies of linkage disequilibrium (LD) in sunflower found that it decayed relatively rapidly (e.g., [62,63]), more recent analyses have revealed the presence of extended islands of LD within the genome [21]. Importantly, none of our mapped genes that exhibit evidence of selection fall within regions that exhibit elevated LD in the sunflower genome. This pattern of selected genes falling in genomic regions with lower overall LD, along with multiple instances of co-localization with QTL for crop-related traits supports the notion that they were indeed targeted by selection. Through the joint application of molecular population genetic analyses and trait-based mapping approaches, we have thus identified a set of promising loci for future functional studies aimed at understanding the molecular basis of sunflower evolution.

Acknowledgments

We would like to thank Eleni Bachlava and Jeff Roeder for assistance in the laboratory and Mark Chapman for help with the selection analyses. We would also like to thank two anonymous reviewers for comments on a previous version of this manuscript.

Author Contributions

Conceived and designed the experiments: JRM EVM SN EG-N JMB. Performed the experiments: JRM EVM SN EG-N. Analyzed the data: JRM EVM SN EG-N JMB. Wrote the paper: JRM EVM SN EG-N JMB.

References

1. Hammer K (1984) Das Domestikationssyndrom. Die Kult 32: 11–34.
2. Doebley JF, Gaut BS, Smith BD (2006) The molecular genetics of crop domestication. Cell 127: 1309–1321.
3. Nesbitt TC, Tanksley SD (2001) fw2.2 directly affects the size of developing tomato fruit, with secondary effects on fruit number and photosynthate distribution. Plant Physiol 127: 575–583.
4. Zhang N, Brewer MT, van der Knaap E (2012) Fine mapping of fw3.2 controlling fruit weight in tomato. Theor Appl Genet 125: 273–284.
5. Chakrabarti M, Zhang N, Sauvage C, Muños S, Blanca J, et al. (2013) A cytochrome P450 regulates a domestication trait in cultivated tomato. Proc Natl Acad Sci U S A 110: 17125–17130.
6. Doebley J, Stec A, Gustus C (1995) Teosinte Branched1 and the Origin of Maize: Evidence for Epistasis and the Evolution of Dominance. Genetics 141: 333–346.
7. Li C, Zhou A, Sang T (2006) Rice domestication by reducing shattering. Science 311: 1936–1939.
8. Lewontin RC, Krakauer J (1973) Distribution of gene frequency as a test of the theory of the selective neutrality of polymorphisms. Genet 74: 175–195.
9. Nielsen R (2005) Molecular signatures of natural selection. Annu Rev Genet 39: 197–218.
10. Wright SI, Gaut BS (2005) Molecular population genetics and the search for adaptive evolution in plants. Mol Biol Evol 22: 506–519.
11. Biswas S, Akey JM (2006) Genomic insights into positive selection. Trends Genet 22: 437–446.
12. Ross-Ibarra J, Morrell PL, Gaut BS (2007) Colloquium Papers: Plant domestication, a unique opportunity to identify the genetic basis of adaptation. Proc Natl Acad Sci 104: 8641–8648.
13. Burke JM, Burger JC, Chapman M a (2007) Crop evolution: from genetics to genomics. Curr Opin Genet Dev 17: 525–532.
14. Tanksley SD, McCouch SR (1997) Seed Banks and Molecular Maps: Unlocking Genetic Potential from the Wild. Science 277: 1063–1066.
15. Innan H, Kim Y (2004) Pattern of polymorphism after strong artificial selection in a domestication event. Proc Natl Acad Sci U S A 101: 10667–10672.
16. Soleri D, DAC (1993) Hopi crop diversity and change. J Ethnobiol 13: 203–231.
17. Putt ED (1997) Early history of sunflower. In: Schneiter A, editor. Sunflower production and technology. Madison, WI: American Society of Agronomy. pp. 1–19.
18. Burke JM, Tang S, Knapp SJ, Rieseberg LH (2002) Genetic analysis of sunflower domestication. Genetics 161: 1257–1267.
19. Wills DM, Burke JM (2007) Quantitative trait locus analysis of the early domestication of sunflower. Genetics 176: 2589–2599.
20. Wills DM, Abdel-Haleem H, Knapp SJ, Burke JM (2010) Genetic architecture of novel traits in the Hopi sunflower. J Hered 101: 727–736.
21. Mandel JR, Nambeesan S, Bowers JE, Marek LF, Ebert D, et al. (2013) Association mapping and the genomic consequences of selection in sunflower. PLoS Genet 9: e1003378.
22. Chapman MA, Pashley CH, Wenzler J, Hvala J, Tang S, et al. (2008) A genomic scan for selection reveals candidates for genes involved in the evolution of cultivated sunflower (Helianthus annuus). Plant Cell 20: 2931–2945.
23. Blackman BK, Rasmussen DA, Strasburg JL, Raduski AR, Burke JM, et al. (2011) Contributions of flowering time genes to sunflower domestication and improvement. Genet 187: 271–287.
24. Chapman M, Burke J (2012) Evidence of selection on fatty acid biosynthetic genes during the evolution of cultivated sunflower. TAG Theor Appl Genet 125: 897–907.
25. Bachlava E, Taylor CA, Tang S, Bowers JE, Mandel JR, et al. (2012) SNP discovery and development of a high-density genotyping array for sunflower. PLoS One 7: e29814.
26. Mandel JR, Dechaine JM, Marek LF, Burke JM (2011) Genetic diversity and population structure in cultivated sunflower and a comparison to its wild progenitor, *Helianthus annuus* L. Theor Appl Genet 123: 693–704.
27. Doyle JL DJ (1987) A rapid DNA isolation procedure for small quantities of fresh leaf tissue. Phytochem Bull 19: 11–15.
28. Yamasaki M, Tenaillon MI, Vroh Bi I, Schroeder SG, Sanchez-Villeda H, et al. (2005) A large-scale screen for artificial selection in maize identifies candidate agronomic loci for domestication and crop improvement. Plant Cell 17: 2859–2872.
29. Chapman MA, Mandel JR, Burke JM (2013) Sequence validation of candidates for selectively important genes in sunflower. PLoS One 8: e71941.
30. Rozas J, Sanchez-DelBarrio JC, Messeguer X, Rozas R (2003) DnaSP, DNA polymorphism analyses by the coalescent and other methods. Bioinformatics 19: 2496–2497.
31. Watterson GA (1975) Number of Segregating Sites in Genetic Models without Recombination. Theor Popul Biol 7: 256–276.
32. Wright SI, Charlesworth B (2004) The HKA test revisited: a maximum-likelihood-ratio test of the standard neutral model. Genet 168: 1071–1076.
33. Hudson RR, Kreitman M, Aguadé M (1987) A test of neutral molecular evolution based on nucleotide data. Genet 116: 153–159.
34. Kimura M (1983) The Neutral Theory of Molecular Evolution. Cambridge, UK: University Press.
35. Wright SI, Charlesworth B (2004) The HKA Test Revisited. Genet 168: 1071–1076.
36. Baack EJ, Sapir Y, Chapman MA, Burke JM, Rieseberg LH (2008) Selection on domestication traits and quantitative trait loci in crop-wild sunflower hybrids. Mol Ecol 17: 666–677.
37. Bowers JE, Bachlava E, Brunick RK, Rieseberg LH, Knapp SJ, et al. (2012) Development of a 10,000 locus genetic map of the sunflower genome based on multiple crosses. G3 GENES, GENOMES, GENET 2: 721–729.
38. Renaut S, Grassa CJ, Yeaman S, Moyers BT, Lai Z, et al. (2013) Genomic islands of divergence are not affected by geography of speciation in sunflowers. Nat Commun 4: 1827.
39. Dechaine JM, Burger JC, Chapman MA, Seiler GJ, Brunick R, et al. (2009) Fitness effects and genetic architecture of plant-herbivore interactions in sunflower crop-wild hybrids. New Phytol 184: 828–841.
40. Brunick RL (2007) Dissertation for the degree of Doctor of Philosophy in Horticulture presented on November 29, 2007. Title: Seed Dormancy in Domesticated and Wild Sunflowers (*Helianthus*).
41. Filiault DL, Wessinger CA, Dinneny JR, Lutes J, Borevitz JO, et al. (2008) Amino acid polymorphisms in Arabidopsis phytochrome B cause differential responses to light. Proc Natl Acad Sci U S A 105: 3157–3162.
42. Neff MM (2012) Light-mediated seed germination: connecting phytochrome B to gibberellic acid. Dev Cell 22: 687–688.
43. Casal JJ, Sánchez RA (1998) Phytochromes and seed germination. Seed Sci Res 8: 317–329.
44. Goyne PJ, Schneiter AA, Cleary KC, Creelman RA, Stegmeier WD, et al. (1989) Sunflower Genotype Response to Photoperiod and Temperature in Field Environments. Agron J 81: 826.
45. Blackman BK (2013) Interacting duplications, fluctuating selection, and convergence: the complex dynamics of flowering time evolution during sunflower domestication. J Exp Bot 64: 421–431.
46. Heiser CB (1954) Variation and subspeciation in the common sunflower, Helianthus annuus. Am Midl Nat 51: 287–305.
47. Blackman BK, Michaels SD, Rieseberg LH (2011) Connecting the sun to flowering in sunflower adaptation. Mol Ecol 20: 3503–3512.
48. Heiser CB (1951) The sunflower among the North American Indians. Proc Am Philos Soc 95: 432–448.
49. Goyne PJ, Schneiter AA (1987) Photoperiod Influence on Development in Sunflower Genotypes. Agron J 79: 704–709.
50. Heiser CB (1945) The Hopi sunflower. Missouri Bot Gard Bull 3: 163–166.
51. Rogers C, Thompson T, Seiler G (1982) Sunflower species of the United States. Bismark, ND: National Sunflower Association.
52. Fick G, Miller J (1997) Sunflower breeding. In: AA S, editor. Sunflower production and technology. Madison: American Society of Agronomy. pp. 395–440.
53. Bachlava E, Tang S, Pizarro G, Schuppert GF, Brunick RK, et al. (2010) Pleiotropy of the branching locus (B) masks linked and unlinked quantitative trait loci affecting seed traits in sunflower. Theor Appl Genet 120: 829–842.
54. Schumacher K, Schmitt T, Rossberg M, Schmitz G, Theres K (1999) The Lateral suppressor (Ls) gene of tomato encodes a new member of the VHIID protein family. Proc Natl Acad Sci U S A 96: 290–295.
55. Schmitz G, Tillmann E, Carriero F, Fiore C, Cellini F, et al. (2002) The tomato Blind gene encodes a MYB transcription factor that controls the formation of lateral meristems. Proc Natl Acad Sci U S A 99: 1064–1069.
56. Greb T, Clarenz O, Schafer E, Muller D, Herrero R, et al. (2003) Molecular analysis of the LATERAL SUPPRESSOR gene in Arabidopsis reveals a conserved control mechanism for axillary meristem formation. Genes Dev 17: 1175–1187.
57. Stirnberg P, van De Sande K, Leyser HMO (2002) MAX1 and MAX2 control shoot lateral branching in Arabidopsis. Development 129: 1131–1141.

58. Miyawaki K, Tarkowski P, Matsumoto-Kitano M, Kato T, Sato S, et al. (2006) Roles of Arabidopsis ATP/ADP isopentenyltransferases and tRNA isopentenyltransferases in cytokinin biosynthesis. Proc Natl Acad Sci U S A 103: 16598–16603.
59. Svistoonoff S, Creff A, Reymond M, Sigoillot-Claude C, Ricaud L, et al. (2007) Root tip contact with low-phosphate media reprograms plant root architecture. Nat Genet 39: 792–796.
60. Huang J, Gu M, Lai Z, Fan B, Shi K, et al. (2010) Functional analysis of the Arabidopsis PAL gene family in plant growth, development, and response to environmental stress. Plant Physiol 153: 1526–1538.
61. Piskurewicz U, Jikumaru Y, Kinoshita N, Nambara E, Kamiya Y, et al. (2008) The gibberellic acid signaling repressor RGL2 inhibits Arabidopsis seed germination by stimulating abscisic acid synthesis and ABI5 activity. Plant Cell 20: 2729–2745.
62. Liu A, Burke JM (2006) Patterns of nucleotide diversity in wild and cultivated sunflower. Genet 173: 321–330.
63. Kolkman JM, Berry ST, Leon AJ, Slabaugh MB, Tang S, et al. (2007) Single nucleotide polymorphisms and linkage disequilibrium in sunflower. Genet 177: 457–468.

Assessment of the Extirpated Maritimes Walrus Using Morphological and Ancient DNA Analysis

Brenna A. McLeod[1,2]*, Timothy R. Frasier[1], Zoe Lucas[2,3]

1 Biology Department, Saint Mary's University, Halifax, Nova Scotia, Canada, **2** Nova Scotia Museum of Natural History, Halifax, Nova Scotia, Canada, **3** The Friends of the Green Horse Society, Halifax, Nova Scotia, Canada

Abstract

Species biogeography is a result of complex events and factors associated with climate change, ecological interactions, anthropogenic impacts, physical geography, and evolution. To understand the contemporary biogeography of a species, it is necessary to understand its history. Specimens from areas of localized extinction are important, as extirpation of species from these areas may represent the loss of unique adaptations and a distinctive evolutionary trajectory. The walrus (*Odobenus rosmarus*) has a discontinuous circumpolar distribution in the arctic and subarctic that once included the southeastern Canadian Maritimes region. However, exploitation of the Maritimes population during the 16th-18th centuries led to extirpation, and the species has not inhabited areas south of 55°N for ~250 years. We examined genetic and morphological characteristics of specimens from the Maritimes, Atlantic (*O. r. rosmarus*) and Pacific (*O. r. divergens*) populations to test the hypothesis that the first group was distinctive. Analysis of Atlantic and Maritimes specimens indicated that most skull and mandibular measurements were significantly different between the Maritimes and Atlantic groups and discriminant analysis of principal components confirmed them as distinctive groups, with complete isolation of skull features. The Maritimes walrus appear to have been larger animals, with larger and more robust tusks, skulls and mandibles. The mtDNA control region haplotypes identified in Maritimes specimens were unique to the region and a greater average number of nucleotide differences were found between the regions (Atlantic and Maritimes) than within either group. Levels of diversity (h and π) were lower in the Maritimes, consistent with other studies of species at range margins. Our data suggest that the Maritimes walrus was a morphologically and genetically distinctive group that was on a different evolutionary path from other walrus found in the north Atlantic.

Editor: Ulrich Joger, State Natural History Museum, Germany

Funding: The research was funded by ExxonMobil (www.exxonmobil.com), The Friends of the Green Horse Society (http://www.greenhorsesociety.com), and the Nova Scotia Museum of Natural History (http://naturalhistory.novascotia.ca). The funders had no role in study design, data collection and analysis, decision to publish, or preparation of the manuscript.

Competing Interests: Regarding any competing interests due to funding from ExxonMobil, the authors confirm that this does not alter the authors' adherence to PLOS ONE policies on sharing data and materials.

* E-mail: bmysticetus@gmail.com

Introduction

The biogeography of a species is a result of a complex series of past and current climate changes, ecological interactions, anthropogenic impacts, physical geography, and evolution. To fully understand the contemporary biogeography of a species, and to make projections for future survival, it is necessary to understand the species' history. It is important to examine specimens from regions of localized extinction, as the loss of a species from such regions may represent the loss of unique adaptations and the loss of a potentially distinct evolutionary trajectory for the species. Of particular importance are margins of the species range, which are often identified as areas of increased genetic differentiation and isolation as well as morphological adaptation to 'new' habitats, or niche evolution [1].

The walrus (*Odobenus rosmarus*) is a large pinniped with a discontinuous circumpolar distribution in the arctic and subarctic. The species is easily recognizable by its large tusks and robust size, and is the only species within its family, Odobenidae. There are currently three recognized subspecies. The Atlantic walrus (*O. rosmarus rosmarus*, Illiger, 1815) is found throughout the Eastern Canadian Arctic to Franz Josef Land, the Barents and Kara Seas. The Pacific Walrus (*O. rosmarus divergens*, Linnaeus, 1758) is found in the Bering and Chukchi Seas. The third subspecies (*O. rosmarus laptevi*, Chapskii, 1940) is found in the Laptev Sea. However, the distinction of this last group has been debated and recent morphological and molecular data suggest that *O. r. laptevi* be considered synonymous with *O. r. divergens* [2]. Following centuries of extensive exploitation, the current status and population size of each subspecies are poorly known and currently being investigated. Hunting still occurs in most regions (Canada, US, Russia, and Greenland), and the species may be threatened by habitat disturbance, pollution and climate change [3].

Contemporary walrus distribution and diversity are a result of the species shifting its range through time, which has been tightly associated with climate changes over the past 5–8 million years. While *Odobenus* likely originated in the Pacific Ocean, approximately 5–8 million years ago (mya), individuals from the Pacific founded the Atlantic stock through the Panama seaway [4]. Subsequently, the Pacific stock is thought to have become extinct ~2mya [5] and it is suggested that the current Pacific subspecies originated by recolonization from the Atlantic, through either the

Canadian or Russian Arctic during a subsequent interglacial period [6].

In the North Atlantic, fossil walrus specimens identified throughout the eastern North American seaboard (Virginia, Maine, North Carolina, New Jersey) [7] suggest a more southerly distribution during the Wisconsinan Glaciation/early Pleistocene. However, with climate warming following the last glacial maximum (LGM), the Atlantic walrus distribution shifted northward reaching the Bay of Fundy and the Grand Banks by 12,500–12,800 before present (BP), southern Labrador by 11,500 BP, and the central Canadian Arctic by 9,700 BP [8].

Today ~8 subpopulations of the Atlantic walrus remain. Five of these are found west of Greenland and three are found east of Greenland [9] (but see [10,11] for recommendations on further subdivision)(Fig. 1). Support for the recognition of subpopulation status of these groups (e.g. for management purposes) has been found in both morphological and genetic data for most of these groups (e.g. [9], [12], [13], [2]).

The walrus once inhabited the Canadian Maritimes (waters of the Eastern Canadian provinces of Quebec, New Brunswick, Nova Scotia, and Prince Edward Island), with a population suggested to have been greater than 100,000 during the 17th century [14–16]. Because it was prized for its ivory-like tusks, thick hide and blubber [7], it was heavily hunted during the 16th–18th centuries, particularly around Sable Island, Nova Scotia; the Magdalene Islands; Prince Edward Island; other Gulf of St. Lawrence islands; and off the coast of New Brunswick and Cape Breton [8,17,18]. The species is now extirpated from these areas and has not been common south of 55°N for ~250 years [19,20]. Although there have been very occasional sightings of apparent strays in the region (e.g. [21], [20], [22], [23]), there have been no signs of recolonization and recovery is considered unfeasible [24].

Some data suggest that the Maritimes walrus was distinct from other Atlantic walrus. Although there are no living examples, numerous postglacial walrus specimens have been identified in New Brunswick, Nova Scotia, Newfoundland, and throughout the Gulf of St. Lawrence [19,25–27]. C. R. Harington and the late F. H. Fay conducted preliminary analyses of 92 adult (72 male, 20 female) postglacial walrus from the Maritimes region. They found that this walrus was larger overall, and had very large upper incisors (I3) relative to the Atlantic walrus (pers. comm. to C. R. Harington, 1992, reported by [8]). They suggested the skull specimens, though different in shape, appeared similar to the Pacific walrus which tends to be ~3% taller and ~10% heavier than the Atlantic walrus, and is recognizable by its longer tusks and broader snout [4,28]. In addition, larger size was evident in large average rostral width and condylobasal length (F. H. Fay pers. comm. to ZL).

We assessed the Maritimes walrus within what is known about global and regional walrus stock structure. The walrus, like several other arctic and subarctic marine mammal species, such as the bowhead (*Balaena mysticetus*) and beluga whales (*Delphinapterus leucas*), appears to have recently inhabited the waters of southeastern Canada. The fact that the area has not been recolonized by walrus may indicate the previous inhabitants were either a distinct, isolated (and perhaps specially adapted) group, or climatic changes over the past 250 years have rendered the habitat now unsuitable. If the group was relatively isolated from more northerly groups, it may have been more genetically and/or morphologically differentiated as well. If this group was adapted for a warmer habitat, its loss represents the loss of evolutionary potential for the species.

For this research, morphological and genetic characteristics of cranial specimens of the extirpated Maritimes walrus collected in Nova Scotia, New Brunswick and Quebec, Canada were compared to those of specimens from current Atlantic and Pacific populations to examine the relationship between the Maritimes group and other walrus subspecies.

Materials and Methods

Morphological Analysis

Morphological data were collected from cranial and mandibular bones held in both public and private collections (n = 278). This included 116 Maritimes (MAR)(*Odobenus rosmarus rosmarus*) specimens (65 cranial/51 mandibular elements), 156 specimens from the Eastern Canadian Arctic (the Atlantic subspecies, ATL) (*Odobenus r. rosmarus*)(82 cranial/74 mandibular elements), and 6 from the North Pacific (the Pacific subspecies, PAC) (*O. r. divergens*) (4 cranial/2 mandibular elements) (Table 1, Fig. 2, Table S1). Most specimens from the ATL have tusk length-at-age and sex information (after [29]). The cranial sample includes 41 males and 30 females while the mandibular sample contains 34 males and 27 females. Ages, as determined by Mansfield [29] ranged from 2.1 – 26+ years. The authors confirm that access to these specimens can be granted to other researchers upon request.

Morphological data were collected using measures previously outlined for seals by the Committee on Marine Mammals [30] and those used previously in *Odobenidae* (after [31], [32], and [2]). Additional measures were used that 1) are commonly used in assessing mammalian cranial morphology; 2) served to capture additional information from skulls in cases where parts of the skull were missing due to fracture or wear (e.g. occipital condyles) and 3) added additional morphological information that we thought was appropriate and informative. Where specimen condition allowed, 18 metric characteristics were examined (Table 2; Fig. 3). These included 5 mandibular measures and 13 cranial measures. Measurements greater than 30.48 cm were taken using vernier calipers (to the nearest 1.0 mm). Measurements less than 30.48 cm were taken using digital calipers (Mitotoyu) (to the nearest 0.01 mm). Tusk circumference and curvilinear tusk length were measured using a flexible measuring tape (to the nearest 1.0 mm). All measurements were taken by one of the authors (BAM).

Morphological Analysis – Data Standardization and Organization

For several of the mandibular and tusk measures, both left and right sides were examined (measures #9–17). Prior to analysis, we examined whether there was any lateral asymmetry by regressing measures from one side on the other and examining the slope of the regression line as well as whether its standard error encompassed 1.0. Lateral asymmetry was identified in most of the measurements, therefore analyses were carried out using the side with the most data available. To quantify the amount of average individual percent directional asymmetry (DA) present we used the formula *%DA = (Right - Left)/(Average of Right and Left)×100* (e.g. [33–35]). This measure allowed us to directly compare the amount of DA present irrespective of dimensions of very different sizes. We also calculated percent absolute asymmetry (%AA) (e.g. [33]); %AA = (maximum value – minimum value)/(average of the two values)×100. This measure allowed us to examine the amount of asymmetry present regardless of directionality, or the amount of "random" asymmetry. To examine whether the %DA and %AA were significantly different between regions, we used student's t-tests with a Bonferroni correction for multiple tests [36].

The goal of this research was to determine if there are differences between animals of ATL and MAR populations; however, such a comparison may be difficult when the sample set

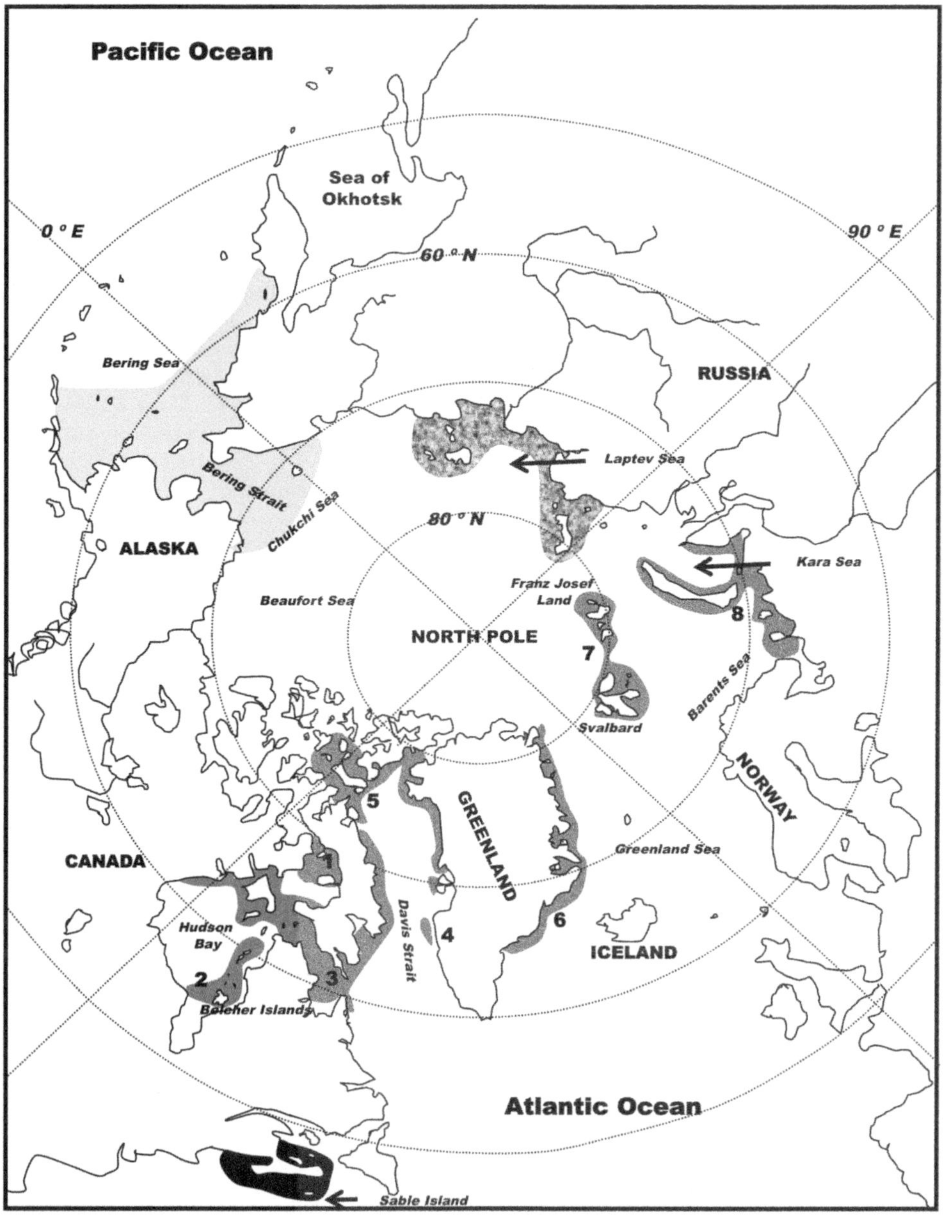

Figure 1. Distribution of the walrus. PAC, ATL, and LAP groups are indicated in light gray, dark grey and mottled gray, respectively. Area of MAR historical specimen sampling for this study is shown in black. Atlantic subpopulations are: (1) Foxe Basin, (2) southern and eastern Hudson Bay, (3) northern Hudson Bay/Hudson Strait/northern Labrador, (4) western Greenland, (5) in the 'North Water' (Baffin Bay/northeastern Canadian Arctic, (6) eastern Greenland, (7) Franz Josef Land/Svalbard, and (8) Kara Sea/Barents Sea.

is comprised of individuals of varied ages and sexes. Therefore, all measures were standardized by a morphological measure that has a relationship with individual age and sex. To determine an appropriate measure for this standardization, we examined the ATL specimens of known age and sex. We found that mastoidal width (#3) and mandible width (#18) were appropriate measures because they showed the strongest relationship with sex and age (Fig. 4). To standardize the analyses, each mandibular measure (#14–#17) was divided by that specimens corresponding mandible width, and all skull and tusk measures (#1, #2, #4–13) were divided by corresponding mastoid widths. Standardized values were then used in all further analyses.

Morphological Analysis - Tests of normalcy, equal variance, and equal means across regions

We evaluated whether the data collected were normally distributed and whether the variances were equal using the Shapiro-Wilk normality test [37] and Fisher's F Test. The means of variables that were normally distributed with equal variances were compared using a Welch 2-sample t-test [38], while all other means were compared using a non-parametric Wilcoxon rank-sum test [39]. These analyses were conducted using only ATL and MAR samples, as PAC sample sizes were too small ($n = 1–4$, depending on measure evaluated). All tests were conducted using the program R [40].

Figure 2. Collection sites for walrus specimens providing morphological data. Black, gray, and white centered circles denote sites of MAR, PAC, and ATL samples respectively. Table S1 gives specific sample origins.

Discriminant Function Analysis of Principal Components

Standardized skull, mandible, and tusk data were used to conduct a Principal Component Analysis (PCA) in R using the MASS library [41]. The resulting principal components were then used to carry out discriminant function analysis (DFA) with the ADE4 package [42]. Prior to this analysis, specimens that were missing any data/measures were removed. Using PCs in a DFA allows us to examine differentiation between the regions, while decreasing the total number of variables into a small number of principal components or 'factors' that account for most of the variance present in the datasets. In this way, we can then examine which factors/variables explain most of the differences between the datasets. We are interested in the differences in skulls from geographic regions, with measures as a proxy for skull shape and size. PCA captures uncorrelated patterns of skull shape and size that could be more informative than the measures themselves. Conducting PCA prior to DFA serves to transform data into uncorrelated variables, which is a requirement for DFA. DFA then allows us to compare the groups. Discriminant analysis of principal components (DAPC) has been shown previously to be a highly effective and relatively rapid method for assessing differentiations between groupings (e.g. [43,44]).

Genetic Analysis

Bone shavings were collected from historical specimens using a cordless drill following the sampling procedure of McLeod et al. [45]. This procedure minimizes DNA contamination and maximizes specimen preservation and integrity. In addition, tissue specimens (tongue and tendon) and DNA from contemporary walrus populations from the Eastern Canadian Arctic were provided by the Makivik Corporation and the Department of Fisheries and Oceans (DFO), respectively. DNA provided from DFO was extracted previously using the Sigma DNA extraction kit or Qiagen DNEasy tissue kit (Qiagen). Contemporary samples were originally collected during aboriginal hunts (Makivik Corporation), or through biopsy sampling (DFO).

A total of 125 samples were collected for DNA analysis (Table S2, Fig. 5) including 37 from the Maritimes (public and private collections) and 88 comparative samples from the Eastern Canadian Arctic. The latter included 74 samples of DNA from southeast Baffin Island (DFO) and 14 tissue samples of contemporary walrus from Nunavik (Akpatok Island in Ungava Bay; Salisbury Island and Notthingham Islands in Hudson Strait, and Sleepers Island in Eastern Hudson Bay)(Makivik Corporation).

DNA was extracted from contemporary specimens using the DNEasy blood and tissue kit (Qiagen) then quantified using a NanoDrop 2000 spectrophotometer (Thermo Scientific Inc.). DNA was extracted from historical specimens using the protocols of Rastogi et al. [46] following stringent ancient DNA (aDNA) protocols to prevent contamination of materials (as per [45] and [47]). All pre-polymerase chain reaction (PCR) aDNA work was conducted in a separate room and in isolation from any contemporary specimens or materials that have been used in association with contemporary materials. In addition, samples were extracted in small batches (n = 5–8) as a means to minimize cross-contamination. Blank samples (samples not containing DNA) were included in each step of the contemporary and historical DNA analyses as a means to identify contamination.

It has been previously shown that the different Atlantic subpopulations of walrus can be distinguished based on analysis of mitochondrial DNA (e.g. [2,9]). Therefore, we PCR amplified all 125 samples for the mitochondrial control region using the primers DL-2F and DL-3R of Lindqvist et al. [2] which amplify a region that is approximately 428 bp. For contemporary DNA samples, PCR cycling conditions consisted of an initial five-minute denaturation step at 94°C; 30 cycles of 94°C for 30 seconds, 57°C for one minute, and 72°C for one minute; and a final extension step at 60°C for 45 minutes. PCR cocktail conditions were as follows within a 15 µl reaction: 10 ng DNA, 1X PCR buffer (20 mM Tris-HCl pH 8.4, 50 mM KCl) (Invitrogen, Burlington, ON), 1.5 mM MgCl2 (Invitrogen, Burlington, ON), 0.2 mM each dNTP (Amersham Biosciences, Piscataway, NJ), 0.3 µg/µl BSA

Table 1. Number of cranial and mandibular samples for which morphological data were collected from public and private institutions/collections.

Specimen type/Region	Collection				
	NSMNH	*NBM*	*CMN*	*Private*	**Total**
Mandibular bone(s) – ATL	0	0	74	0	74
Cranial bones – ATL	0	0	82	0	82
Mandibular bone(s) – MAR	32	5	10	4	51
Cranial bones – MAR	38	4	7	16	65
Mandibular bone(s) – PAC	0	0	2	0	2
Cranial bones - PAC	0	0	4	0	4
Total	**70**	**9**	**179**	**20**	**278**

'NSMNH', 'NBM', and 'CMN' denote Nova Scotia Museum of Natural History, New Brunswick Museum, and Canadian Museum of Nature, respectively.

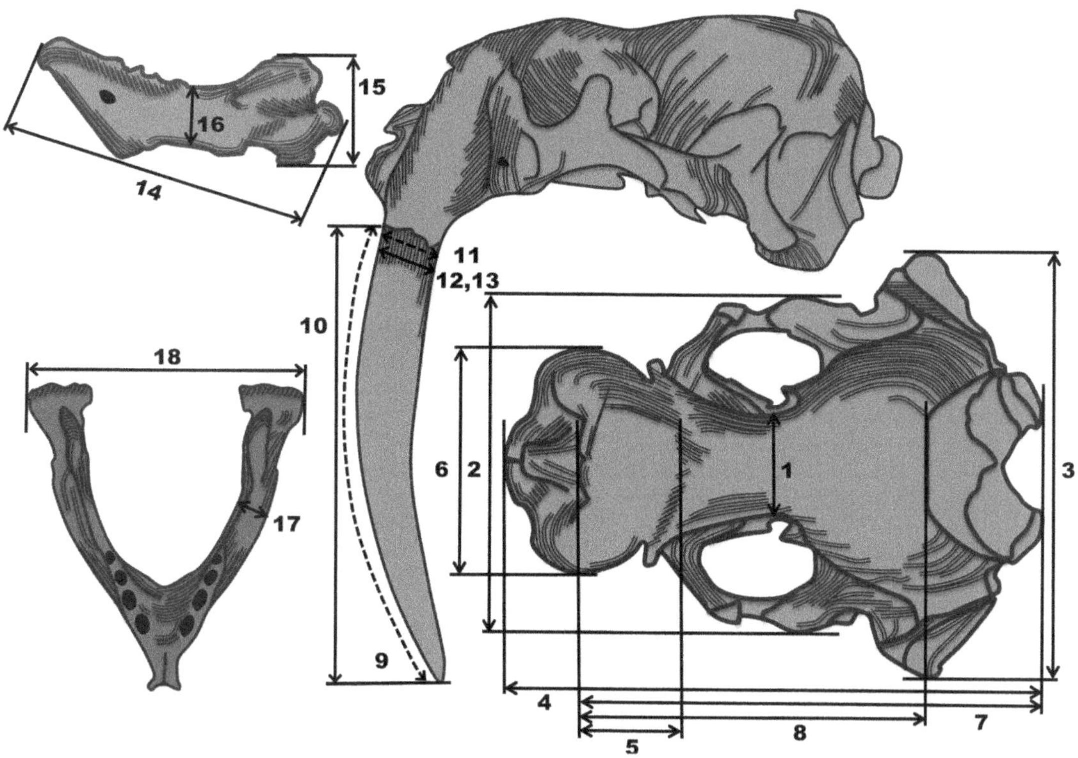

Figure 3. Cranial, tusk and mandibular measurements collected from walrus crania where possible. Numbers indicate measures taken as described in Table 2. Dashed lines indicate measures taken with a measuring tape. Measure 13 is not shown but is a mediolateral equivalent of measure 12. Images modified from Allen [7].

Table 2. Morphological measurements taken and corresponding definitions.

Measure	Name	Definition
Cranial Measures		
1	*Interorbital width (IW)**	least distance between the orbital fossae
2	*Zygomatic width (ZW)**	greatest width, at right angle to the axis of the skull, across the zygomatic arches
3	*Cranial width/mastoid width (CW)**	greatest transverse width of the skull posterior to the zygomatic arches
4	*Condylobasal length (CBL)**	measured from a transverse line touching the most posterior points on the occipital condyles to a transverse line touching the most anterior points on the premaxillary bones
5	*Nasal length (NL)*	greatest length from the anteriormost to posteriormost points of the nasal bones
6	*Rostral width (RW)**	greatest width of the maxillary bones at the level of the canines
7	*Occiptonasal length (ONL)*	greatest length from the tip of the nasals to the posterior of the occipital condyles
8	*Nasal-occipital crest (N-Oc)*	measured from the most anterior tip of the nasals to the most anterior tip of the occipital crest
9	*Curvilinear tusk length (TL-C)****	curvilinear length from alveolar margin to the tip along the anterior edge of the tusk
10	*Straight tusk length (TL-S)****	straight line length from the outer alveolar margin to the tusk tip
11	*Tusk circumference (TC)****	tusk circumference at the alveolar margin
12	*Tusk diameter (anterposterior) (TD-AP)*	maximum anterposterior diameter at alveolar margin
13	*Tusk diameter (mediolateral) (TD-ML)*	maximum mediolateral diameter at alveolar margin
Mandibular Measures		
14	*Mandible length (ML)***	distance between most anterior point on mandible and midpoint on posterior surface of articular condyle
15	*Mandible height (MH)***	distance between most dorsal point on coronoid process and most ventral point on angular process
16	*Least mandible depth (MD)***	minimal distance between dorsal and ventral mandibular surfaces, posterior to the last post-canine
17	*Least mandible thickness (MT)***	minimal lateral distance between medial and lateral mandibular surfaces, posterior to the last post-canine
18	*Mandible width (MW)*	maximum mediolateral distance between the left and right mandibular condyles

*After Committee on Marine Mammals (1967), **Wiig and Gjertz (1996) and ***Wiig et al. (2007).

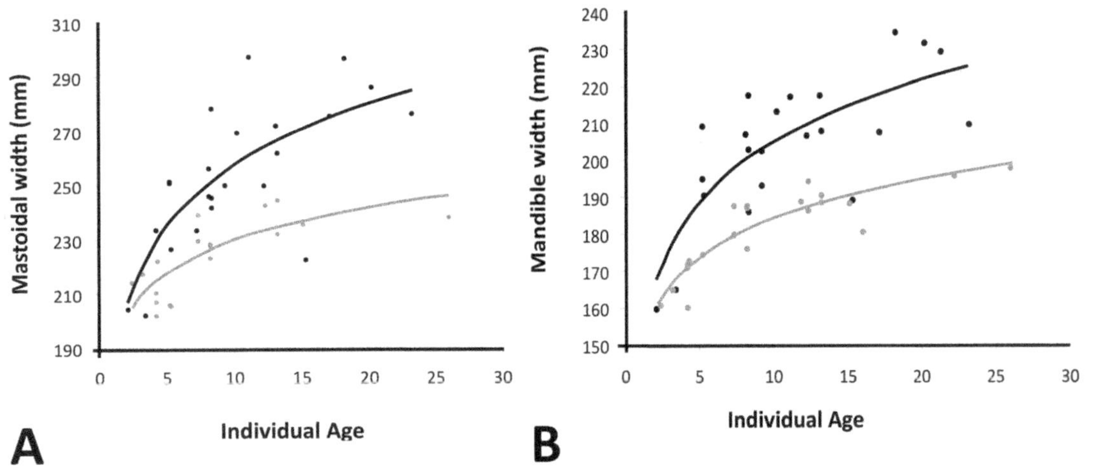

Figure 4. Mastoid width (a) and mandible width (b) versus age in known age males (black) and females (gray) from ATL.

(Sigma, Oakville, ON), 0.05 U/vl Taq polymerase (Invitrogen, Burlington, ON), and 0.3 μM of each primer. For historic DNA samples, conditions were identical with the exception of a 20 μl reaction volume, 2 μl DNA (of unknown concentration), and 50 PCR cycles. To examine amplification success and DNA quantity, amplified DNA was visualized under UV light following electrophoresis through 1.5% agarose gels stained with ethidium bromide.

To prepare DNA sequences for analysis, dNTPs and excess primers were first degraded. We incubated samples for 15 minutes at 37°C, then 15 minutes at 80°C within a 5.78 μl cocktail containing 5 μl amplified DNA, 1.29X Antarctic phosphatase buffer (50 mM Bis-Tris-Propane-HCl, 1 mM MgCl2, 0.1 mM ZnCl2, pH 6.0), 0.1 U/μl Antarctic phosphatase (New England Biolabs), and 0.123 U/μl exonuclease I. Samples were then sequenced using an ABI BigDye Terminator v3.1 Cycle Sequencing Kit (Applied Biosystems), de-salted using ethanol precipitation [48], and size separated and visualized using capillary electrophoresis on an ABI 3500xl Genetic Analyzer (Applied Biosystems). All MAR samples were sequenced using both primers. Sequences were then examined visually using 4Peaks [49] and edited in BioEdit ver. 7.1.3 [50]. A multiple sequence alignment was carried out using ClustalX ver. 2.1 [51]. All historical samples were sequenced in both directions (using each primer), while contemporary samples were sequenced in one direction only. Unique mitochondrial control region haplotypes were then determined by eye and assigned to samples. Any samples with unique haplotypes (found only once within the sample set) were reamplified and sequenced for confirmation.

Haplotypes from this study were then compared to those available from contemporary populations in the Pacific, Laptev Sea and Atlantic regions by Lindqvist et al. [2]. As first steps towards evaluating the amount of evolutionary divergence both within and between groups, we examined nucleotide diversity (π) and haplotype diversity (*h*) using DnaSP ver. 5.10.01 [52]. To visualize relationships between individual haplotypes, a median joining network was constructed using the program Network 4.610 [53]. We used the HKY [54] model of molecular evolution with gamma distributed rate variation across sites and a proportion of invariant sites (HKY+G+I), as indicated by MODELGENERATOR to be appropriate for our data set [55] to construct a phylogenetic tree using Bayesian inference of phylogeny approach implemented in MrBayes ver. 3.2 [56]. Analysis incorporated running 4 chains for 5,000,000 generations, with a sample taken

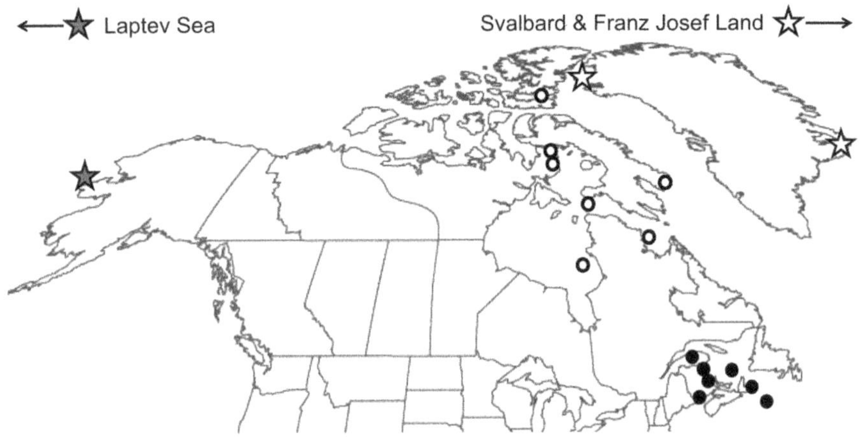

Figure 5. Collection sites for walrus specimens providing DNA data. Black, gray, and white centered shapes denote MAR, PAC, and ATL regions respectively. Circles indicate new data collected in this study, stars indicate data presented previously [2]. Not shown on the map are samples from Laptev Sea, Svalbard and Franz Josef Land of Lindqvist et al. [2].

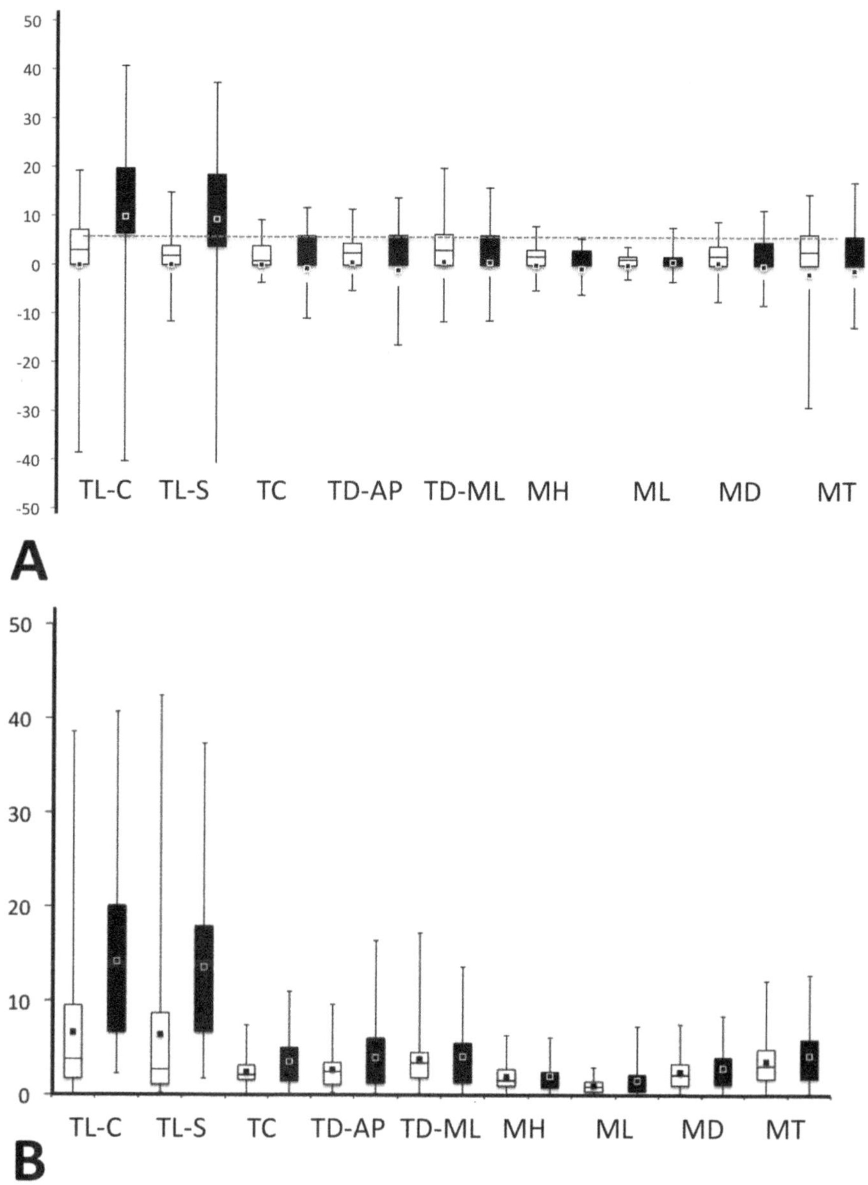

Figure 6. Boxplots of (a) %DA and (b) %AA. Whiskers denote minimum and maximum of the data. Boxes denote the 2nd and 3rd quartiles of the data and the squares denote the data mean. For %DA, negative values indicate a left-side asymmetry while positive values indicate a right-side asymmetry.

every 500 generations, and 1,250,000 (25%) initial steps discarded as burn-in. This tree was visualized using FigTree ver. 1.3.1 [57].

Results

Morphological Analysis

Regression analysis suggested that lateral asymmetry was evident in 6 of the 9 tusk and mandible measures for which left and right sides were examined. Both ATL and MAR showed lateral asymmetry for all tusk measures. While MAR did not show signs of asymmetry in mandibular measures, ATL had slight asymmetry for mandible thickness. Because of this asymmetry, right sides were not exchanged for missing left sides when standardizing the datasets. As a result, the sides with a higher count (left sides for all measures) were used. To further examine asymmetry in the specimens, we also examined %DA and %AA separately for each of the regions (Table 3; Fig. 6) and whether average %DA and %AA within each region was significantly different. After incorporating a Bonferroni correction, only two measures were significantly different between the regions: straight and curvilinear tusk length (Measures #9 and #10). We found that MAR average individual %DA was ~10%, with right tusks longer than left. For the remaining measures, %DA ranged from 0.017%–1.79% while %AA ranged from ~0.978%–4.181% with the MAR exhibiting greater %AA for all measures.

Parametric and non-parametric tests of means of each of the morphological variables showed that 13 of the 16 comparisons were significantly different ($p<0.05$) between the ATL and MAR groups (Fig. 7). Skull measurements corresponding to overall skull length (CBL, ONL) and width (RW, IW) were significantly different across regions, with MAR longer and wider than ATL, while other measures (NL, N-Oc) were not. Mandible length,

Table 3. Regional average percent individual directional asymmetry (%DA) and percent individual absolute asymmetry (%AA) in the MAR and ATL specimens.

Measure:	%DA		Significance	%AA		Significance
	MAR	ATL		MAR	ATL	
TL-C (#9)	**9.781**	**−0.033**	**t = 3.8403, p = 0.0003**	**14.63**	**5.628**	**t = 3.9893, p = 0.0002**
TL-S (#10)	**9.278**	**0.084**	**t = 3.0414, p = 0.0036**	**15.07**	**3.912**	**t = 5.64, p = 0.0001**
TC (#11)	−0.699	0.068	t = 0.8450, p = 0.4013	3.591	2.363	t = 2.3530, p = 0.0218
TD-AP (#12)	−1.026	0.559	t = 0.8421, p = 0.4028	3.977	2.671	t = 1.8982, p = 0.0620
TD-ML (#13)	0.568	0.697	t = 0.1002, p = 0.9205	4.181	3.732	t = 0.5295, p = 0.5983
ML (#14)	0.617	−0.017	t = 1.7243, p = 0.0882	1.544	0.978	t = 2.2098, p = 0.0297
MH (#15)	−0.733	−0.095	t = 1.133, p = 0.2603	2.003	1.616	t = 1.2916, p = 0.1999
MD (#16)	−0.261	0.462	t = 1.0622, p = 0.2907	2.854	2.451	t = 0.9785, p = 0.3302
MT (#17)	−1.09	−1.79	t = 0.6751, p = 0.5010	4.165	3.962	t = 0.2809, p = 0.7794

depth and thickness were significantly different between the two regions, with MAR values being greater, while mandible height was not significantly different. Finally, all tusk measures were significantly different between the regions, with MAR tusks longer and more robust than ATL tusks. Although four PAC specimens were included in the boxplots (Fig. 7) the sample size is too small for these data to be used in any interpretation of the relationships between these regional groupings. It should be noted again that all these results are based on the standardized data, and not original data, therefore we have accounted for sex and age differences in the datasets.

We conducted discriminant function analysis (DFA) using the principal components (PCs) of the PCA. For the skull data, the first two (of seven) PCs accounted for 73.7% of the variance in the dataset. For the mandible data, the first two PCs (of four) accounted for 75.2% of the variance, and for the tusk data the first two (of five) accounted for 99.2% of the variance. In Fig. 8, the DFA scatterplots from each of the three regions are plotted for the skull, mandible and tusk data. In all three graphs there is differentiation between the ATL and MAR regions, and sometimes the PAC. Although differentiation is evident, particularly between tusk and skull measures, there is still considerable overlap in the mandible and tusk distributions. However, there was complete isolation of the ATL and MAR groups within the DFA analysis of skull characteristics.

DNA Extraction, Amplification, Sequencing and Phylogenetic Analysis

Mitochondrial DNA was successfully amplified and sequenced from all 88 of the extracted contemporary specimens and 28 of the 37 extracted historic specimens. We obtained clean and readable sequences that were 344 bp in length. The final haplotype length was largely determined by the sequence length obtained from contemporary specimens, since they were sequenced in a single direction. Among these sequences, 28 unique haplotypes were identified within the contemporary ATL (Genbank accession

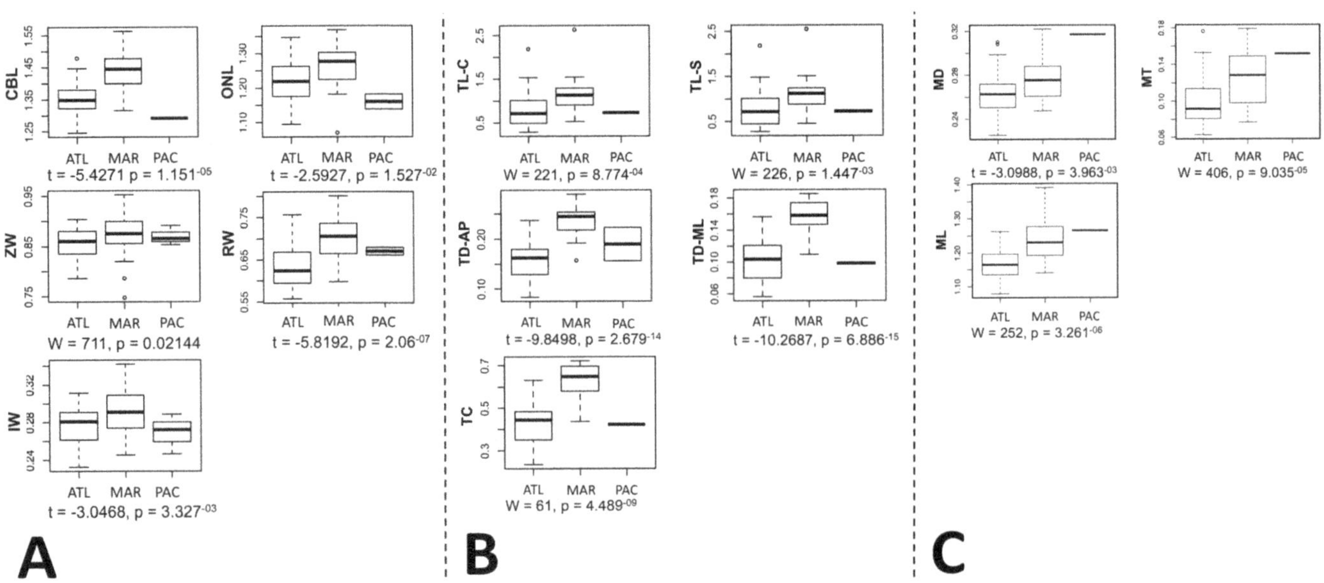

Figure 7. Boxplots for standardized measures of (a) skull, (b) tusk, and (c) mandible data with corresponding *t* or *W* values for the comparison of ATL and MAR regions. Measurement abbreviations correspond to those described in Table 2.

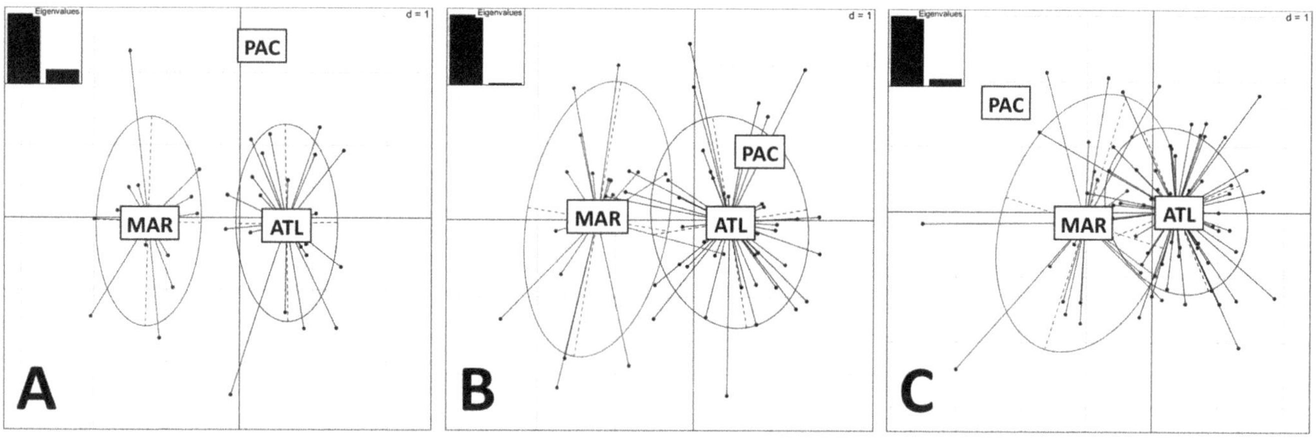

Figure 8. Discriminant function analysis scatterplots for skull (a), tusk (b), and mandible (c) data. Embedded is also a plot of the first two eigenvalues of the discriminant analysis. Circles encompass the range of 2/3 of the data points.

KJ522887 - KJ522911, KJ522920 - KJ522922) and 8 within the historic MAR specimens (Genbank accession KJ522912 – KJ522919). Haplotypes from this study were then compared to those available from Lindqvist et al. [2]; however, because our sequences were shorter than those of Lindqvist et al. [2], three of their haplotypes were collapsed into one (Sequence IDs ATL13/ATL14, a unique haplotype, became identical to ATL11 and ATL21, two other haplotypes). Across the two studies, 3 of our ATL haplotypes were shared with the Lindqvist et al. [2] Atlantic specimens. These were the same as the following sequence IDs from Thule, NW Greenland: ATL07/ATL17, ATL09/ATL10/ATL16, and ATL6/ATL08. However, no haplotypes were shared between historic MAR specimens and any other Atlantic specimens.

Haplotype diversity (*h*), nucleotide diversity (π), and the average number of nucleotide differences both within and between regions were calculated for samples from PAC, Laptev Sea (LAP), ATL, and MAR (Table 4). Both *h* and π were found to be the lowest in the MAR, while they were the highest in the PAC. In addition the MAR had the lowest number of nucleotide differences between sequences.

Haplotypes from PAC, LAP, and ATL (both east and west of Greenland) appear to cluster together within the network, with the exception of two PAC haplotypes that fall out in the network closer to ATL haplotypes from east of Greenland (Fig. 9). Although the samples from the MAR did not share any haplotypes with the PAC/LAP group or the ATL group, the MAR sequences are more similar to the ATL than the PAC/LAP specimens. Many of the ATL haplotypes are highly similar, with most differing from each other by only a single basepair. In addition, the MAR haplotypes appear to branch off in two separate lineages from a single ATL haplotype. Haplotypes that are most similar (and closest) to the MAR haplotypes within the network originate from animals sampled around Southeast Baffin Island and Nottingham Island.

Similar to the median joining network, the phylogenetic tree shows the ATL and PAC/LAP subspecies grouping within separate and distinct clades, with the MAR specimens grouping within the ATL clade (Fig. 10).

Discussion

The morphological analyses presented here indicate that the Maritimes walrus was a physically distinctive group from other populations of walrus in the North Atlantic, west of Greenland. Comparisons of morphological measures across ATL and MAR indicated that most of the cranial and mandibular measures were significantly different between the two groups, with MAR values being greater in all cases. The discriminant analysis of principal components (DAPC) also identified the ATL and MAR as distinctive groups, with complete isolation of skull features. The MAR group appears to have been comprised of larger animals, with larger and more robust tusks, skulls and mandibles. This agrees with previous suggestions that the group was morphologically distinct, and more robust in overall size (pers. comm. to

Table 4. Haplotype (*h*) and nucleotide diversity (π) and corresponding standard error (in parentheses) for PAC, LAP, ATL and MAR regions.

Region	PAC	ATL	MAR	LAP
	$h = 0.967$ (0.036)	$h = 0.844$ (0.024)	$h = 0.439$ (0.114)	$h = 0.800$ (0.164)
	$\pi = 0.02574$ (0.00340)	$\pi = 0.00666$ (0.00031)	$\pi = 0.00332$ (0.00121)	$\pi = 0.00468$ (0.00096)
PAC	8.675			
ATL	13.980	2.230		
MAR	14.469	4.644	1.119	
LAP	7.463	12.957	13.850	1.600

Below this is the average number of nucleotide differences within (on diagonal) and between regions.

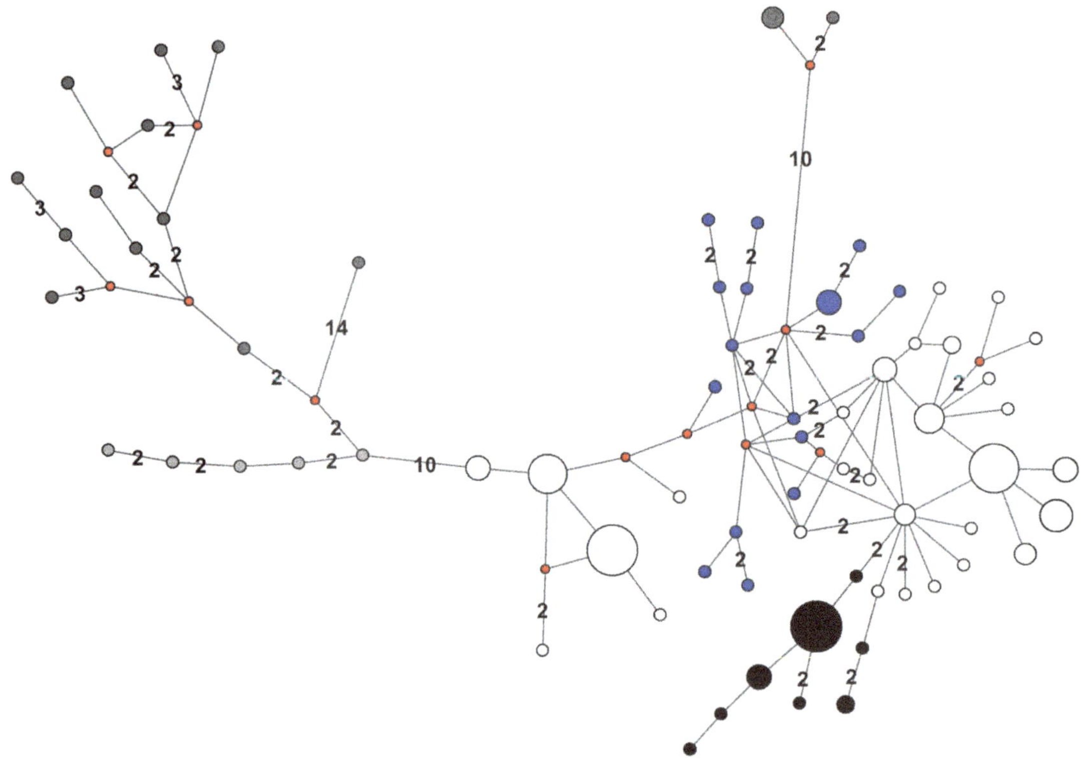

Figure 9. Median joining network of walrus haplotypes from PAC (dark gray), LAP (light gray), ATL (white) and MAR (black) regions as found in Lindqvist et al. [2] **and this study.** In addition, ATL samples from east of Greenland are indicated in blue. Each line segment denotes one sequence difference, except where numbers indicate differences between more divergent sequences. The circle sizes are indicative of haplotype frequencies across the two studies. Small red circles indicate inferred median vectors.

CRH, 1992, reported by [8]). More thorough morphological comparison of the MAR and PAC groups is required.

Previous studies of marine mammals have identified morphological variation in relation to environmental characteristics such as latitude, average water temperature (*Tursiops* spp.) [58], and primary productivity (*Phocoenoides dalli*) [59]. It is not known what selective pressures may have resulted in the differences in size that have been identified here. Morphological differences identified in the MAR group may be a result of genetic divergence and/or environmental effects on phenotype. At this southerly habitat, at the margin of historical walrus distribution, MAR individuals may have been subjected to warmer conditions, as well as different prey and substrate types from those of other walrus populations.

We cannot exclude the possibility that the ATL walrus is less robust than the extirpated MAR walrus as a result of selective hunting pressures towards larger, more robust animals in the Canadian Arctic. The walrus has a long history of exploitation, both aboriginal and commercial. Prized for its blubber and ivory, larger individuals (with larger tusks) were likely preferentially hunted. If exploitation was extensive and skewed towards the hunt of more robust individuals with larger tusks this may have resulted in a phenotypic shift towards smaller individuals with smaller tusks (e.g. [60]).

The DNA analysis indicates that the extirpated MAR group was most genetically similar to contemporary ATL population(s). The MAR haplotypes identified are not monophyletic, as would be expected following long isolation and subsequent genetic divergence, and instead fall within a larger clade including ATL haplotypes. However, in support of the hypothesis that the MAR group was distinctive, we found no shared haplotypes between the regions. All haplotypes identified within the MAR are located within two distinctive and closely related lineages. As well, there are a greater average number of nucleotide differences between the ATL and MAR than within either group suggesting some degree of divergence. Finally, levels of diversity (h and π) were lower in the MAR. This finding is consistent with many previous studies indicating lower levels of genetic variation at range edges, especially in areas where greater isolation has occurred; these include arctic marine mammals (e.g. beluga, bowhead, Saimaa ringed seal (*Pusa hispida saimensis*) (reviewed by [61])). Unfortunately, we are currently limited in the types of additional analyses that can be conducted (e.g. mismatch distribution, etc.) to estimate a divergence time of the MAR and ATL groups because we do not have radiocarbon ages for the MAR specimens. It is therefore assumed that the samples represent a variety of ages, which could range from ~250–12,800 BP [8,27,62].

Samples from the ATL and PAC exhibit strikingly different haplotype divergence patterns (Figure 10). While haplotypes with the PAC clade are relatively divergent from each other, with long branches, the ATL clade has very short comb-like branching. In addition, ATL exhibits lower levels of haplotype and nucleotide diversity (Table 4). This may suggest a shorter evolutionary history for the ATL clade and is consistent with more recent colonization of the area [6]. The similarity of ATL haplotypes (most branching from a single limb) may be indicative of a population bottleneck event, perhaps resulting from a relatively small number of founding individuals. The fact that the MAR haplotypes exist within this clade suggests that the two groups originated from the same stock of founding individuals, which likely shared southerly habitat during the LGM [7,8].

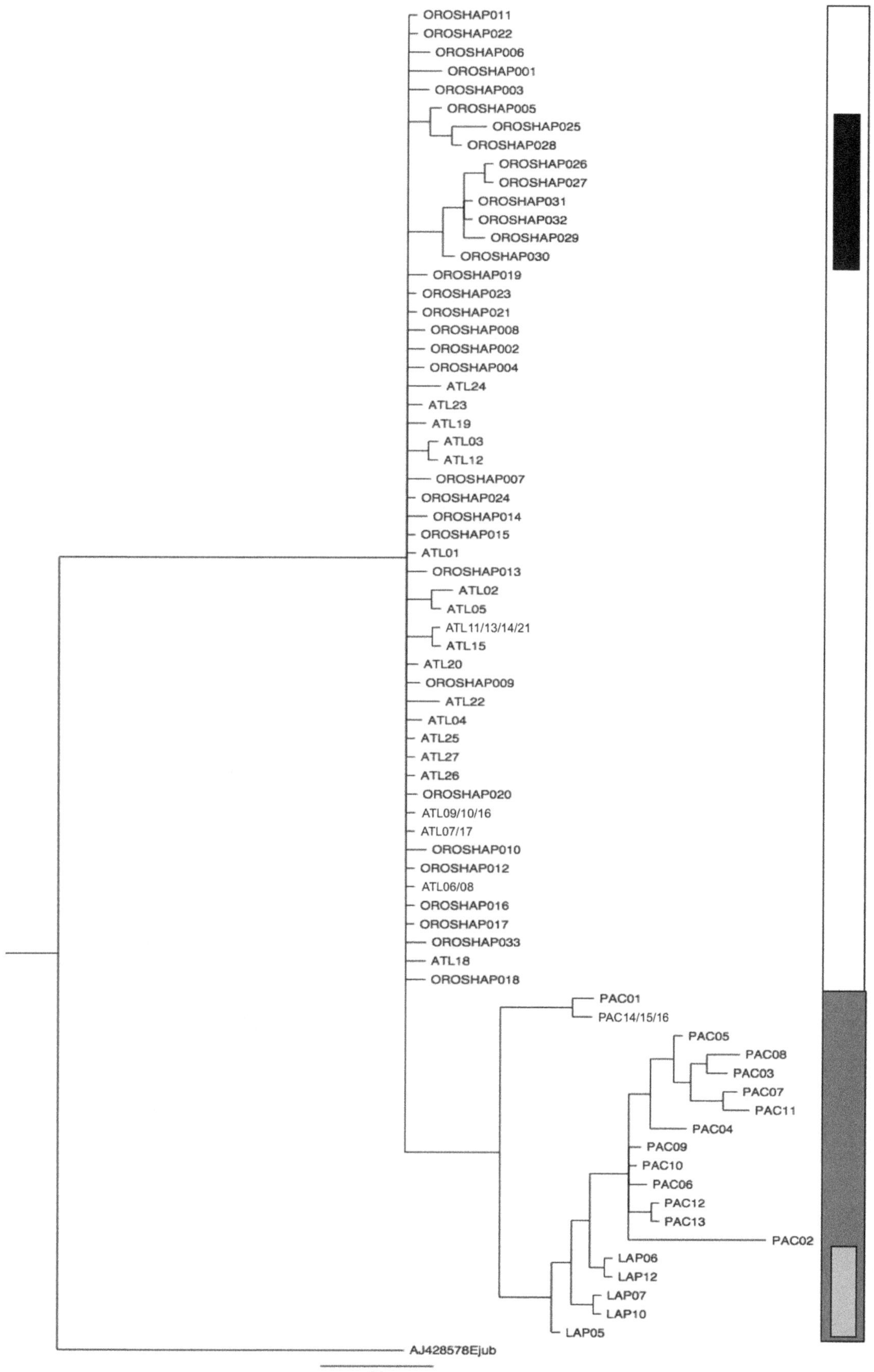
OROSHAP011
OROSHAP022
OROSHAP006
OROSHAP001
OROSHAP003
OROSHAP005
OROSHAP025
OROSHAP028
OROSHAP026
OROSHAP027
OROSHAP031
OROSHAP032
OROSHAP029
OROSHAP030
OROSHAP019
OROSHAP023
OROSHAP021
OROSHAP008
OROSHAP002
OROSHAP004
ATL24
ATL23
ATL19
ATL03
ATL12
OROSHAP007
OROSHAP024
OROSHAP014
OROSHAP015
ATL01
OROSHAP013
ATL02
ATL05
ATL11/13/14/21
ATL15
ATL20
OROSHAP009
ATL22
ATL04
ATL25
ATL27
ATL26
OROSHAP020
ATL09/10/16
ATL07/17
OROSHAP010
OROSHAP012
ATL06/08
OROSHAP016
OROSHAP017
OROSHAP033
ATL18
OROSHAP018
PAC01
PAC14/15/16
PAC05
PAC08
PAC03
PAC07
PAC11
PAC04
PAC09
PAC10
PAC06
PAC12
PAC13
PAC02
LAP06
LAP12
LAP07
LAP10
LAP05
AJ428578Ejub
0.4

Figure 10. Phylogenetic tree of mtDNA control region haplotypes identified in this study and by Lindqvist et al. [2]. Samples from LAP, PAC, ATL and MAR are shown highlighted by light gray, dark gray, white, and black, respectively. The Stellar sea lion (*Eumetopias jubatus*) is used as an outgroup.

The duration and extent to which the MAR region was inhabited by walrus is not known. Prior to ~12,500–12,880 BP walrus in the North Atlantic inhabited their LGM refugium; south of the Bay of Fundy [8]. By 9,700 BP the species had entered the Central Canadian Arctic [8]. Thus, occupation of the MAR region by walrus was feasible between <12,880 BP ~1750 [8,19,20]. It is not known whether occupation of the coast of southeastern Canada was continual over this ~11,000 yr period. Radiocarbon dates for walrus specimens found around southeastern Canada are scattered throughout the period [8,27,63,64], with a cluster of bones that date to ~9,000–10,000 BP [27]. Historical records suggest that the walrus inhabited the area in large numbers between the late 16th century and mid-18th century [7,14,15,19,20] at which time they were extirpated. It is also not known whether the animals were seasonal or year-round inhabitants. Though highly gregarious, walrus are separated by sex and age classes for most of the year [65] and sites may have been used by different sex or age groups at various times of the year. In historical times, walrus were (at least) present in the spring to give birth on land [7]. Despite the aforementioned uncertainties, our results suggest that individuals inhabiting the area were isolated enough from northerly groups to allow for the accumulation of both morphological and genetic adaptations.

Sable Island has provided an abundance of MAR walrus bones (>190), making it the most productive site. While such a large sample set may partially be a result of dedicated survey efforts (by ZL), it also appears that Sable Island and its coastal waters hosted significantly more walrus relative to other sites. Did Sable Island represent a refugium haulout site during glacial periods? Sable Island is an emerged part of a ~30,000 km^2 sand and gravel bank (Sable Island Bank) that was deposited by a Wisconsinan ice front. Lower postglacial sea levels in Atlantic Canada resulted in the emergence of several large banks between 12,000–6000 BP [66]. It is possible that during this time period exposed banks (e.g. Sable Island Bank, Georges Bank, and Grand Bank) provided ideal walrus habitat: haul out sites surrounded by large areas of shallow waters and substrates appropriate to house large communities of bivalve mollusks, the preferred prey of walrus (e.g. *Mya* sp., *Hiatella* sp., and *Serripes* sp.) [67,68]. Certainly, by the 16th century, when most of Sable Island was again submerged, it, and other islands in Eastern Canada continued to provide adequate habitat for walrus.

We identified directional asymmetry in several measures. In most cases the degree of asymmetry appeared to be small, with no significant difference between ATL and MAR regions. The only significant difference in degree of asymmetry (both %AA and %DA) was tusk length, with right tusks ~9–10% longer than left tusks in the MAR. We speculate that the directional asymmetry (DA) may indicate lateralized side use (a 'handedness'), during feeding. This phenomenon has been shown in several marine mammal species (e.g. walrus [69], fin whale (*Balaenoptera physalus*) [70], humpback whales (*Megaptera novaeangliae*) [71], gray whales (*Eschrichtius robustus*) [72]). In some cases, lateralized side use is hypothesized as a cause for morphological asymmetry (e.g. walrus [69], harbour porpoise (*Phocoena phocoena*) [73], gray whales [72]). Although walrus have several feeding methods, one method is to beat one flipper to remove sediments while suction feeding with the muzzle along the sea floor. Levermann et al. [69] found that walrus in northeastern Greenland have a significant preference for beating the right flipper while feeding. The dimensions of right forelimbs (humerus, scapula, ulna) were significantly longer than those on the left, suggesting the animals in that region have a tendency of dextrality ['right handedness'] in feeding [69]. Within our specimens, it is possible that if animals exhibited a preference for beating with a particular flipper they also had a tilt in body axis while feeding that applied additional pressure and/or abrasion to the left side, which slowed growth and/or abraded the length (of tusks). Indeed, many of the tusks from the MAR region showed extensive wear on the anterior anterior/distal edges, presumably a result of abrasion during feeding. Interestingly, while the MAR specimens show this 'skew', the same relationship is not evident in the ATL specimens. Curvilinear tusk length of the right tusk was on average 35 mm (DA = 9.78%) longer than the left for the MAR specimens, while for the ATL specimens the right tusk was on average 2.2 mm (DA = 1.1%) longer than the right. This difference across the two regions may thus reflect different feeding strategies and/or environments in the two regions.

We assumed that our standardization of measures using variables associated with age and sex have minimized any as-yet unidentified age-based affects from differences in the ATL and MAR sample sets. Yet, it is noteworthy that there appeared to be age-dependent differences between the ATL and MAR samples. Many of the ATL individuals were estimated to range from ~2–26+ years old. The majority of these specimens, despite being from mature animals, still had several unfused cranial sutures, which are known to close in a sequential pattern with age. In contrast, most samples from the MAR were heavily ossified and had few open sutures. This likely suggests that the MAR specimens were much older and more physically mature animals than most from the ATL samples. This apparent skew towards older specimens from the MAR may be a result of 1) the older skulls, that were bound more strongly by sutures and have denser bone, making them more resistant to breakage and erosion; and/or 2) some as yet unknown behavioural/ecological characteristics of the population that inhabited Sable Island (e.g. habitat that was used by older individuals). In addition, it is likely that the modern ATL population from which the bone specimens were taken may have had older individuals selectively removed.

Although questions remain regarding the duration and extent to which the MAR region was inhabited by walrus, our data suggest that the walrus that once inhabited the region was morphologically and genetically distinctive. The MAR group appears to have accumulated unique genetic mutations and was morphologically distinct, suggesting that it was on a different evolutionary path than other walrus found in the north Atlantic. The loss of this group following extensive commercial exploitation represents the loss of adaptive potential of the species.

Supporting Information

Table S1 Table of walrus (Odobenus) bone sample specimen information including: sample ID at corresponding institution, origin, bone type as well as radiocarbon date, sex, and age (if known). Museum abbreviations are as follows: Canadian Museum of Nature (CMN), New Brunswick Museum (NBM), New Brunswick Museum (geology collection) (NBMG), NSM (Nova Scotia Museum), PRIV (private collection).

Table S2 Table of samples included in DNA analysis and corresponding results including DNA lab ID, sample ID at corresponding institution, site of origin, tissue type, and mtDNA haplotype. Museum abbreviations are as follows: Department of Fisheries and Oceans (DFO), Makivik Corporation (MAK), New Brunswick Museum (NBM), New Brunswick Museum (geology collection) (NBMG), NSM (Nova Scotia Museum), PRIV (private collection).

Acknowledgments

Logistical support was provided by the Friends of the Green Horse Society, ExxonMobil Canada, Ltd., Saint Mary's University, the Nova Scotia Museum of Natural History, the Canadian Museum of Nature, and the New Brunswick Museum. Contemporary walrus samples (tissue and/or DNA) were provided by Dr. R. E. A. Stewart (Department of Fisheries and Oceans, Canada) and M. Simard (Nunavik Research Centre, Makivik Corporation). This work would not have been possible without the kind assistance of many people during specimen sampling and measurement. We would like to thank D. Bowen, D. Colville, S. Crowell, M. Currie, G. Forbes, C. R. Harington, A. Hebda, K. Khidas, M. Strong, Y. Sivret, D. McAlpine, A. Macfarlane, J. McMillan, R. Miller, K. Ogden, S. Petersen, R.E.A. Stewart, D. Skilliter, D. Tenkula and the walrus hunters of Nunavik. We extend our appreciation to the reviewers of this manuscript, Ulrich Joger and Michael Hofreiter, whose comments greatly improved the manuscript.

We dedicate this paper to the late Francis H. 'Bud' Fay (1927–1994).

Author Contributions

Conceived and designed the experiments: BAM TRF ZL. Performed the experiments: BAM. Analyzed the data: BAM. Contributed reagents/materials/analysis tools: BAM TRF ZL. Wrote the paper: BAM.

References

1. Sexton JP, McIntyre PJ, Angert AL, Rice KJ (2009) Evolution and ecology of species range limits. Annu Rev Ecol Evol Syst 40: 415–436.
2. Lindqvist C, Bachmann L, Andersen LW, Born EW, Arnason U, et al. (2009) The Laptev Sea walrus *Odobenus rosmarus laptevi*: an enigma revisited. Zool Scr 38: 113–127.
3. Lowry L, Kovacs K, Burkanov V (2009) *Odobenus rosmarus*. IUCN Red List of Threatened Species: www.iucnredlist.org.
4. Fay FH (1982) Ecology and biology of the Pacific walrus, *Odobenus rosmarus divergens* Illiger. North American Fauna 74: 1–279.
5. Repenning CA, Ray CE, Grigorescu D (1979) Pinniped biogeography. In: Gray J, Boucot AJ, editors. Historical biogeography, plate tectonics, and the changing environment. Corvallis: Oregon State University Press. pp. 357–369.
6. Fay FH (1985) *Odobenus rosmarus*. Mamm Species 238: 1–7.
7. Allen JA (1880) History of North American pinnipeds: a monograph of the walruses, sea-lions, sea-bears and seals of North America. Washington: G.P.O. 814p.
8. Dyke AS, Hooper J, Harington CR, Savelle JM (1999) The Late Wisconsinan and Holocene record of walrus (*Odobenus rosmarus*) from North America: a review with new data from Arctic and Atlantic Canada. Arctic 52: 160.
9. Born E, Andersen L, Gjertz I, Wiig Ø (2001) A review of the genetic relationships of Atlantic walrus (*Odobenus rosmarus rosmarus*) east and west of Greenland. Polar Biol 24: 713–718.
10. Outridge PM, Davis WJ, Stewart REA, Born EW (2003) Investigation of the stock structure of Atlantic walrus (*Odobenus rosmarus rosmarus*) in Canada and Greenland using dental Pb isotopes derived from local geochemical environments. Arctic 56: 82–90.
11. Stewart REA (2008) Redefining walrus stocks in Canada. Arctic 61: 292–308.
12. Andersen LW, Born EW, Gjertz I, Wiig Ø, Holm LE, et al. (1998) Population structure and gene flow of the Atlantic walrus (*Odobenus rosmarus rosmarus*) in the eastern Atlantic Arctic based on mitochondrial DNA and microsatellite variation. Mol Ecol 7: 1323–1336.
13. Andersen LW, Born EW (2000) Indications of two genetically different subpopulations of Atlantic walruses (*Odobenus rosmarus rosmarus*) in west and northwest Greenland. Can J Zool 78: 1999–2009.
14. Mowat F (1984) Sea of Slaughter. Toronto: McClelland and Stewart. 438p.
15. Born EW, Gjertz I, Reeves RR (1995) Population assessment of Atlantic walrus (*Odobenus rosmarus rosmarus L.*). Oslo: Norsk Polarinstitutt. Meddelelser Nr. 138. 100p.
16. Naughton D (2012) The Natural History of Canadian Mammals. Toronto: University of Toronto Press. 824p.
17. Allen JA (1876) The former range of some New England carnivorous mammals. Am Nat 10: 708–715.
18. Martijn CA (2003) Early Mi'kmaq presence in Southern Newfoundland: an ethnohistorical perspective, c.1500–1763. NLS 19: 44–102.
19. Allen GM (1930) The walrus in New England. J Mammal 11: 139–145.
20. Manville RH, Favour PG Jr (1960) Southern distribution of the Atlantic walrus. J Mammal 41: 499–503.
21. Wright BS (1951) A walrus in the Bay of Fundy. Can Field Nat 65: 61–64.
22. Richer S (2003) Nova Scotians awake to find a rare walrus in their midst. Globe and Mail 12 Jun 2003.
23. Sobey DG (2010) An analysis of the historical records for the native mammalian fauna of Prince Edward Island. Can Field Nat 121: 384–396.
24. Department of Fisheries and Oceans (2008) Recovery Strategy for the Atlantic walrus (*Odobenus rosmarus rosmarus*), Northwest Atlantic population, in Canada. Ottawa: Species at Risk Act Recovery Strategy Series, Department of Fisheries and Oceans. x +11p.
25. Harington CR, Anderson TW, Rodrigues CG (1993) Pleistocene walrus (*Odobenus rosmarus*) from Forteau, Labrador. Geogr Phys Quatern 47: 111–118.
26. Miller RF (1990) New records of postglacial walrus and a review of Quaternary marine mammals in New Brunswick. Atlantic Geol 26: 97–107.
27. Miller RF (1997) New records and AMS radiocarbon dates on Quaternary walrus (*Odobenus rosmarus*) from New Brunswick. Geogr Phys Quatern 51: 107–112.
28. Fay FH (1981) Walrus *Odobenus rosmarus* (Linnaeus, 1758). In: Harrison RJ, Ridgeway SH, editors. Handbook of Marine Mammals. 1. The walrus, sea lions, fur seals and sea otter. London: Academic Press. pp. 1–23.
29. Mansfield AW (1958) The biology of the Atlantic walrus, *Odobenus rosmarus rosmarus* (Linnaeus) in the eastern Canadian Arctic. Fish Res Board Can MS Rept Ser (Biology), 653:146p.
30. Committee on Marine Mammals (1967) Standard measurements of seals. J Mammal 48: 459–462.
31. Wiig Ø, Gjertz I (1996) Body size of male Atlantic walruses (*Odobenus rosmarus rosmarus*) from Svalbard. J Zool 240: 495–499.
32. Wiig Ø, Born EW, Gjertz I, Lydersen C, Stewart REA (2007) Historical sex-specific distribution of Atlantic walrus (*Odobenus rosmarus rosmarus*) in Svalbard assessed by mandible measurements. Polar Biol 31: 69–75.
33. Auerbach BM, Ruff CB (2006) Limb bone bilateral asymmetry: variability and commonality among modern humans. J Hum Evol 50: 203–218.
34. Steele J, Mays S (1995) Handedness and directional asymmetry in the long bones of the human upper limb. Int J Osteoarchaeol 5: 39–49.
35. Mays S (2002) Asymmetry in metacarpal cortical bone in a collection of British post-mediaeval human skeletons. J Archaeol Sci 29: 435–441.
36. Hochberg Y (1988) A sharper Bonferroni procedure for multiple tests of significance. Biometrika 75: 800–802.
37. Shapiro SS, Wilk MB (1965) An analysis of variance test for normality (complete samples). Biometrika 52: 591–611.
38. Welch BL (1951) On the comparison of several mean values: an alternative approach. Biometrika 38: 330–336.
39. Wilcoxon F (1945) Individual comparisons by ranking methods. Biometrics Bulletin 1: 80–83.
40. R Core Development Team (2012) R: A language and environment for statistical computing. R Foundation for Statistical Computing. Available: http://www.R-project.org. Accessed: 2014 May 20.
41. Venables WN, Ripley BD (2002) Modern applied statistics with S. New York: Springer. xi + 495p.
42. Chessel D, Dufour AB, Thioulouse J (2004) The ade4 package. I. One-table methods. R News 4: 5–10.
43. Jombart T, Devillard S, Balloux F (2010) Discriminant analysis of principal components: a new method for the analysis of genetically structured populations. BMC Genet 11: 1–15.
44. Sariyar-Akbulut B (2010) Discriminant function analysis based on principal components for rapid discrimination of metabolic capabilities of new isolates. Chem Biochem Eng Q 24: 119–127.
45. McLeod BA, Brown MW, Moore MJ, Stevens W, Barkham SH, et al. (2008) Bowhead whales, and not right whales, were the primary target of 16th-to 17th-century Basque Whalers in the Western North Atlantic. Arctic 61: 61–75.
46. Rastogi T, Brown MW, McLeod BA, Frasier TR, Grenier R, et al. (2004) Genetic analysis of 16th-century whale bones prompts a revision of the impact of Basque whaling on right and bowhead whales in the western North Atlantic. Can J Zool 82: 1647–1654.
47. McLeod BA, Frasier TR, Dyke AS, Savelle JM, White BN (2012) Examination of ten thousand years of mitochondrial DNA diversity and population demographics in bowhead whales (*Balaena mysticetus*) of the Central Canadian Arctic. Mar Mamm Sci 28: E426–E443.
48. Irwin DL, Mitchelson KR, Findlay I (2003) PCR product cleanup methods for capillary electrophoresis. Biotechniques 34: 932–4, 936.

49. Griekspoor A, Groothuis T. 4Peaks website. Available: http://nucleobytes.com/index.php/4peaks. Accessed 2014 May 20.
50. Hall T. BioEdit Sequence Alignment Editor for Windows 95/98/NT/XP website. Available: http://www.mbio.ncsu.edu/BioEdit/page2.html. Accessed 2014 May 20.
51. Larkin MA, Blackshields G, Brown NP, Chenna R, McGettigan PA, et al. (2007) Clustal W and Clustal X version 2.0. Bioinformatics 23: 2947–2948.
52. Rozas J, Sánchez-DelBarrio JC, Messeguer X, Rozas R (2003) DnaSP, DNA polymorphism analyses by the coalescent and other methods. Bioinformatics 19: 2496–2497.
53. Bandelt H-J, Forster P, Röhl A (1999) Median-joining networks for inferring intraspecific phylogenies. Mol Biol Evol 16: 37–48.
54. Hasegawa M, Kishino H, Yano T (1985) Dating of the human-ape splitting by a molecular clock of mitochondrial DNA. J Mol Evol 22: 160–174.
55. Keane TM, Creevey CJ, Pentony MM, Naughton TJ, McInerney JO (2006) Assessment of methods for amino acid matrix selection and their use on empirical data shows that ad hoc assumptions for choice of matrix are not justified. BMC Evol Biol 6: 1–17.
56. Huelsenbeck JP, Ronquist F (2001) MRBAYES: Bayesian inference of phylogeny. Bioinformatics 17: 754–755.
57. Rambaut A. Molecular Evolution, Phylogenetics and Epidemiology website, program FigTree. Available: http://tree.bio.ed.ac.uk/software/figtree/). Accessed 2014 May 20.
58. Ross GJB, Cockroft VG (1990) Comments on Australian bottlenose dolphins and the taxonomic status of *Tursiops aduncas* (Ehrenberg, 1832). In: Leatherwood S, Reeves RR, editors. The Bottlenose Dolphin. London: Academic Press. 101–128.
59. Amano M, Miyazaki N (1992) Geographic variation and sexual dimorphism in the skull of Dall's porpoise *Phocoenoides dalli.* Mar Mamm Sci 8: 240–261.
60. Coltman DW, O'Donoghue P, Jorgenson JT, Hogg JT, Strobeck C, et al. (2003) Undesirable evolutionary consequences of trophy hunting. Nature 426: 655–658.
61. O'Corry-Crowe G (2008) Climate change and the molecular ecology of Arctic marine mammals. Ecol Appl 18: 56–76.
62. Harington CR (2003) Quaternary vertebrates of Québec: a summary. Geogr Phys Quatern 57: 85–94.
63. Bouchard MA, Harington CR, Guilbault J-P (1993) First evidence of walrus (*Odobenus rosmarus L.*) in late Pleistocene Champlain Sea sediments, Quebec. Can J Earth Sci 30: 1715–1719.
64. Harington CR (2003) Annotated bibliography of Quaternary vertebrates of northern North America. Toronto: University of Toronto Press. 360p.
65. Perrin WF, Wursig B, Thewissen JGM (2009) Encyclopedia of Marine Mammals. San Diego: Academic Press. 1352p.
66. Shaw J, Gareau P, Courtney RC (2002) Palaeogeography of Atlantic Canada 13-0 kyr. Quat Sci Rev 21: 1861–1878.
67. Fisher KI, Stewart REA (1997) Summer foods of Atlantic walrus, *Odobenus rosmarus rosmarus*, in northern Foxe Basin, Northwest Territories. Can J Zool 75: 1166–1166.
68. Sheffield G, Grebmeier JM (2009) Pacific walrus (*Odobenus rosmarus divergens*): Differential prey digestion and diet. Mar Mamm Sci 25: 761–777.
69. Levermann N, Galatius A, Ehlme G, Rysgaard S, Born EW (2003) Feeding behaviour of free-ranging walruses with notes on apparent dextrality of flipper use. BMC Ecol 3: 1–13.
70. Tershy BR, Wiley DN (1992) Asymmetrical pigmentation in the fin whale: a test of two feeding related hypotheses. Mar Mamm Sci 8: 315–318.
71. Clapham PJ, Leimkuhler E, Gray BK, Mattila DK (1995) Do humpback whales exhibit lateralized behaviour? Anim Behav 50: 73–82.
72. Kasuya T, Rice DW (1970) Notes on baleen plates and on arrangement of parasitic barnacles of gray whale. Whales Research Institute 22: 39–43.
73. Galatius A, Jespersen S (2005) Bilateral directional asymmetry of the appendicular skeleton of the harbor porpoise (*Phocoena phocoena*). Mar Mamm Sci 21: 401–410.

15

Impact of Gene Molecular Evolution on Phylogenetic Reconstruction: A Case Study in the Rosids (Superorder Rosanae, Angiosperms)

Khidir W. Hilu*, Chelsea M. Black¤a, Dipan Oza¤b

Department of Biological Sciences, Virginia Tech, Blacksburg, Virginia, United States of America

Abstract

Rate of substitution of genomic regions is among the most debated intrinsic features that impact phylogenetic informativeness. However, this variable is also coupled with rates of nonsynonymous substitutions that underscore the nature and degree of selection on the selected genes. To empirically address these variables, we constructed four completely overlapping data sets of plastid *matK*, *atpB*, *rbcL*, and mitochondrial *matR* genes and used the rosid lineage (angiosperms) as a working platform. The genes differ in combinations of overall rates of nucleotide and amino acid substitutions. Tree robustness, homoplasy, accuracy in contrast to a reference tree, and phylogenetic informativeness are evaluated. The rapidly evolving/unconstrained *matK* faired best, whereas remaining genes varied in degrees of contribution to rosid phylogenetics across the lineage's 108 million years evolutionary history. Phylogenetic accuracy was low with the slowly evolving/unconstrained *matR* despite least amount of homoplasy. Third codon positions contributed the highest amount of parsimony informative sites, resolution and informativeness, but magnitude varied with gene mode of evolution. These findings are in clear contrast with the views that rapidly evolving regions and the 3rd codon position have inevitable negative impact on phylogenetic reconstruction at deep historic level due to accumulation of multiple hits and subsequent elevation in homoplasy and saturation. Relaxed evolutionary constraint in rapidly evolving genes distributes substitutions across codon positions, an evolutionary mode expected to reduce the frequency of multiple hits. These findings should be tested at deeper evolutionary histories.

Editor: Ming Zhang, University of Georgia, United States of America

Funding: This work was supported by grant from the National Science Foundation, USA, Assembling Tree of Life (grant numbers EF-043105 and REU-4776833) to KH. The funders had no role in study design, data collection and analysis, decision to publish, or preparation of the manuscript.

Competing Interests: The authors have declared that no competing interests exist.

* Email: hilukw@vt.edu

¤a Current address: Department of Cellular and Molecular Medicine, 1830 Monument Street, John Hopkins School of Medicine, Baltimore, Maryland, United States of America

¤b Current address: 3707 St. George's University, University Centre, St. George's, Grenada, West Indies

Introduction

Genomic regions vary considerably in their potential phylogenetic informativeness and subsequent contribution to resolving a given set of taxa over specified time points. Among the properties inherent to genomic regions, rate of nucleotide substitution is regarded to have a profound impact in phylogenetic reconstruction [1–8]. This has led to a debate over the "optimal" rate of substitution for a choice genomic region within a contextual historic depth [1,2,6,9–15]. Rate of substitution in this ideal genomic region should not be too low to generate inadequate signal or too high to inflate noise. Consequently, the prevalent approach promotes the use of rapidly evolving regions in shallow evolutionary histories and slowly evolving regions for deeper epochs [3,16–20]. The exclusion of rapidly evolving regions from deep phylogenetic reconstruction is generally based on the premise that multiple hits confounded by extended time scale could be significant enough to conceal phylogenetic signals and elevate homoplasy, with saturation reaching levels that can negatively impact tree structure [9,11,14,15]. It was further suggested that accumulation of multiple hits in rapidly evolving regions can obscure potential synapomorphies and may also result in long-branch attraction [6,21,22]. However, opposing points have been raised that promote the effectiveness of rapidly evolving and less constrained genomic regions in deep-level phylogenetics [2,13,23–26].

Similarly, the 3rd codon position was down weighted or excluded in phylogenetic analyses due to higher rates of substitution compared with the 1st and 2nd positions [27–33]. It has been argued [22] that the greater average of substitution rate for 3rd codon positions might reduce phylogenetic signal due to long-branch attraction in parsimony reconstructions. Li and Graur [34] indicated that unlike 1st and 2nd codon positions, the 3rd codon position is highly saturated and contributes higher levels of homoplasy. However, these views have been disputed [2,10,35–43]. The 3rd codon position is under relaxed selection since mutations in it are less likely to lead to nonsynonymous substitutions. Theoretically, mutations in the 1st, 2nd and 3rd codon positions translate into 96%, 100%, and 31% nonsynonymous substitutions, respectively [34].

These opposing notions have not been tested within a well-defined experimental design. We evaluate here the overall phylogenetic informativeness of four protein-coding genes (*rbcL*, *atpB*, *matK*, *matR*) with contrasting combinations of mode and tempo of evolution in the rosids lineage (angiosperms) using PhyDesign [6] in conjunction with various statistical measures. The *matR* gene is mitochondrial whereas the others are plastid genes. Functionally, *matK* and *matR* are group II intron maturases, *rbcL* codes for the large subunit of RuBisCo, an essential photosynthesis enzyme, while *atpB* encodes the beta subunit of the plastid ATPase [44–46]. Substitution rate in *matK* is about twice that of *rbcL*, and 6.5 times at nonsynonymous sites [25,47]. The synonymous substitution rate in *matR* is approximately four times slower than those of *rbcL* and *atpB*, whereas the rate of nonsynonymous mutations is considerably higher [3,48,49]. The substantially higher rates of nonsynonymous mutations in *matR* and *matK* imply reduced selectional constraints compared with *rbcL* and *atpB* [3,25,46,48,50]. Thus, the four genes represent spectra of nucleotide substitution rates from the rapidly evolving *matK* to *rbcL*, *atpB*, and *matR*, and amino acid substitution rates from *matR* to *matK*, *rbcL*, and *atpB*. For convenience, we use the terms *rapidly* and *slowly* evolving for overall rates of substitutions and *constrained* and *unconstrained* for degrees of nonsynonymous substitution.

The choice of the rosids (Superorder Rosanae APG III [51]) is based on current availability of a robust multi-gene phylogeny [19], the documented monophyly of the group with well defined lineages (e.g. rosids, fabids, core malvids), and the detection of rapid radiation in parts of its ~108 million years (MY) of evolution [19]. The rosid clade, as circumscribed by APG III [52], includes 140 families placed in 18 orders (Vitales placement is uncertain [53]). Two recent rosids-focused phylogenetic studies exist based on four genes [48], and 36 genes plus intervening spacer sequences [19]. The latter study provides the most robust tree for the rosids, and will thus be used as reference tree here. The rosid families fall into two large subclades, Fabidae (fabids) and Malvidae (malvids) (Figure 1). Within the fabids, the Zygophyllales is sister to two clades: the nitrogen fixing clade (NFC; Rosales, Fabales, Cucurbitales, Fagales), and the COM clade (Celastrales, Malpighiales, Oxalidales). The malvids clade includes the historically difficult to place orders Myrtales, Crossosomatales, Geraniales, and Picramniales, in a grade sister to the core malvids (Brassicales, Malvales, Sapindales, Huerteales).

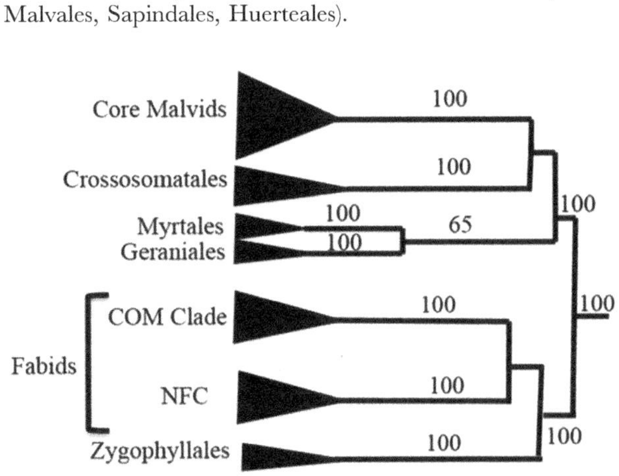

Figure 1. A summarized backbone relationships among the rosids lineages based on Wang et al. [19] phylogenetic study. A total evidence maximum likelihood rosid tree redrawn from Wang et al. (2009) with percent bootstrap support values ≥50% noted on branches. COM: Celastrales, Malpighiales, and Oxalidales. NFC (Nitrogen Fixing Clade): Rosales, Fabales, Cucurbitales, and Fagales. The malvids clade includes the basal three orders Myrtales+Geranials/Crossosomatales as sister to the core group.

We demonstrate here that the rapidly evolving and unconstrained *matK* provides substantially higher phylogenetic information and structure throughout the rosids history, whereas slowly evolving genes exhibit lower degrees of informativeness regardless of rates of nonsynonymous mutations. The 3rd codon position consistently superseded the 1st and 2nd positions in phylogenetic signal, but the disparity in informativeness is accentuated in the constrained genes. The four genes informativeness profiles combined provide complementary phylogenetic signal across the rosids phylogenetic tree.

Materials and Methods

Genomic regions, taxon sampling, and data partitioning

To empirically evaluate the phylogenetic informativeness of four single-copy orthologs selected for this study (plastid *atpB*, *rbcL*, and *matK*, and mitochondrial *matR*), we generated overlapping data sets of 60 species from the rosid clade. The data set includes newly-generated complete or partial sequences for *matK* and sequences obtained from GenBank (Table S1). The data set covers 43 of the 140 families and represents 16 of the 17 orders currently assigned to the rosid clade *sensu* APG III [51]. The order Rafflesiales sensu APG III [51] is not represented, however, the APG III [51] places the family in the Malpighiales. The Vitaceae (Vitales) is included as part of the core eudicots since its placement in the rosids is equivocal [53,54]. Taxon sampling was based on achieving strong overlap with the sample used in the Wang et al. [19] study of the rosids. In addition to the rosid taxa, 21 species were included to represent the remaining core eudicot lineages (Solanales, Lamiales, Cornales, Ericales, Caryophyllales, Phyllanthaceae, Berberidopsidales, Vitales, and Dilleniaceae). The data set was partitioned into individual genomic regions and respective codon positions. Trees were rooted in all analyses with representatives of the early diverging eudicots Buxaceae and Trochodendraceae and the first diverging core eudicot Gunneraceae [13,55]. Information on the species used and sources of the sequences are noted in Table S1.

Materials, DNA isolation and sequencing

Genomic DNA for the 54 new *matK* sequences was either isolated from material collected in the field or was obtained from various sources (Table S1). Genomic DNA was isolated following Doyle and Doyle [56] procedure as modified in M'ribu and Hilu [57]. The *matK* gene was amplified as described in Crawley and Hilu [8]. Sequencing was performed at the Virginia Bioinformatics Institute at Virginia Tech or Duke University using a Big Dye Terminator Cycle Sequencing Ready Reaction Kit (Applied Biosystems, Foster City, California).

Sequence alignment and phylogenetic reconstruction

Sequences were manually aligned using the QuickAlign program [58]. Gaps were inserted at the cost of two or more substitutions. However, all data sets were analyzed without the inclusion of gaps as characters in order to avoid character bias among genes since *atpB*, *rbcL* and *matR* either lack or have few gaps, and to render the data comparable to the Wang et al. [19] study that excluded gaps.

The data sets were analyzed using both Maximum Parsimony (MP) and Maximum Likelihood (RAxML). The MP analyses were conducted in PAUP* version 4.0b [59] and consisted of a heuristic tree search with TBR branch swapping and 1000 random stepwise addition replicates with indels treated as missing data. In the MP analyses of codon partition, the runs did not reach completion for the 1st and 2nd data sets in some of the slowly evolving genes. To avoid subjective inflations in the number of most parsimonious trees and the subsequent impact on consensus tree resolution, we opted to analyze concatenated 1st and 2nd data sets. However, results from individual codon position analyses that reached completion will be noted wherever relevant. A strict consensus tree was generated when multiple most parsimonious trees were recovered. Bootstrap support (BS; Felsenstein [60]) was calculated in PAUP* with 1000 replicates each with 10 random sequence addition replicates using the same conditions as in MP analysis. Due to time constraints, the data sets were analyzed using the Ratchet algorithm [61] as implemented in PRAP2 [62] and executed in PAUP*. The RAxML analyses were conducted in CIPRES portal (http://www.phylo.org) applying the default setting and conducting 1000 replicates. Bootstrap support was calculated for the 50% majority trees. The default Model GTR+I+G was used.

Measurements of phylogenetic informativeness

PhyDesign (http://phydesign.townsend.yale.edu; [63]) was used to estimate phylogenetic informativeness of genomic regions across rosid evolutionary history to assess their effectiveness in phylogenetic reconstruction in the context of their mode and tempo of evolution. This software program computes the amount of phylogenetic information in a genomic region across the history of the group based on character evolutionary rates. The Townsend [6] metric phylogenetic informativeness has been used effectively in predicting informativeness profiles in various groups [8,64–70]. Although Klopfstein et al. [14] argued that it could lead to overestimation of informativeness in rapidly evolving genes when taxon sampling increased beyond the 4-taxa case, Townsend and Luenberger [71] refuted that notion. Both net and per-site informativeness were computed and contrasted to assess cost-effectiveness of the genes. The concatenated nucleotide data set was partitioned by genes and codon positions. We used these alignments to generate ML trees with RAxML. A fixed age of 108 MY was selected for the rosids divergence, and 6 minimum ages (*Phytolacca/Polygonum* 83.5 MY; *Galax/Sarracenia* 91.2 MY; *Leea/Vitis* 57.9 MY; *Citrus/Bursera* 65 MY; *Malpighia/Passiflora* 49 MY; *Populus/Salix* 48 MY) were designated following Wang et al [19]. The RaxML best tree file and the designated dates were used to reconstruct ultrametric trees in PATHd8 (www.math.su.se/PATHdh; [72]). The ultrametric tree files were executed in MEGA 4.0 [73] to generate a Newick format. These tree files and their corresponding data sets were used as input files in PhyDesign to extract the phylogenetic informativeness for genes and their codon positions.

Measures of phylogenetic structure and accuracy

Phylogenetic structure encompasses tree resolution and support for depicted relationships. With the focus on the backbone of the rosids (the major clades, their subclades, and the orders), the total number of nodes in a fully resolved tree is 41. We used the number and percentages of nodes resolved in the MP strict consensus tree derived from partitioned (genes and codons) data sets as a measure of resolution. Bootstrap [60] values obtained from RAxML were used as statistical measure of support. Since the Townsend [6] phylogenetic informativeness does not account for homoplasy [63], ensemble consistency index (CI; Kluge and Farris [74]) and ensemble retention index (RI; Farris [75]) were used as measures of homoplasy to evaluate signal vs. noise for the four genomic. Phylogenetic accuracy was assessed by comparing the reconstructed trees from the partition analyses with a model tree for the rosids, namely the total-evidence tree of Wang et al. [19] for incongruences. This latter tree was based on >43,000 base pair (bp) from two nuclear and 34 plastid genes plus some intervening spacers of plastid inverted repeat. It is fully resolved, strongly supported, and topologically highly congruent with relationships recovered in phylogenetic studies on angiosperms, e.g. [24,55].

Assessments of molecular evolution and statistical tests

The four protein coding genes display different rates and modes of evolution as reflected in the rates of nonsynonymous substitution and its subsequent impact on amino acid mutations. We estimated the ratio of nonsynonymous substitution per nonsynonymous site to synonymous substitution per synonymous site, dN/dS, using SNAP (www.hiv.lanl.gov; [76]). Statistical testing was carried out in JMP 9 (www.jmp.com). To assess degree of variation in phylogenetic signal at different eras of rosids evolution, the 108 MY time scale was divided into four equal epochs and both net informativeness and its standard deviation for the genes were calculated for each epoch.

Results

To be consistent in contrasting phylogenetic tree reconstructed in this study with the reference tree of Wang et al. [19], we will focus on the RaxML trees. However, consensus trees and tree statistics obtained from the MP analyses will also be discussed. The RAxML trees for the four gene partitions are summarized to highlight the major rosid clades (Figures 2–5); the detailed trees are provided in Figures S1–S4.

Phylogenetic structure

Three tree parameters will be underscored here, namely resolution, node support, and accuracy as compared with the reference tree. Degree of resolution is presented as number and percentage of nodes resolved in the MP strict consensus tree out of a total 41 expected nodes (Table 1). Degree of resolution is highest in *matK* (40, 98%) and lowest in *atpB* (29, 70%). The BS support for 17 major nodes that describe the backbone of the rosids is plotted for analyses based on gene partitions (Figure 3). In the *matK* partition (rapidly evolving/unconstrained), the rosids' monophyly receive 100% BS support (Figure 2), and core malvids and fabids are recovered with 97% and 87% BS support, respectively. Of the three basal malvids orders, the Myrtales appear sister to core malvids (70% BS), whereas the Crossosommatales and Geraniales emerge in a clade sister to remaining rosids (<50% BS). The core malvids structure (Figure S1) is topologically identical to that retrieved in Wang et al. [19]. Within the fabids, Zygophyllaceae emerge as sister to the NFC (99% BS) plus the COM clade (100% BS). Phylogenetic relationships among the orders within the NFC and COM clades differ from those obtained by Wang et al. [19], but incongruences are weakly supported. All orders received 100% BS support (Table S2). The CI and RI values are 0.374 and 0.429, respectively.

The slowly evolving/unconstrained mitochondrial *matR* recovers the rosids (58% BS) as well as the core malvid clade (72% BS), but the fabid clade as traditionally defined [19,24,53,55] is not retrieved (Figure 2). Instead, the COM clade of the fabid appears sister to the malvids, albeit BS support for this relationship is 58%. Two of the early diverging malvids lineages (Myrtales and Geraniales) form a clade sister to all rosids, whereas the third lineage, the Crossosomatales, form along with the Zygophyllales a weakly supported grade sister to the malvids plus COM (Figure 2). All these nodes received 51% BS support at best. The topology of the COM clade is congruent with that of Wang et al. [19], whereas those of the malvid and NFC clades are not (Figure S2). BS support for the monophyly of the orders range from 95–100%

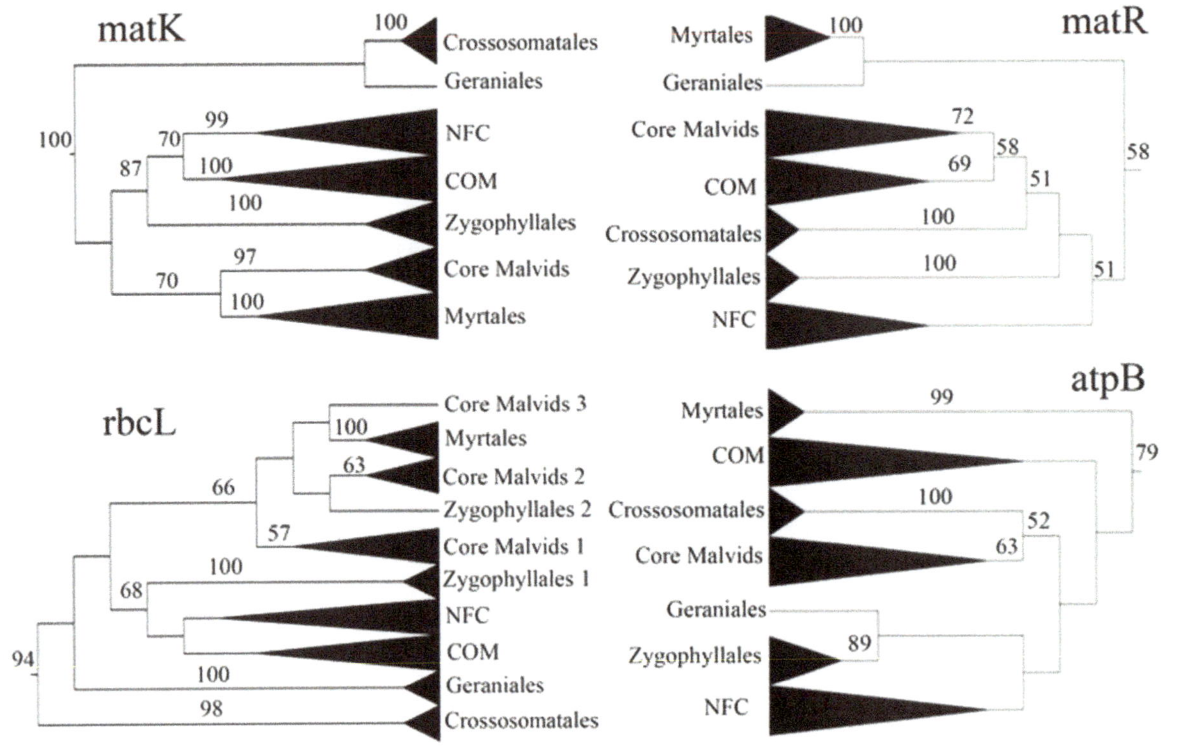

Figure 2. Summaries of the rosids RAxML trees based on the four gene partitions. The backbone relationships among the rosids lineages based on partitioned data sets of *matK*, *matR*, *rbcL*, and *atpB*. Percent bootstrap support values ≥50% are noted. COM: Celastrales, Malpighiales, and Oxalidales. NFC (Nitrogen Fixing Clade): Rosales, Fabales, Cucurbitales, and Fagales. The core malvids group and the Zygophyllales were not recovered as monophyletic lineages in the *rbcL* partition analysis.

(Table S2). The CI and RI values are 0.692 and 0.638, respectively.

The *rbcL* (slow/constrained) data provided 94% BS support for the rosids monophyly and 68% for the fabids. However, core malvids are not monophyletic, with Myrtales and *Tribulus* (Zygophyllaceae) nested in them with 66% BS support, a topology that depicts polyphyletic Zygophyllaceae (Figure 2). The Crossosomatales and Geraniales appear as consecutive sisters to all rosids but their placements receive <50% BS support. The ordinal relationships within the malvid and fabid clades (Figure S3) are incongruent with those obtained by Wang et al. [19], but BS support is weak. Monophyly for the orders, except paraphyletic Rosales, receive 58–100% BS support (Table S2). The CI and RI values are 0.353 and 0.416, respectively.

The *atpB* (slow/constrained) data recover the rosids with 79% BS, but fail to recover the fabids as COM + NFC. Instead, the

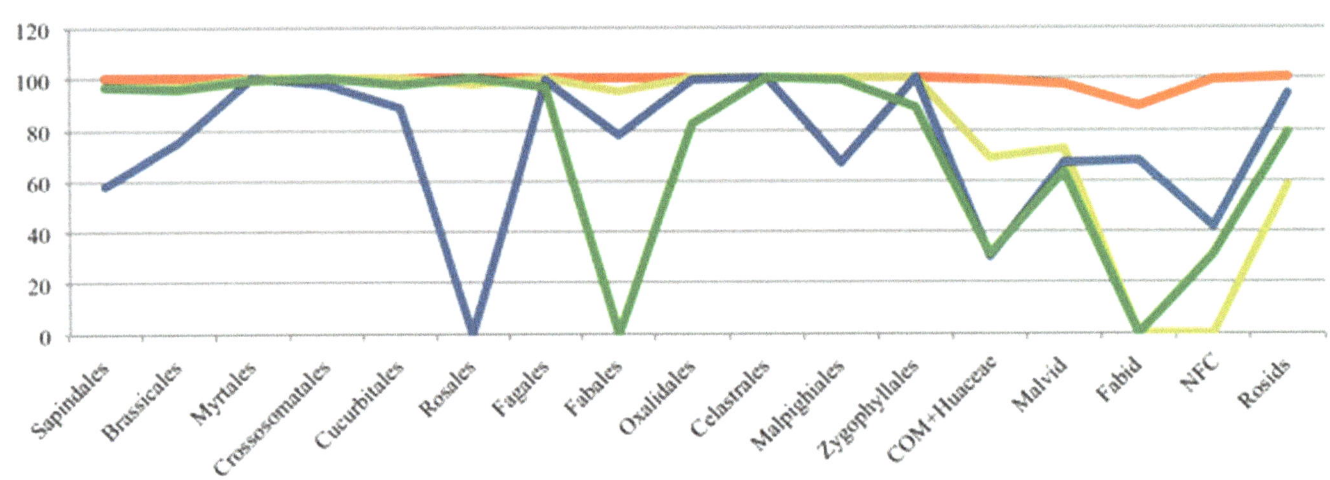

Figure 3. Variation in bootstrap support for the major rosids lineages derived from gene partition analyses. Illustration of the differential bootstrap support for the rosids orders and the major clades calculated from the RAxML analyses based on partitioned gene data sets of *matK* (red), *matR* (yellow), *rbcL* (blue), and *atpB* (green).

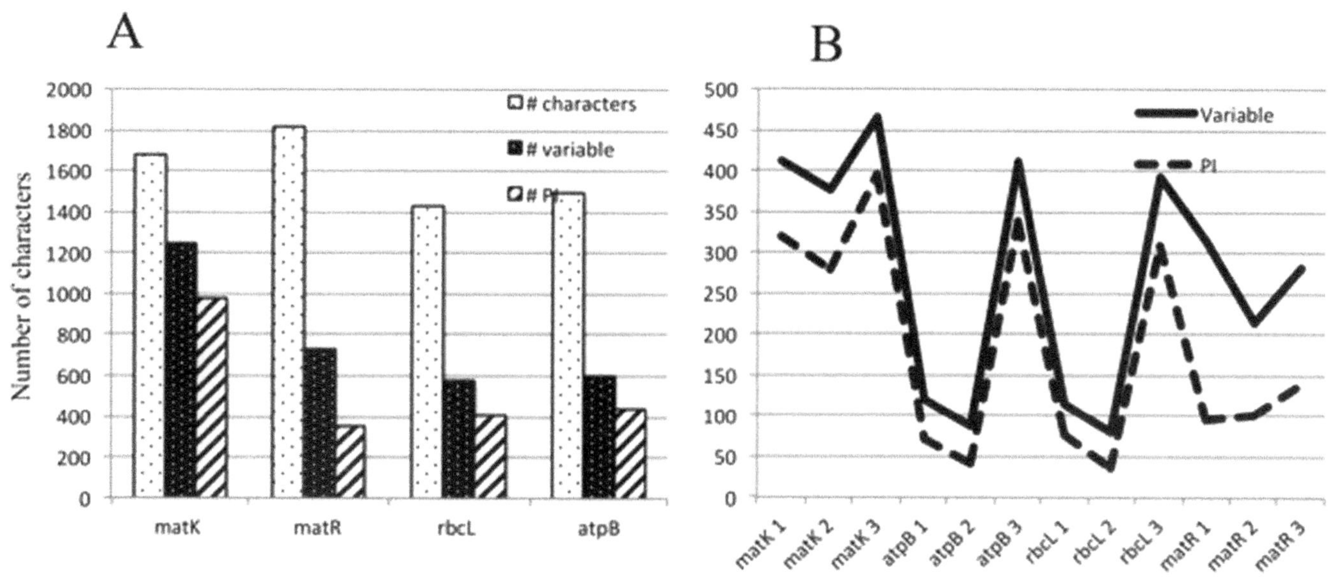

Figure 4. Diagrammatic representation of information on molecular characters from the four gene data sets. The number of total, variable and parsimony informative (PI) characters calculated from the maximum parsimony analyses of *matK, matR, rbcL,* and *atpB.* (A) gene partitions, (B) the three codon positions partitions (B).

COM clade diverge after Myrtales as sister to remaining rosids, albeit BS support for this topology is <50%. The Crossosomatales and Geraniales are scattered across the tree (Figure 2, Figure S4). BS Support for the monophyly of the rosids orders is 82–100% (Table S2). The CI and RI values are 0.390 and 0.426, respectively.

Phylogenetic informativeness of the four genes

Measures of phylogenetic information are based on numbers of PI characters as well as gene informativeness profiles computed in PhyDesign (table 1). The number of PI characters and their proportion out of total characters is lowest for the slowly evolving/ unconstrained *matR* (350, 19%) and highest for the rapidly evolving/unconstrained *matK* (975, 58%); the slowly evolving/ constrained *atpB* and *rbcL* are intermediate, providing 438 (29%) and 402 (28%) PI characters, respectively (Table 1, Figure 4). Similar trend is also notable in the phylogenetic informativeness estimated in PhyDesign with rapidly evolving *matK* superseding the other genes in net and per-site informativeness across the rosids history (Figure 5). The *matR* gene stands at the lower end of the informativeness spectrum, and *rbcL* and *atpB* are intermediates (Table 1, Figure 5). Rapidly evolving *matK* displays a relative decline in informativeness at deeper histories (≥50MY; (Figure 5). In contrast, net informativeness of the slowly evolving but constrained *rbcL* and *atpB* remain constant at deeper histories but declined in recent epochs (Figure 5). The informativeness profile of *matR* is elevated at deeper epochs than recent ones. The Standard deviation of informativeness for all four genes was quite low across the rosids' history except for the most recent epoch (Figure 5). Curious spikes in the informativeness profiles of unconstrained *matK* and *matR* are notable in modern era; *rbcL* and *atpB* lack these spikes. These spikes are intriguing since both genes evolve under relaxed selection but differ considerably in rates of substitution. However, this phenomenon has been addressed on the PhyDesign website (http://phydesign.townsend.yale.edu/), stating that "those few sites all are estimated to evolve at one very fast rate, leading to a spike that has little biological meaning".

Per-site informativeness profiles followed the same trend as net informativeness profiles in all four genes (Figure 5).

Phylogenetic informativeness at codon partitions

The 3rd codon positions account for larger proportions of PI characters compared with their respective 1st and 2nd positions, but degree of disparity varied with gene mode of evolution (Figure 4; Table 1). Net and per-site informativeness was highest in the 3rd codon positions, followed by the 1st codon position, except for the per-site informativeness in *matR* where it is highest in the 1st codon position followed by the 3rd (Table 1). Overall, the three codon partitions of *matK* exhibit higher magnitudes of PI characters than corresponding codon positions of the other three genes (Table 1, Figure 4). Both unconstrained *matK* and *matR* display more uniformity in number of PI characters across codon positions with standard deviation (SD) being 59 and 23, respectively. This is juxtaposed with *atpB* and *rbcL* where the number of PI characters are disproportionally skewed towards the 3rd codon positions (SD = 164 and 147, respectively) (Table 1, Figure 4). Codon position informativeness calculated in PhyDesign varies with gene mode than tempo of evolution, mirroring the patterns noted in the PI data (Table 1; Figure 6), with *matR* showing the highest uniformity among codon positions (SD = 0.019). The 3rd codon position profile of *matK* experienced a relatively higher degree of decline in informativeness at deeper histories (≥50 MY; Figure 6). When contrasted with slight decline displayed by the 3rd codon positions of *atpB* and *rbcL* (≥85 and 75 MY ago, respectively). The informativeness profiles of all three codon positions in *matR* remain elevated at deeper rosids history (Figure 6). Similar recent spikes in informativeness profiles are evident in *matK* and *matR* codon positions.

In the phylogenetic analyses of codon partitions, degrees of resolution are comparable among trees based on the individual codon partitions in the unconstrained *matK* and *matR*, but increase resolution is skewed toward the 3rd codon position in the constrained *rbcL* and *atpB* (Table 1). In *matR*, the 3rd codon position provides less tree resolution than the 1st and 2nd (Table 1).

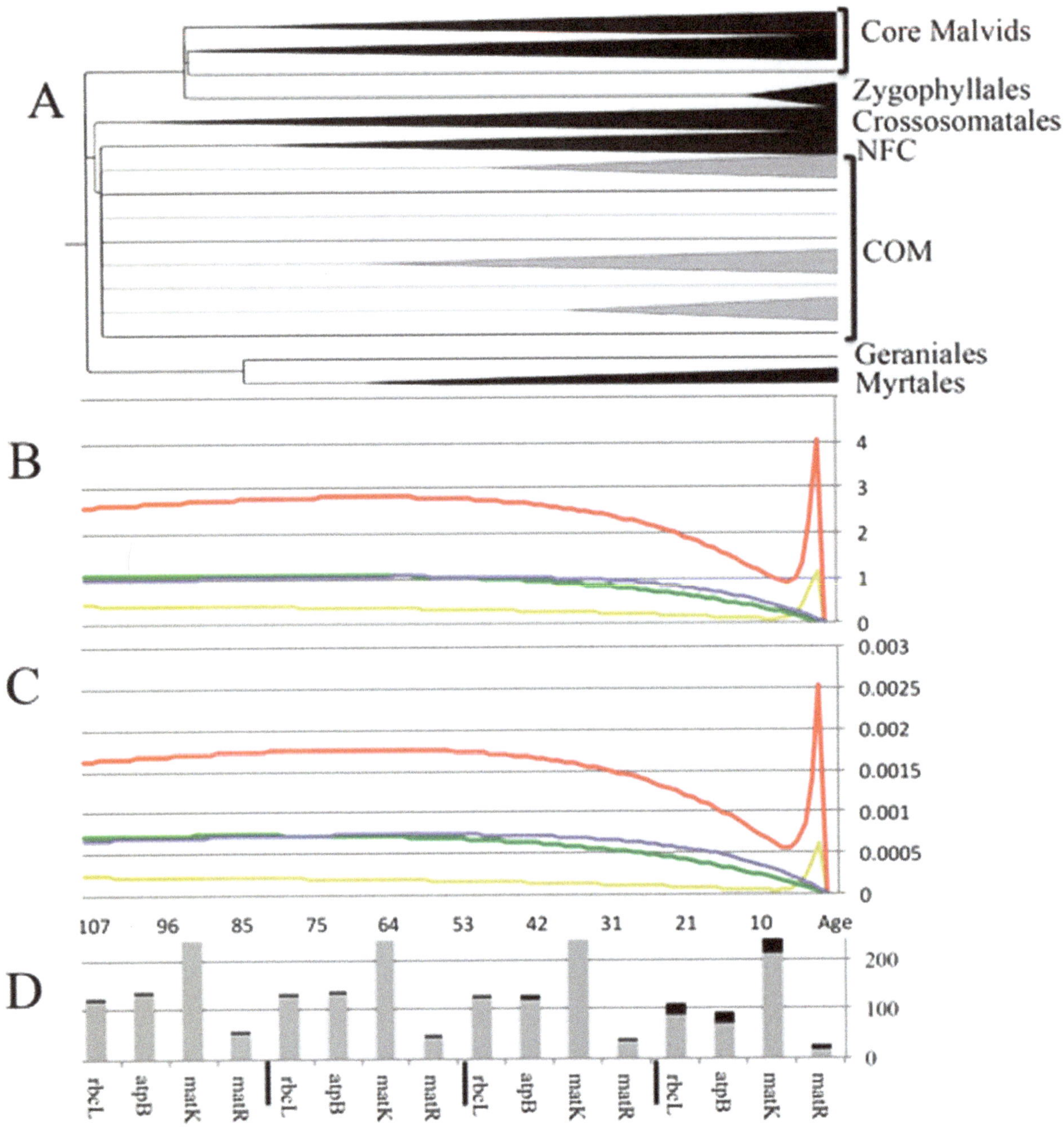

Figure 5. Rosids ultrametric reconstructed in PATHd8 and the informativeness profiles estimated in PhyDesign along with their variability at different epochs of rosids history. (A) Ultrametric tree for the rosids calculated in PATHd8 based on the concatenated four-gene data set. (B) Net informativeness profiles for *matK* (red), *matR* (yellow), *rbcL* (blue), and *atpB* (green) gene partitions; (C) Per-site informativeness profiles for the same genes; (D) The mean (gray) and standard deviation (black) for gene informativeness at four equally divided epochs in the evolutionary history of the rosids. COM: Celastrales, Malpighiales, and Oxalidales. NFC: Nitrogen Fixing Clade, Rosales, Fabales, Cucurbitales, and Fagales. Both *matK* and *matR* display spikes in their informativeness profiles at very recent times.

Phylogenetic noise

We used ensemble CI and RI values as measures of homoplasy. PhyDesign does not profile phylogenetic noise or provide estimates of overall noise for gene or codon partitions. The values for the CI and RI measures are highly correlated ($r^2 = 0.938$, $F_{1,16} = 212$, $P < 0.0001$) throughout the partitioned analyses of genes and codon positions. Consequently, we arbitrarily opted to use the CI values. Homoplasy (1-CI) was lowest in *matR*, whereas it is quite comparable for the other three genes despite striking difference in rates of nucleotide and amino acid substitutions (Table 1). Homoplasy is highest in the 3rd codon position and lowest in the 2nd for all four genes. Again, homoplasy is lowest in *matR* for all three codon positions (Table 1). Homoplasy in the 3rd codon position is quite similar for *matK*, *atpB* and *rbcL* (0.672, 0.654 and 0.668, respectively) despite marked differences in rates of substitutions.

Table 1. Phylogenetic information and informativeness (net and per-site) of genes and their codon partitions.

Genes/	Character included	Variablesites	PI[a]	CI[b]	RI[c]	MP[d] node # (%)	Net	Per-site
matK	1635	1250	975	0.374	0.429	40 (98)	2.765	0.187
atpB	1496	600	438	0.39	0.426	29 (71)	0.885	0.074
rbcL	1427	580	402	0.353	0.416	36 (88)	0.912	0.079
matR	1832	732	350	0.692	0.638	33 (80)	0.764	0.021
matK 1/2	1090	789	597	0.400	0.455	40 (98)	0.753/0.596	0.003/0.002
matK 3	545	466	394	0.328	0.435	40 (98)	1.155	0.004
atpB 1/2	998	209	114	0.519	0.478	18[f] (44)	0.100/0.069	0.000/0.000
atpB 3	498	410	339	0.346	0.438	23 (56)	0.681	0.001
rbcL 1/2	952	195	112	0.414	0.465	15 (37)	0.129/0.061	0.000/0.000
rbcL 3	475	392	308	0.332	0.448	36 (88)	0.571	0.001
matR 1/2	1221	467	216	0.683	0.642	34 (83)	0.203/0.194	0.004/0.000
matR 3	610	281	138	0.604	0.622	23 (56)	0.231	0.001

a: Parsimony Informative.
b: Consistency Index.
c: Retention Index.
d: Maximum Parsimony.
f: Analysis did not reach completion.
Net informativeness could be calculated only for partitioned codon positions individually. Numbers associated with genes refer to codon positions.

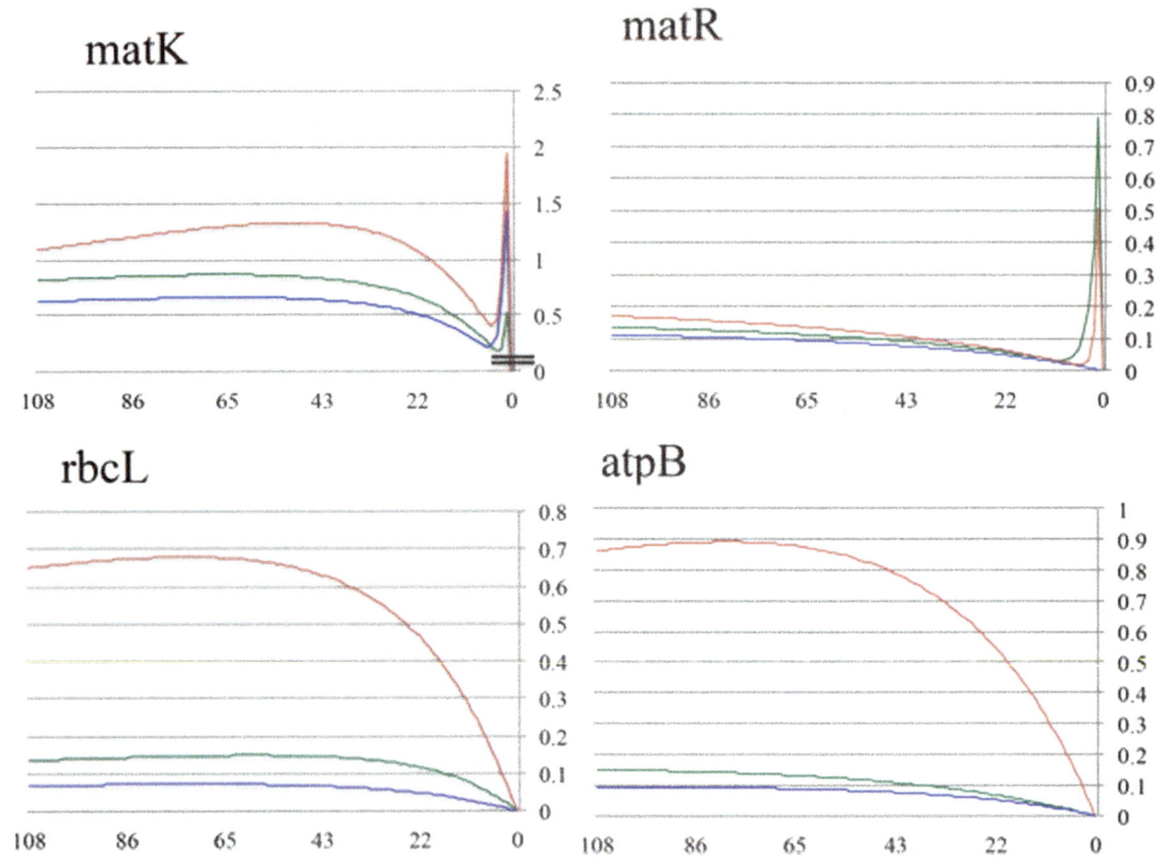

Figure 6. Informativeness profiles for codon positions measured in PhyDesign. Net informativeness profiles for 1st (green), 2nd (blue) and 3rd (red) codon positions of *matK, matR, rbcL*, and *atpB* computed in PhyDesign. Note the informativeness values for *matK* codon positions starting at 0.5, highlighted by double lines on the y-axis. Both *matK* and *matR* exhibit spikes in their informativeness profiles at very recent times.

Discussion

Gene informativeness

The disparity in tempo of gene evolution is reflected by the marked differences in the proportions of variable characters, ~40% in *rbcL, atpB* and *matR* compared with 75% in *matK* (Table 1, Figure 4). However, the likelihood of a variable character to be parsimony informative was comparable ($X^2 = 6.591$, $P = 0.037$) among the rapidly evolving/unconstrained *matK* and the slowly evolving/constrained *rbcL* and *atpB* (78% vs. 69% and 73%, respectively). As a consequence, the 2.2–2.4 folds in number of PI characters contributed by *matK* compared with *rbcL* and *atpB* could be ascribed in part to the availability of a larger pool of variable characters in *matK* (Table 1, Figure 4). In a marked contrast, the slowly evolving but unconstrained *matR* exhibits the lowest proportion of PI characters (48%) despite its similarity to the other two slowly evolving genes, *rbcL* and *atpB*, in proportion of variable characters. This pattern may imply lower probability for a *matR* variable characters to be translated into PI characters in the rosids.

Net informativeness of *matK* is 2.6 fold those of *atpB* and *rbcL*, and 7.8 that of *matR* (Table 1). A strong correlation ($r^2 = 0.994$, $F_{1,4} = 329$, $P < 0.003$) is found between the per-site informativeness and the number of PI characters for the four genes (Table 1, Figure 5). These two measures together provide a reliable estimation of phylogenetic signal in the rosids, and point to *matK* as being by far the most informative gene across the evolutionary history of the rosids compared with the three slowly evolving genes (Figure 5). The high performance of *matK* is likely due to expanding character state choices (character-state space [77]) and subsequent minimization of the impact of homoplasy [43,77–79]. In the rosids, the degree of substitution in the 3rd codon position in *matK* is quite comparable to those in the 3rd codon positions of *atpB* and *rbcL* (466 vs. 410 and 392, respectively, Fig. 4B). Should this rapidly evolving *matK* be as evolutionarily conserved as *rbcL* and *atpB*, the anticipated skewed substitution rate toward its 3rd codon position would augment the likelihood of multiple hits and homoplasy due to an expected 41% surge in nucleotide mutations in its 3rd codon positions (calculated from Table 1). Therefore, superimposing elevation in rate of nonsynonymous mutations on the overall accelerated rate of substitution in *matK* helps in reducing the mutation load on its 3rd codon position. Mossel and Steel [80], Steel and Penny [81] and Townsend et al. [15] have demonstrated that increased accessibility of characters to multiple states reduces the potential noise stemming from multiple hits. The decline in the informativeness profiles of the *matK*'s 3rd codon position from the rosids crown group divergence to ~50 MY ago is probably a function of its substitution rate. Despite this decline, its overall informativeness and the profile remain above those of the other two codon positions (Figure 6; Table 1).

The informativeness profile of *matK* declined deep in rosids history (≥50 MY ago) in contrast with that of *matR* (Figure 5); both are evolutionary unconstrained (dN/dS for *matR* = 0.975, *matK* = 0.427 compared with *rbcL* = 0.074, *atpB* = 0.064). Considering the differences in tempo of evolution between *matK* and *matR*, the relative decline in informativeness in *matK* at deep historic levels could be a function of elevation in multiple hits that could obscure phylogenetic signal, but nonstationarity may be another factor. The low rate of substitution in *matR* and its 2.3–15 fold dN/dS are expected to increase signal to noise ratios by curbing the overall proportions of multiple hits [80,81]. This mode and tempo of evolution in *matR* may account for maintaining higher informativeness profiles at deeper history in the rosids (Figs. 5, 6).

The decline in the informativeness profile of *matK* ≥~54 MY ago (Figure 5) corresponds to the epoch at which the problematic orders Myrtales, Geraniales and Crossosomatales diverged, possibly causing their uncertain placement in the rosids tree (Figure 2). In contrast, the decline in recent epochs in the *rbcL* and *atpB* informativeness (≤~60 MY; Figure 5) may account for their low utility in discerning pattern of divergences for orders emerging at that evolutionary period (Figure 2). Notable is the considerably low standard deviation values for informativeness at deeper evolutionary history for all four genes (Figure 5). These values suggest minimum stochasticity in phylogenetic signal for the genes despite the substantial differences in their tempo and mode of evolution.

Phylogenetic informativeness should be considered alongside noise caused by homoplasy and multiple hits, particularly in rapidly evolving genes. Homoplasy (1-CI) for *matK*, *rbcL* and *atpB* values are quite comparable (0.626, 0.647 and 0.610). In contrast, *matR* deviates from these three genes by having substantially lower homoplasy, 0.308. This is possibly an outcome of a compounding effect of low rate of nucleotide substitution and high dN/dS ratio (0.975), which points to evenness in mutations across the three codon positions, an evolutionary mode that minimizes the potential of saturation and multiple hits. The homoplasy value for *matK*, being intermediate between the two slowly evolving genes *rbcL* and *atpB*, does not reflect the common notion that rapidly evolving genes are highly homoplasic.

Tree robustness and accuracy

The question is how these varied informativeness profiles and homoplasy are translated into tree robustness. The *matK* MP strict consensus tree resolved 98% of the expected nodes (Table 1), approaching the full resolution achieved in the Wang et al. [19] multigene tree. This resolution is contrasted with 88%, 71%, and 80% in the *rbcL*, *atpB* and *matR* trees, respectively (Table 1). Bootstrap values for major nodes are much higher in *matK* compared with the other genes (Figure 3), with means and standard deviation being 99%±3 for *matK* vs. 63%±36, 74%±36, and 84%±28 for *rbcL*, *atpB*, and *matR*, respectively. Thus, not only notable differences in magnitude of resolution and support exist, but also a substantial variation in node support across the rosids for the three latter genes (Figure 1, 3).

The ultimate goal in phylogenetic reconstruction is accuracy in depicting true patterns of historic divergences. The Wang et al. [19] ML tree represents the best available hypothesis for the rosids phylogeny. Compared to this reference tree, the backbone of the *matK* ML tree displays one topological incongruence by depicting the phylogenetically difficult-to-place Geraniales and Crossosomatales sister to remaining rosids instead of being basal in a malvids clade, but BS support is <50% (Figure 2). Accuracy was reduced in the *matR* tree as a consequence of the inability to recover the fabid clade and in the placement of the COM clade, Zygophyllales, Myrtales, and Geraniales (Figure 2). These topological inconsistencies receive weak support (Figure 2). Homoplasy cannot account for the incongruences since *matR* displays the lowest degree of homoplasy among all four genes (Table 1). It is likely that such shortcomings are a consequence of the low signal in *matR* across the rosids tree (Figure 5, Table 1). Fong and Fujita [67] have shown in three data sets of vertebrates genes that phylogenetic signal was greatly reduced although they were less subject to homoplasy. Zhu et al. [48] recovered the same topology for the rosids in their *matR* partition, which they attributed to a difference in history or evolutionary phenomena for *matR*.

Accuracy varies between the slowly evolving/constrained *rbcL* and *atpB* ML trees (Figure 2), despite the comparable amounts of PI characters and net informativeness (Figure 4, 5, Table 1). The pronounced differences in topological accuracy between the *atpB* and *rbcL* trees and that of the *matK* cannot be explained by homoplasy since CI values of three genes are comparable, and that of *matK* being intermediate between *atpB* and *rbcL* (*atpB* = 0.390, *matK* = 0.374, *rbcL* = 0.353). Therefore, it follows that tree accuracy and robustness are impacted by disparities in phylogenetic signal. Net phylogenetic informativeness in *matK* is ≥3× of the slowly evolving genes (Table 1), and its phylogenetic informativeness consistently surpassed them across rosids evolution (Table 1, Figure 5). Further, the number of PI characters in *matK* is 2.2–2.4 times that of the other genes (Table 1).

Phylogenetic informativeness at codons level

The informativeness profile of the 3rd codon positions consistently superseded those of the 1st and 2nd in all genes across the rosids evolutionary history, but disparity in informativeness is inversely proportional to their respective dN/dS ratio (Table 1, Figure 6). The dN/dS for constrained *matR* (0.968) and *matK* (0.427) are 6–16 times the dN/dS of unconstrained *rbcL* (0.074) and *atpB* (0.062), reflecting the differential rates of substitution at the three codon positions. The high uniformity in informativeness profiles across codon positions for unconstrained *matR* and the 10 fold difference in net informativeness between the 3rd codon position and the 2nd of constrained *atpB* and *rbcL* reflect the two extremes in dN/dS ratios (Table 1). In *matK*, the 3rd codon position is twice as informative as the 2nd, in concordance with its dN/dS ratio. In general, the 2nd codon position exhibited the least amount of informativeness (Figure 6). Further, homoplasy was lowest in the 2nd codon positions for all genes except for *matR* where it was comparable for 1st and 2nd codon positions. Nucleotide substitutions in the 2nd codon position are translated into 100% nonsynonymous mutations [34], and consequently it is highly constrained evolutionarily. These findings are in agreement with the Björklund's [39] partitioned phylogenetic analyses of cytochrome *b* codon partitions in vertebrate where performance is lowest in 2nd codon and highest in the 3rd.

Tree resolution in codon position partition analyses followed closely the inherent dN/dS for the genes. The 3rd codon positions of constrained *rbcL* and *atpB* provided higher resolution than 1st+ 2nd partition, whereas the amount of resolution was lower in the *matR* 3rd codon position than the equally-informative 1st and 2nd codon positions combined. In *matK*, the amount of resolution was the same for the two partitions (Table 1). Similar patterns are notable when the number of PI characters are considered (Table 1). Strong correlation exists between phylogenetic informativeness and PI characters for the codon partitions ($r^2 = 0.933$, $F_{1,12} = 477$, $P<0.0001$).

The higher rate of substation in the 3rd codon position has been negatively construed since such an attribute is expected to increase the likelihood of site saturation and to elevate the degree of homoplasy [34]. Although this might be a cautionary points for genomic regions with excessively higher rates of substitution, such animal mitochondrial DNA, it has been overly generalized, leading to the tendency of excluding or down weighting of the 3rd codon position in phylogenetic reconstruction, e.g.

[7,27,29,30,32]. Simmons et al. [42] contended that PI characters of 3rd codon position in *rbcL* and *atpB* angiosperms data set outperformed the 1st and 2nd combined in phylogenetic signal, and that Jackknife support was 14% higher with the 3rd codon position compared with the tree based on 1st and 2nd combined in seed plants phylogenetic study. Imposing a 4:17:1 weighing criteria for the 1st, 2nd and 3rd codon positions in an analyses of cytochrome *b* [28] reduced resolution and increased probabilities of support for erroneous trees. Similarly, it has been found [82] that exclusion of the 3rd codon position led to a substantial problem in recovering the true tree.

Conclusions

Our study demonstrates that tree robustness and phylogenetic informativeness for the four genes work in concert with their mode and tempo of evolution. Phylogenetic signal from rapidly evolving and unconstrained *matK* provides by far the most structure and accuracy, whereas slowly evolving, constrained and unconstrained, genes display decreasing degrees of informativeness and tree structure. The 3rd codon positions consistently supersede the 1st and 2nd positions in phylogenetic signal, and its differential informativeness is accentuated in the constrained genes. The study underscores the need for assessments of phylogenetic informativeness of genomic regions for a given biological lineage within the framework of overall rates of nucleotide as well as nonsynonymous substitutions across their historic divergence. A priori judgments on performances of genomic regions without empirical data may hinder efforts aiming at achieving the best phylogenetic hypothesis. Specifically, our findings in the rosids argues against the notion that arbitrarily discourages the use of rapidly evolving genomic regions in deep phylogenetics due to potential multiple hits, homoplasy and saturation [9,11,14,15,71,83]. Simmons et al. [43] demonstrated that increasing rates of evolution in a simulation model consistently improved resolution. Yang [2] concluded that optimal limits for sequence divergence are higher than previously suggested for saturation of substitutions and, consequently, the problem of saturation may have been exaggerated. In a phylogenetic analyses of an *rbcL* data set for green plants, Källersjö et al. [10] asserted that homoplasy can provide phylogenetic structure. We have demonstrated in a study of early diverging angiosperms [25] that a PI site for *matK* provides more structure than that of *rbcL*, and that homoplasy in *matK* has less negative impact on phylogenetic structure than it does in *rbcL*. In a phylogenetic analysis of 1070 genes in a yeast data set, Salichos and Rokas [84] found that using slowly evolving genes and conserved sites increased incongruence across many internodes. Recently, Magallón et al. [84] demonstrated in an assessment of land plant phylogeny that *matK* provides phylogenetic signal and structure matching those derived from a concatenated, three slowly evolving genes data.

The per-site informativeness profiles, which excludes gene length bias, mirrors closely their corresponding net informativeness profiles and the two are highly correlated ($r^2 = 0.903$, $F_{1,3} = 19$, $P < 0.0499$). The number of nucleotides sequenced for *matR* (1822 nucleotides) exceeds those of *matK* (1672), *atpB* (1496) and *rbcL* (1427). Nevertheless, *matR* is at disadvantage in terms of number of PI characters and net-informativeness and, thus, lessening its cost-effectiveness per nucleotide sequenced. The *matK* gene stands at the other end of the cost-effectiveness spectrum.

Although the study promotes the consideration of rapidly evolving regions in phylogenetic reconstruction, homology assessment of sequence alignments at deep histories should not be compromised since this step represents a crucial foundation in molecular phylogenetics. The rosids divergence and diversification spans some 108 million years, and thus it would be useful to carry out similar detailed studies for groups with substantially deeper evolutionary histories.

Supporting Information

Figure S1 Detailed *matK* RAxMl tree for the rosids and representatives of remaining core eudicots.

Figure S2 Detailed *matR* RAxMl tree for the rosids and representatives of remaining core eudicots.

Figure S3 Detailed *rbcL* RAxMl tree for the rosids and representatives of remaining core eudicots.

Figure S4 Detailed *atpB* RAxMl tree for the rosids and representatives of remaining core eudicots.

Table S1 Taxa used in this study. The species used, their family and order affiliation and the GenBank accessions numbers.

Table S2 Bootstrap support for rosids orders represented by more than one taxon.

Acknowledgments

We thank Sunny Crawley, Shelli Newman, Jaimin Patel and Patrick Mason for help in the lab, and Royal Botanic Garden (Kew) and Pamela and Douglas Soltis for providing some of the DNA samples.

Author Contributions

Conceived and designed the experiments: KH. Performed the experiments: CB. Analyzed the data: DO. Contributed reagents/materials/analysis tools: KH. Wrote the paper: KH.

References

1. Goldman N (1998) Phylogenetic information and experimental design in molecular systematics. Proceedings of the Royal Society of London Series B: Biological Sciences 265: 1779–1786.
2. Yang Z (1998) On the best evolutionary rate for phylogenetic analysis. Systematic Biology 47: 125–133.
3. Graham SW, Olmstead RG (2000) Utility of 17 Chloroplast genes for inferring the phylogeny of the basal angiosperms. American Journal of Botany 87: 1712–1730.
4. Sanderson MJ, Shaffer HB (2002) Troubleshooting molecular phylogenetic analyses. Annual Review of Ecology and Systematics 33: 49–72.
5. Mueller RL (2006) Evolutionary rates, divergence dates, and the performance of mitochondrial genes in Bayesian phylogenetic analysis. Systematic Biology 55: 289–300.
6. Townsend JP (2007) Profiling phylogenetic informativeness. Systematic Biology 56: 222–231.
7. Regier JC, Shultz JW, Ganley ARD, Hussey A, Shi D, et al. (2008) Resolving arthropod phylogeny: Exploring phylogenetic signal within 41 kb of protein-coding nuclear gene sequence. Systematic Biology 57: 920–938.
8. Crawley SS, Hilu KW (2012) Caryophyllales: Evaluating phylogenetic signal in trnK intron versus matK. Journal of Systematics and Evolution 50: 387–410.
9. Graybeal A (1994) Evaluating the phylogenetic utility of genes: A search for genes informative about deep divergences among vertebrates. Systematic Biology 43: 174–193.
10. Kälersjö M, Albert VA, Farris JS (1999) Homoplasy increases phylogenetic structure. Cladistics 15: 91–93.
11. Wenzel JW, Siddall ME (1999) Noise. Cladistics 15: 51–64.

12. Shpak M, Churchill GA (2000) The information content of a character under a Markov model of evolution. Molecular Phylogenetics and Evolution 17: 231–243.
13. Hilu KW, Black C, Diouf D, Burleigh JG (2008) Phylogenetic signal in matK vs. trnK: A case study in early diverging eudicots (angiosperms). Molecular Phylogenetics and Evolution 48: 1120–1130.
14. Klopfstein S, Kropf C, Quicke DLJ (2010) An evaluation of phylogenetic informativeness profiles and the molecular phylogeny of Diplazontinae (Hymenoptera, Ichneumonidae). Systematic Biology 59: 226–241.
15. Townsend JP, Su Z, Tekle YI (2012) Phylogenetic signal and noise: predicting the power of a data set to resolve phylogeny. Syst Biol 61: 835–849.
16. Hillis DM (1987) Molecular versus morphological approaches to systematics. Annual Review of Ecology and Systematics 18: 23–42.
17. Sogin ML (1991) Early evolution and the origin of eukaryotes. Current Opinion in Genetics & Development 1: 457–463.
18. Brown WM, Prager EM, Wang A, Wilson AC (1982) Mitochondrial DNA sequences of primates: Tempo and mode of evolution. Journal of Molecular Evolution 18: 225–239.
19. Wang H, Moore MJ, Soltis PS, Bell CD, Brockington SF, et al. (2009) Rosid radiation and the rapid rise of angiosperm-dominated forests. Proceedings of the National Academy of Sciences 106: 3853–3858.
20. Brockington Samuel F, Alexandre R, Ramdial J, Moore Michael J, Crawley S, et al. (2009) Phylogeny of the Caryophyllales Sensu Lato: Revisiting hypotheses on pollination biology and perianth differentiation in the core Caryophyllales. International Journal of Plant Sciences 170: 627–643.
21. Felsenstein J (1978) Cases in which parsimony or compatibility methods will be positively misleading. Systematic Zoology 27: 401–410.
22. Magallón S, Sanderson MJ (2002) Relationships among seed plants inferred from highly conserved genes: Sorting conflicting phylogenetic signals among ancient lineages. American Journal of Botany 89: 1991–2006.
23. Hilu KW, Liang H (1997) The matK gene: Sequence variation and application in plant systematics. American Journal of Botany 84: 830–839.
24. Hilu KW, Borsch T, Müller K, Soltis DE, Soltis PS, et al. (2003) Angiosperm phylogeny based on matK sequence information. American Journal of Botany 90: 1758–1776.
25. Müller KF, Borsch T, Hilu KW (2006) Phylogenetic utility of rapidly evolving DNA at high taxonomical levels: Contrasting matK, trnT-F, and rbcL in basal angiosperms. Molecular Phylogenetics and Evolution 41: 99–117.
26. Worberg A, Quandt D, Barniske A-M, Lohne C, Hilu KW, et al. (2007) Phylogeny of basal eudicots: insights from non-coding and rapidly evolving DNA. Org Divers Evol 7: 55–77.
27. Edwards SV, Arctander P, Wilson AC (1991) Mitochondrial resolution of a deep branch in the genealogical tree for perching birds. Proceedings: Biological Sciences 243: 99–107.
28. Milinkovitch MC, LeDuc RG, Adachi J, Farnir F, Georges M, et al. (1996) Effects of character weighting and species sampling on phylogeny reconstruction: a case study based on DNA sequence data in cetaceans. Genetics 144: 1817–1833.
29. Swofford DL, Olsen G, Waddell PJ, Hillis DM (1996) Phylogenetic inference. In: Hillis DM, Moritz C, Mable BK, editors. Molecular Systematics.Sunderland (MA): Sinauer Associates. pp. 407–514.
30. Blouin MS, Yowell CA, Courtney CH, Dame JB (1998) Substitution bias, rapid saturation, and the use of mtDNA for nematode systematics. Molecular Biology and Evolution 15: 1719–1727.
31. Olmstead RG, Reeves G, Yen AC (1998) Patterns of sequence evolution and implications for parsimony analysis of chloroplast DNA. In: Soltis DE, Soltis PS, Doyle JJ, editors. Molecular Systematics of Plants II: DNA Sequencing. Boston (MA): Kluwer. pp. 164–187.
32. Goremykin VV, Hirsch-Ernst KI, Wölfl S, Hellwig FH (2003) Analysis of the Amborella trichopoda chloroplast genome sequence suggests that Amborella is not a basal angiosperm. Molecular Biology and Evolution 20: 1499–1505.
33. Criscuolo A, Michel CJ (2009) Phylogenetic inference with weighted codon evolutionary distances. Journal of Molecular Evolution 68: 377–392.
34. Li W-H, Graur D (1991) Fundamentals of molecular evolution. Sunderland, MA: Sinauer.
35. Lewis LA, Mishler BD, Vilgalys R (1997) Phylogenetic relationships of the liverworts (Hepaticae), a basal embryophyte lineage, inferred from nucleotide sequence data of the chloroplast GenerbcL. Molecular Phylogenetics and Evolution 7: 377–393.
36. Yang Z (1996) Maximum-likelihood models for combined analyses of multiple sequence data. Journal of Molecular Evolution 42: 587–596.
37. Yoder AD, Vilgalys R, Ruvolo M (1996) Molecular evolutionary dynamics of cytochrome b in strepsirrhine primates: the phylogenetic significance of third-position transversions. Molecular Biology and Evolution 13: 1339–1350.
38. Kallersjo M, Farris JS, Chase MW, Bremer B (1998) Simultaneous parsimony jackknife analysis of 2538 rbcL DNA sequences reveals support for major clades of green plants, land plants, seed plants and flowering plants. Plant systematics and evolution.
39. Björklund M (1999) Are third positions really that bad? A test using vertebrate cytochrome b. Cladistics 15: 191–197.
40. Campbell DL, Brower AVZ, Pierce NE (2000) Molecular evolution of the wingless gene and its implications for the phylogenetic placement of the butterfly family Riodinidae (Lepidoptera: Papilionoidea). Molecular Biology and Evolution 17: 684–696.
41. Sennblad B, Bremer B (2000) Is there a justification for differential a priori weighting in coding sequences? A case study from rbcL and Apocynaceae s.l. Systematic Biology 49: 101–113.
42. Simmons MP, Ochoterena H, Freudenstein JV (2002) Amino acid vs. nucleotide characters: challenging preconceived notions. Molecular Phylogenetics and Evolution 24: 78–90.
43. Simmons MP, Zhang L-B, Webb CT, Aaron R (2006) How can third codon positions outperform first and second codon positions in phylogenetic inference? An empirical example from the seed plants. Systematic Biology 55: 245–258.
44. Woessner JP, Gillham NW, Boynton JE (1986) The sequence of the chloroplast atpB gene and its flanking regions in Chlamydomonas reinhardtii. Gene 44: 17–28.
45. Hasebe M, Omori T, Nakazawa M, Sano T, Kato M, et al. (1994) rbcL gene sequences provide evidence for the evolutionary lineages of leptosporangiate ferns. Proceedings of the National Academy of Sciences 91: 5730–5734.
46. Barthet MM, Hilu KW (2008) Evaluating evolutionary constraint on the rapidly evolving gene matK using protein composition. Journal of Molecular Evolution 66: 85–97.
47. Olmstead RG, Palmer JD (1994) Chloroplast DNA systematics: A review of methods and data Analysis. American Journal of Botany 81: 1205–1224.
48. Zhu X-Y, Chase M, Qiu Y-L, Kong H-Z, Dilcher D, et al. (2007) Mitochondrial matR sequences help to resolve deep phylogenetic relationships in rosids. BMC Evolutionary Biology 7: 217.
49. Qiu Y-L, Li L, Wang B, Xue J-Y, Hendry TA, et al. (2010) Angiosperm phylogeny inferred from sequences of four mitochondrial genes. Journal of Systematics and Evolution 48: 391–425.
50. Young ND, dePamphilis CW (2000) Purifying selection detected in the plastid gene matK and flanking ribozyme regions within a group II intron of nonphotosynthetic plants. Molecular Biology and Evolution 17: 1933–1941.
51. The Angiosperm Phylogeny G (2009) An update of the Angiosperm Phylogeny Group classification for the orders and families of flowering plants: APG III. Botanical Journal of the Linnean Society 161: 105–121.
52. Chase MW, Reveal JL (2009) A phylogenetic classification of the land plants to accompany APG III. Botanical Journal of the Linnean Society 161: 122–127.
53. Cantino PD, Doyle JA, Graham SW, Judd WS, Olmstead RG, et al. (2007) Towards a phylogenetic nomenclature of Tracheophyta. Taxon 56: 822–846.
54. Moore MJ, Soltis PS, Bell CD, Burleigh JG, Soltis DE (2010) Phylogenetic analysis of 83 plastid genes further resolves the early diversification of eudicots. Proceedings of the National Academy of Sciences.
55. Soltis DE, Smith SA, Cellinese N, Wurdack KJ, Tank DC, et al. (2011) Angiosperm phylogeny: 17 genes, 640 taxa. American Journal of Botany 98: 704–730.
56. Doyle JJ, Doyle JL (1990) Isolation of plant DNA from fresh tissue. Focus 12: 13–25.
57. M'Ribu KH, Hilu KW (1996) Application of random amplified polymorphic DNA to study genetic diversity in Paspalum scrobiculatum L. (Kodo millet, Poaceae). Genetic Resources and Crop Evolution 43: 203–210.
58. Müller J, Müller K (2003) QuickAlign: A new alignment editor. Plant Molecular Biology Reporter 21: 5–5.
59. Swofford DL (2003) PAUP*. Phylogenetic analysis using parsimony (*and other methods. Sunderland (MA): Sinauer Associates.
60. Felsenstein J (1985) Confidence limits on phylogenies: an approach using the bootstrap. Evolution 39: 783–791.
61. Nixon KC (1999) The parsimony ratchet, a new method for rapid parsimony analysis. Cladistics 15: 407–414.
62. Müller K (2004) PRAP—computation of Bremer support for large data sets. Molecular Phylogenetics and Evolution 31: 780–782.
63. Lopez-Giraldez F, Townsend J (2011) PhyDesign: an online application for profiling phylogenetic informativeness. BMC Evolutionary Biology 11: 152.
64. Townsend JP, López-Giráldez F, Friedman R (2008) The phylogenetic informativeness of nucleotide and amino acid sequences for reconstructing the vertebrate tree. Journal of Molecular Evolution 67: 437–447.
65. Schoch CL, Sung G-H, López-Giráldez F, Townsend JP, Miadlikowska J, et al. (2009) The ascomycota tree of life: A phylum-wide phylogeny clarifies the origin and evolution of fundamental reproductive and ecological traits. Systematic Biology 58: 224–239.
66. Moeller AH, Townsend JP (2011) Phylogenetic informativeness profiling of 12 genes for 28 vertebrate taxa without divergence dates. Molecular Phylogenetics and Evolution 60: 271–272.
67. Fong JJ, Fujita MK (2011) Evaluating phylogenetic informativeness and data-type usage for new protein-coding genes across Vertebrata. Molecular Phylogenetics and Evolution 61: 300–307.
68. Dornburg A, Brandley MC, McGowen MR, Near TJ (2012) Relaxed clocks and inferences of heterogeneous patterns of nucleotide substitution and divergence time estimates across whales and dolphins (Mammalia: Cetacea). Molecular Biology and Evolution 29: 721–736.
69. Raja H, Schoch CL, Hustad VP, Shearer CA, Miller AN (2011) Testing the phylogenetic utility of MCM7 in the Ascomycota. MycoKeys 1: 63–94.
70. Shen X-X, Liang D, Zhang P (2012) The development of three long universal nuclear protein-coding locus markers and their application to Osteichthyan phylogenetics with nested PCR. PLoS ONE 7: e39256.
71. Townsend JP, Leuenberger C (2011) Taxon sampling and the optimal rates of evolution for phylogenetic inference. Systematic Biology 60: 358–365.

72. Britton T, Anderson CL, Jacquet D, Lundqvist S, Bremer K (2007) Estimating divergence times in large phylogenetic trees. Systematic Biology 56: 741–752.
73. Tamura K, Dudley J, Nei M, Kumar S (2007) MEGA4: Molecular Evolutionary Genetics Analysis (MEGA) Software Version 4.0. Molecular Biology and Evolution 24: 1596–1599.
74. Kluge AG, Farris JS (1969) Quantitative phyletics and the evolution of Anurans. Systematic Zoology 18: 1–32.
75. Farris JS (1989) The retention index and the rescaled consistency index. Cladistics 5: 417–419.
76. Korber B (2000) HIV sequence signature and similarities. In: Rodrigo AG, Learn GHJ, editors. Computational and Evolutionary Analysis of HIV Molecular Sequences.Dordrecht (Netherlands): Kluwer Academic Publishers. pp. 52–72.
77. Simmons MP, Reeves A, Davis JI (2004) Character-state space versus rate of evolution in phylogenetic inference. Cladistics 20: 191–204.
78. Naylor GJP, Collins TM, Brown WM (1995) Hydrophobicity and phylogeny. Nature 373: 565–565.
79. Steel M, Penny D (2005) Maximum parsimony and the phylogenetic information in multistate characters. In: Albert VA, editor. Parsimony, phylogeny, and genomics.Oxford (UK): Oxford University Press. pp. 163–178.
80. Mossel E, Steel M (2004) A phase transition for a random cluster model on phylogenetic trees. Mathematical Biosciences 187: 189–203.
81. Steel M, Penny D (2000) Parsimony, likelihood, and the role of models in molecular phylogenetics. Molecular Biology and Evolution 17: 839–850.
82. Håstad O, Björklund M (1998) Nucleotide substitution models and estimation of phylogeny. Molecular Biology and Evolution 15: 1381–1389.
83. Collins TM, Fedrigo O, Naylor GJP (2005) Choosing the best genes for the job: The case for stationary genes in genome-Sscale phylogenetics. Systematic Biology 54: 493–500.
84. Magallón S, Hilu KW, Quandt D (2013) Land plant evolutionary timeline: Gene effects are secondary to fossil constraints in relaxed clock estimation of age and substitution rates. American Journal of Botany 100: 556–573.

16

Evolutionary Divergence of Geographic Subspecies within the Scalloped Spiny Lobster *Panulirus homarus* (Linnaeus 1758)

Shane D. Lavery[1,2], Ahmad Farhadi[3]*, Hamid Farahmand[3], Tin-Yam Chan[4], Ashkan Azhdehakoshpour[5], Vibhavari Thakur[1], Andrew G. Jeffs[2]

1 School of Biological Sciences, University of Auckland, Auckland, New Zealand, **2** Institute of Marine Science, University of Auckland, Auckland, New Zealand, **3** Dept. of Fisheries, University of Tehran, Karaj, Iran, **4** Institute of Marine Biology and Center of Excellence for the Oceans, National Taiwan Ocean University, Keelung 202, Taiwan, R.O.C., **5** IFRO- Chabahar Branch, Chabahr, Iran

Abstract

Panulirus homarus is an economically important spiny lobster that is widespread through the Indo-West Pacific Region, but has an uncertain taxonomic status, with three or four geographic subspecies having been described. This study used mitochondrial (16S, COI and control region) and nuclear (18S, ITS-1) DNA sequences to examine specimens of all putative subspecies and forms from throughout their range, in order to determine their genetic validity, and understand the evolutionary history of this species. Despite the range of diversity present in the loci examined, the results were consistent across genes. *P. h. rubellus* from the SW Indian Ocean comprised the most divergent lineage that was reciprocally monophyletic with respect to all other *P. homarus* (approx. 9% divergence in COI), and has likely evolved reproductive barriers. The putative *P. h.* "Brown" subspecies from the Marquesas Is in the central Pacific also comprised a somewhat divergent monophyletic lineage (approx. 3% in COI), but may simply be an allopatric population. The widespread *P. h. homarus* was not diverged at all from the described *P. h. megasculpta* from the NW Indian Ocean. The degree of evolutionary divergence of populations at the extremes distribution of the species is somewhat surprising, given the long pelagic larval stage, but suggests that allopatric speciation has been an important driver in the evolution of the genus.

Editor: James Bell, Victoria University Wellington, New Zealand

Funding: The study was conducted with the financial support of the Australian Research Council's Industrial Transformation Research Hub funding scheme (project number IH 120100032) and the Worldwide Universities Network (AJ), the University of Auckland (SL), grants from the National Science Council, Taiwan, R.O.C. (TYC) and the University of Tehran. The funders had no role in study design, data collection and analysis, decision to publish, or preparation of the manuscript.

Competing Interests: The authors have declared that no competing interests exist.

* E-mail: farhadia63@yahoo.com

Introduction

Panulirus homarus (Linnaeus, 1758) is an economically important spiny lobster from the Indo-Pacific region [1]. Despite considerable scientific attention over the years [2], its taxonomic status remains somewhat uncertain, with three or four geographic subspecies or forms having been described [1,3,4]. This tropical species is distributed very widely, ranging from the Natal coast of South Africa in the west, to French Polynesia in the east [3] (Figure 1). Recent reports outline a serious decline in the fisheries status of this species throughout its distribution [5,6] making it more urgent to clarify the ambiguity in its taxonomic status and to assist management processes. The species is also attractive for aquaculture because it is hardy, amenable to culture in sea cages, and grows quickly [7], for which its taxonomic and regional variation is of considerable importance as the basis for selective breeding. Furthermore, an understanding of the origins, evolution and maintenance of any distinct *P. homarus* subspecies will likely provide considerable insight into the important oceanographic and evolutionary forces acting on lobsters and other marine organisms throughout the Indo-West Pacific (IWP) [8–10].

Based principally on the pattern of sculpturing on the abdomen and colouration, three subspecies have been described for *P.homarus* previously [1,3,4]. The nominotypical form *P. homarus homarus* (Linnaeus, 1758) with type locality in Amboina, Indonesia, has small squamae on the transverse abdominal groove, and a mainly green colour, and is believed to occur widely throughout the species' distribution, from South Africa through to the Pacific [3,4]. *Panulirus h. megasculpta* Pesta 1915 has the type locality in South Yemen, and is distributed in the Arab Sea region. It has large squamae on the abdomen, with yellowish spots on the abdomen, and more or less continuous yellowish lines along the margins of the tergites and pleura. *Panulirus h. rubellus* Berry, 1974 with type locality off the eastern coast of South Africa, with large squamae and brick red colouration, is known only from the South-east African coast and Madagascar [3,11]. A fourth form has also been proposed from the Marquesas Islands as *P. homarus* "Brown", which has small squamae and brown colouration (Figure 2). Thus *P. h. rubellus*, *P. h. megasculpta* and *P. h.* "Brown" [4] have geographically discrete distributions, while *P. h. homarus* is described as being widespread (Figure 1). The only genetic investigation so far of variation within this species is that of Ptacek

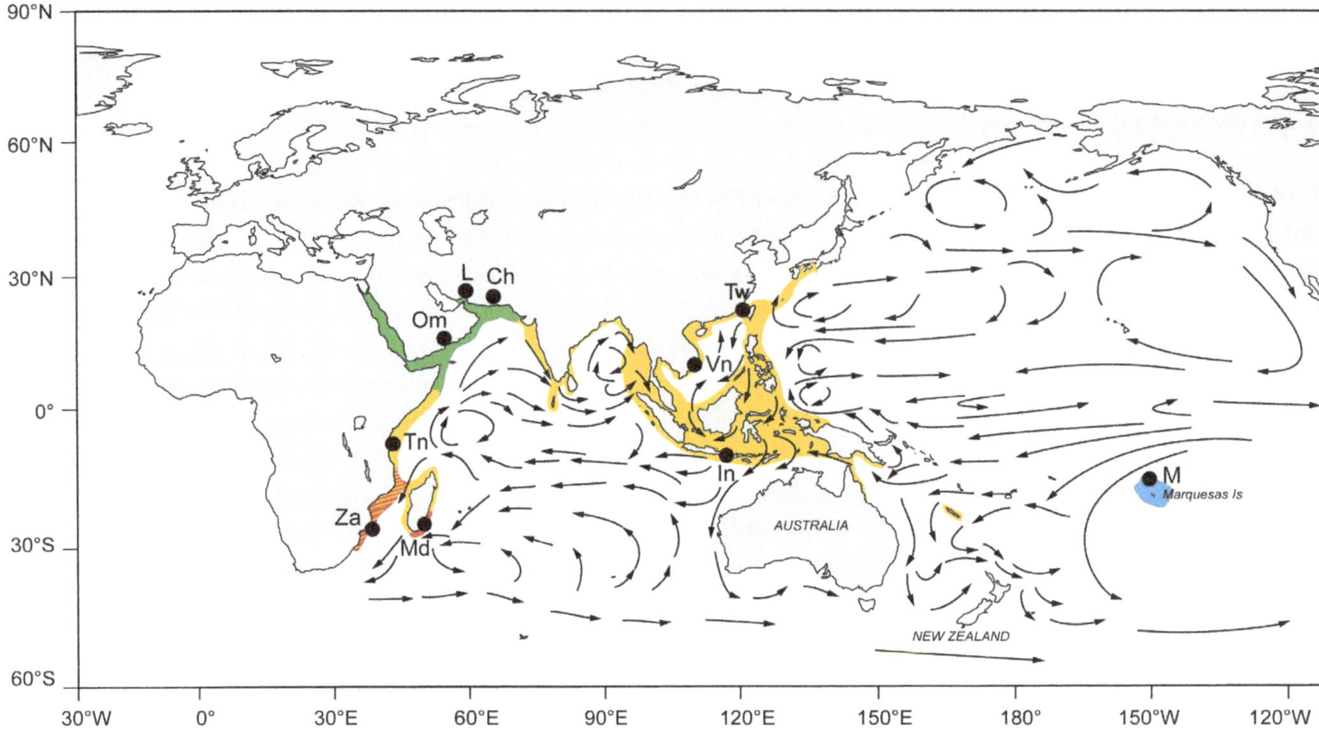

Figure 1. Sampling locations and reported distribution of *Panulirus homarus* and its described subspecies and forms. Yellow: *P. h. homarus*; Red: *P. h. rubellus*; Green: *P. h. megasculpta*; Blue: *P. h.* "Brown". Sampling location abbreviations are listed in Table 1. The eastern and western limits of *P. h. megasculpta* have not been clearly described. In the SW Indian Ocean, there is uncertainty about the exact distributions of *P. h. rubellus* and *P. h. homarus*, as well as their degree of overlap. Approximate geographic coordinates of sampling locations; Za (−31.880002N, 29.262922E), Md (−25.036189N, 46.983933E), Tn (6.1333SN, 39.3167E), Om (16.947833N, 54.705391E), L (26.854245N, 56.313686E), C (25.348491N 60.512433E), In (−8.942053N, 116.199646E), Vn (16.198191N, 108.247604E), Tw (25.122284N, 121.936569E), M (−8.884732N, −139.998436E).

et al (2001), who examined two potential subspecies using partial sequences from the 16S rRNA and COI genes. That study described one specimen from the Marquesas Islands and one from Singapore as belonging to *P. h. homarus* and one specimen from Oman as belonging to *P. h. megasculpta*. They found less than 1% sequence difference between these specimens from different subspecies at the 16S rRNA locus but reported 14% sequence divergence in the COI gene between *P. h. megasculpta* and the Marquesas *P. h. homarus* [12]. They concluded that more study was required on the subspecies status of *P. homarus*.

Genetic techniques have successfully been applied to resolving the taxonomic status of many crustacean species [13,14–19]. In particular, mitochondrial 16S rRNA and COI sequences have been used widely in lobster phylogenetic studies [2], and have shown that some other *Panulirus* species are comprised of more than one genetic lineage [16].

To date there has been no comprehensive genetic study focusing on the different subspecies of *P. homarus*. Recent population genetic analysis of samples from the north-west Indian Ocean has revealed some genetic population differentiation within this region [20]. Here, we use DNA sequences from a range of mitochondrial and nuclear genes in all *P. homarus* putative subspecies and forms collected throughout the species' geographic distribution, to test the validity of these previously described subspecies and forms, and to be better understood the species' evolutionary history. We address the following questions: (1) Do the recognized morphological subspecies and forms *P. h. homarus*, *P. h. megasculpta*, *P. h. rubellus* and *P. h.* "Brown" form reciprocally monophyletic lineages indicative of reproductively isolated types? (2) Do any distinct lineages have overlapping distributions, indicating the formation of reproductive barriers between them? (3) Are distinct lineages suggestive of subspecies status, or have they diverged in genetic and morphological characters sufficiently to suggest distinct species status? (4) What do the geographic patterns of genetic divergence suggest about the past evolutionary divergence within *P. homarus*?

Methods

Sampling

Specimens of *P. homarus* were collected during 2011–2012 from a range of geographic locations throughout the species' distribution (Figure 1, Table 1). The tissue samples were taken from either identified museum specimens or identified fisheries-caught individuals and were preserved in ethanol until subsequent DNA extraction. The lobster samples from Tanzania and Vietnam were purchased from fishermen, and elsewhere were donated (see Table S1 for details of fisheries). Details of museum specimens sequenced, including institution names, holding locations and specimen catalogue numbers, are provided in Table 1. All other specimens sequenced were sub-sampled from normal fisheries-caught animals, and did not require animal ethics permits, or collecting permits. All fisheries had the appropriate permissions for collections. The species is not endangered or protected. Specimens previously identified by taxonomic experts and/or obtained from Muséum National d'Histoire Naturelle, Paris (MNHN), Natur-Museum Senckenberg, Frankfurt (SMF), University of Florida, Gainesville (FLMNH) and National Taiwan Ocean University,

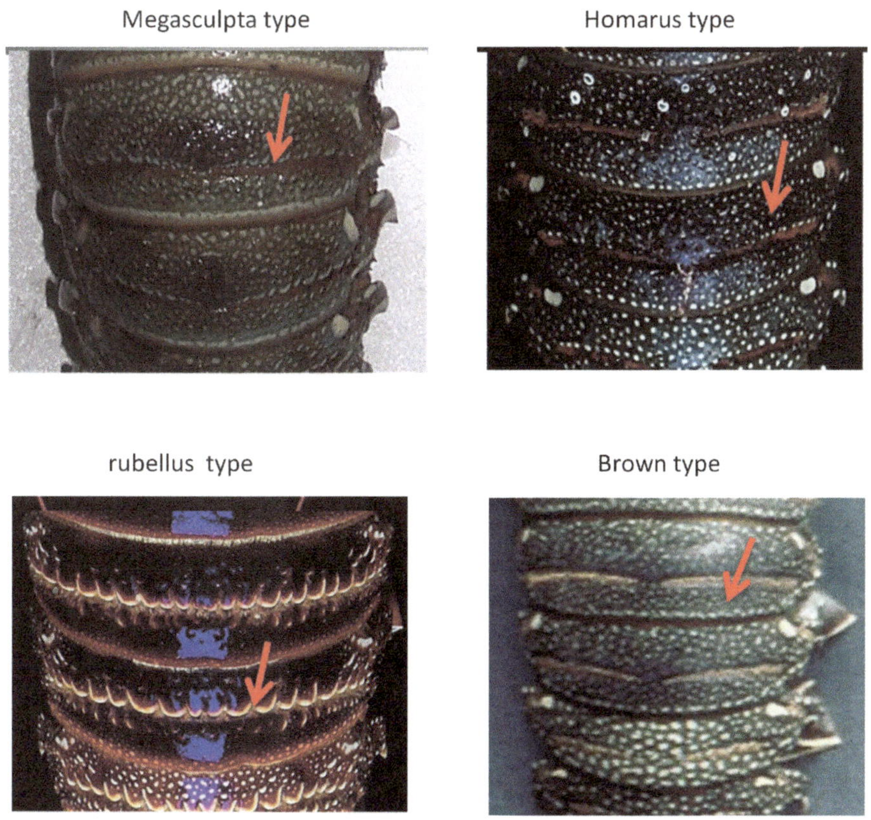

Figure 2. Distinguishing morphological appearances of the subspecies and forms of P. homarus. The colour and sculptus patter are the main morphological identification criteria. The red arrows shows squamae pattern. Rubellus (Madagascar), homarus (Taiwan) and Brown (Marquesas Island) photo; by TYC, megasculpta photo (Chabahar-Iran) by AF.

Keelung (NTOU) were also examined by T. Y. Chan and their identity determined in relation to existing morphological descriptions. All museums provided their consent for the use of samples. In the same manner, whole specimens or those with sufficient-quality photographs available were identified by T.Y Chan or Andrew Jeffs based on colour and sculptus pattern. Each putative subspecies and form is represented by multiple individuals from multiple locations, except *P. h.* "Brown", which is reported from only the Marquesas Islands.

DNA extraction and PCR

Whole DNA was extracted from small subsamples of muscle tissue (approx. 50 mg) using a modified phenol-chloroform method [21]. PCR fragments were amplified from two mitochondrial loci 16S rRNA (16S) and cytochrome oxidase subunit I (COI), using the primers 16Sar/16Sbr [22] and LCO1490/HCO2198 [23] respectively. An improved pair of primers for amplifying *P. homarus* COI (LCO-Ph: 5′- TCGGAGCATGAGCTGGGATAGT -3′ and HCO-Ph 5′-ACTTCTGGGTTGTCGAGGACTC-3′) was designed for more consistent amplification and sequencing. Approximately 800 bp of the mtDNA Control Region (CR) was amplified with CRL-F and CRL-R primers [24]. A 1800 bp fragment of the nuclear 18S rRNA (18S) gene was amplified using the primers 18e [25] and 18p [26]. An approximately 800 bp fragment of the nuclear ribosomal complex including the internal transcribed spacer 1 (ITS-1) was amplified using the primers Sp1-5: 5′-CACACCGCCCGTCGCTACTA-3′ and Sp1-3: 5′-ATTTAGCTGCGGTCTTCATC-3′ [27].

All PCR amplifications were undertaken in 25 µl reactions. The reactions contained 2.5 µl of 10× PCR reaction buffer, 2 mM MgCl2, 200 µM of dNTP mix, 0.4 mM primer, 0.125 units of Taq Ti polymerase (Fisher Thermoscientific) and 10–20 ng of total DNA. Reactions were exposed to initial denaturation of 94°C for 4 min, following by 35 cycles of 94°C for 10 s, the respective annealing temperature (59°C for 16S, 62°C for COI, 63°C for ITS-1 and 59°C for CR) for 20 s and 72°C for 30 s, followed by a final extension of 72°C for 5 min. 18S rRNA was amplified with the same conditions except with an annealing temperature of 63C and with 45S of extension in 72°C. All reactions were accompanied by a negative control.

DNA Sequencing

Free nucleotides and primers were removed from PCR products using a SAP-ExoI protocol [28]. The cleaned products were directly sequenced using the standard protocols of BigDye terminator sequencing chemistry on an ABI PRISM 3100 Genetic Analyser (Perkin-Elmer, Foster City, CA) automated capillary sequencer. Unincorporated dye-labelled nucleotides were removed using the CleanSEQ (Agencourt Bioscience Corporation, Beverly, MA) magnetic bead protocol under recommended conditions. Sequence fragments were generated from approximately 700 bp of the 5′ end of the mitochondrial mtDNA control region (CR), 630 bp of COI, 450 bp of 16S and 1800 bp of 18S, all using the forward primer. Approximately 800 bp of sequence including the ITS-1 were generated using both forward and reverse primers. The heterozygous sequences from individual PhomZa13 were confirmed from multiple PCR products from multiple DNA

Table 1. *P. homarus* and outgroup sampling details and number of individuals (N) sequenced for each locus in this study.

Sample geographic location	Sequence Abbrev.	Collected/Identified by	Sample reference	Morphotype	N				
					COI	16S	CR	ITS-1	18S
South Africa	PhomZa	Fielding, Jeffs	Photos of some	rubellus	5	4	6	6	1
Madagascar	PhomMd	Chan	MNHN[1]	rubellus	3	3	4	2	1
Tanzania	PhomTn	Jeffs	-	homarus	3	1	3	1	-
Oman	PhomOm, PhomChl, Phom06031	Chan, Ptacek	FLMNH[2] and Ptacek	megasculpta	5	6	6	2	-
Iran-Larak	PhomL	Farhadi, Jeffs	Photos	megasculpta	2	2	3	1	1
Iran Chabahar	PhomC	Farhad, Jeffs	Photos	megasculpta	4	6	6	2	1
Indonesia	PhomIn	Victor Estilo	-	homarus			1		
Vietnam	PhomVn	Ngoc Nguyen Thi	-	homarus	3	3	3	1	-
Taiwan	PhomTw	Chan	NTOU[3]	homarus	2	3	3	1	1
Marquesas Islands	PhomM PhomMLC	Jeffs, Farhadi Chan	Photos MNHN[1]	"Brown"	7	4	7	6	-
Marquesas island	Phom06021	Childress, George	Photos	"Brown"	1	1	1	-	-
Iran – Hangam Isl	PversIr	Farhadi	Photos	*P.versicolor*	-	-	1	-	-
Western Australia	PornChl	Childress	-	*P.ornatus*	-	-	1	-	-

[1]Catalogue nos; MNHN 2009-2111, MNHN 2010-4705, MNHN-2010-4706, MNHN 2010-5074. Muséum National d'Histoire Naturelle, Paris, France.
[2]Catalogue no; AR15182, Florida Museum of Natural History, University of Florida, Gainesville, FL.
[3]Catalogue nos; NTOU M01782, NTOU M01783, NTOU M01784. National Taiwan Ocean University, Keelung, Taiwan.

extractions from that individual. All sequence calls had phred quality scores above 30. Sequences have been submitted to GenBank under accession nos (CR: 906454–906484, COI: KJ802748–KJ802782, 16S: KF923507–KF923532, ITS-1: KJ802725–KJ802747).

Sequence alignment and phylogenetic analysis

Previously published sequences from additional *Panulirus* species (Table 2) were included in the sequence alignments and phylogenetic analyses to enable comparison of inter- and intra-species divergences. Sequence alignment and editing was undertaken using Geneious v5.6 [29] (MUSCLE alignment [30]) and confirmed by eye. For the CR, sequence divergences beyond the most closely related species were too great to enable confident alignment. For the ITS-1, sequence divergence was too great to successfully align any outgroup species.

It is necessary to note that some previously published DNA sequences from *P. homarus* have been found here to be artefacts, and therefore have not been included in the current analyses, and are clearly not reliable for future analyses. The COI sequence of Ptacek *et al.* (2001) from the Marquesas Is (identified there as *P. h. homarus*) (GenBank accession #AF339457) is incorrect, and may be a nuclear pseudogene (or numt), as has been reported from other decapods COI sequences [31]. It is around 22% divergent from all other *P. homarus*, including our samples from the Marquesas, and does not match any other *Panulirus* species. This sequence thus gave a false impression in their work of *P. h. homarus* being very divergent from their *P. h. megasculpta* (Oman) sequence. Similarly, the 18S sequences of *P. homarus* from India reported in [32] appear to be incorrect. These sequences are quite variable among several sampled locations in India, however, they are all quite divergent from our 18S sequences, which show no variation within *P. homarus*.

The best DNA substitution model for each marker was determined by Jmodeltest2 [33,34]. The models selected were: 16S - HKY+I+G, $I = 0.27$, $G = 0.40$; COI - GTR+I+G, $I = 0.54$, $G = 0.62/0.11$; CR - GTR+G, $G = 0.35$; ITS-1 - GTR. Neighbor-joining, maximum likelihood and Bayesian phylogenetic analyses were performed in Geneious, using PHYML [35] and MrBayes [36] with the best substitution models. Between-group average divergences were calculated in Mega 5.2.2 [37].

Results

As expected, there are considerable differences among genes in the levels of nucleotide diversity within *P. homarus* (Table 3). Nucleotide diversity among all *P. homarus* specimens ranged from 0.0 in 18S to 1.1% in 16S, 5.9% in COI, 8.7% in CR and 2.6% in ITS-1. Although each gene region provides varying detail of the divergence between distinct lineages, all results within *P. homarus* are consistent among loci.

Table 2. Details of previously published sequences of *Panulirus* species used in phylogenetic analyses.

Species	COI	16S	Location	Ref
P. h. homarus	AF339457	AF337963	Singapore, Marq. Isl	[12]
P. h. megasculpta	AF339458	AF337961	Oman	[12]
P. homarus	JN & JQ[1]	HM & JQ[2]	India	GenBank
P. argus	GU476041	JQ412154	North Atlantic	[16]
P.a. westonii	GU476045	JQ412156	Brazil	[16]
P. interruptus	JN701682	AF337959	California, USA	[12]
P. gutatus	AF339456	AF337963	Florida, USA	[12]
P. penicillatus-WP	AB610678	AB610710	West Pacific	[56]
P. penicillatus-EP	AB610698	AB610717	East Pacific	[56]
P. echinatus	AF339454	AF337965	Brazil	[12]
P. pascuensis	AF339466	AF337973	Easter Isl	[12]
P. l. longipes	AF339464	AF339464	Phillippines	[12]
P. l. bispinosa	AF339463	AF339463	Singapore	[12]
P. cygnus	AF339453	AF337967	Western Australia	[12]
P. japonicus	AF339461	AB610735	Japan	[56]
P. polyphagus	AF339469	AF337975	Singapore	[12]
P. laevicuda	AF339462	AF337969	Brazil	[12]
P. regius	AF339470	AF337976	Congo	[12]
P. gracilis	AF339455	AF337964	Mexico	[12]
P. inflatus	AF339459	AF337960	Mexico	[12]
P. marginatus	AF339465	AF337972	Hawaii	[12]
P. versicolor	AF339472	AF337978	Phillippines	[12]
P. ornatus	This study	AF337971	Indonesia, Singapore	[12]
P. stimpsoni	AF339471	AF337977	Hong Kong	[12]
Jasus edwardsii	AF339473	AF337979	New Zealand	[12]

[1]-JN418937, JQ229884, JQ229888, JQ229914, JQ229916, JQ229925.
[2]- JQ229869, HM015270, HM015271, HM015272.

Table 3. Average pairwise sequence divergences[1] (%) between putative subspecies of scalloped spiny lobster (*Panulirus homarus*) (below diagonal) and within-subspecies diversities (on diagonal).

Subspecies	Locus	*P. h. homarus*	*P. h. megasculpta*	*P. h.* "Brown"	*P. h. rubellus*
P. h. homarus	16S	0.6			
	COI	1.7			
	CR	6.7			
	ITS-1	0.2			
P. h. megasculpta	16S	0.6 (0.2)[2]	0.2		
	COI	1.3 (0.1)	0.6		
	CR	6.1 (0.3)	4.9		
	ITS-1	1.8 (0.6)	2.2		
P. h. "Brown"	16S	1.0 (0.6)	0.8 (0.6)	0.1	
	COI	3.1 (1.8)	2.6 (1.9)	0.9	
	CR	11.1 (6.5)	10.5 (6.8)	2.6	
	ITS-1	0.5 (0.5)	2.1 (1.0)	0.0	
P. h. rubellus	16S	2.5 (1.9)	2.8 (2.4)	3.0 (2.6)	0.7
	COI	9.0 (7.1)	8.5 (7.2)	9.3 (7.9)	1.9
	CR	31.6 (22.8)	31.4 (23.5)	31.6 (24.8)	9.0
	ITS-1	5.0 (4.8)	3.8 (2.6)	4.4 (4.3)	0.3

[1]% sequence divergences calculated using the best model for that locus described in Methods.
[2]Net pairwise divergences in brackets.

Phylogenetic trees were constructed for each gene, including sequences available from related species as outgroups, where these could be unambiguously aligned (Figures 3–6). The outgroup sequences allow comparison of within- and between-species divergences. The patterns of relationships among the *P. homarus* individuals belonging to the different putative subspecies and forms are remarkably consistent among loci. The most obvious feature is that there are two major well-supported reciprocally monophyletic lineages observed at each variable locus, representing all specimens of *P. h. rubellus* in one clade and all specimens of all other subspecies and forms in the second clade. This is clearly seen in all of the 16S, COI, CR and ITS-1 phylogenies. The average sequence divergences between the two lineages range between 2.5% in 16S to 31% in the CR (estimated from the best-fit substitution model, accounting for multiple substitutions per site; Table 3). The only individual that does not follow this pattern is PhomZa13. This specimen was collected from South Africa, and identified as *P. h. rubellus*, but has mitochondrial 16S, COI and CR sequences that belong to the other lineage. On examining the ITS-1 sequence derived from this individual, it can be seen that it is heterozygous for two alleles (Figure 6), one that falls into a *P. h. rubellus*–only clade, and the other that falls into the second clade containing all other *P. homarus* specimens. Unfortunately we do not have the whole specimen, or photographs of it, to confirm the morphological appearance of this specimen.

The second most obvious pattern seen in the phylogenies is that all specimens from the Marquesas Islands, and described as *P. h.* "Brown", also fall into a distinct monophyletic lineage within, and paraphyletic to, the major clade. This lineage is well-supported in all phylogenies except that of ITS-1, where these individuals are all identical to one another, but do not form a distinct well-supported monophyletic lineage. The average sequence divergences between the *P. h.* "Brown" lineage and the *P. h. homarus*/*P. h. megasculpta* lineage range between 0.5% in ITS-1 to 11% in the CR (Table 3).

The remaining sequences of all genes from specimens identified as belonging to the *P. h. homarus* and *P. h. megasculpta* subspecies are found completely mixed within one lineage, and do not show any distinct separation into monophyletic clades. This pattern is repeated in all of the 16S, COI, CR and ITS-1 phylogenies. The net sequence divergences between all *P. h. homarus* specimens and all *P. h. megasculpta* specimens are insignificant for all genes (Table 3).

Discussion

Reproductive isolation of subspecies

All the available genetic data shown here from *P. homarus* is consistent in interpretation across the loci. It is apparent that the most distinct lineage within *P. homarus* is that of *P. h. rubellus*. Morphologically, it is readily distinguished from all other subspecies and forms by its red colouring (Figure 2) [3]. The considerably larger squamae on the abdomen also separate *P. h. rubellus* from *P. h. homarus* and *P. h.* "Brown". Genetically, it clearly belongs to a reciprocally monophyletic lineage that is quite diverged from the remaining *P. homarus* specimens at all variable loci examined (approx. 30% in the CR and 9% in COI). Importantly, this includes divergence at a nuclear locus, ITS-1.

The specimens from the other extreme of the species' range, the Marquesas, also form a distinct monophyletic lineage in 16S, COI and CR phylogenies, but it falls within the major *P. homarus* lineage (i.e., the remaining *P. h. homarus* and *P. h. megasculpta* are paraphyletic with respect to the Marquesas lineage). This Marquesas population has recently been described as another potential subspecies (*P. h.* "Brown"), based on its brownish colouration and remote location [4]. Genetically, it is diverged from the remaining *P. homarus* at most loci examined, but to a much lesser degree than the *P. h. rubellus* lineage (approx. 10% in the CR and 3% in COI). There appears to be insufficient variation at the nuclear ITS-1 locus to unequivocally distinguish the

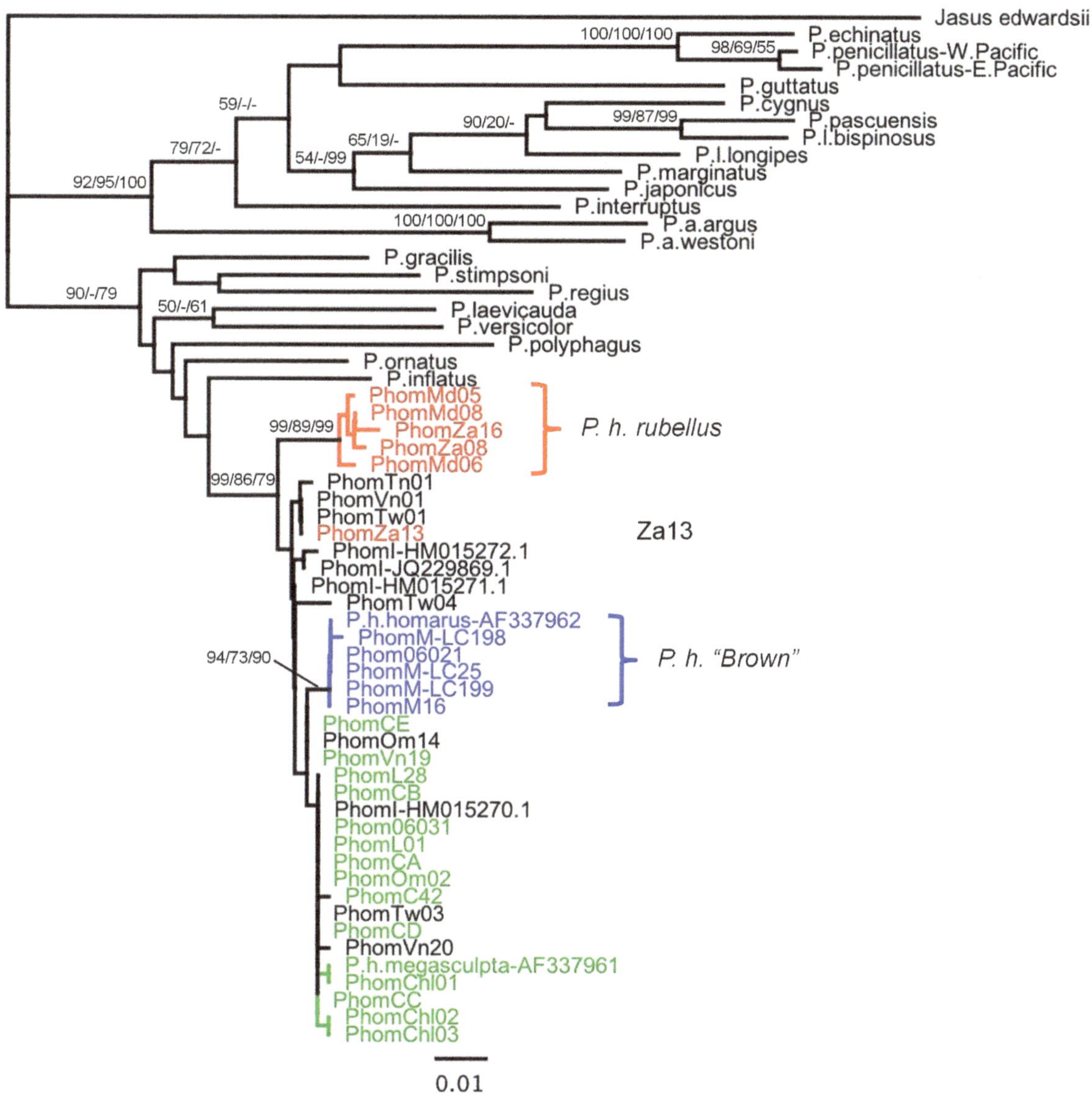

Figure 3. Maximum likelihood phylogenetic tree of 16S sequences from *Panulirus homarus* and other *Panulirus* species. Percent support values (Neighbor-Joining bootstrap/Maximum Likelihood bootstrap/Bayesian probability) shown for nodes with more than 50% support. Sequence location codes are as given in Table 1. Colours are correlated with those in Fig. 1 - Black: *P. h. homarus*; Red: *P. h. rubellus*; Green: *P. h. megasculpta*; Blue: *P. h.* "Brown".

Marquesa specimens as a distinct lineage. Also, there is no clear fixed morphological difference between the Marquesas material and the *P. h. homarus* from Taiwan. Moreover, with additional colouration information now available for both *P. h. homarus* and the Marquesas population, it is clear that there are actually no fixed colouration differences between them. The colouration of the Marquesas specimens is generally olive-green and is simply sometimes lighter (i.e. more greenish) and sometimes darker (i.e. more brownish).

It is also quite clear that the genetic data do not support the separation of the proposed subspecies *P. h. homarus* and *P. h. megasculpta*. Specimens allocated by morphology and geography to these subspecies do not appear to be genetically distinct at all, with the specimens' sequences found mixed throughout the main genetic lineage in each locus examined. There is negligible net sequence divergence between the two putative subspecies at all loci examined. Thus it appears *P. h. megasculpta* should be synonymised with *P. h. homarus*.

These results clearly support the existence of some degree of reproductive isolation among *P. h. rubellus*, *P. h.* "Brown" and the remaining *P. h. homarus*, based on characteristic genetic divergence at mtDNA loci. There is also support from a nuclear locus (ITS-1) for the distinctiveness of *P. h. rubellus*. Existing morphological evidence also supports the distinctiveness of *P. h. rubellus*. The morphological distinctiveness of *P. h. homarus* and *P. h. megasculpta* (based largely on sculpture pattern of abdomen) does not appear to have a clear genetic basis, and there is no evidence of reproductive isolation between these forms.

Overlapping distributions and reproductive barriers

The four putative subspecies and forms previously described from *P. homarus* have largely non-overlapping geographical

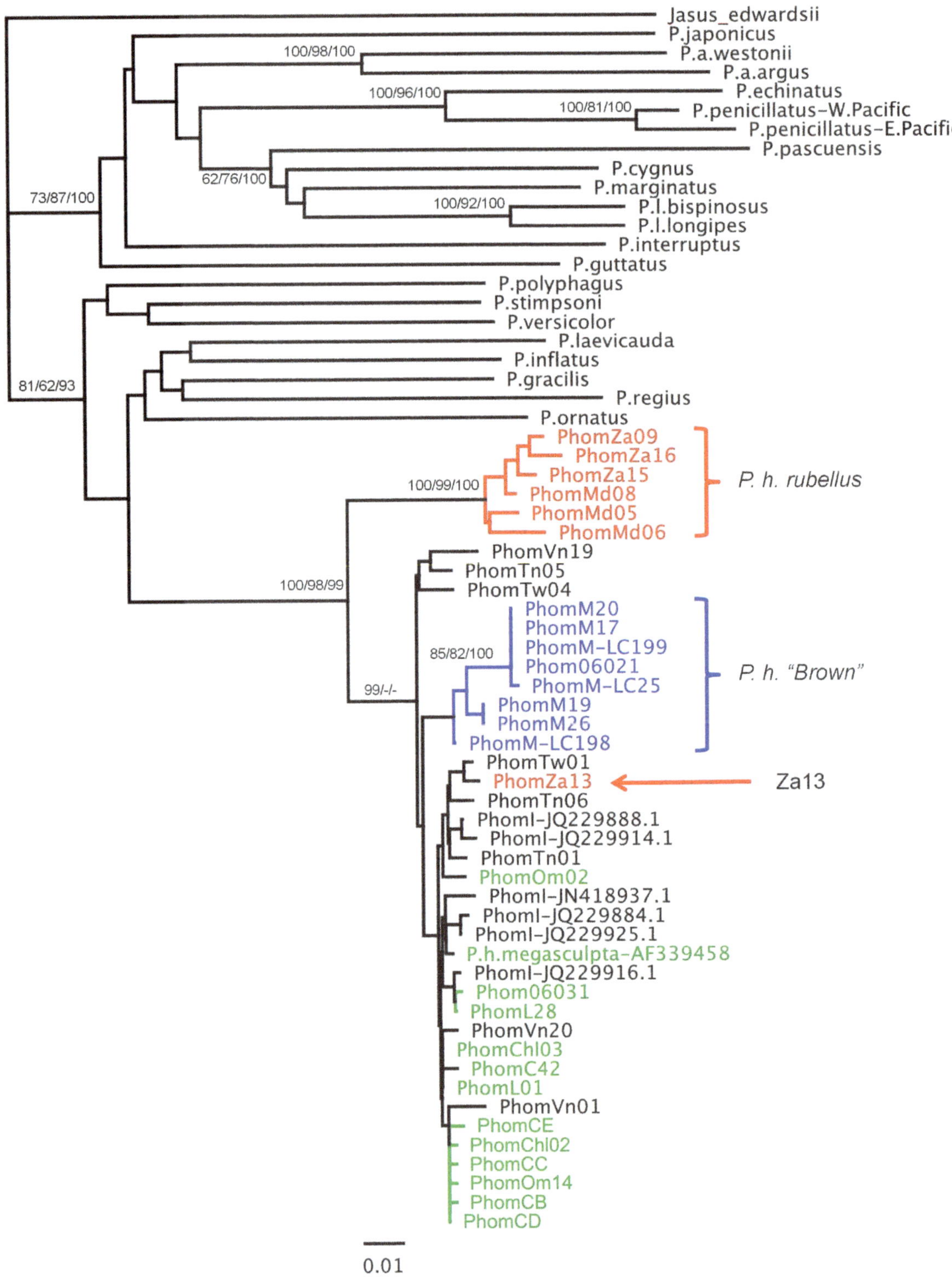

Figure 4. Maximum Likelihood phylogenetic tree of COI sequences from *Panulirus homarus*, and other *Panulirus* species. Percent support values (Neighbor-Joining bootstrap/Maximum Likelihood bootstrap/Bayesian probability) shown for nodes with more than 50% support. Sequence location codes and colours are as in Fig. 2.

distributions. *P. h. rubellus* is reported to occur only in the southwest Indian Ocean (mainly South Africa and Madagascar), *P. h. megasculpta* in the northwest Indian Ocean, and *P. h.* "Brown" from the eastern extreme of the range in the Marquesas. Only *P. h. homarus* has been reported as being widespread through the Indo-West Pacific. Populations of a species can become reproductively isolated over time simply through long-term geographic allopatric separation. Populations reproductively isolated in this way do not necessarily have any biological reproductive barriers between them, and if not, could still successfully reproduce with each other

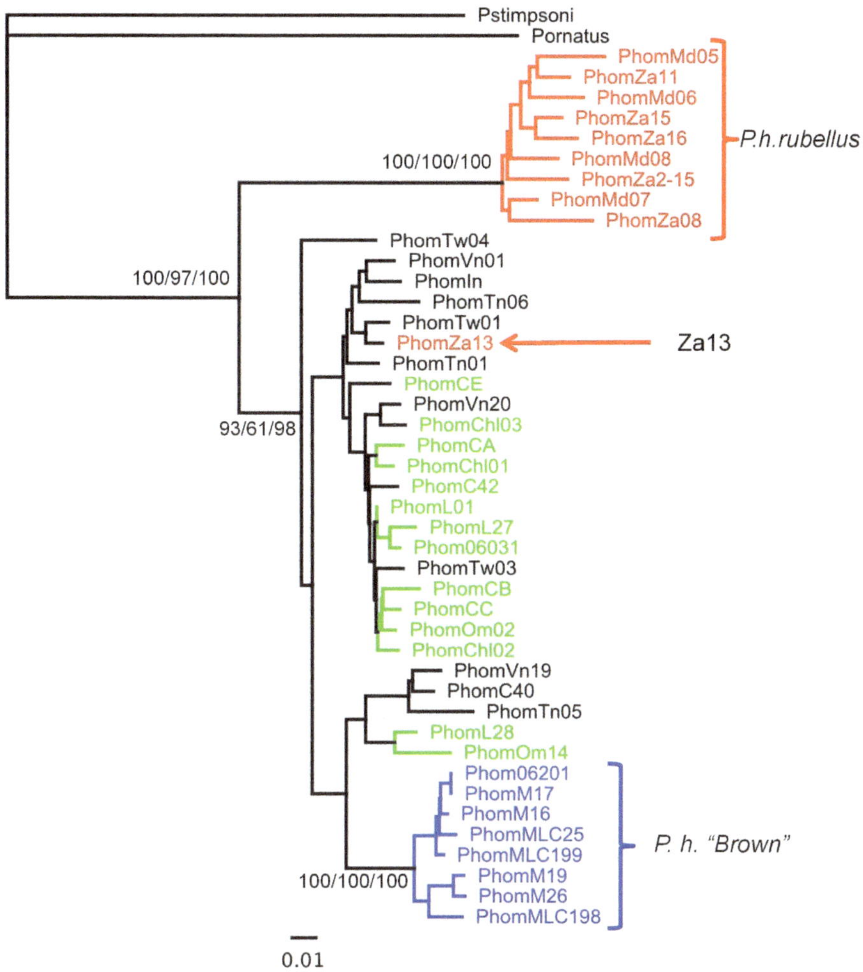

Figure 5. Maximum Likelihood phylogenetic tree of CR sequences from *Panulirus homarus* and outgroup. Percent support values (Neighbor-Joining bootstrap/Maximum Likelihood bootstrap/Bayesian probability) shown for nodes with more than 50% support. Sequence location codes and colours are as in Fig. 2.

if their gametes meet [38]. If, however, forms that have diverged morphologically and genetically occur in geographic sympatry, then it is likely that there must exist some form of biological reproductive barrier between them at the pre-or post-zygotic stage [39].

Evidence in support of a reproductive barrier with *P. h. rubellus* comes from the maintenance of genetic distinctiveness at both mitochondrial and nuclear genes, in the face of geographic sympatry with the more widespread *P. h. homarus* in at least part of its range. It has previously been reported that along the east African coast (including our sampling site in Tanzania) the common *P. h. homarus* form occurs exclusively as far as the northern Mozambique coast [3]. On the southeast Madagascar coast the *P. h. rubellus* form occurs exclusively [3,40], whereas along the nearby southeast African coast (including our South African sampling site) both these forms have been reported. Berry recorded that *P. h. rubellus* increases in frequency from approximately 80% of *P. homarus* specimens in southern Mozambique, to over 98% in Natal [3]. The sympatric distribution of two morphological forms that are genetically divergent at both mitochondrial and nuclear genes suggests that these two forms have developed some level of biological reproductive barrier. Their evolutionary divergence appears to have progressed further than merely the stochastic divergence of neutral loci through allopatric separation. Unless there were some reproductive barriers slowing the mixing of their gene pools, the complete monophyletic divergence at neutral genetic loci would rapidly erode with the inevitable gene flow that would arise from sympatric or parapatric distributions. This is particularly so for species such as *P. homarus* with very widely dispersed larvae. The divergent evolution of reproductive structures has previously been suggested to play a role in the development of reproductive barriers between spiny lobster species [41]. Further research could more closely examine differences in reproductive structures between *P. h. rubellus* and *P. h. homarus*.

The extent of the evolutionary divergence between the forms has been clarified to some degree by our genetic studies. One *P. homarus* specimen from South Africa (Za13) has been identified as belonging to the *P. h. homarus* mtDNA lineage (unlike our remaining specimens from that location that belong to the *P. h. rubellus* lineage), whereas this individual is heterozygous in its nuclear ITS-1 alleles, with one allele from the *P. h. rubellus* lineage and one from the *P. h. homarus* lineage. This clearly suggests that this individual is either an F1 hybrid between the two forms, or at least a descendent of one. Thus, although there must likely exist some level of biological reproductive barrier between the forms,

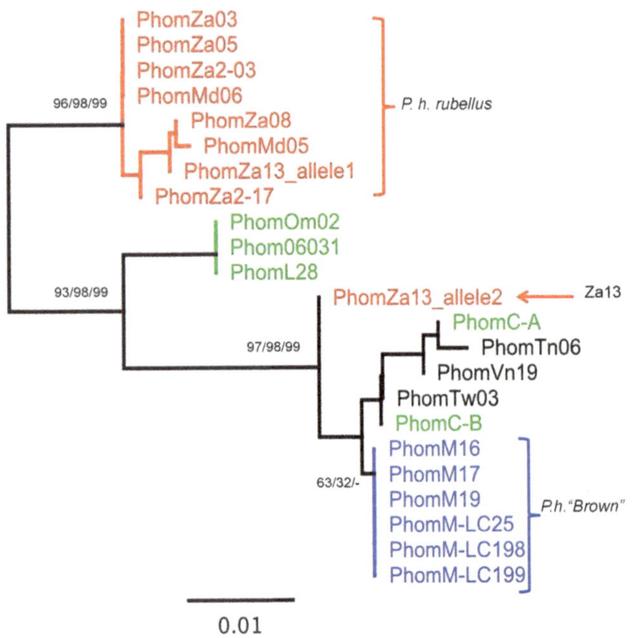

Figure 6. Unrooted Maximum Likelihood phylogenetic tree of ITS-1 partial sequences from *Panuliraus homarus*. Percent support values (Neighbor-Joining bootstrap/Maximum Likelihood bootstrap/ Bayesian probability) shown for nodes with more than 50% support.

for them to maintain their distinctiveness, it clearly must be somewhat permeable. The fact that the hybrid individual possesses *P. h. homarus* mtDNA, indicates that a male *P. h. rubellus* has successfully mated promiscuously with a female *P. h. homarus* when depositing its spermatophores. The extent of hybridisation between the two forms, and whether male *P. h. homarus* can also successfully mate promiscuously with *P. h. rubellus* requires further sampling to clarify. The fact that only one hybrid has been found from our sampling of 44 individuals from South Africa, East Africa and Madagascar suggests an extremely low level of gene flow between *P. h. rubellus* and *P. h. homarus*. The discovery of this hybrid does confirm the suspicions of Berry, who presumed that the occurrence of rare "intermediate" morphotypes in this region indicated the presence of hybrids. He found the frequency of these "hybrids" dropped from 5.1% in southern Mozambique to only 0.1% in Natal [3].

It is also possible that the presence of two divergent alleles (from the *P. h. rubellus* and *P. h. homarus* lineages) in the one individual may be due to either incomplete lineage sorting or past introgression. We consider hybridization to be a more likely cause in this instance, but clearly further investigation is warranted.

Status of lineages as subspecies or species

The rigorous taxonomic determination of species or subspecies generally requires evidence of distinct morphological and genetic characters, indicating reproductive isolation [42,43]. We have shown firstly that *P. h. rubellus* is distinctive for morphological colour and genetic characters, and, due to its overlapping distribution with *P. h. homarus*, appears to have some degree of biological reproductive barrier from this form. Furthermore, there is a substantial degree of genetic divergence between *P. h. rubellus* and the other *P. homarus* forms (approx. 30% in the CR and 9% in COI). Thus *P. h. rubellus* warrants at least a distinct subspecies status, and perhaps even a distinct species status. Its genetic divergence from the other *P. homarus* forms is of the same order as that of other described *Panulirus* species. The genetic divergence at COI between species of the genus *Panulirus* varies from approximately 10%, to about 32% [16].

Nevertheless, *P. h. rubellus* still forms a monophyletic clade with the other *P. homarus* forms, indicating its close evolutionary relationship with *P. homarus*. Moreover, the present study shows that *P. h. megasculpta* cannot be separated from *P. h. homarus*. Thus, the larger squamae in *P. h. rubellus* are not useful in differentiating it from *P. h. homarus* and the only reliable distinguishing character is its brick red colouration. However, where both *P. h. rubellus* and *P. h. homarus* occur sympatrically, the two can be readily separated by colouration and size of abdominal squamae. In view of the only distinguishing character of *P. h. rubellus* being colouration (Figure 2), and that natural hybridization occurs between *P. h. rubellus* and *P. h. homarus*, we propose to continue treating *P. h. rubellus* as a subspecies of *P. homarus* until more information on the hybrids is known (e.g. degree of hybridization, fertility of the hybrids, etc.).

Panulirus h. megasculpta has previously been reported to occur exclusively in the NW Indian Ocean where no *P. h. rubellus* or *P. h. homarus* morphotypes were found [3]. The Arabian Sea specimens obtained in this study (including four additional specimens from Yemen and Oman deposited in SMF but which could not be successfully sequenced) are all of the *P. h. megasculpta* morphotype. This *Panulirus h. megasculpta* morphotype differs from the *P. h. homarus* morphotype in having bigger squamae and more distinctive yellowish spots on the abdomen (with those on the margins of tergites and pleura often somewhat continuous as a line). However, there is no distinct genetic difference between the Arabian Sea material and *P. h. homarus* in the other Indo-West Pacific localities. Thus the differences in colouration and squamae size do not indicate different genetic lineages or subspecies, and the subspecific status of *P. h. megasculpta* is invalid. It would be very informative to map the exact geographical limitations of the "*Panulirus h. megasculpta*" morphotype as this form appears to be absent in India in the east and Kenya in the west [3], and if both the "*Panulirus h. megasculpta*" and *P. h. homarus* morphotypes can occur sympatrically or not.

The taxonomic status of the *P. h.* "Brown" form is less clear. It is still clearly reproductively isolated from other forms based on its genetic distinctiveness. However, this could be due entirely to its geographic isolation, because there is no fixed morphological and colouration difference between the Marquesas material and other Indo-West Pacific *P. h. homarus*. It is quite possible that no biological reproductive barriers have developed in this form, and it may represent simply a geographically and genetically distinct population of *P. homarus* that has diverged on the eastern periphery of the species' distribution. The fact that *P. h. homarus* mtDNA is paraphyletic with respect to the *P. h.* "Brown" lineage supports the latter being a distinct population rather than a subspecies. The degree of genetic divergence (approx. 10% in the CR and 3% in COI) is similar to that between both distinct subspecies and populations of other *Panulirus* that have been previously described [16]. The taxonomic status of *P. h. "Brown"* can only be finally determined when a greater number of specimens from intermediate localities in the South Pacific islands are included in the analysis.

The gene most used in recent studies for calibrating the age of divergence between crustacean lineages is COI. The most recent and apparently reliable *Panulirus* divergence rate estimate is approximately 1% COI sequence divergence per million years [16], although other estimated rates have been as high as 4% [44]. Using the rate of 1% gives an approximate divergence time for *P. h. rubellus* of about 9 million years ago (MYA), and a divergence time for *P.h.*"Brown" of approximately 3MYA. Using the rate of 4% per MY, gives minimum divergence times of approximately 2.25 MYA and 0.75 MYA, respectively.

Past evolutionary divergence within *P. homarus*

The *Panulirus* genus is somewhat extraordinary in that most species have relatively long oceanic pelagic larval stages [45], with consequent relatively wide geographic distributions, yet at the same time have undergone extensive evolutionary speciation and radiation [2]. In the Indo-West Pacific alone, there are 12 recognised species of *Panulirus*, many with relatively wide distributions [46]. Given that their long pelagic larval durations would predict relative genetic homogeneity throughout the IWP, the substantial evolutionary radiation in this genus suggests that other mechanisms are involved that restrict successful long-range gene flow through this region. *Panulirus homarus* has one of the wider Indo-West Pacific distributions (although one of the shorter larval durations, at around 6 months) [7], and the evolutionary mechanisms dividing this species into distinct lineages may offer insight into the mechanisms operating throughout this genus to make it such a diverse and successful group.

There are several factors that may potentially limit successful long-range dispersal and/or gene flow in this species. Continental plate movement appears to not have played a role here, as the maximum estimated dates of divergence of the lineages (9MYA) are too recent to be impacted by this, unlike older species divergences in the Palinuridae [47]. Firstly, it is feasible, although not entirely obvious, that the direction of flow of the major currents in the region may have acted to isolate the *P. homarus* populations at the extremes of the species' distribution (Figure 2). Certainly, the predominantly westward flowing currents from the Marquesas would restrict ongoing gene flow in the opposite direction from the distant populations of the western Pacific. The Marquesas are recognised as having a high degree of endemism, particularly in fish and molluscs [9], and its prevailing currents and upwellings have previously been implicated in its genetic isolation [48]. However there is no clear major current directions in the SW Indian Ocean that would isolate this region. It may be more local, less-understood water circulation patterns that play a crucial role here. Of relevance is the apparent genetic divergence of the spiny lobster *Palinurus delagoae* into two distinct lineages (approx. 5% divergent at mtCR) separated by the Mozambique Channel [49]. Discrete SW water movements of the Agulhas and East Madagascar currents may act to minimise larval movement across the Channel [50]. Perhaps more importantly, at both extremes of the *P. homarus* distribution, it is likely that it is the long-term changes in water-circulation patterns that have played the pivotal role. Currents must have been favourable in the past to permit colonisation of these regions by *P. homarus*, and may have subsequently changed for a sufficient period of time to isolate these populations and permit genetic divergence through random drift. Recent currents in the SW Indian Ocean appear to have reduced isolation of the region to some extent, as evidenced by the subspecies' now overlapping distributions, and the occasional hybrid formation. In the Marquesas, perhaps the crucial currents there are eddies and upwelling that permit local retention of larvae after many months [4].

Apart from the influence of currents, other factors are also likely to be important in restricting gene flow in this species. If the larval stages acted as completely passive particles, it is likely that (over their long larval period) they would easily disperse throughout the SW Indian Ocean, breaking down any geographic isolation of populations in this region. It does seem that some level of active movement of the larvae (possibly in response to species-specific orientation cues), in conjunction with the highly active coastward swimming of the postlarval puerulus stage, [51,52] must restrict dispersal and gene flow to be lower than that expected from currents alone. Several potential behavioural adaptations have been proposed that would help palinurid larvae recruit to local habitat, including diurnal vertical movement [41].

Regardless of the exact mechanisms involved, it is clear that it is only the most geographically extreme populations of *P. homarus* that have diverged to the extent that they may be subspecies. It appears that the most eastern and western populations have diverged relatively recently from the ancestral, central population. This points to allopatric divergence being the dominant evolutionary mechanism in this species. It also suggests that allopatric speciation may be the dominant evolutionary mechanism throughout the genus *Panulirus*. Recent genetic studies have revealed that several described *Panulirus* species are actually comprised of more than one distinct lineage, which have characteristic morphological differences and are now described as either distinct subspecies or species. Most of these distinct lineages are allopatrically distributed. Mitochondrial DNA sequence data was used to propose the divergence of *P. argus argus* from *P. a. westonii* as two subspecies with distinct geographic distributions in the Caribbean and south-west Atlantic respectively [53]. These two clades have recently been confirmed as two monophyletic lineages for both mtDNA genes (16S and COI) and a nuclear gene (ANT) [16]. *P. penicillatus* has been shown recently to consist of two very divergent allopatric lineages from the eastern Pacific and the central and western Pacific [51]. Morphological variants of *P. longipes* have also been confirmed to belong to distinct lineages by mitochondrial COI [54,55], with two subspecies named *P. l. longipes* and *P. l. bispinosus* (~3% divergent) that partially overlap in distribution. The closely-related species of the "*japonicus*" group, *P. longipes*, *P. cygnus*, *P. marginatus*, *P. pascuensis*, and *P. japonicus*, all have largely allopatric distributions [55]. This expanding evidence of many closely-related *Panulirus* species and subspecies having non-overlapping distributions highlights the great potential importance of allopatric speciation of peripheral populations in the formation of new genetic lineages within this genus. In the case of *P. h. rubellus*, we appear to have caught this peripheral speciation process in action, as this subspecies has not yet developed a complete reproductive barrier from its sibling, *P. h. homarus*.

Conclusions

We have shown here using both mtDNA and nDNA sequencing that two of the four putative subspecies in *P. homarus* (*P. h. rubellus* and *P. h.* "Brown") belong to genetically distinct lineages and are valid taxa, while the remaining two (*P. h. homarus* and *P. h. megasculpta*) are genetically indistinguishable from each other. The partially overlapping distributions of the *P. h. rubellus* & *P. h. homarus* lineages, and the existence of morphological hybrids confirmed genetically by one apparently hybrid specimen in this study, suggest there exists a semi-permeable biological reproductive barrier between these forms. The taxonomic status *P. h. rubellus* is thus valid while that of *P. h. megasculpta* is not. The status of *P. h.* "Brown" remains to be further investigated, particularly with more sampling in the South Pacific, but is clearly reproductively isolated from south-east Asian *P. h. homarus*. Finally, the evolutionary history of this species suggests that it has been dominated by allopatric divergence of populations at the extreme of its distribution. This may point to the principal evolutionary mechanism operating throughout the diverse and successful *Panulirus* genus.

Acknowledgments

The molecular analyses for this study were carried out in the Molecular Ecology and Evolution lab in the School of Biological Sciences at the University of Auckland. Sincere thanks are extended to the Iranian Fisheries Research Bureau in Chabahar and the University of Tehran, Iran, Dr Matthew Slater, Dr Peter Fielding, Renee Albertson, Nguyen Thi BichNgoc, Victor Estilo, L. Corbari of the Muséum National d'Histoire Naturelle, Paris (MNHN), M. Türkay of the Natur-Museum Senckenberg, Frankfurt, and G. Paulay of the University of Florida, Gainesville for kindly providing us with samples. Samples from Madagascar were collected during the ATIMO VATAE Expedition, which is a one of a cluster of Mozambique-Madagascar expeditions (La Planète Revisitée/Our Planet Reviewed) funded by the Total, Prince Albert II of Monaco, and Stavros Niarchos foundations, and conducted by MNHN and Pro-Natura International (PNI). The ATIMO VATAE Expedition was undertaken in partnership with Institut d'Halieutique et des Sciences Marines, University of Toliara, and the Madagascar Bureau of the Wildlife Conservation Society. We thank all these collectors for their generous assistance with providing samples. Special thanks go to Yang C. W and members of the Molecular Ecology and Evolution Lab for experimental help. Michael Childress kindly provided tissue subsamples from specimens previously sequenced in an earlier phylogenetic study [12]. We also thank the University of Auckland Centre for Genomics and Proteomics for sequencing services, and Vivian Ward for Figure 2. Special thanks to Ray George and anonymous reviewers of this manuscript.

Author Contributions

Conceived and designed the experiments: AF HF AA AJ SL. Performed the experiments: AF VT TYC. Analyzed the data: AF VT SL TYC. Contributed reagents/materials/analysis tools: SL AJ TYC. Wrote the paper: SL AF TYC AJ.

References

1. Holthius LB (1991) Marine lobsters of the world. An annotated and illustrated catalogue of species of interest to fisheries known to date. FAO Species Catalogue (FIR/S125 vol 13). Rome: FAO. pp. 125.
2. Patek SN, Feldmann RM, Porter M, Tshudy D (2007) Phylogeny and evolution. in: B. F. . Phillips, editor. Lobsters: biology, management, aquaculture and fisheries: 113–145.
3. Berry PF (1974) A revision of the *Panulirus homarus*-group of spiny lobsters (Decapoda, Palinuridae). Crustaceana 27 (1): 31–42.
4. George RW (2005) Tethys sea fragmentation and speciation of *Panulirus* spiny lobsters. Crustaceana 78: 1281–1309.
5. FAO (2010) FAO yearbook, Fishery and aquaculture statistics. Rome. pp. 100.
6. Al-Marzouqi A, Al-Nahdi A, Jayabalan N, Groeneveld JC (2007) An asssessment of the spinylobster *Panulirus homarus* fishery in Oman — Another decline in the Western Indian Ocean? Western Indian Ocean J Mar Sci 6: 159–174.
7. Phillips B, Matsuda H (2011) A global review of spiny lobster aquaculture. In: R. . Fotedar and B. F. . Phillips, editors. Recent advances and new species in aquaculture. Wiley-Blackwell. pp. 22–84.
8. Pollock DE (1992) Palaeoceanography and speciation in the spiny lobster genus *Panulirus* in the Indo-Pacific. Bull Mar Sci 51: 135–146.
9. Springer VG (1982) Pacific plate biogeography, with special reference to shorefishes. Smithsonian Contribution to Zoology 367: 1–182.
10. Ekman S (1953) Zoogeography of the sea London: Sidgwick and Jackson.
11. Berry PF (1971) The biology of the spiny lobster *Panulirus homarus* (Linnaeus) off the east coast of southern Africa. Oceanographic Research Institute. Investigational Report, No 28, pp. 75.
12. Ptacek MB, Sarver SK, Childress MJ, Herrnkind WF (2001) Molecular phylogeny of the spiny lobster genus *Panulirus* (Decapoda: Palinuridae). Mar Freshw Res 52: 1037–1047.
13. Groeneveld JC, Gopal K, George RW, Matthee CA (2007) Molecular phylogeny of the spiny lobster genus *Palinurus* (Decapoda: Palinuridae) with hypotheses on speciation in the NE Atlantic/Mediterranean and SW Indian Ocean. Mol Phylogenet Evol 45: 102–110.
14. Palero F, Crandall K, Abelló P, Macpherson E, Pascual M (2009) Phylogenetic relationships between spiny, slipper and coral lobsters (Crustacea, Decapoda, Achelata). Mol Phylogenet Evol 50: 152–162.
15. Li CP, De Grave S, Chan TY, Lei HC, Chu KH (2011) Molecular systematics of caridean shrimps based on five nuclear genes: implications for superfamily classification. Zoologischer Anzeiger-A J Com Biol 250: 270–279.
16. Tourinho JL, Solé-Cava AM, Lazoski C (2012) Cryptic species within the commercially most important lobster in the tropical Atlantic, the spiny lobster *Panulirus argus*. Mar Biol 159: 1897–1906.
17. Jimoh AA, Anetekhai MA, Cummings S, Abanikanda OTF, Turner GF, et al. (2013) Mismatch between molecular (mtDNA) and morphological classification of *Macrobrachium* prawns from Southern Nigeria: Cryptic freshwater species and brackish water morphotypes. Aquaculture 410–411: 25–31.
18. Yang C-H, Bracken-Grissom H, Kim D, Crandall KA, Chan T-Y (2012) Phylogenetic relationships, character evolution, and taxonomic implications within the slipper lobsters (Crustacea: Decapoda: Scyllaridae). Mol Phylogenet Evol 62: 237–250.
19. Chan T-Y, Ho KC, Li CP, Chu KH (2009) Origin and diversification of the clawed lobster genus Metanephrops (Crustacea: Decapoda: Nephropidae). Mol Phylogenet Evol 50: 411–422.
20. Farhadi A, Farhamand H, Nematollahi MA, Jeffs A, Lavery SD (2013) Mitochondrial DNA population structure of the scalloped lobster *Panulirus homarus* (Linnaeus 1758) from the West Indian Ocean. ICES J Mar Sci 70 (7): 1491–1498.
21. Baker CS, Slade RW, Bannister JL, Abernethy RB, Weinrich MT, et al. (1994) Hierarchical structure of mitochondrial DNA gene flow among humpback whales *Megaptera novaeangliae*, world-wide. Mol Ecol 3: 313–327.
22. Palumbi SR (1996) Nucleic acids II: the polymerase chain reaction. Mol Syst 2: 205–247.
23. Folmer O, Black M, Hoeh W, Lutz R, Vrijenhoek R (1994) DNA primers for amplification of mitochondrial cytochrome c oxidase subunit I from diverse metazoan invertebrates. Mol Mar Biol Biotechnol 3: 294–299.
24. Diniz FM, Maclean N, Ogawa M, Cintra IHA, Bentzen P (2005) The hypervariable domain of the mitochondrial control region in Atlantic spiny lobsters and its potential as a marker for investigating phylogeographic structuring. Marine Biotechnology 7: 462–473.
25. Hillis DM, Dixon MT (1991) Ribosomal DNA: Mol Evol Phylogenet Inference. Q Rev Biol 66: 441–446.
26. Halaynch KM, Bacheller JD, Aguinaldo AM, Liva SM, Hillis DM, et al. (1995) Evidence from 18S ribosomal DNA that lophophorates are protostome animals. Science 267: 1641–1643.
27. Chu KH, Li CP, Ho HY (2001) The first internal transcribed spacer (ITS-1) of ribosomal DNA as a molecular marker for phylogenetic and population analyses in Crustacea. Mar Biotechnol 3: 355–361.
28. Werle E, Schneider C, Renner M, Völker M, Fiehn W (1994) Convenient single-step, one tube purification of PCR products for direct sequencing. Nucleic Acids Res 22: 4354–4355.
29. Drummond AJ, Ashton B, Buxton S, Cheung M, Cooper A, et al. (2011) Geneious Auckland: Biometra.
30. Edgar RC (2004) MUSCLE: multiple sequence alignment with high accuracy and high throughput. Nucleic Acids Res 32: 792–797.
31. Buhay JE (2009) "COI-like" sequences are becoming problematic in molecular systematic and DNA barcoding studies. J Crust Biol 29: 96–110.
32. Huxley VAJ, Rejinie Mon TS, Ramesh R (2011) 18S rRNA gene polymorphisms of *Panulirus homarus* populations from different geographic regions of Peninsular India. J Theor Exp Biol 8: 58–93.
33. Darriba D, Taboada GL, Doallo R, Posada D (2012) jModelTest 2: more models, new heuristics and parallel computing. Nat Methods 9: 772–772.
34. Posada D (2008) jModelTest: Phylogenetic model averaging. Mol Biol Evol 25: 1253–1256.
35. Guindon S, Gascuel OA (2003) Simple, fast, and accurate algorithm to estimate large phylogenies by maximum likelihood. Syst Biol 52: 696–704.
36. Ronquist F, Teslenko M, van der Mark P, Ayres DL, Darling A, et al. (2012) MrBayes 3.2: Efficient bayesian phylogenetic inference and model choice across a large model space. Syst Biol 61: 539–542.
37. Tamura K, Peterson D, Peterson N, Stecher G, Nei M, et al. (2011) MEGA5: Molecular evolutionary genetics analysis using maximum likelihood, evolutionary distance, and maximum parsimony methods. Mol Biol Evol 28: 2731–2739.
38. Knowlton N (1993) Sibling Species in the Sea. Annu Rev of Ecol Sys 24: 189–216.
39. Quenouille B, Hubert N, Bermingham E, Planes S (2011) Speciation in tropical seas: Allopatry followed by range change. Mol Phylogenet Evol 58: 546–552.
40. Chan TY. Personal observation.
41. George RW (2005) Comparative morphology and evolution of the reproductive structures in spiny lobsters, *Panulirus*. N Z J Mar Freshw Res 39: 493–501.
42. Avise JC (2000) Phylogeography, the history and formation of species Boston: Harvard university press. 228 p.
43. Ryder OA (1986) Species conservation and systematics: the dilemma of subspecies. Trends Ecol Evol 1: 9–10.
44. Palero F, Abelló P, Macpherson E, Gristina M, Pascual M (2008) Phylogeography of the European spiny lobster (*Palinurus elephas*): Influence of current

oceanographical features and historical processes. Mol Phylogenet Evol 48: 708–717.

45. Phillips BF, Booth JD, Cobb JS, Jeffs AG, McWilliam P (2006) Larval and postlarval ecology. In: B. F. . Phillips, editor editors. Lobsters: biology, management, aquaculture and fisheries. Oxford: Blackwell Scientific Press. pp. 231–262.
46. Phillips BF, Melville-Smith R (2006) *Panulirus* species. In: B. F. . Phillips, editor. Lobsters: Biology, Management, Aquaculture and Fisheries. Oxford: Blackwell Scientific Press. pp. 30.
47. George RW (2006) Tethys origin and subsequent radiation of the spiny lobsters (Palinuridae). Crustaceana 79: 397–422.
48. Gaither MR, Toonen RJ, Robertson DR, Planes S, Bowen BW (2010) Genetic evaluation of marine biogeographical barriers: perspectives from two widespread Indo-Pacific snappers (*Lutjanus kasmira* and *Lutjanus fulvus*). J Biogeogr 37: 133–147.
49. Gopal K, Tolley KA, Groeneveld JC, Matthee CA (2006) Mitochondrial DNA variation in spiny lobster *Palinurus delagoae* suggests genetically structured populations in the southwestern Indian Ocean. Mar Ecol-Prog Ser 319: 191–198.
50. Lutjeharms J (1988) On the role of the East Madagascar Current as a source of the Agulhas Current. South African J Sci 84: 236–238.
51. Jeffs A, Montgomery J, Tindle C (2005) How do spiny lobster post-larvae find the coast? N Z J Mar Freshw Res 39: 605–617.
52. Yeung C, Jones DL, Criales MM, Jackson TL, Richards WJ (2001) Influence of coastal eddies and counter-currents on the influx of spiny lobster, *Panulirus argus*, postlarvae into Florida Bay. Mar Freshw Res 52: 1217–1232.
53. Sarver SK, Silberman JD, Walsh PJ (1998) Mitochondrial DNA sequence evidence supporting the recognition of two subspecies or species of the Florida Spiny Lobster *Panulirus argus*. J Crust Biol 18: 177–186.
54. Chan TY, Ng PKL (2001) On the nomenculture of the comercially important spiny lobster *Panulirus longipes femorstriga* (Von Martin, 1872), *P.bispinosus* Borradaile, 1899, and *P.albiflagellum* Chan and Chu, 1996 (Decapoda, Palinuridae). Crustaceana 74: 123–127.
55. Ravago RG, Menez MAJ (2002) Phylogenetic position of the striped-legged forms of *Panulirus longipes* (A. Milne-Edwards, 1868) (Decapoda, Palinuridae) inferred from mitochondrial DNA sequences. Crustaceana 75: 1047–1059.
56. Chow S, Jeffs A, Miyake Y, Konishi K, Okazaki M, et al. (2011) Genetic Isolation between the Western and Eastern Pacific Populations of Pronghorn Spiny Lobster *Panulirus penicillatus*. PLoS One 6: e29280.

Support Vector Machine Classification of Streptavidin-Binding Aptamers

Xinliang Yu[1,2]*, Yixiong Yu[3], Qun Zeng[4]

1 College of Chemistry and Chemical Engineering, Hunan Institute of Engineering, Xiangtan, Hunan, China, 2 State Key Laboratory of Chemo/Biosensing and Chemometrics, College of Chemistry and Chemical Engineering, Hunan University, Changsha, Hunan, China, 3 G1302, Lushan International Experimental School, Changsha, Hunan, China, 4 Department of Neurological Surgery, Xiangtan Central Hospital, Xiangtan, Hunan, China

Abstract

Background: Synthesizing and characterizing aptamers with high affinity and specificity have been extensively carried out for analytical and biomedical applications. Few publications can be found that describe structure–activity relationships (SARs) of candidate aptamer sequences.

Methodology: This paper reports pattern recognition with support vector machine (SVM) classification techniques for the identification of streptavidin-binding aptamers as "low" or "high" affinity aptamers. The SVM parameters *C* and γ were optimized using genetic algorithms. Four descriptors, the topological descriptor *PW4* (path/walk 4 - Randic shape index), the connectivity index *X3A* (average connectivity index chi-3), the topological charge index *JGI2* (mean topological charge index of order 2), and the free energy *E* of the secondary structure, were used to describe the structures of candidate aptamer sequences from SELEX selection (Schütze *et al.* (2011) PLoS ONE (12):e29604).

Conclusions: The predicted fractions of winning streptavidin-binding aptamers for ten rounds of SELEX conform to the aptamer evolutionary principles of SELEX-based screening. The feasibility of applying pattern recognition based on SVM and genetic algorithms for streptavidin-binding aptamers has been demonstrated.

Editor: Mark Isalan, Imperial College London, United Kingdom

Funding: This work was supported by the Hunan Provincial Natural Science Foundation of China (Contract No. 12JJ6011). The funder had no role in study design, data collection and analysis, decision to publish, or preparation of the manuscript.

Competing Interests: The authors have declared that no competing interests exist.

* E-mail: yxliang5602@sina.com.cn

Introduction

Aptamers are structured single-stranded oligonucleotides that show an affinity toward a variety of targets, including proteins, viruses, and whole cells [1,2]. Compared to antibodies, aptamers are economical and easy to synthesize and modify, possess long-term stability, and display low immunogenicity, fast blood clearance, rapid tissue and tumor penetration. Furthermore, unlike antibodies developed *in vivo*, aptamers can be developed *in vitro* [3–5]. Aptamers, a promising class of compounds, both for target recognition and therapy, can be derived from a process termed Systematic Evolution of Ligands by EXponential enrichment (SELEX). This is a reiterative process of partitioning of aptamer candidates from non-binding sequences by an affinity method, followed with amplification of the bound sequences by polymerase chain reaction (PCR) [5].

The conventional SELEX method usually takes 10–15 cycles of selection and amplification, which are labor-intensive and time-consuming [6]. Furthermore, the consumption of samples/reagents is relatively high. By applying the relationship between the chemical structure of a molecule and its biological activity or properties, i.e., the structure–activity relationship (SAR) model, pattern recognition can be used to screen a series of candidate molecules, including those not yet synthesized, on the computer in order to select the structures having the desired set of predicted activities/properties [7]. It is then possible to select the most promising candidates for synthesis, laboratory testing, and optimization. Thus, the pattern recognition approach based on such SARs as classification models can conserve resources and accelerates the process of selecting candidates for any purpose.

Streptavidin is widely used as a detection tool in biology research because of the strongest non-covalent interaction known in nature between streptavidin and biotin [8,9]. In addition, biotin-streptavidin system can be explored for detection of infection and tumor in clinical medicine [10]. Support vector machines (SVMs) are a popular technique for classification. The aim of this paper is to develop a pattern recognition model for aptamers against streptavidin, using SVM as the classification technique. A genetic algorithm is employed to find suitable parameters for SVM model that has relatively optimal prediction performance.

Materials and Methods

Data set

Table S1 shows the candidate aptamer sequences, which were taken from Reference [9]. These candidate aptamers were obtained through ten rounds of SELEX selection and the 100 most frequent clones of each round were listed. Each of the clones contains 40 bases with a tolerance of two bases. The candidate aptamers were split into two sets: a training set and a test set. The training set includes the sequences from the 1st and 10th rounds of SELEX selection. The sequence R10#86 in the 10th round recurs in the 1st selection round, i.e., R1#1. In addition, the sequences R1#11 and R1#60 have no pair structure. Therefore, the three sequence, R10#86, R1#11 and R1#60 are removed from the data set of the 1st round selection. Totally, 197 different sequences were obtained to generate the training set. The test set contains the sequences from the rest of SELEX rounds, i.e., the 2nd, 3rd, 4th, 5th, 6th, 7th, 8th, and 9th rounds. Similarly, the sequences R2#55, R2#63 and R2#94 without pair structure are deleted from the test set.

In general, the binding affinity of aptamers with targets increases exponentially with increasing selection rounds. Such exponential increase in binding affinity is obvious during the first few rounds. In the end, the affinity is close to a saturation point after a certain number of rounds are carried out, and subsequently the binding affinity does not increase obviously [11–13]. Thus, the class labels (or target values) of sequences from the 1st round of SELEX were set as 1, denoting the low affinity and specificity aptamer candidates. The class labels of sequences in the 10th round were defined as 2, denoting the high affinity and specificity aptamer candidates. The training set was used to train and optimize algorithm parameters of SVM models. The test set was used later to evaluate predictive performance of the developed model.

Molecular descriptor calculation

The RNAstructure package (version 5.3) was used for prediction of secondary structures of candidate aptamers by minimizing the free energy (*E*) [14]. The loop structures were adopted to calculate molecular descriptors for corresponding candidate aptamers. Since the size of the loop is important for binding, the priority was given to the loops with 5–7 nucleotides, which were selected to calculate descriptors. Besides the free energy descriptor (*E*), three groups of molecular descriptors were calculated with Dragon software [15], which are 119 topological descriptors, 21 topological charge indices, and 33 connectivity indices. Totally, 174 molecular descriptors were calculated for each sequence. To calculate molecular descriptors, the loop of each aptamer candidate was sketched using ChemBioDraw Ultra 11.0 [16], and optimized using molecular mechanics (MM2) in ChemBio3D Ultra 11.0 until the *rms* of gradient value became smaller than 0.1 kcal/mol Å. The energy minimized molecules were then used as the inputs for Dragon software [15].

Topological descriptors based on a graph representation of a molecule can describe one or more such chemically interesting features as size, shape, symmetry, branching and cyclicity. They can also reflect chemical information like atom type and bonding environments [15]. Connectivity indices are calculated from the non- hydrogen part of a molecule where each vertex (non-hydrogen atom) is weighted by the vertex degree, i.e., the number of connected non-hydrogen atoms [17]. Topological charge descriptors are derived from an unsymmetrical matrix CT, whose single element CT_{ij} is equal to the vertex degree δ_i of the *i*th atom under the condition of $i = j$. Otherwise, CT_{ij} equals to the difference of m_{ij} and m_{ji}. Here m_{ij} and m_{ji} are elements of the matrix obtained by multiplying the adjacency matrix by the reciprocal square distance matrix [15,18]. For each path of length *k*, a topological charge index *GGIk* is defined as the half-sum of all charge terms CT_{ij} (absolute values) corresponding to pair of vertices with topological distance equal to *k* [15,18].

Principle of support vector machine classification

SVMs are known as maximum-margin classifiers, since they find the optimal hyperplane that is equidistant from two classes and defined by a number of support vectors. In general, the larger the margin or distance between these parallel hyperplanes, the smaller the generalization error of the classifier will be [19–22].

Let $(\mathbf{x}_i; y_i)$ be a set of training examples, where $i = 1, 2 \ldots, n$, $\mathbf{x}_i \in R^d$ is an input vector, $y_i \in \{-1, 1\}$ is its corresponding desired output, i.e., a constant denoting the class to which that point $\mathbf{x}_i$ belongs, n is the number of training data, and d denotes the number of dimensions of input data.

The SVM requires the solution of the following optimization problem

$$\min_{w,b,\xi} J(w,\xi,b) = \frac{1}{2} w^T w + C \sum_{i}^{l} \xi_i \quad (1)$$

subject to:

$$y_i(w^T \phi(x_i) + b) \geq 1 - \xi_i \quad (2)$$

$$\xi_i \geq 0 \quad (3)$$

Here $\mathbf{w}$ is the weight vector, b is the bias, ξ is a non–negative slack variable for the data points, and C is a penalty factor that controls the tradeoff between the complexity of the decision function and the number of training examples misclassified. SVM maps input vectors $\mathbf{x}_i$ into a higher (may be infinite) dimensional space, where a margin hyperplane with the maximal margin is constructed. Under constraints $\sum_i \alpha_i y_i = 0$ and $0 \leq \alpha_i \leq C$, the optimization problem becomes

$$\min W(\alpha) = -\sum_i \alpha_i + \frac{1}{2} \sum_{i,j} \alpha_i \alpha_j y_i y_j (x_i, x_j) \quad (4)$$

Quadratic programming method can be adopted to solve the above extreme problem. The points $\mathbf{x}_i$ with $\alpha_i > 0$ are called support vectors. Patterns with $0 < \alpha_i < C$ are called unbounded support vectors, while those with $\alpha_i = C$ are called bounded support vectors [22]. SVM can be easily generalized to non–linear decision surfaces by replacing the inner product $(x_i \cdot x_j)$ with a kernel function $K(x_i, x_j)$. The Gaussian radial basis function kernel (RBF) is a popular kernel function used in SVM classification and can be expressed with

$$K(x_i, x_j) = \exp(-\gamma \|x_i - x_j\|^2) \quad (5)$$

Here γ is a kernel parameter. For the SVM classification models based on the RBF kernel, two parameters, C and γ, should be carefully tuned to the problem at hand. If the factor C is too large, a large penalty is assigned to non-separable points, which leads to store too many support vectors and thus over fit. On the other hand, if C is too small, an under fitting can occur. The γ parameter specifies the radius of the RBF, also exerting a strong impact on model performance [23].

In this paper, we used the genetic algorithm to optimize the SVM parameters, C and γ. Genetic algorithm belongs to the family of evolutionary algorithms, which generate solutions to optimization problems using techniques inspired by the evolution of species emphasizing the law of survival of the strongest [24]. It uses random mutation, crossover and selection procedures to breed better models or solutions from an originally random starting population or sample [25]

The parameter values used in our experiments were set as following: the population size of genetic algorithm being 20, evolutionary generation being 200, and both the SVM parameter, C and γ, being selected in the ranges [1, 1000]. The 5-fold cross validation procedure was carried out for the training set during the optimization of SVM parameters, C and γ. LibSVM package [26] was used to develop the SVM models.

Results and Discussion

The selection of appropriate theoretical descriptors is crucial to obtain good classification performance. With the number of selection rounds being the dependant variable and 174 molecular descriptors being the independent variables, the correlation between independent and dependent variables was analyzed with stepwise multiple linear regression (MLR) in SPSS 11.5, to select an optimal subset of variables used as the inputs of the SVM model. Four descriptors, the topological charge index *JGI2* (mean topological charge index of order 2), the topological descriptor *PW4* (path/walk 4 - Randic shape index), the connectivity index *X3A* (average connectivity index chi-3), and the free energy (*E*) of the secondary structure were obtained to describe the structure features of each candidate aptamer, which are listed in Table S2.

Molecular connectivity indices are used widely in various areas of physical, chemistry, biology, pharmacology, polymer, and environmental science. One of the most important reasons about their successful applications is that these indices are based on sound chemical, structural (topologic and geometrical), and mathematical grounds. The descriptor *X3A* (a connectivity index; average connectivity index chi-3) belongs to average connectivity indices *XkA*, which are obtained by dividing each connectivity index by the number of paths involved in its calculation [15]

$$XkA = \sum_{k=1}^{k} \left(\prod_{a=1}^{n} \delta_a \right)_k^{-1/2} \tag{6}$$

Where δ_a represents the corresponding vertex degree, k is the total number of *m*th order subgraphs, and n is the number of vertices in the subgraph. The descriptor *X3A* ($k = 3$ for *XkA*) relates the characteristic dimension of the molecule to the atomic parameters (quantum number, bond indexes, etc.) [27]. *X3A* can denote the molecular size and the electronic distribution of a loop, which is related to the induced fitting behavior and molecular recognition of an aptamer.

Mean topological charge indices *JGIk* (from order 1 to order 10) are obtained by dividing the corresponding topological charge index (*GGIk*) by the total number of summation terms in *GGIk*. The mean topological charge index *JGI2* ($k = 2$ for *JGIk*) is related to the topological valence of the atoms and the net charge transferred from the atom j to the atom i [15].

Path/walk Randic shape indices (*PWk*) are calculated by summing the ratios of the atomic path count over the atomic walk count of the same order k and then dividing by the number of non-H atoms [15,28]. DRAGON calculates path/walk shape indices from order 2 up to 5. The index of first order is not provided as the counts of the paths and walks of length one are equal and, therefore, the corresponding molecular index always equals one for all molecules [15]. Since path/walk count ratio is independent of molecular size, the topological descriptor *PW4* (path/walk 4 - Randic shape index) can be considered as a shape descriptor.

The free energy E of an aptamer secondary structure is approximated as the sum of individual contributions from loops, stacked base pairs, and other secondary structure elements. Aptamer molecules fold by intramolecular base pairs and are stabilized by hydrogen bonds that form between the base pairs along the DNA or RNA molecule. In addition, base pair stacking in a helix also stabilizes the molecule and decreases the free energy of the folded aptamer. Thus the free energy (*E*) of the secondary structure reflects the conformational stability of an aptamer. Generally, an aptamer with a high free energy E (absolute value) does not mean a stronger binding with its targets. Because the binding is an aptamer-target interaction provided by different intermolecular interactions such as electrostatic interactions between charged groups, stacking of aromatic structures contained in organic compounds and the nucleobases, hydrogen bonds, and the complementary in three-dimensional shape [29], while the free energy here is the property of an aptamer only. For streptavidin-binding aptamers, a longer stem of an aptamer secondary structure leads to a larger descriptor E (absolute value). On the other hand, a longer stem can more effectively maintain the loop and bulge structures for binding [8]. Therefore, for a streptavidin DNA aptamer, its free energy E is correlated with its binding ability to streptavidin. Figure 1 shows the correlation of the mean free energy (*E*) of the secondary structure and the number of rounds of SELEX [9]. As can be seen, the descriptor E decreases with the increasing number of performed rounds.

Selecting appropriate values for parameters, C and γ, is also important for SVM performance. The optimization results show that the relatively optimal SVM model possesses parameters C of 705.933 and γ of 749.802. The model based on $C = 705.933$ and $\gamma = 749.802$ has classification accuracy of 97.98% for the training set. To evaluate the model, we calculated prediction values for the test set, which are listed in Table S2. In Reference [9], at identical concentration (1 mM), the binding affinity of eight sequences of R10#1, R10#2, R10#4, R10#6, R10#10, R10#17, R10#62, and R10#86, were studied. Only the sequence R10#62 has low

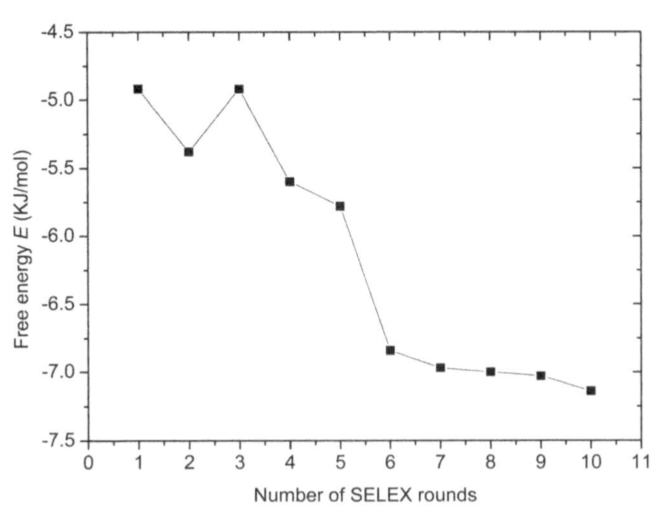

Figure 1. Correlation of the mean free energy and the number of performed rounds.

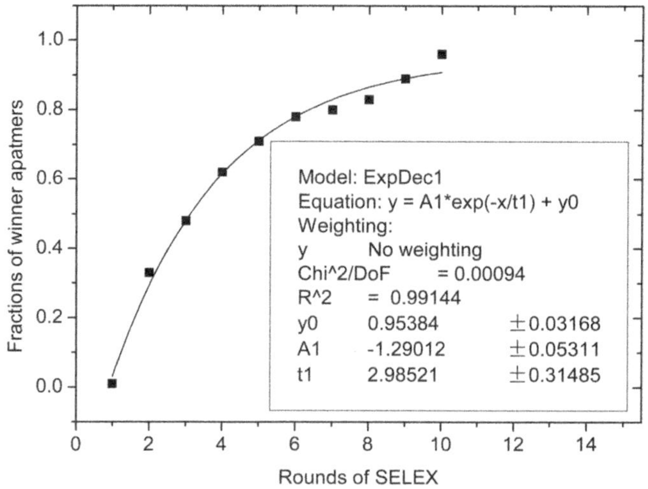

Figure 2. Evolutionary curve of streptavidin-binding aptamers.

prediction class labels being 1, which is acceptable since R10#62 really has a weak binding ability to streptavidin [9].

For *in vitro* DNA aptamer selection procedures based on SELEX technology, starting point of each SELEX process is a synthetic random DNA oligonucleotide library consisting of a multitude of single-stranded DNA fragments (approximately 10^{15}) with different sequences. After the library and the target molecules are incubated for binding, unbound oligonucleotides are removed by several stringent washing steps of the binding complexes. The target-bound oligonucleotides being eluted and subsequently amplified by PCR are then used to generate a new enriched pool for the next selection round. During the SELEX process, the average binding affinity of the selected sequence against specific targets can increase exponentially with the number of selection rounds, just as the word "exponential" suggests in the term of Systematic Evolution of Ligands by EXponential enrichment (SELEX) [11–13]. The fractions of winning aptamers (i.e. their prediction class labels being 2) from the first round to 10th rounds of experiments are 0.01, 0.33, 0.48, 0.62, 0.71, 0.78, 0.80, 0.83, 0.89, and 0.96, respectively. We can obtain the following fitting curve and the corresponding exponential equation in Figure 2, which shows the average binding affinity of the candidate sequences towards streptavidin increases exponentially during the first six rounds of SELEX (the experimental results also show that the binding signal increased strongly until round six [9]). After that, the affinity is close to the saturation point, the fractions of winning aptamers being 08, although the binding affinity increases gently in subsequent rounds. Obviously, the prediction result consists with the aptamer evolutionary principles of SELEX based screening, as stated above [11–13].

Conclusions

The four descriptors, *PW4*, *X3A*, *JGI2*, and *E*, reflecting the structures of candidate aptamer sequences, were used as the input variables of SVM model. Genetic algorithms were chosen to optimize the SVM parameters, C and γ. The relatively optimal SVM model with parameters C of 705.933 and γ of 749.802 has classification accuracy of 97.98% for the training set. Furthermore, the prediction fractions of winning aptamers from the 1st round to 10th round are 0.01, 0.33, 0.48, 0.62, 0.71, 0.78, 0.80, 0.83, 0.89, and 0.96, respectively. The prediction result consists with the aptamer evolutionary principles of SELEX based screening, which shows that pattern recognition for streptavidin-binding aptamers is successful. The investigation may encourage the application of pattern recognition methods to the designs of candidate aptamers.

Author Contributions

Conceived and designed the experiments: XY YY. Performed the experiments: XY YY. Analyzed the data: XY YY. Contributed reagents/materials/analysis tools: XY QZ. Wrote the paper: XY QZ.

References

1. Wilson DS, Szostak JW (1999) In vitro selection of functional nucleic acids. Annu Rev Biochem 68: 611–647. doi: 10.1146/annurev.biochem.68.1.611
2. Bunka DH, Stockley PG (2006) Aptamers come of age – at last. Nat Rev Microbiol 4: 588–596. doi:10.1038/nrmicro1458
3. Hwang DW, Ko HY, Lee JH, Kang H, Ryu SH, et al. (2010) A nucleolin-targeted multimodal nanoparticle imaging probe for tracking cancer cells using an aptamer. J Nucl Med 51: 98–105. doi: 10.2967/jnumed.109.069880
4. Shangguan D, Meng L, Cao ZC, Xiao Z, Fang X, et al. (2008) Identification of liver cancer-specific aptamers using whole live cells. Anal Chem 80: 721–728. doi: 10.1021/ac701962v
5. Tan WH, Donovan MJ, Jiang JH (2013) Aptamers from cell-based selection for bioanalytical applications. Chem Rev 113: 2842–2862. doi: 10.1021/cr300468w
6. Levine HA, Nilsen-Hamilton M (2007) A mathematical analysis of SELEX. Comput Biol Chem 31: 11–35. doi: 10.1016/j.compbiolchem.2006.10.002
7. Karelson M, Lobanov VS, Katritzky AR (1996) Quantum-chemical descriptors in QSAR/QSPR studies. 96: 1027–1043. doi: 10.1021/cr950202r
8. Bing T, Yang XJ, Mei HC, Cao ZH, Shangguan D (2010) Conservative secondary structure motif of streptavidin-binding aptamers generated by different laboratories. Bioorg Med Chem 18: 1798–1805. doi: 10.1016/j.bmc.2010.01.054
9. Schütze T, Wilhelm B, Greiner N, Braun H, Peter F, et al. (2011) Probing the SELEX Process with Next-Generation Sequencing. PLoS ONE 6: e29604. doi: 10.1371/journal.pone.0029604
10. Yomogida K, Chou Y, Pang J, Baravati B, Maniaci BJ, et al. (2012) Streptavidin suppresses T cell activation and inhibits IL-2 production and CD25 expression. Cytokine 58: 431–436. doi: 10.1016/j.cyto.2012.02.007
11. Djordjevic M (2007) SELEX experiments: new prospects, applications and data analysis in inferring regulatory pathways. Biomol Eng 24: 179–189. doi: 10.1016/j.bioeng.2007.03.001
12. Vant-Hull B, Payano-Baez A, Davis RH, Gold L (1998) The mathematics of SELEX against complex targets. J Mol Biol 278: 579–597. doi: 10.1006/jmbi.1998.1727
13. Djordjevic M, Sengupta AM (2006) Quantitative modeling and data analysis of SELEX experiments. Phys Biol 3: 13–28. doi: 10.1088/1478-3975/3/1/002
14. Reuter JS, Mathews DH (2010) RNAstructure: software for RNA secondary structure prediction and analysis. BMC Bioinformatics 11: 129:1–129:9. doi: 10.1186/1471-2105-11-129
15. Talete SRL (2006) DRAGON for Widows (Software for the Calculation of Molecular Descriptors), Version 5.4. Milan, Italy.
16. Cambridge Soft Inc (2008) ChemBioOffice Ultra Version 11.0. Cambridge, USA.
17. Mandloi M, Sikarwar A, Sapre NS, Karmarkar S, Khadikar PV (2000) A comparative QSAR study using Wiener, Szeged, and molecular connectivity indices. J Chem Inf Comput Sci 40: 57–62. doi: 10.1021/ci980139h
18. Gálvez J, Garcia-Domenech R, Salabert-Salvador MT, Soler R (1994) Charge indexes. New topological descriptors. J Chem Inf Comput Sci 34: 520–525. doi: 10.1021/ci00019a008

19. Birzele F, Kramer S (2006) A new representation for protein secondary structure prediction based on frequent patterns. Bioinformatics 22: 2628–2634. doi: 10.1093/bioinformatics/btl453
20. Wang B, Chen JW, Li XH, Wang YN, Chen L, et al. (2009) Estimation of Soil organic carbon normalized sorption coefficient (Koc) using least squares-support vector machine. QSAR Comb Sci 28:561–567. doi: 10.1002/qsar.200860065
21. Li S, Fedorowicz A, Andrew ME (2007) A new descriptor selection scheme for SVM in unbalanced class problem: a case study using skin sensitisation dataset. SAR QSAR Environ Res 18: 423–441. doi: 10.1080/10629360701428474
22. Afantitis A, Melagraki G, Sarimveis H, Koutentis PA, Igglessi-Markopoulou O, et al. (2010) A combined LS-SVM & MLR QSAR workflow for predicting the inhibition of CXCR3 receptor by quinazolinone analogs. Mol Divers 14: 225–235. doi: 10.1007/s11030-009-9163-7
23. Plewczynski D, Tkacz A, Wyrwicz LS, Godzik A, Kloczkowski A, et al. (2006) Support-vector-machine classification of linear functional motifs in proteins. J Mol Model 12: 453–461– doi: 10.1007/s00894-005-0070-2
24. Yu X, Yu R (2013) Setschenow Constant Prediction Based on the IEF-PCM Calculations. Ind Eng Chem Res 52:11182–11188. doi: 10.1021/ie400001u
25. Turabekova MA, Rasulev BF (2004) A QSAR Toxicity Study of a Series of Alkaloids with the Lycoctonine Skeleton. Molecules 9: 1194–1207. doi: 10.3390/91201194
26. Chang CC, Lin CJ (2011) LIBSVM: A library for support vector machines. ACM Trans Intell Syst Technol 2: 27:1–27: 27. doi: 10.1145/1961189.1961199
27. Petrova T, Rasulev BF, Leszczynski J, Toropov AA, Leszczynska D (2011) Improved model for fullerene C 60 solubility in organic solvents based on quantum-chemical and topological descriptors. J Nanopart Res 13: 3235–3247. doi: 10.1007/s11051-011-0238-x
28. Randic M (2001) Novel shape descriptors for molecular graphs. J Chem Inf Comput Sci 41: 607–613. doi: 10.1021/ci0001031
29. Nikolaus N, Strehlitz B (2014) DNA-Aptamers Binding Aminoglycoside Antibiotics. Sensors, 14(2): 3737–3755. doi:10.3390/s140203737

The Roles of Standing Genetic Variation and Evolutionary History in Determining the Evolvability of Anti-Predator Strategies

Daniel R. O'Donnell[1,2*⁹], Abhijna Parigi[1,2⁹], Jordan A. Fish[2,3⁹], Ian Dworkin[1,2,4], Aaron P. Wagner[4]

1 Program in Ecology, Evolutionary Biology, and Behavior, Michigan State University, East Lansing, Michigan, United States of America, 2 Department of Zoology, Michigan State University, East Lansing, Michigan, United States of America, 3 Center for Microbial Ecology, Michigan State University, East Lansing, Michigan, United States of America, 4 BEACON Center for the Study of Evolution in Action, Michigan State University, East Lansing, Michigan, United States of America

Abstract

Standing genetic variation and the historical environment in which that variation arises (evolutionary history) are both potentially significant determinants of a population's capacity for evolutionary response to a changing environment. Using the open-ended digital evolution software Avida, we evaluated the relative importance of these two factors in influencing evolutionary trajectories in the face of sudden environmental change. We examined how historical exposure to predation pressures, different levels of genetic variation, and combinations of the two, affected the evolvability of anti-predator strategies and competitive abilities in the presence or absence of threats from new, invasive predator populations. We show that while standing genetic variation plays some role in determining evolutionary responses, evolutionary history has the greater influence on a population's capacity to evolve anti-predator traits, i.e. traits effective against novel predators. This adaptability likely reflects the relative ease of repurposing existing, relevant genes and traits, and the broader potential value of the generation and maintenance of adaptively flexible traits in evolving populations.

Editor: Nadia Singh, North Carolina State University, United States of America

Funding: Funding was provided by BEACON Center for the Study of Evolution in Action (NSF Cooperative Agreement DBI-0939454), beacon-center.org; and Michigan State University Institute for Cyber Enabled Research, icer.msu.edu. The funders had no role in study design, data collection and analysis, decision to publish, or preparation of the manuscript.

Competing Interests: The authors have declared that no competing interests exist.

* Email: odonn146@msu.edu

⁹ These authors contributed equally to this work.

Introduction

The diversity and complexity of any biological system reflects past evolutionary responses to environmental conditions. Underlying those responses are a number of extrinsic and intrinsic factors influencing populations' capacities to evolve responses through the generation or utilization of genetic variation [1,2]. Because evolutionary responses rely largely on available heritable variation, mechanisms that influence the generation and maintenance of variation [3] can strongly shape the evolutionary trajectories of populations [4]. Among the potential factors involved, standing genetic variation (SGV) and evolutionary history (EH) are likely to be significant determinants of adaptability to novel environments [5]. Accordingly, understanding how these critical factors either promote or constrain population evolutionary potential provides insight into the realized pathways that led to historical evolutionary outcomes, as well as those that will shape future populations.

Standing genetic variation is the presence of alternative forms of a gene (alleles) at a given locus [5] in a population. While an allele may be mildly deleterious or confer no fitness advantage over other forms under one set of environmental conditions [6], that allele may become beneficial if the environment changes. As selection can act only on available variation, SGV provides a potential means for more rapid adaptive evolution ([7–20]; reviewed in [21]) compared with the *de novo* mutations [5,22], particularly if the environment changes (e.g. if a new predator or competitor invades the system, or if abiotic conditions change).

In addition to SGV, a population's historical selection environment (i.e. evolutionary history) may play a strong role in determining the speed and the extent to which populations can adapt to directional environmental change [23–25]. In particular, EH will have influenced the genetic variation and genetic architecture of traits in contemporary populations. If changes in the environment alter the strength of selection on a trait, populations with an evolutionary history of adaptation to similar pressures may be mutationally "closer" to the discovery of new [26], or rediscovery of historical beneficial traits [27]. For example, Gillings and Stokes [25] suggest that historical exposure to antibiotics may confer greater adaptability of bacteria when exposed to novel antibiotics.

With the contemporary rise of experimental evolution as a means of testing evolutionary and ecological hypotheses [28,29], a valuable and untapped opportunity now exists to elucidate the roles SGV and EH have played in determining historical, realized rates of adaptive evolution. Furthermore, while SGV is known to be an important determinant of the speed of evolution [5,11], it is less clear how levels of SGV and EH, alone or in combination, impact the overall evolutionary potential of populations. Such an understanding could allow population geneticists to evaluate how

SGV and EH may contribute to or constrain the future evolvability of populations, particularly in human-modified environments [1,30,31].

To evaluate the individual and interactive effects of SGV and EH on evolutionary outcomes, we measured their independent as well as their combined contributions toward evolutionary potential in populations using the digital evolution software Avida [32]. Digital organisms in Avida inhabit a controlled, two-dimensional environment, and experience evolutionary adaptation. The complexity of evolutionary dynamics possible in Avida can be compared to that of bacteria or viruses [33]. In Avida, as in biological systems, three necessary and sufficient conditions for producing evolution via natural selection are met: replication, heritable variation, and differential fitness resulting from that variation [34].

Digital evolution experiments carry several significant advantages for addressing evolutionary questions requiring systematic manipulation and highly controlled environments. Among these, generation times are rapid, the experimenter has full control over initial environmental conditions, and detailed genetic, demographic, and behavioral trait data can be recorded perfectly. Organisms in Avida can also engage in ecological competition and other complex interactions, and the system can allow for co-evolution in predator-prey systems [35–37]. Furthermore, unlike in evolutionary models and evolutionary simulations that impose artificial selection via explicit selection functions, Avida uniquely allows for unrestricted, unsupervised, and non-deterministic evolution via natural selection [32]. With biological organisms, it is difficult to know the exact number of unique genotypes present at a given time, especially when population sizes are large. Hence, manipulating levels of SGV in populations of biological organisms is difficult (but see [19]). Because our question required precise knowledge of the amount of genetic diversity present within each treatment, experimental evolution with biological organisms would have been impractical, given the scale of our study.

Predation is an ecologically important agent of selection [38,39] as demonstrated by the diverse array of prey defenses that have evolved in nature [40–43]. Accordingly, we used Avida to test whether historical exposure to predation influenced how prey populations responded to pressures from new, invasive predators. We then further examined which factors (SGV, EH, and their interaction) were important in determining the future defensive and competitive abilities of those populations. While both factors played significant roles, we show that evolutionary history has the stronger influence on evolutionary potential, with the strength of SGV's role being contingent on a population's evolutionary history. Irrespective of the level of SGV, populations with *predator* EH almost always outcompeted populations without predator EH. In *predator* EH populations, SGV appeared to have no effect on evolvability, whereas, in *no predator* EH populations, higher SGV increased evolvability.

Materials and Methods

Avida

The Avida digital evolution platform is a tool for conducting evolutionary experiments on populations of self-replicating computer programs, termed "digital organisms" [32]. Digital organisms are composed of a set of instructions constituting their "genome". Organisms execute genome instructions in order to perform actions such as processing information and interacting with their environment, and for reproduction. Additionally, predefined combinations of instructions allow organisms to consume resources from the simulated environment. Sufficient consumption of resources, to a level defined by the experimenter, is a prerequisite for organisms to copy their genomes and divide (i.e. reproduce). During the copy process, there is an experimenter-defined probability that genomic instructions will be replaced with a different instruction (substitution mutation) randomly selected from the full set of all 60 possible instructions (instruction set files are in the skel directory in the GitHub repository [https://github.com/fishjord/avida_predation_scripts]). Separately, there are also set probabilities for the copy of an instruction to fail (creating a deletion mutation) or for a chance insertion of a new instruction into the copy's genome. In real-time, generation times in Avida are typically only a few seconds, and experimental environments are fully definable.

Environment. We ran all Avida trials in a 251×251 bounded grid-cell environment. Each cell started with one unit of resource. If a prey organism fed in a given cell (by executing an 'output' instruction), the resource in that cell was completely consumed and that cell was replenished with resources after 100 updates (an update is the measurement of time in Avida, reflecting the number of genomic instructions executed across the population).

Timing. All Avida trials were run for a set number of updates, as indicated in each of the Phase descriptions below. Here, each organism executed 30 instructions per update. Organisms had a maximum lifespan of 15,000 instructions executed (500 updates). Realized average number of generations for the populations in each treatment is given in their respective sections below.

Reproduction. Organisms, both predator and prey, were required to have consumed 10 resource units (in one "bite", individuals could consume a maximum of 1 resource unit) and be at least 100 updates old to successfully reproduce. Genome copying and reproduction occurred when the organism executed a single reproduction ('repro') instruction. In most cases, biological organisms must consume copious resources (and thus perform a number of other feeding-related behaviors, e.g. moving, searching, handling) in order to execute one reproductive event; thus, execution of the reproduction instruction took one update, compared to 1/30th of an update for other instructions (except for predation handling time, as noted below). When an organism reproduced, its offspring was placed in the cell the parent was facing. This new organism's genome was a copy of the parent's genome, with the exception any substitutions, insertions, and deletions that occurred during the copy process. Substitutions were independent and only one insertion or deletion was allowed per reproduction (at rates specified in the 'Phase 1' and 'Phase 2' descriptions, below). In addition to the birth of the new organism upon reproduction, the parent organism was effectively reborn, with all internal states (e.g. stored values such as those describing objects seen, genomic marker positions, and execution pointer position) reset. Unlike in legacy Avida experiments, multiple organisms could occupy the same cell (as in [35–37]). Thus, newborn organisms did not replace existing occupants of the cells into which they were born.

Predation. In Avida, predation occurs via the evolution of an 'attack-prey' instruction, the execution of which allows an organism to attack and kill a non-predator in the cell it is facing. Predation in Avida and the base predator Avida configuration used are described in [32]. Here, as in handling time, the attack-prey instruction costs the predator 10 cycles (one-third of an update) if the attack is successful. Predators receive 10% of each prey individual's consumed resources after a successful kill, which is then applied toward the consumption threshold required for reproduction. An organism is classified as a predator in Avida if it attacks a prey organism, not simply by evolving the 'attack-prey' instruction. Predators can only consume prey organisms, and cannot consume resources.

In simple (i.e. lacking topographic features like refuges) and confined environments like the one used here, Avida trials containing predators typically require a minimum prey population level below which predator attacks fail. In top-down limited populations like those used here, minimum prey levels prevent population extinctions and serve to standardize prey population sizes and thus intra-specific competitive pressures. In practice, because prey are constantly being born into populations, prey kills are prevented for very short periods of time and failures simply require predators to make multiple, rapid attacks. Here, attacks on prey were always fatal if there were more than 1,000 prey in the environment (i.e., 1,000 was a minimum prey level below which fatal attacks were prevented). Similarly, in trials without predators, prey population sizes were controlled via a preset maximum population cap (1,000 organisms), as described for each Phase below. When a birth caused that limit to be exceeded, a random prey organism was removed from the population.

Phase One: Evolutionary History

Our study was divided into three experimental phases (Figure 1). In the first phase ("Phase 1"), we evolved two sets of 30 base populations. Each of these populations was initiated via the placement of 9 identical prey organisms into a new grid environment (Figure 1, Phase 1). These ancestors had simple genomes that allowed them only to blindly move around the environment, repeatedly attempting to feed and reproduce. In one set of populations, the attack-prey instruction was allowed to appear in the organisms' genomes through random mutations, thus allowing for the potential evolution of predators (Figure 1, Phase 1). Henceforth we refer to these populations as "*predator* EH" treatments. In the second set of populations, the "attack" instruction was prevented from mutating into organisms' genomes. This second set of "*no predator* EH" base populations thus evolved in the absence of predators in Phase 1. Each population evolved for two million updates (on average, 20,000 generations). During this phase, reproduction carried a 25% probability of a single, random, genomic instruction substitution (maximum one substitution per genome, conditions identified in [44] as being optimal to facilitate evolution of complexity in Avida), and 5% probabilities (each) of single insertion and deletion mutations.

Phase Two: Standing Genetic Variation

The second phase (Figure 1, Phase 2) of the experiments allowed us to evaluate how the amount of SGV as well as EH influenced the evolutionary potential of prey in the presence of novel predators. For all Phase 2 trials, prey populations (compositions described below) were evolved for 200,000 updates in the presence of identical copies of a single static predator population (i.e. predators in Phase 2 were prevented from evolving further, dying, or reproducing). We used static predators in Phase 2 trials in order to prevent predator-prey coevolution dynamics from causing predation pressure to vary among replicates. Taking in sensory information is a prerequisite for behaviorally responding to and targeting prey, and we used visual input rates as a proxy for predator effectiveness. If organisms evolve to use visual sensors, these sensors provide them with information about their environment, allowing them to "look" at the objects (food, prey or other organisms) in the 45-degree area in front of them. Because predator populations varied across Phase 1 replicates, we selected the Phase 2 predator population from the Phase 1 replicate with highest realized usage of visual sensors (i.e. highest per-predator 'look-ahead-intercept' instruction executions) to ensure that an effective predator population was used in Phase 2. Ultimately, the predator population selected and copied for use in all Phase 2 trials had 209 predator organisms.

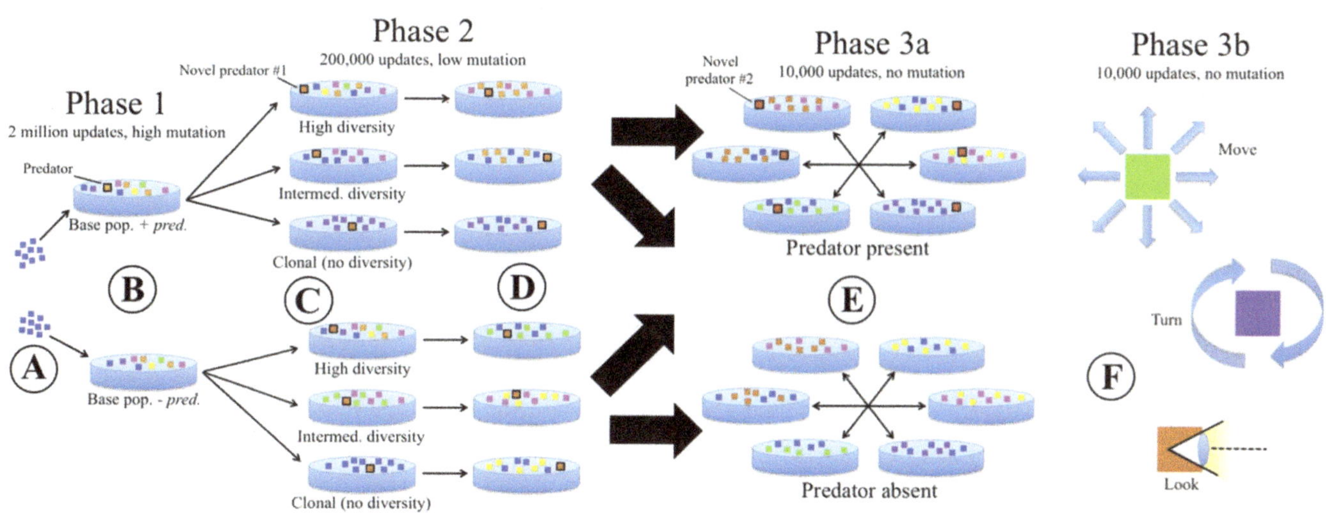

Figure 1. Design of the experiment to test the effects of SGV and EH on evolution of anti-predator traits. A) 9 identical prey are injected to initialize two sets of 30 base populations. B) Over the course of 2 million updates of evolution, populations diversified, including the *de novo* evolution of predators in half of these EH populations (top). C) One fully evolved prey population from each EH treatment (with and without evolving predators) was used to create 30 identical sets of *high*, *intermediate*, and no diversity (*clone*) SGV populations (totaling $2 \times 30 \times 3 = 180$ populations), with clones of the best evolved predator from Phase 1 added to each. D) After a further 2 million updates of evolution at a low mutation rate, replicate populations converged on an intermediate level of diversity, but SGV and EH differences result in trait variation among populations. E) To evaluate the effectiveness of evolved anti-predator traits, each of the 180 fully evolved populations from D were introduced, in turn, into separate competition assays with one randomly selected replicate from each of the other EH×SGV combination populations, both in the presence and in the absence of a new novel predator. F) Separately, prey rates of executing "anti-predator" actions (moves, turns, looks) were recorded in the presence and absence of predators, to quantify expressed behavioral responses to predators for all fully evolved populations.

To create Phase 2 Evolutionary History (EH) treatment source populations, we first selected a single prey population from a random Phase 1 replicate in each EH treatment (*predator* and *no predator* EH), excluding the population from which the Phase 2 predator population was drawn. For each of the two selected populations, we excluded any organisms that were classified as predators or had parents that were predators (as determined by an internal state, see [35,36]), to ensure that Phase 2 prey populations consisted only of prey organisms. An organism must execute an attack instruction to be designated a predator, and some prey organisms may have attack instructions in their genomes, even if they have not yet used them; we therefore replaced any attack instructions in the remaining prey populations with a 'do nothing' (nop-X) instruction, and prevented any new copies of the attack instruction from mutating into offspring genomes.

To create Phase 2 SGV treatment populations, we created separate "*clone*", "*intermediate*", and "*high*" SGV populations by sampling each of the two EH treatment source populations ("seed populations"; one *predator* EH population and one *no predator* EH population: Figure 1, Phase 2). First, for *high* SGV populations, we simply used duplicates of the selected EH treatment source populations. Because all possible genomes are present in this treatment, evolutionary trajectories of replicates within this treatment were not constrained by the lack of genetic variation. Second, we created each *clone* SGV population by randomly selecting a single genotype from a seed population and making as many duplicates of that genotype as there were organisms in the seed population. A genotype could be selected for multiple Phase 2 clone replicates. Given that most evolution experiments with microorganisms begin with a single clone, we wanted to compare the evolutionary trajectory of a clonal population to the high SGV populations. Third, we created the *intermediate* SGV populations by randomly sampling genotypes (with replacement) from a seed population, creating up to 55 identical copies of each sampled genotype. We repeated this process until the new *intermediate* SGV population was the same size as the seed population; this sampling scheme yielded *intermediate* SGV prey populations the same size as their *high* SGV seed populations, but with ~50% lower diversity (Shannon's diversity index), on average. Evolution experiments with diploid, sexually reproducing organism like *Drosophila melanogaster* start with large, wild caught natural populations. However, sampling from a natural population cannot capture all possible genotypes. Intermediate SGV populations are analogous to such "sampled" populations that have some, but not all the available SGV. In all, there were six (2 EH×3 SGV) Phase 2 treatments, each with 30 replicates (Figure 1).

For each of the six combinations of SGV (*clone, intermediate, high*) and EH (*predator, no predator* EH), we evolved 30 replicate populations for 200,000 updates (500 generations, on average), each in the presence of a copy of the constructed Phase 2 predator population. The mutation rates in Phase 2 were lowered to 0.1% substitution probability, and 0.5% insertion and deletion probability per generation. We found empirically that these mutation rates yielded, on average, 5% divergence (see below) between ancestor and final organisms in Avida after 500 generations (200,000 updates using the Phase 2 settings), matching the expected divergence of a bacterial genome over 500 generations [45]. Other configuration settings were identical to those used in Phase 1.

At the end of Phase 2, we calculated how different the resulting population was from the starting population by using a dynamic programming algorithm [46]. We measured the genetic divergence of each prey organism in each population by aligning the current genome sequence with that of its ancestor from the beginning of Phase 2 (alignment was necessary to account for changes in genome length due to insertion/deletion). Each organism from the seed population was tagged with a unique Lineage ID, which is shared with all progeny of the organism. The Lineage ID of the organisms present in the population at the end of Phase 2 could then be used to identify their ancestor from the beginning of Phase 2. We then calculated the divergence as the percent identity between the two aligned genome sequences.

Phase 3: Competitive Evaluation of prey populations and assessment of trait evolution

For the third and final phase (Figure 1, Phase 3a), we used a set of both "ecological" evaluation simulations to measure the fitness of the final prey populations in the presence and absence of predators (referred to as "PT" for "predator treatment" from now on), and trait assays to evaluate the evolution of anti-predator traits during Phase 2. For all Phase 3 evaluations, substitution, insertion and deletion mutation rates were set to zero. For all types of Phase 3 trials (described below), in order to create an uneven resource landscape (i.e. in the absence of consumption by prey, all cells would contain resources that could be consumed), populations were introduced into their test environments, run for an initial 1,000 updates, and then reintroduced a second time. Once prey were reintroduced, we ran each trial for an additional 10,000 updates, and recorded population sizes for the two competing populations.

As in Phase 2, Phase 3 trials used copies of a single Phase 1 predator population. The Phase 3 predator was intended to exhibit novel strategies relative to the Phase 2 predator population. We selected the Phase 3 predator population by first eliminating the three Phase 1 replicates used for creating Phase 2 predator and prey populations, and then selecting one predator population that exhibited an average level of visual sensor usage (about half that of the predator population selected for Phase 2). The final selected predator population for Phase 3 contained 253 predators.

Phase 3 involved two sets of trials: the first set allowed for evaluating competitiveness, while the second set allowed for evaluating the effects of SGV and EH on expressed levels of evolved anti-predator behavioral traits. In the first set of Phase 3 trials (Figure 1, Phase 3a), we conducted pairwise competitions in which each population from each of the six Phase 2 treatments was paired with a randomly selected population from each of the five other treatments (for a total of 30 pairings ×30 replicates = 900 competitions). For each pairing, the two populations were injected once into an environment with the Phase 3 predator population ("*predators present* PT") and once into an environment without predators ("*predators absent* PT"). For each of these trials, we enforced a prey population level of 2,000 and recorded the relative abundance of each of the two competing populations every 1,000 updates over the course of the 10,000-update trial.

The second set of Phase 3 trials was to evaluate the degree to which traits likely involved in defense evolved in prey populations during Phase 2 (Figure 1, Phase 3b), thus elucidating proximate causes of the outcomes of ecological competition in Phase 3a. Each of the starting ("pre-Phase 2") and final ("post-Phase 2") Phase 2 prey populations was exposed to a copy of the new Phase 3 predator population and reintroduced into a fresh environment. We then recorded the number of moves and turns executed and the usage of visual sensors ("look" instructions), as proxies for anti-predator responses. Each prey population was then introduced a second time in a separate evaluation in a *predators absent* PT environment. For each of these trials, we kept the number of prey at a constant 1,000 individuals via enforcement of population caps (i.e. a random prey was killed if a new birth would bring the prey population above 1,000) and minimum thresholds (below which predator attacks would fail).

Statistical analysis

Due to the anticorrelation of the relative abundances of lineages in competition, as well as the extreme heterogeneity of variances among competition scenarios (Figure 2), we used a permutational multiple analysis of variance (PerMANOVA) (Table S3) with SGV, EH, and PT as predictors and the relative abundances of lineages in competition as response variables (Table S3) to analyze the effects of SGV, EH, and PT outcomes of ecological competition (Phase 3; Figure 2). The SGV×EH×PT interaction was far from significant ($F = 0.60$; $df = 2$; $P = 0.56$), so we dropped this term from the model and used a reduced model containing only main effects and all first-order interactions (Table S3).

To generate response variables for models describing change in traits due to Phase 2 evolution, we divided instruction counts by total instructions performed (e.g., number of moves/total lifetime instructions) to derive the proportion of instructions represented by each instruction type. Trait assays were conducted pre- and post-Phase 2. Thus, to determine the effect of evolution specifically during Phase 2, we subtracted pre-Phase 2 instruction proportions from post-Phase 2 proportions, yielding the raw change in instruction proportions.

We used linear mixed effects models (Equation 1; Appendix S1) fit by maximum likelihood to determine the effects of SGV, EH, and Phase 3 predator treatment (PT) on change in total instructions and in instruction proportions due to Phase 2 evolution. SGV, EH, and PT were included as fixed effects. Since both predator treatment (PT) levels (*predators present, predators absent*)

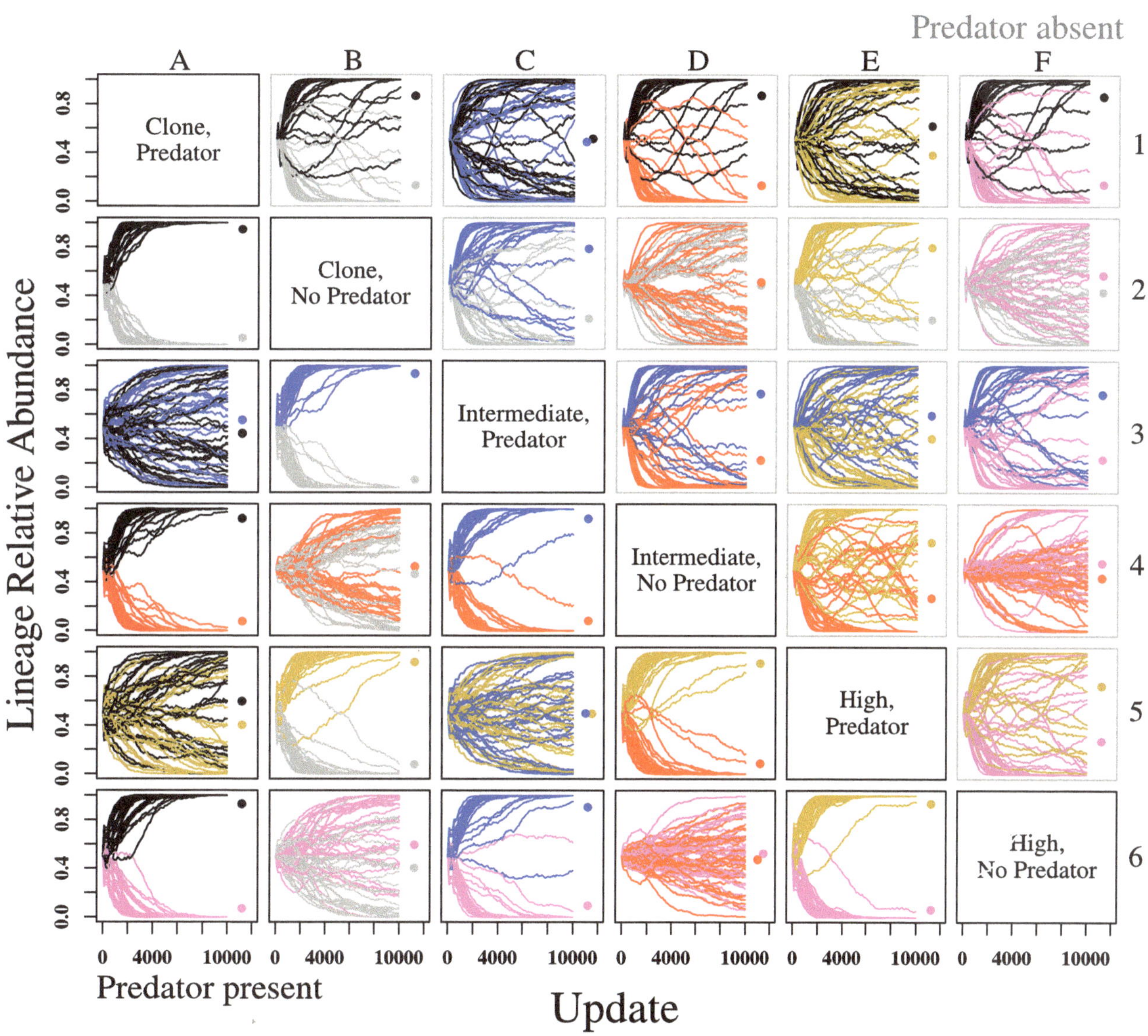

Figure 2. Ecological competition. Populations from all SGV and EH treatment combinations are randomly paired in competition in the absence (gray boxes, above diagonal) and presence (black boxes, below diagonal) of a novel predator (Phase 3). EH had a far greater effect on competitive outcomes than SGV in both the presence and absence of predators, with *predator* EH populations competitively dominant in most cases. Y-axis: relative abundance of each lineage in competition. Colors for each lineage indicate SGV and EH treatment history. Black: *clone, predator*; Gray: *clone, no predator*; Blue: *intermediate, predator*; Red: *intermediate, no predator*; Gold: *high, predator*; Pink: *high, no predator*. Dots indicate mean relative abundance for lineages of each treatment after 10,000 updates of competition.

were applied to each population, we fitted replicate as a random intercept, and included a random slope across PT levels to account for non-independence of PT levels. Where appropriate, we also allowed for unequal variances among factor levels using the "varIdent()" function from the "nlme" [47] package (Table S4). Equation 1 describes the change in total prey instructions due to Phase 2 evolution. Models describing change in prey moves, turns, and looks varied in complexity, but were of the same general form (Appendix S1). For each trait, we used a likelihood ratio test for model selection, starting with the full model (Equation 1) and dropping interaction terms that did not improve the model fit (Table S4).

EQUATION 1

$$y \sim \beta_{0ij} + \beta_{1ij} + \beta_1 PT + \beta_2 EH + \beta_3 SGV + \beta_4 (PT \times EH) +$$
$$\beta_5 (PT \times SGV) + \beta_6 (EH \times SGV) + \beta_7 (PT \times EH \times SGV) +$$
$$\epsilon_{ijk}, \beta_{0ij} \sim \left(\beta_{0j}, \sigma^2_{\beta_0}\right), \beta_{1ij} \sim \left(\beta_{1j}, \sigma^2_{\beta_1}\right), \sigma^2_{\beta_2} \sim EH$$

where β_{0ij} are the random intercepts, β_{1ij} are the randomly varying slopes across PT levels, and $\sigma^2_{\beta 2}$ ~EH is the variance function allowing variances to differ between EH levels. We used a method recently developed by [48] to calculate marginal R^2 (proportion of variation accounted for by fixed effects alone) and conditional R^2 (proportion of variation accounted for by fixed and random effects together). R^2_m and R^2_c for each model are given in the Table S5.

To test the effects of SGV and EH on predator attack rates during trait assays, we used a general linear model with SGV and EH included as fixed effects, and the raw difference in attack rates between pre- and post-Phase 2 evolution as the response variable. A single replicate population was excluded from analyses of change in total instructions, as this replicate had an unusually high pre-Phase 2 instruction count, was highly unrepresentative of prey populations in general, and greatly increased the variance of the *clone* SGV + *no predator* EH treatment group (Figure S6).

Software and Hardware

We used Avida version 2.14 and the Heads-EX hardware [44] for all experiments. Avida did not originally support immortal, sterile predators as used in Phase 2 and Phase 3, so we added an additional option to Avida that, when set, causes predators to 'reset' (akin to being reborn) instead of reproducing or dying of old age. These modifications are available as a patch file against Avida 2.14, along with all analysis scripts, in the GitHub repository (https://github.com/fishjord/avida_predation_scripts).

We conducted all experiments using the Michigan State University's High Performance Computing Cluster (HPCC). The MSU HPCC contains three general purpose computing clusters containing 2944 cores. Individual Avida runs were submitted as compute jobs to the HPCC general processing queue in parallel where possible. The Avida jobs took between 30 minutes and seven days to run, depending on the number of updates (2,000 to 2,000,000 updates).

We performed all statistical analyses and constructed all figures using the R statistical programming language version 3.0.2. The PerMANOVA was conducted using the "adonis()" function in the "vegan" package (last updated by [49]). Linear mixed effects models were constructed using the "lme" function from the "nlme" package (last updated by [47]), and we used the "allEffects" function from the "effects" package (last updated by [50]) to extract marginalized fixed effects. To calculate marginal and conditional R^2, we used the "r.squaredGLMM" function from the "MuMIn" package (last updated by [51]).

Results

Ecological competitions

Evolutionary history, i.e. historical exposure to predation in Phase 1, was the most important selective agent shaping competitive abilities of prey both in the presence and in the absence of a novel predator. Prey that evolved with *predator* EH were, in general, stronger competitors than those that evolved with

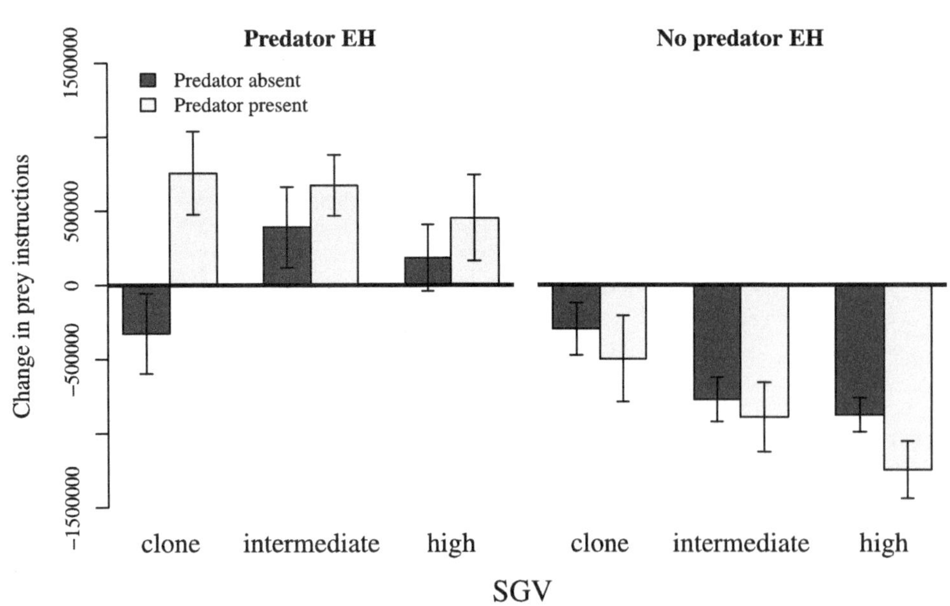

Figure 3. Change in total prey instructions executed (start of Phase 2 minus end of Phase 2), reflecting change in gestation time (or lifespan) resulting from Phase 2 evolution. *"Predator present/absent"* refers to the presence or absence of a novel predator during trait assays before and after Phase 2 evolution. Total instructions either increased or decreased very little in *predator* EH populations, and universally decreased in *no predator* EH populations. Bars are ±95% CI.

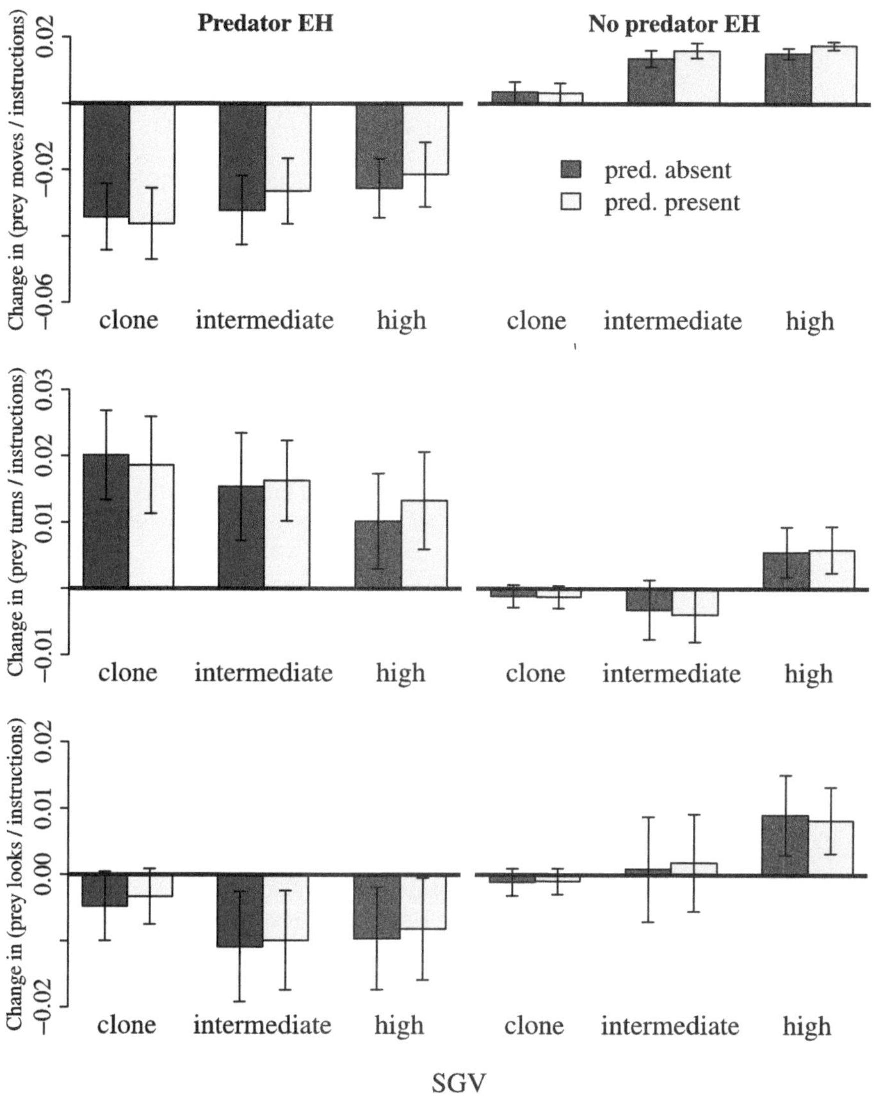

Figure 4. Change in (top) prey moves, (middle) prey turns, and (bottom) prey looks as proportions of total prey instructions, resulting from Phase 2 evolution with a novel predator. *"Predator present/absent"* refers to the presence/absence of the Phase 3 novel predator during the trait assays. Traits in *predator* EH populations generally evolved to a greater extent than did traits in *no predator* EH populations, and traits often evolved in different directions between the two EH treatments. Bars are ±95% CI.

no predator EH. In competitions between *predator* EH and *no predator* EH treatments, *predator* EH treatments always had the higher mean final relative abundance (Figure 2; Table S3). Competitive exclusion of a *predator* EH lineage by a *no predator* EH lineage after 10,000 updates in competition was extremely rare (but see e.g. Figure 2, panel D3, in which 1 out of 30 *predator* lineages was excluded). Competitive outcomes were not affected by the presence of a novel predator, regardless of SGV or EH (Figure 2; Table S3). While there was no main effect of SGV on mean final relative abundances (SGV effect: Table S3), higher SGV seems to have conferred some benefit to *no predator* EH populations (Figure 2; SGV×EH effect: Table S3). Similarly, there appears to be a subtle interaction between PT and EH, wherein in the presence of a predator, the difference in relative abundances between predator EH and no predator EH populations during competition assays increases in favor of predator EH populations.

Evolution of prey traits during Phase 2

Lifetime instruction counts for prey do not reflect higher rates of instruction execution, but rather longer lifespans of prey organisms. Changes in instruction counts varied greatly among treatments (PT×EH×SGV interaction; Figure S2). During Phase 2 evolution, regardless of presence or absence of a novel predator during trait assays, *predator* EH populations generally evolved to increase total instructions executed per lifetime, except in the case of *clone* populations in the absence of predators (Figure 3). Conversely, *no predator* EH populations universally decreased in lifetime instruction counts. In *predator* EH populations, SGV had a markedly positive effect on the magnitude of the decrease in instruction counts, while SGV had little or no effect on instruction counts in *predator* EH populations. Prey generally changed more in total instructions executed with *predators present* than with *predators absent* (Figure 3).

Prey moves as proportions of total prey instructions universally decreased in *predator* EH populations as a result of Phase 2

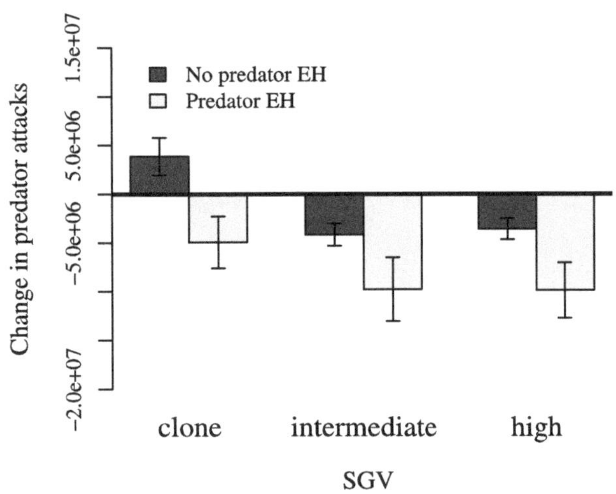

Figure 5. Change in attacks in behavioral assays pairing the Phase 3 predator with pre- and post-Phase 2 prey populations. Predator attack frequencies on *predator* EH populations universally decreased, and generally changed more than did attack frequencies on *no predator* EH populations. Bars are ±95% CI.

evolution, and largely increased in *no predator* EH populations (EH effect, Figure 4; Figure S3). Effects of SGV on change in moves were subtle, though slightly less so with *predators absent.* There was no main effect of PT on change in moves (Figure 4; Figure S3).

Prey turns as proportions of total prey instructions increased dramatically in *predator* EH populations, but changed very little in *no predator* EH populations (Figure 4; Figure S4). Effects of SGV were subtle, but were most pronounced in *no predator* EH populations: *clone* and *intermediate* SGV populations decreased slightly on average in proportion moves, while *high* SGV populations increased (EH×SGV interaction, Figure S4). PT did not affect change in turns (no PT main effect, Figure 4; Figure S4), nor was there a main effect of SGV (Figure S4).

Proportions of instructions that were looks generally decreased in *predator* EH populations, but changed little in *no predator* EH populations; EH effects varied among levels of SGV, with the smallest EH effects in *clone* SGV populations (EH×SGV interaction, Figure 4; Figure S5). Main effects of SGV and PT were not statistically significant (Figure S5).

Effects of Phase 2 prey evolution on predator attack rates

Phase 2 evolution led to a reduction in predator attacks in nearly all cases, and EH and SGV both affected predator attacks on prey (Figure 5; Table S2). *High* and *intermediate* SGV populations experienced qualitatively similar reductions in predator attacks as a result of Phase 2 evolution, and prey evolved with *predator* EH experienced similar reductions in predator attacks across SGV treatments (Figure 5). Only *clone* SGV + *no predator* EH populations were attacked more after Phase 2 evolution.

Discussion

Though the role of SGV in aiding rapid adaptive evolution has been established [5,7–12,22], few studies have empirically tested the effect of varying levels of SGV on evolvability. Additionally, how a population's EH affects evolvability, both independently and in combination with SGV is still largely unexplored [26]. For example, Lohbeck et al. 2012 demonstrated that clonal and genetically diverse populations of the marine coccolithophore *Emiliana huxleyi* adapted similarly to increased levels of CO_2. In genetically diverse populations clone sorting occurred, such that only a subset of the clones persisted in each CO_2 treatment. However, all *E. huxleyi* clones were isolated from the same location, so the effect of EH on adaptation to elevated CO_2 was not tested in this case.

Here, we were able to separately control for and vary the background levels of SGV and the EH, and subsequently test the roles of each in shaping the evolvability of populations in a predator-prey context. Avida is meant to be analogous to biological systems, and the hypotheses testable in Avida could, in theory, apply to biological systems as well. However, the greatest challenge in conducting evolution experiments is the time required to measure significant evolution. Digital organisms reproduce much more quickly than even bacteria, making large-scale experiments covering millions of generations practical. Evolution experiments on the scale of our study are only practical in digital systems, as the labor and time required to conduct such experiments in biological systems would be prohibitive.

Role of evolutionary history

Evolutionary history (the combination of selective pressures under which populations evolve) likely determines which mutations become fixed in populations [23]. We found that having an evolutionary history with a predator improves a population's ability to evolve defenses in response to novel pressures from novel invading predator populations: in general, *predator* EH populations experienced greater changes in trait expression compared to *no predator* EH populations (Figure 4) when evolving defenses against invading (Phase 2) predators. This occurs regardless of the level of SGV, suggesting that the importance of EH eclipses that of evolvability originating from SGV. Furthermore, in pairwise competitions (Figure 2), populations with *predator* EH demonstrated competitive superiority over *no predator* EH populations, as in [35]. If a novel predator was present during the competition, the advantage was even more pronounced (also see Table S3), with the vast majority of *predator* EH populations competitively excluding their *no predator* EH competitors (Figure 2). Additionally, when introduced into separate ecological trials with the novel predator, populations that evolved with *predator* EH experienced a greater decline in predator attacks over Phase 2 than populations evolved with *no predator* EH (Figure 5).

Because populations in both EH treatments were further evolved in the presence of predators (in Phase 2) before they were competed against each other, the ecological advantage enjoyed by *predator* EH populations in competition with *no predator* EH populations was due, in part, to their historical exposure to predators (i.e. in Phase 1), even though that initial evolution was in response to an entirely different predator population than that used in Phase 2. It is clear that evolution in the presence of predators yields not only improved antipredator traits, but also improved ecological efficiency. It is likely that traits such as enhanced moving, turning, and looking improve both predator resistance and foraging success [36].

Large differences in behavioral traits or competitive ability in the presence v. absence of predators (Phase 3) would indicate plasticity. However, while some plasticity is apparent, differences in trait values and competitive outcomes between PT levels are extremely subtle. The observed changes across Phase 2 were thus primarily in constitutively expressed traits, not plastic ones.

Both the mutation rates used for the evolution of Phase 1 predator EH populations and the duration of the evolutionary trials will have enhanced the effect of evolutionary history in each trial (i.e. the degree to which prey populations were adapted to predators). Because we did not vary Phase 1 mutation rates or duration, we cannot comment on how experimentally altering the strength of historical selection could impact the observed trends in trait evolution and competitive outcomes.

Role of standing genetic variation

Many studies have pointed to SGV as a key component of a population's evolutionary potential [6]. When there is greater genetic variation in a population, there is more raw material upon which natural selection can act. Thus, we expected SGV alone to play a significant role in the rapid evolution of anti-predator traits. However, we found marginal effects of historical SGV in determining the final outcomes of ecological competitions (Figure 2). Furthermore, the effects of SGV on the evolution of specific anti-predator traits (moves, turns, looks) were detectable only within EH treatments (Figure 4).

Standing genetic variation (SGV) vs Evolutionary history (EH)

Across the *predator* EH populations, we found no detectable effects of SGV on anti-predator trait values (Figure 4) or competitive outcomes (Figure 2). We suggest that a lack of an SGV effect on *predator* EH populations may be due, in part, to the historical predator weeding out unfit prey genotypes. For example, predators reduced prey diversity (Figure S1; Table S1) because they operate as a very strong agent of selection against less fit phenotypes. However, we would expect SGV to have an effect within *no predator* EH populations (as seen in Figure 4). Given the lack of an evolutionary history with predators, in *no predator* EH populations, the odds of one sampled genotype (*clone* SGV) being able to quickly acquire a beneficial mutation are low. On the other hand, if the full suite of all discovered genotypes (*high* SGV) remains in the population, there is a greater potential for providing precursory material for rapidly generating anti-predator traits. However, if EH leads to a large number of genotypes having the prerequisites for rapid adaptive response, the odds of sampling a beneficial genotype are high, and differences in SGV will not significantly affect evolvability.

As we have shown, in order to adaptively address novel predation pressures, it appears to be easier to modify historically realized and relevant adaptations than to repurpose unrelated genes, or evolve effective traits *de novo*. Thus, the greater evolvability of the *predator* EH populations would have arisen out of their ability to adapt anti-predator strategies across predation contexts. More broadly, evolvability may commonly be determined by a population's ability to use strategies across environmental contexts, i.e. their adaptive complexity [52].

Supporting Information

Figure S1 Shannon's diversity across 200,000 updates of Phase 2 evolution in (A) *high*, (B) *intermediate*, and (C) *clone* SGV treatments. Filled circles represent *no predator* EH treatments, open circles represent *predator* EH treatments. Bars are ±95% CI.

Figure S2 Change in instructions. Shown are the marginalized fixed effects of a generalized linear mixed effects model of scaled change in total prey instructions across Phase 2 evolution as a function of SGV (A. clone; B. intermediate; C. high), EH, and PT. Bars are ±95% CI. EH had the greatest effect on change in prey instructions, though PT also affected change in instructions, especially in *clone* SGV populations. Formal model, random effects output, and marginal and conditional R^2 can be found in Table S5.

Figure S3 Change in prey moves. Shown are the marginalized fixed effects of a generalized linear mixed effects model of scaled change across Phase 2 evolution in the proportion of prey instructions that were moves as a function of **A.** SGV and PT and **B.** EH. Bars are ±95% CI. Both EH and SGV affected change in proportion moves, with SGV increasing more dramatically across SGV treatments with *predator present*. Formal model, random effects output, and marginal and conditional R^2 can be found in Table S5.

Figure S4 Change in prey turns. Shown are the marginalized fixed effects of a generalized linear mixed effects model of scaled change across Phase 2 evolution in the proportion of prey instructions that were turns as a function of A. PT and B. SGV×EH. Bars are ±95% CI. *Predator* EH populations increased turns as a results of Phase 2 evolution; change in turns was marginal for *no predator* EH populations, and SGV affected *predator* and *no predator* populations differently. PT had no detectable effect. Formal model, random effects output, and marginal and conditional R^2 can be found in Table S5.

Figure S5 Change in prey looks. Shown are the marginalized fixed effects of a generalized linear mixed effects model of scaled change across Phase 2 evolution in the proportion of prey instructions that were looks as a function of A. PT and B. SGV×EH. Bars are ±95% CI. *Clone* and *high* SGV populations with *predator* EH increased looks as a result of Phase 2 evolution and change in turns was marginal to negative for *no predator* EH populations. PT had no detectable effect. Formal model, random effects output, and marginal and conditional R^2 can be found in Table S5.

Figure S6 Total instructions executed by each replicate prey population before and after Phase 2 evolution. The slope of each line represents the evolutionary change in instructions executed per lifetime (i.e. gestation time). The same replicate populations were assayed in the absence (top) and in the presence (bottom) of the Phase 3 predator. Arrows indicate the replicate population that was excluded from all statistical analyses, as it is not representative of its treatment group (*clone* SGV, *no predator* EH).

Table S1 Analysis of Variance (calculated using the anova() function in the R 3.0.2 base package) on a linear model*† testing the effects of SGV and EH and their interaction on Shannon's diversity among populations at the 200,000th update of Phase 2 evolution.

Table S2 Analysis of Variance (calculated using the anova() function in the R 3.0.2 base package) on a linear model*† testing the effects of SGV and EH and their interaction on change in predator attacks as a result of Phase 2 evolution.

Table S3 Permutational multiple analysis of variance (PerMANOVA) output describing effects on SGV, EH, and PT (and all first-order interactions) on outcomes of

ecological competitions (see Figure 2). 999 permutations were used.

Table S4 AIC, LRT output, and model selection processes for linear mixed-effects models describing change in A. total prey instructions; proportion moves; C. proportion turns and D. proportion looks. Variance functions are shown with traits modeled (in italics).

Table S5 Random intercepts and slopes (linear contrasts) between levels of PT with parametric bootstrap 95% confidence intervals for models describing change in A. total prey instructions; B. proportion moves; C. proportion turns; and D. proportion looks. Also included are marginal and conditional R^2 and residual standard deviations with parametric bootstrap 95% CI.

Appendix S1 S1.1: Genetic diversification during Phase 2 evolution. S1.2: Linear mixed effects models describing evolutionary change in traits.

Acknowledgments

We thank Chris Adami for comments and feedback on an earlier draft of this manuscript, and Max McKinnon for contributions to early stages of this research. Finally, we thank two anonymous reviewers for comments on the earlier versions of this work.

Author Contributions

Conceived and designed the experiments: AP DRO JAF APW. Performed the experiments: JAF APW. Analyzed the data: DRO. Wrote the paper: DRO AP JAF APW ID. Designed all figures and tables, aided in experimental design, contributed significantly to interpretation of results, and conducted significant revision of the manuscript: DRO. Significant revision/editing of manuscript: ID.

References

1. Frankham R, Lees K, Montgomery ME, England PR, Lowe EH, et al. (1999) Do population size bottlenecks reduce evolutionary potential? Anim Conserv 2: 255–260.
2. Colegrave N, Collins S (2008) Experimental evolution: experimental evolution and evolvability. Heredity 100: 464–470.
3. Mitchell-Olds T, Willis JH, Goldstein DB (2007) Which evolutionary processes influence natural genetic variation for phenotypic traits? Nat Rev Genet 8: 845–856.
4. Futuyma DJ (1979). Evolutionary Biology. 1st ed. Sinauer Associates, Sunderland, Massachusetts. ISBN 0-87893-199-6.
5. Barrett R, Schluter D (2008) Adaptation from standing genetic variation. Trends Ecol Evol 23: 38–44.
6. Orr HA, Betancourt AJ (2001) Haldane's sieve and adaptation from the standing genetic variation. Genetics 157: 875–884.
7. Innan H, Kim Y (2004) Pattern of polymorphism after strong artificial selection in a domestication event. Proc Natl Acad Sci USA 101: 10667–10672.
8. Colosimo PF, Hosemann KE, Balabhadra S, Villarreal Jr G, Dickson M, et al. (2005) Widespread parallel evolution in sticklebacks by repeated fixation of ectodysplasin alleles. Science 307: 1928–1933.
9. Prezeworski M, Coop G, Wall JD (2005) The signature of positive selection on standing genetic variation. Evolution 59: 2312–2323.
10. Pennings PS (2012) Standing genetic variation and the evolution of drug resistance in HIV. PLoS Comput Biol 8: e1002527.
11. Elmer KR, Meyer A (2011) Adaptation in the age of ecological genomics: insights from parallelism and convergence. Trends Ecol Evol 6: 298–306.
12. Yoshida T, Jones LE, Ellner SP, Fussmann GF, Hairston NG (2003) Rapid evolution drives ecological dynamics in a predator-prey system. Nature 424: 303–306.
13. Litchman E, Edwards KF, Klausmeier CA, Thomas MK (2012) Phytoplankton niches, traits and eco-evolutionary responses to global environmental change. Mar Ecol Prog Ser 470: 235–248.
14. Jerome JP, Bell JA, Plovanich-Jones AE, Barrick JE, Brown CT, et al. (2011) Standing Genetic Variation in Contingency Loci Drives the Rapid Adaptation of Campylobacter jejuni to a Novel Host. PLoS ONE 6: e16399
15. Dionne M, Miller KM, Dodson JJ, Bernatchez L (2009) MHC standing genetic variation and pathogen resistance in wild Atlantic salmon. Philos T Roy Soc B 364: 1555–1565.
16. Dettman JR, Anderson JB, Kohn LM (2008) Divergent adaptation promotes reproductive isolation among experimental populations of the filamentous fungus Neurospora. BMC Evol Biol 8: 35.
17. Bell G, Collins S (2008) Adaptation, extinction and global change. Evol Appl 1: 3–16.
18. McGuigan K, Sgrò CM (2009) Evolutionary consequences of cryptic genetic variation. Trends Ecol Evol 24: 305–311.
19. Lohbeck KT, Riebesell U, Reusch TBH (2012) Adaptive evolution of a key phytoplankton species to ocean acidification. Nat Geosci 5: 346–351.
20. Hayden EJ, Ferrada E, Wagner A (2011) Cryptic genetic variation promotes rapid evolutionary adaptation in an RNA enzyme. Nature 474: 92–95.
21. Messer PW, Petrov DA (2013) Population genomics of rapidadaptation by soft selective sweeps. Trends Ecol Evol 28: 659–669.
22. Anderson CJ (2013) The role of standing genetic variation in adaptation of digital organisms to a new environment. Artif Life 13: 3–10.
23. Hermisson J (2005) Soft sweeps: molecular population genetics of adaptation from standing genetic variation. Genetics 169.4: 2335–2352.
24. Levins R (1963) Theory of fitness in a heterogeneous environment. II. Developmental flexibility and niche selection. Am Nat 97: 75–90.
25. Gillings MR, Stokes HW (2012) Are humans increasing bacterial evolvability? Trends Ecol Evol 27: 346–352.
26. Blount ZD, Borland CZ, Lenski RE (2008) Historical contingency and the evolution of a key innovation in an experimental population of *Escherichia coli*. Proc Natl Acad Sci USA 105: 7899–7906.
27. Hedrick PW (1976) Genetic variation in a heterogeneous environment. II. Temporal heterogeneity and directional selection. Genetics 84: 145–157.
28. Kawecki TJ, Lenski RE, Ebert D, Hollis B, Olivieri I, et al. (2012) Experimental evolution. Trends Ecol Evol 27: 547–560.
29. Fry JD (2001) Direct and correlated responses to selection for larval ethanol tolerance in *Drosophila melanogaster*. J Evolution Biol 14: 296–309.
30. Bonnell ML, Selander RK (1974) Elephant seals: genetic variation and near extinction. Science 184.4139: 908–909.
31. Lande R (1988) Genetics and demography in biological conservation. Science 241: 1455–1460.
32. Ofria C, Bryson DM, Wilke CO (2009) Avida: A Software Platform for Research in Computational Evolutionary Biology. In: Adamatzky A, Komosinski M, editors. Artificial Life Models in Software. London, UK: Springer Verlag, London, UK. Second edition.
33. Wilke CO, Adami C (2002) The biology of digital organisms. Trends Ecol Evol 17: 528–532.
34. Lenski RE, Ofria C, Pennock RT, Adami C (2003) The evolutionary origin of complex features. Nature 423: 139–144.
35. Fortuna MA, Zaman L, Wagner AP, Ofria C (2013) Evolving Digital Ecological Networks. PLoS Comput Biol 9: e1002928.
36. Wagner AP, Zaman L, Dworkin I, Ofria C (2013) Coevolution, behavioral chases, and the evolution of prey intelligence. arXiv:1310.1369v2.
37. Lehmann K, Goldman BW, Dworkin I (2013) From cues to signals: evolution of interspecific communication via aposematism and mimicry in a predator-prey system. PLoS One 9: e91783.
38. Lima SL, Dill LM (1990) Behavioral decisions made under the risk of predation: a review and prospectus. Can J Zool 68: 619–640.
39. Sih A (1987) Predator and prey lifestyles: an evolutionary and ecological overview. Predation: direct and indirect impacts on aquatic communities (Kerfoot WC, Sih A, eds). Hanover, New Hampshire: University Press of New England; 203–224.
40. McCollum SA, Van Buskirk J (1996) Costs and benefits of a predator-induced polyphenism in the gray treefrog *Hyla chrysoscelis*. Evolution 50: 583–593.
41. Dodson S (1989) Predator-induced reaction norms. BioScience 39: 447–452.
42. Rundle HD, Nosil P (2005) Ecological speciation. Ecol Lett 8: 336–352.
43. Schmitz OJ (2008) Effects of predator hunting mode on grassland ecosystem function. Science 319: 952–954
44. Bryson DM, Ofria C (2013) Understanding evolution potential in virtual CPU architectures. PLoS ONE 8: e83242.
45. Drake JW, Charlesworth B, Charlesworth D, Crow JF (1998) Rates of spontaneous mutation. Genetics 148: 1667–1686.
46. Needleman SB, Wunsch CD (1970) A general method applicable to the search for similarities in the amino acid sequence of two proteins. J Mol Biol 48: 443–453.
47. Pinheiro J, Bates D, DeRoy S, Sarkar D (R Core Development Team) (2014) Linear and nonlinear mixed effects models. Version 3.1–117. Updated 31 March 2014.
48. Nakagawa S, Schielzeth H (2013) A general and simple method for obtaining R2 from generalized linear mixed-effects models. Methods Ecol Evol 4: 133–142.
49. Oksanen J, Blanchet FG, Kindt R, Legendre P, Minchin P, et al. (2013) vegan: Community ecology package. Version 2.0–10. Updated 12 December 2013.

50. Fox J, Weisberg S, Friendly M, Hong J, Andersen R, et al. (2013) R package effects: Effects displays for linear, generalized linear, multinomial-logit, proportional-odds logit models and mixed-effects models. Version 2.3–0. Updated 7 November 2013.

51. Bartón K (2013) R package MuMIn: Multi-modal inference. Version 1.9.13. Updated 29 October 2013.

52. Potts R (1996) Evolution and climate variability. Science 273: 922–923.

19

Rapid Birth-and-Death Evolution of Imprinted snoRNAs in the Prader-Willi Syndrome Locus: Implications for Neural Development in Euarchontoglires

Yi-Jun Zhang[1¤], Jian-Hua Yang[1], Qiao-Su Shi[2], Ling-Ling Zheng[1], Jun Liu[3], Hui Zhou[1], Hui Zhang[3], Liang-Hu Qu[1]*

1 Key Laboratory of Gene Engineering of the Ministry of Education, State Key Laboratory of Biocontrol, and School of Life Sciences, Sun Yat-sen University, Guangzhou, P. R. China, **2** Laboratory of Liver Disease Hospital, The Third Affiliated Hospital, Sun Yat-sen University, Guangzhou, P. R. China, **3** Institute of Human Virology, Zhongshan School of Medicine, Sun Yat-sen University, Guangzhou, P. R. China

Abstract

Imprinted small nucleolar RNAs (snoRNAs) are only found in eutherian genomes and closely related to brain functions. A complex human neurological disease, Prader-Willi syndrome (PWS), is primarily attributed to the deletion of imprinted snoRNAs in chromosome 15q11-q13. Here we investigated the snoRNA repertoires in the PWS locus of 12 mammalian genomes and their evolution processes. A total of 613 imprinted snoRNAs were identified in the PWS homologous loci and the gene number was highly variable across lineages, with a peak in Euarchontoglires. Lineage-specific gene gain and loss events account for most extant genes of the HBII-52 (SNORD115) and the HBII-85 (SNORD116) gene family, and remarkable high gene-birth rates were observed in the primates and the rodents. Meanwhile, rapid sequence substitution occurred only in imprinted snoRNA genes, rather than their flanking sequences or the protein-coding genes located in the same imprinted locus. Strong selective constraints on the functional elements of these imprinted snoRNAs further suggest that they are subjected to birth-and-death evolution. Our data suggest that the regulatory role of HBII-52 on 5-HT_{2C}R pre-mRNA might originate in the Euarchontoglires through adaptive process. We propose that the rapid evolution of PWS-related imprinted snoRNAs has contributed to the neural development of Euarchontoglires.

Editor: Robert Feil, CNRS, France

Funding: This research was supported by the National Natural Science Foundation of China (No. 30830066) to Liang-Hu Qu. National Natural Science Foundation of China (No. 81301431), China Postdoctoral Science Foundation (No. 2012M511866 and No. 2013T60824) to Yi-Jun Zhang. The National Natural Science Foundation of China (No.31370791); the funds from Guangdong Province (No. S2012010010510, S2013010012457); the project of Science and Technology New Star in ZhuJiang Guangzhou city (No. 2012J2200025) to Jian-Hua Yang. The funders had no role in study design, data collection and analysis, decision to publish, or preparation of the manuscript.

Competing Interests: The authors have declared that no competing interests exist.

* E-mail: lssqlh@mail.sysu.edu.cn

¤ Current address: Institute of Human Virology, Zhongshan School of Medicine Sun Yat-sen University, Guangzhou, P. R. China

Introduction

Small nucleolar RNAs (snoRNAs) belong to a large group of small, metabolically stable RNAs that primarily direct the post-transcriptional modification of ribosomal RNA (rRNA) nucleotides. Based on common sequence motifs involved in the assembly of sno-ribonucleoprotein (snoRNP) particles, snoRNAs fall into two major classes, the C/D box and H/ACA box groups, which account for the 2′-O-ribose methylation and pseudouridylation modifications respectively of rRNA or snRNA [1,2]. These snoRNAs are called 'guide' snoRNAs, and there is a fraction of snoRNAs, the so called 'orphan' snoRNAs, which do not target to any rRNA or snRNA. In recent years, hundreds of snoRNAs have been discovered in imprinted regions of the eutherian genomes (imprinted snoRNAs) [3–8]. The expressions of imprinted genes are depending on the parent of origin. Unlike other snoRNAs, which show broad expression profiles, imprinted snoRNAs are mainly expressed in the brain and are closely associated with brain function [9,10]. Another feature of imprinted snoRNAs is that they usually fall within large gene clusters containing several gene families. One of such family usually contains dozens of snoRNA genes. Two large paternally expressed C/D box snoRNA families, HBII-85 (SNORD116) and HBII-52 (SNORD115), are located in the 15q11-q13 imprinted region of the human genome. Large interstitial deletions of this region underlie ~70% of cases of Prader-Willi syndrome (PWS, [MIM176270]), a complex human neurological disease [11]. Duplication of the same region is the only recurrent cytogenetic aberration associated with autism, occurring in up to 5% of autism cases [12–14]. The clinical symptoms of PWS include neonatal hypotonia, feeding difficulties and failure to thrive during infancy, excessive weight gain after 18 months, hyperphagia, hypogonadism, global developmental delay and equivocal facial features. Deletion of the HBII-85 snoRNA cluster results in an exhibition of all major clinical symptoms of PWS in humans [15–17], but the role of the HBII-52 cluster in PWS has been difficult to define [11,18]. The neuronal-specific HBII-52 snoRNAs have been reported to participate in the post-transcriptional regulation of the pre-mRNA encoding the 5-hydroxytryptamine 2C receptor (5-HT_{2C}R), an important neuro-transmission protein, including A-to-I RNA editing and alternative

RNA splicing, with *in vitro* experiments [19–21]. The biological importance of the regulatory role of HBII-52 *in vivo*, if any, remains to be demonstrated. Knockout of MBII-52 (murine homologues of HBII-52) in a murine model led to an increase in the editing of 5-HT_{2C}R pre-RNA and alterations in a number of 5HT_{2C}R-related behaviors, including impulsive responses, locomotor activity and reactivity to palatable food [22]. Although the molecular targets of HBII-85 are as yet unknown, an increasing number of studies have investigated the regulatory roles of HBII-85 snoRNAs and their connection to PWS phenotypes [15–17,23,24]. In addition, the PWS locus contains several singleton snoRNA genes of unknown function, e.g., HBII-438, HBII-13 and HBII-436 (Figure 1).

SnoRNA genes have traditionally been regarded as conservative components of the genome. SnoRNAs have been identified in a wide spectrum of organisms spanning eukaryotes to archaebacteria, suggesting that they represent an evolutionary ancient group of non-coding RNAs [25]. Homologues of almost all yeast snoRNAs can be found in the human genome [26,27], and their modification sites in rRNAs are also highly conserved among distantly related eukaryotes [1,28]. However, the discovery of a large number of eutherian-specific imprinted snoRNAs led us to reexamine this supposedly ancient group of RNAs from a new perspective. Further investigations failed to find homologs of these imprinted snoRNAs in any other non-eutherian mammals or vertebrates [29–31]. In addition, a number of lineage-specific snoRNA genes were identified only in rodents rather than humans, and a rat-specific imprinted snoRNA cluster, RBII-36, containing 74 member genes, was not detected in mice (see the review of [32]). These observations suggest that imprinted snoRNA genes represent a class of newly derived small non-coding RNAs that might be subjected to rapid evolution.

Although an increasing number of imprinted snoRNAs have now been identified, they are mainly limited to a few model species, including human, mouse and rat. Since the PWS-related imprinted snoRNAs are closely associated with mammalian brain functions, it is of great interest to investigate the evolution of these highly dynamic non-coding RNAs. Nahkuri et al. suggest that the HBII-52 gene family originates from a non-imprinted snoRNA, SNORD119 [33]. Ogorelkova et al. revealed that the HBII-52 family was subjected to positive selection in Europeans populations, suggesting these gene are still undergoing evolution in human populations [34]. However, these studies did not fully reveal the evolutionary process of these imprinted snoRNA genes along with the differentiation of the eutherian or its significance to mammalian neural development. In this study, using a genome-wide snoRNA identification program developed by our group, snoSeeker [35], we performed comprehensive identification and evolutionary analyses to reveal the pattern and process underlying the evolution of mammalian imprinted snoRNAs.

Materials and Methods

Identification and Classification of Imprinted snoRNA Genes in Mammalians PWS Locus

The genome sequences and related whole genome alignments (WGAs) of 12 placental mammals (Human, Chimp, Rhesus, Rat, Mouse, Dog, Cat, Horse, Cow, Armadillo, Tenrec and Elephant)

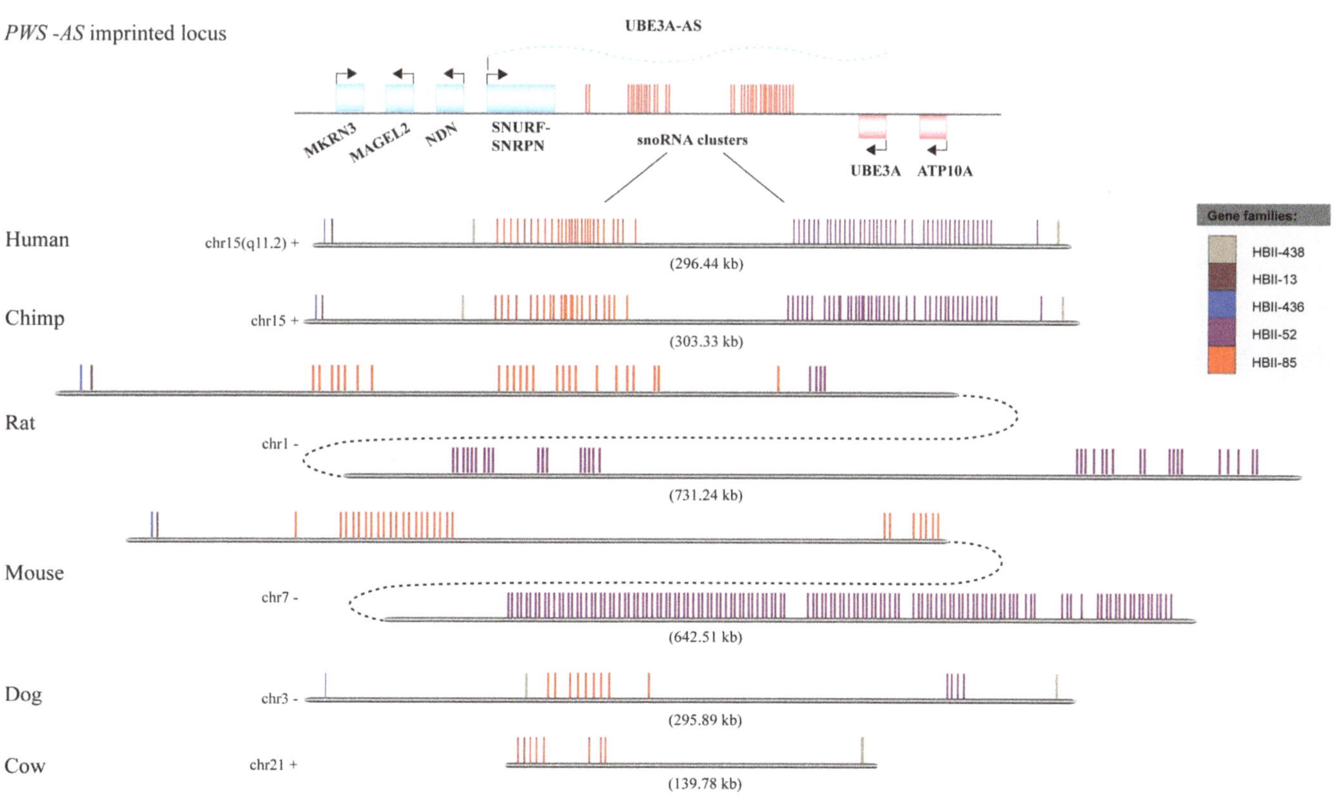

Figure 1. A schematic diagram of the imprinted snoRNA gene cluster in the PWS-related imprinted locus. The top panel depicts the gene distribution across the entire PWS locus. Pink represents maternally expressed genes, blue represents paternally expressed genes, red represents imprinted snoRNA clusters, and the dashed wavy line represents non-coding transcripts. In the lower panels, each snoRNA gene is represented by a vertical bar. The distribution of snoRNA genes was scaled to their genomic location, and the colors indicate their respective gene families. Clusters are scaled relative to their size within the chromosome.

were downloaded from the UCSC genome website (http://genome.ucsc.edu/and http://genome-test.cse.ucsc.edu/). These species represent all major lineages of extant placental mammals. The range of human PWS locus was identified by the gene pair of MKRN3 and ATP10A. The homologous region of PWS in other species was obtained using the "covert" tool of the UCSC genome website and the gene content was used as an index when the result of covert tool is poor. The information of PWS loci in different species is listed in Table S1.

Two independent strategies were employed to search imprinted snoRNA candidates in the PWS loci: (1) The comparative genomics identification strategy with WGAs was performed as described in Yang et al. [35]. (2) The single-genome strategy was performed by searching the PWS homologous regions with snoSeeker program (http://202.116.74.192/snoSeeker/downloads.php) under the cutoff value of >24.0. The non-redundant candidates from these two strategies were combined for further gene family classification. The sequences of non-imprinted snoRNA genes of 15 vertebrates (with the addition of opossums, platypus and chicken) were identified from the WGAs of human and each of other species as described in Yang et al. [35]. All identified imprinted and non-impritned snoRNA gene sequences were provided in the Data S1.

For imprinted snoRNA gene family classification, BLAST search was first conducted to identify homologs to know human or rodents imprinted snoRNA genes, the cutoff is set to 10^{-3}. Nevertheless, due to algorithmic characteristics and sensitivity constraint, this classification method will lead to ambiguities when dealing with the species that are distantly related to the reference. In addition, the genomes of some species were poorly assembled, e.g. the elephant and the tenrec genome, which could not provide accurate gene mapping information to validate the classification results. Thus, phylogenetic methods were further exploited to classify imprinted snoRNA candidates into known gene families. For this purpose, phylogenetic trees containing imprinted snoRNA candidates of each species were constructed with the human imprinted snoRNA genes as the reference for family classification. The threshold value of branch support for the assignment of a candidate to known gene families was >50%. The phylogenetic trees were constructed with MrBayes ver. 3.1 [36] with a HKY+G model.

Phylogenetic Analyses

To reconstruct the phylogenetic relationship of imprinted snoRNA genes in the *PWS* locus, a dataset containing gene sequences from all families in this locus (i.e. HBII-436, HBII-13, HBII-438, HBII-85 and HBII-52) was generated. For gene families containing more than 20 members, 4–6 representative genes were selected based on a gene tree constructed using the neighbor-joining method implemented in MEGA ver. 3.1 [37]; for gene families containing less than 20 members, all sequences were added to the dataset. The sequences were aligned with ClustalX [38]. The sequence alignment is provided in data S1 and S2. Then a phylogenetic network was constructed using the NeighborNet method implemented in SplitsTree ver. 4.10 [39].

To further investigate the phylogenetic relationship within the HBII-52 and HBII-85 multi-gene families, the alignment of each family was built as described above. The phylogenetic relationships were reconstructed using MrBayes ver. 3.1 [36] with a HKY+G model suggested by MrModeltest (ver. 2.2) [40]. Every analysis consisted of two runs each containing four Markov chains that ran for 2×10^6 generations, sampled every 1000 generations, with a random starting tree, default priors and equal branch lengths. The consensus trees were edited with iTOL [41]. The sequence alignments are provided in Data S2.

Evolutionary Changes of the Number of Imprinted snoRNA Genes

To estimate the numbers of gene in ancestral species and gene gains and losses, we used the reconciled tree method [42–44], in which the topology of a gene tree is reconciled with that of a species tree. A simple example is shown by Niimura and Nei [45] and Nam and Nei [44]. To apply this method to imprinted snoRNA genes, we requested the computer program described in Niimura and Nei [45,46] from Y. Niimura. We first constructed a phylogenetic tree using all genes of the two multi-gene families (HBII-52 and HBII-85) together with outgroup genes chosen from other close related families. The reconciled tree method was applied to these two phylogenetic trees and the Euarchontoglires tree topology. Given the relative low bootstrap value caused by short sequence of snoRNA genes, the 50% bootstrap condensed trees were used here.

To investigate the expansion process of imprinted snoRNAs, a combined physical-phylogenetic map for the HBII-52 or HBII-85 gene family of human or mouse was constructed. The phylogenetic trees were built with the Bayesian method as described above, each Markov chain ran for 1×10^7 generations. The clades receiving >50% posterior probabilities were kept and the derived consensus trees were reconciled with gene physical locations in the chromosomes.

Evolutionary Analyses

The number of new genes was calculated as follows. For imprinted snoRNAs, the number of newly derived genes was estimated according to the evolutionary change in gene number described above, whereas the number of newly derived non-imprinted snoRNAs was estimated by determining the difference in gene number between closely related species. Birth rates of new genes were calculated by dividing the number of newly derived genes by the lapsed time since the species diverged from their most recent common ancestor. The time scale was set according to Hedges et al. [47], and the weighted average data were selected.

For each imprinted gene family, the substitution rates between humans and each of the other species were calculated using the "between-group average calculation" method with the Kimura 2-parameter model implemented with MEGA ver. 3.1 [37,48]. A "group" corresponds to a multi-gene family from one species, which contains all gene members in this family. For each between-group average, an arithmetic average is computed for all valid inter-group pairwise comparisons. The standard error for each calculation was estimated by bootstrap analysis with 500 replicates. To calculate the substitution rates of non-imprinted snoRNA genes, we first identified all of the families that were conserved between humans and each of the other species (Table S2, the sequences were provided in data S1 and S2). The guide snoRNAs and orphan snoRNAs were analyzed separately. The orphan snoRNA genes were identified by the absence of targeting sites in rRNA or snRNA according the information in snoRNA-LBME-db (https://www-snorna.biotoul.fr/index.php) [27]. These orphan snoRNAs were listed in Table S2. For multi-gene families, one gene member and its orthologous in the other species s was selected for the analysis. Then gene sequences from one species were linked together in order of gene name and an aliment was built with this integrated sequence of human and each of the other species. Next, the average substitution rates of the conserved families were calculated with the Kimura 2-parameter model. All of the calculations were performed using MEGA version 3.1 [37].

To test whether the rapid rate of evolution was specific to the imprinted snoRNAs, we compared the evolutionary rate between imprinted snoRNA genes and their flanking intergenic sequences. We obtained 500 bp of both upstream and downstream flanking sequences for each analyzed imprinted snoRNA gene from the UCSC genome website. The substitution rates of both imprinted snoRNA genes and their flanking sequences were calculated as described above. Similar analyses were also performed using non-imprinted snoRNA genes as the reference.

According to the secondary structure we divided box C/D snoRNs gene into 5 parts: the stem region, C/D boxes, antisense element 1 (ASE1), antisense element 2 (ASE2) and the remaining sequence. To investigate the evolutionary rates of different parts of the imprinted snoRNA genes, we created an alignment for HBII-52 or HBII-85 family and defined different structural components in the alignment. The evolutionary distances (*K*) between humans and each of the other species were calculated for each of the five parts of the snoRNAs using the Kimura 2-parameter model.

Tajima's *D* test was performed with DnaSP software version 5.10.01[49]. For the Tajima's neutrality test, significant deviations of Tajima's *D* values from zero suggest rejecting the standard neutral model and negative values of Tajima's *D* indicate positive selection. All the gene sequences of HBII-52 or HBII-85 gene family from one species were used for the test. The statistical significance was obtained assuming that *D* follows the beta distribution[50].

The K_a/K_s or d_n/d_s test is one of the most frequently used tests of selection. The test relies on the simple fact that the rate of evolution can exceed that of a neutral sequence only if there is positive selection [51]. Ka/Ks test for protein-coding gene was performed with MEGA 4.0 software [52].

PWS Related snoRNA Targeting Pattern Analysis

To investigate the targeting pattern of HBII-52 gene families in different eutherians species, homologous sequence to human 5-$HT_{2C}R$ mRNA of different species were got by BALSTP program from ensembl genome web site (www.ensembl.org). The binding pattern between ASE2 sequences and 5-$HT_{2C}R$ mRNA in same species was analyzed by BLAST method with some modifications. Two major characters, the length of perfect binding and the distance of perfect binding from D box, were analyzed.

Results

Identification and Characterization of Imprinted snoRNAs in the PWS Homologous Locus of 12 Eutherian Genomes

Both single genomic and comparative genomic strategies were applied to identify imprinted snoRNA genes from 12 placental mammals to achieve high sensitivity and accuracy simultaneously (Table S1). To classify the snoRNA gene families, BLAST searches and phylogenetic analyses were combined to avoid the misassignment of individual gene members to known gene families. A total of 613 imprinted snoRNA genes were identified in the PWS homologous loci (the 15q11-q13 region of the human genome) of 12 eutherian genomes (Table 1, all snoRNA gene sequences were provided in Data S1). In general, primates and rodents have many more imprinted snoRNAs than other mammals, mainly reflected by the big volume of the HBII-52 and the HBII-85 multi-gene families. On average, primates have 77 PWS-related imprinted snoRNAs, rodents have 112, while Laurasiatheri and Xenarthra have only 17. The number of imprinted snoRNAs in elephants was comparable to that of primates. Certain gene families are missing from the ancestral lineages of placental mammals, such as Xenarthra and Afrotheria. For example, the HBII-85 gene family was missing from the tenrec genome, whereas HBII-52 was absent from the cow and the armadillo genomes. These observations may have two explanations: the first is due to the incomplete assembly of these genomes, the second one attributes to particular gene loss events occurred in these lineages. Conversely, extensive expansions of particular gene families have occurred in the mouse and elephant, leading to an excess of HBII-52 genes (130 and 70, respectively). The gene numbers of the multi-gene families (HBII-52 and HBII-85) predicted in this study were slightly lower than those of previous studies on the model species (e.g. human and mouse) [4,5,53]. It may attribute to the more stringent criteria for snoRNA identification used in the snoSeeker program [35].

Based on the genomic location information of each snoRNA gene, we illustrated the precise distribution of the imprinted snoRNA genes in the PWS homologous loci of representative species (Figure 1). The size of the PWS imprinted snoRNA cluster was notably expanded in rodents. The cluster sizes of both mouse and rat were approximately two-fold that of the human. In both the rat and mouse genomes, the PWS-related snoRNA clusters were extended by the insertion of a long heterogeneous fragment containing no snoRNA genes. However, the insertion sites were different, occurring within the HBII-52 gene family in the rat genome and in the HBII-85 gene family in mouse. Interestingly, the size of the imprinted snoRNA cluster does not appear to determine the gene number. Despite similar cluster sizes, the dog has only 25% of the snoRNA genes presented in human. These results imply that the PWS imprinted snoRNA cluster is located in a highly active genomic region, a hot spot for chromosomal rearrangements and gene duplications as observed in other imprinted loci [54].

Phylogenetic Analyses

The phylogenetic network indicated that each gene families could form a single phylogenetic clade (Figure. S1). The phylogenetic relationship among the gene families exhibits consistence with their physical distribution in the chromosome. For example, HBII-436 and HBII-13, which are closely distributed in the locus, showed close phylogenetic relationship. Interestingly, the HBII-438 family, consisting of two gene members HBII-438A and HBII-438B, was separated by the large gene cluster of HBII-52 and HBII-85. This suggests these two gene cluster might origin from an insertion of ancestral gene(s) within the HBII-438 family and followed by extensive expansions [33]. Figure 2 shows the intra-family gene relationship of the HBII-52 and the HBII-85 family. A remarkable feature exhibited by the phylogenetic trees is that gene members from same species always form a single clade, especially in non-primates species, suggesting their close relationship. In primates, an intermixed pattern among species was observed, implying of a progressive evolutionary process of these gene families. This pattern was also suggested by the phylogenetic network shown in Figure S1.

Recent Creation of PWS-related Imprinted snoRNA Genes through a Birth-and-death Process

To further explore the evolutionary history of the PWS-related imprinted snoRNA genes, we estimated the number of imprinted snoRNA genes in all ancestral organisms and their increases and decreases at different stages during the evolution of eutherian. Considering that their functional importance and gene volume, we focused this analysis on the two multi-gene families, HBII-52 and HBII-85. To estimate the gene number in ancestral species as well as gene gains and losses, we used the reconciled tree method [42–44], in which the topology of a gene tree is reconciled with that of a species tree. The analysis was based on the "Euarchontoglires

Table 1. Statistics of imprinted snoRNA genes in the PWS locus of mammalians.

			HBII-436	HBII-13	HBII-438A	HBII-85	HBII-52	HBII-438B	Total
Euarchontoglires	Primates	Human	1	1	1	27	41	1	72
		Chimp	1	1	1	22	44	1	70
		Rhesus	1	1	1	28	58	1	90
	Rodentia	Rat	1	1	0	24	39	0	65
		Mouse	1	1	0	27	130	0	159
Laurasiatheri	Carnivora	Dog	1	0	1	9	4	1	16
		Cat	1	2	0	6	20	0	29
	Perissodactyla	Horse	1	0	1	12	2	1	17
	Cetartiodactyla	Cow	0	0	0	8	0	1	9
Xenarthra		Armadillo	1	2	0	9	0	1	13
Afrotheria		Tenrec	0	0	0	0	0	0	0
		Elephant	0	1	0	1	70	1	73

tree" proposed by Murphy et al. [55], which suggested that primates and rodents are sister groups forming a clade named Euarchontoglires. Figure 3 summarizes the gene numbers estimated for each ancestral species and the associated changes along the different lineages. The numbers of PWS-related imprinted snoRNA genes were extremely low in ancestral organisms as compared with extant species. Interestingly, the evolutionary history of PWS-related snoRNA families is dominated by gene increases (Figure 3). In both the HBII-52 and HBII-85 families (Figure S2 and S3), most (55–98%) of the existing family members arose from lineage-specific gene gain events. In the human genome, 31 of the 41 HBII-52 genes were gained after the species diverged from chimpanzees; for the mouse, 127 of the 130 HBII-52 genes were gained after the species diverged from rats. In contrast, gene loss events were very rare. Similar results were observed for the HBII-85 cluster. Figure 3 provided an overview of the evolutionary dynamics of PWS-related imprinted snoRNA genes in mammals, which reflects the rapid gene turnover in these families. It also suggeststhat the extant genes in each species primarily arose through a lineage-specific gene gaining process occurring in each species' recent history. This trend can also be intuitively inferred by their grouping patterns, i.e., gene members from the same species were always clustered together (Figure S1 and Figure 2).

Since the results suggest rapid intra-species gene expansion occurred during eutherian evolution, especially in primates and rodents, we want to ask how this happened. For this purpose, combined physical-phylogenetic maps of the HBII-52 and HBII-85 gene families in human and mouse were constructed to investigate gene duplication processes (Figure S4). There are several models proposed for gene duplication [56]. The tandem duplication model predicts a close phylogenetic relationship for physically neighboring genes. This pattern was observed in all the investigated gene families, i.e. HBII-52, HBII-85, MBII-52 and MBII-85. However, in the HBII-52 and MBII-52 family, a lot of physically distant genes show phylogenetic connections, which seems like the duplicated genes have jumped or trans-located from their original sites. This pattern could be explained with a 'regional ectopic duplication' model, which usually refers to duplication of individual or a small group of genes to an unlinked locus [57,58]. The mechanism behind this may be recombination during the gametogenesis. These results also suggest different duplication processes were favored by different gene families. The tandem duplication plays a major role in the recent expansion of the HBII-85 and MBII-85 family; while the regional ectopic duplication dominates the expansion process of the HBII-52 and MBII-52 families.

The birth-and-death model of gene evolution predicts the death of genes, or "gene degradation". Consistent with this idea, we found pseudogenes in all species (Table S3, Figure S5). Gene degradation mainly occurs through the loss-of-function mutations that impair snoRNA processing or snoRNP formation, e.g., in the C/D boxes. Primates and rodents generally exhibit much lower pseudogene ratios than other species (Table S3). For example, in the HBII-52 family, the average pseudogene ratios were 0.138 and 0.158 for primates and rodents, respectively, but 0.667 for other species. This observations suggest that a trend toward gene expansion was favored in both primates and rodents and explains why these groups have more PWS-related imprinted snoRNA genes.

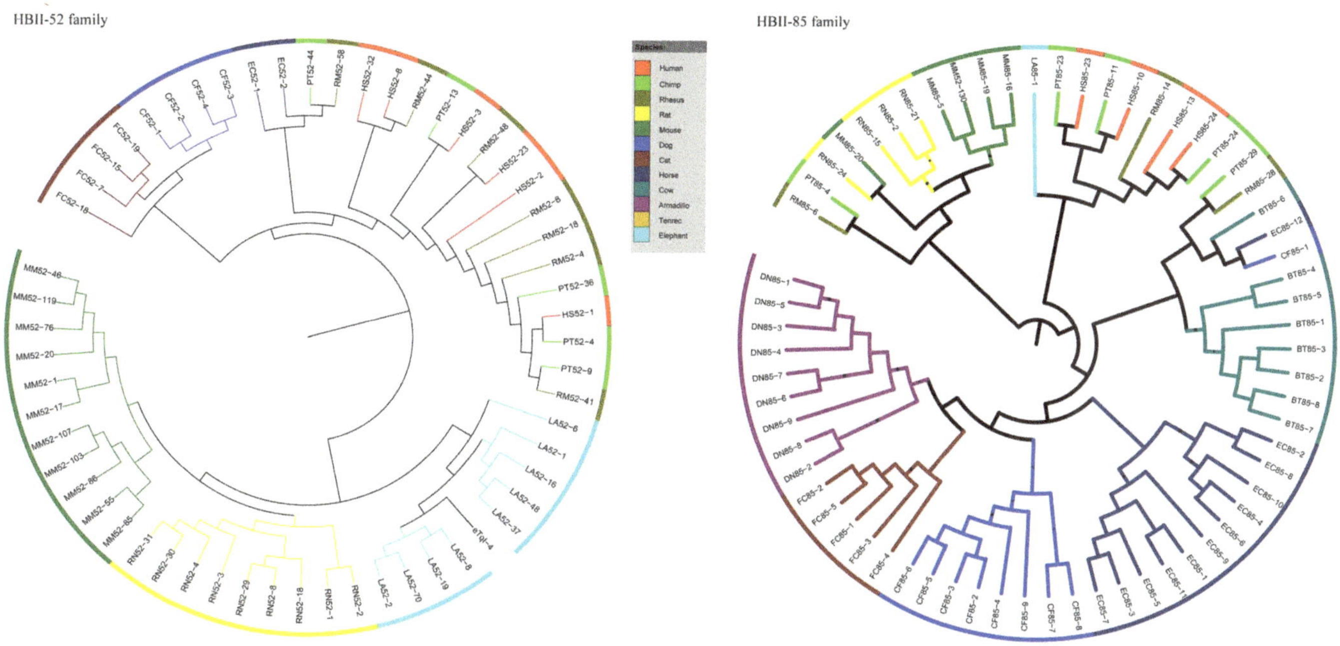

Figure 2. Phylogenetic analysis of HBII-52 and HBII-85 gene family. Abbreviations: HS, human; PT, chimpanzee; RM, rhesus; MM, mouse; RN, rat; CF, dog; FC, cat; EC, horse; BT, cow; DN, armadillo; LA, elephant.

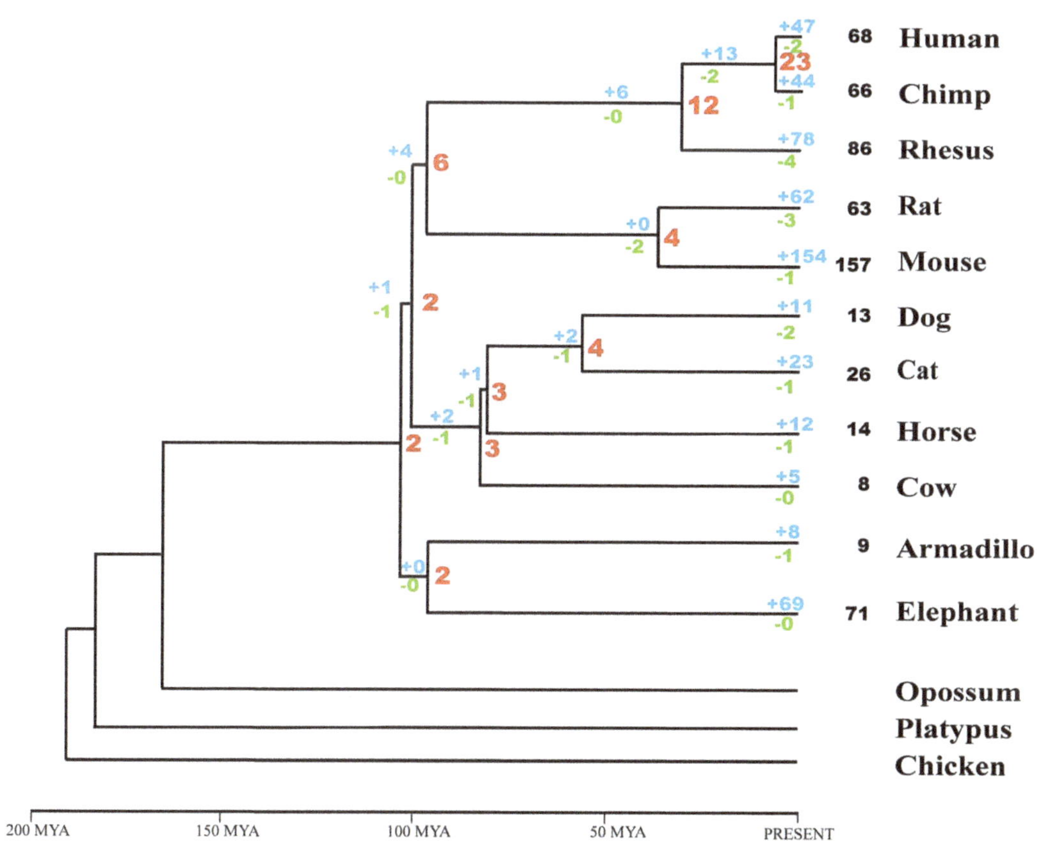

Figure 3. Gene numbers and changes in PWS-related imprinted snoRNA genes with respect to eutherian divergence. Numbers in black indicate the current gene number in each species, whereas red numbers indicate the ancestral gene number, blue numbers denote gained genes, and green numbers indicate lost genes. The major diversification events of eutherians are indicated on the timescale. MYA, a million years ago.

Rapid Evolution of PWS-related Imprinted snoRNAs with Functional Constraints

We next asked whether the PWS-related imprinted snoRNAs have undergone rapid evolution. The evolutionary rates of PWS-related imprinted snoRNAs were investigated by two aspects, the first is the birth rate of new genes and the second is sequence substitution rates. Table 2 presents the birth rate of new genes for both the imprinted and non-imprinted snoRNA gene families. Only the C/D box non-imprinted snoRNAs were included in this comparison because all imprinted snoRNAs belong to this class. In the PWS locus, 6.9 and 4.1 new genes were born per million years (Mys) in the human and the mouse respectively, while 8.4 and 2.9 non-imprinted snoRNAs were born per Mys throughout the rest of the genome. Note that the number of non-imprinted gene families involved in this calculation is much larger than that of the imprinted snoRNAs, i.e., ~180 vs. 2. When the number of gene families was taken into account, the average gene birth rate of the imprinted clusters is over 80-fold faster than that of the non-imprinted snoRNAs. This rate increased to 127 folds in mouse. The higher gene birth rate observed in mouse may account for the emergence of many rodent-specific imprinted snoRNAs. The high gene birth rate was also observed in other small non-coding RNAs, such as microRNA [59,60]. However, the high birth rate of the imprinted snoRNA genes is quite remarkable, even as compared with the microRNAs. A birth rate of 12 new microRNA genes per Mys has been reported in the fruit fly; however, only 0.3 genes/Mys have been preserved over long stretches of time [61]. In fact, only 5 new, consistently expressed microRNA genes have been maintained in the *Drosophila* genus over the last 55 Mys of evolution [62].

We also observed rapid sequence evolution of the PWS-related imprinted snoRNAs. Figure 4A shows the evolutionary rates of non-imprinted (including guide and orphan) and the imprinted snoRNA families between human and the other analyzed mammals (the standard errors are indicated in Table S4). This comparison spanned the full range of mammals, and K represents the number of substitutions per site. For non-imprinted snoRNAs, the mean K value was calculated for all the homologous snoRNA families that are conserved between the human and each of the other species (Table S2); for imprinted snoRNAs, the K value was calculated for the HBII-52 and HBII-85 families. This analysis suggests that PWS-related imprinted snoRNA families evolved with a higher sequence substitution rate than non-imprinted snoRNAs. Since orphan snoRNA have no known targeting site on rRNA or snRNAs, it is speculating that they are subjected to relative rapid evolution than guide snoRNAs due to the lack of selective constrains. But our results did not support this notion (Figure 4A). Depending on the species, the HBII-52 family evolved 1.2 to 2.2-fold faster than non-imprinted snoRNAs, with an average value of 1.6 folds; the HBII-85 family evolved 1.3 to 4.6-fold faster than non-imprinted snoRNAs, with an average value of 2.2 folds. Primate species exhibited the highest relative evolution rates (imprinted vs. non-imprinted) of the HBII-85 family (Figure 4B, Table S4). Assuming that the mean evolutionary rate of non-imprinted snoRNAs is nearly constant over a long time scale, this pattern suggests that the evolution of this snoRNA family may have been accelerated in primates.

To determine whether the rapid sequence evolution is specific to the PWS-related imprinted snoRNAs or spans the entire imprinted locus, we compared the substitution rates of the imprinted snoRNA genes and their flanking sequences. A similar comparison was performed using non-imprinted snoRNAs as the reference. We calculated the between-species substitution rates (human vs. chimpanzee and human vs. rhesus monkey) for the

Table 2. Birth rates of new genes among PWS-related imprinted and non-imprinted snoRNAs.

	Imprinted snoRNA gene clusters					Non-imprinted snoRNA genes	
	Human	Chimp	Rhesus	Mouse	Rat	Human	Mouse
New genes	47	44	78	154	62	57	108
Rate of birth (per Mys)	6.96	6.52	2.52	4.17	1.68	8.44	2.92
Rate of birth per family (per Mys)	3.48	3.26	1.26	2.085	0.84	0.0433	0.0163

Ratios of gene birth rate between imprinted and non-imprinted snoRNAs: human: 80.37; mouse: 127.91.

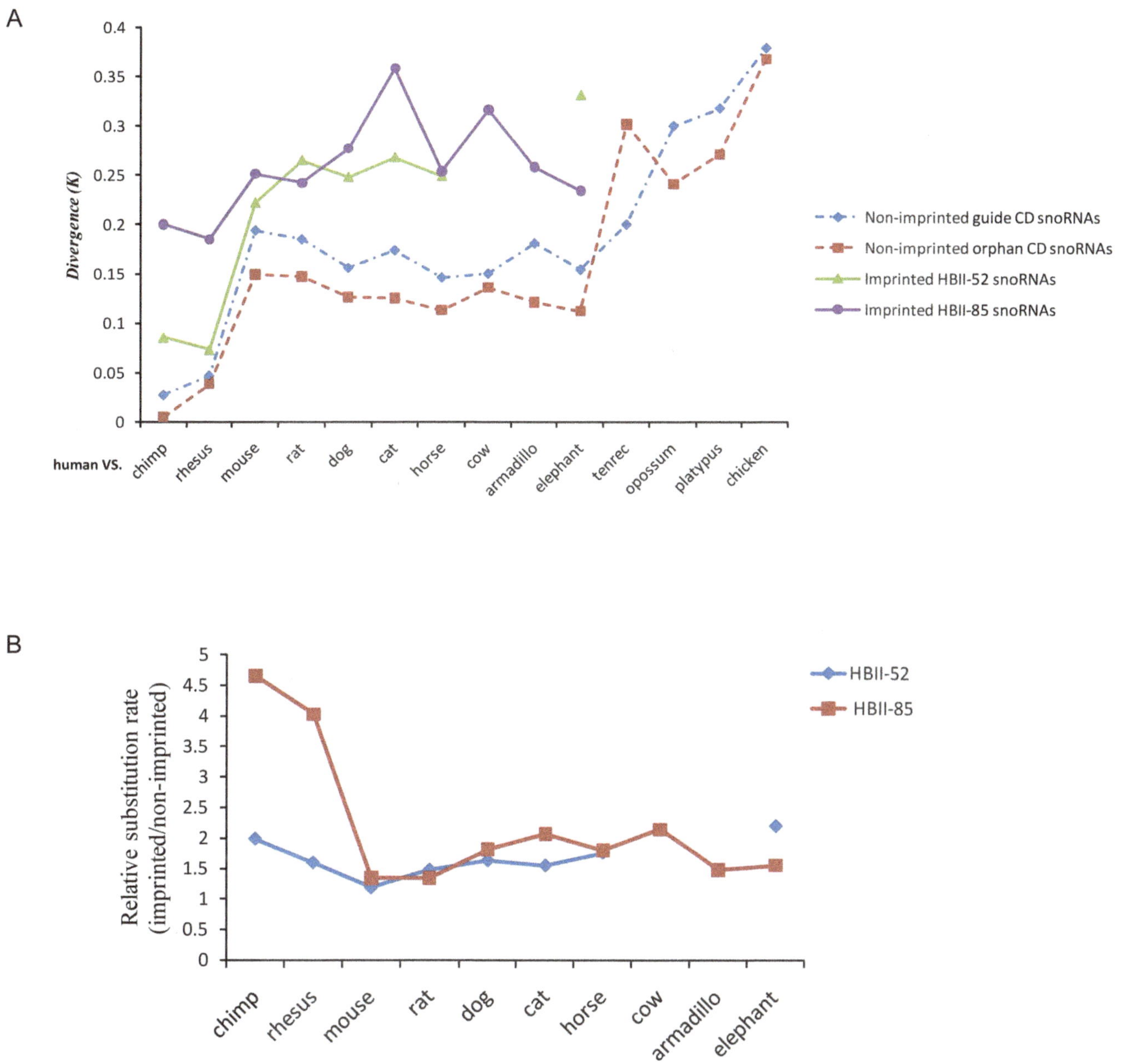

Figure 4. Evolutionary divergence of PWS-related imprinted vs. non-imprinted snoRNA gene families between the human and other mammals. (A) The evolutionary divergence of HBII-52, HBII-85, non-imprinted guide and orphan snoRNA genes between human and other species. *K* denotes the number of substitutions per site. (B) Relative sequence substitution rate between imprinted and non-imprinted snoRNA genes of eutherian mammals.

snoRNAs (K_{sno}) and the flanking genomic sequences (K_f) (presumably nonfunctional, Table 3). The average K_f values were similar between the PWS-related imprinted snoRNAs and the non-imprinted snoRNAs, i.e., 0.011 vs. 0.012 for human vs. chimp and 0.085 vs. 0.078 for human vs. rhesus, indicating nearly equivalent substitution rates. However, the PWS-related imprinted snoRNA cluster had a much higher average K_{sno} as compared with the non-imprinted snoRNAs (0.006 VS. 0.001 and 0.033 VS. 0.013), again supporting the proposed rapid evolution of the PWS-related imprinted snoRNAs. These results suggest that the rapid evolution of PWS-related imprinted snoRNAs was not conferred by a similar process operating in the entire imprinted region but was specific to the snoRNA genes.

We further investigated the evolutionary patterns of different parts of the imprinted snoRNA genes. As mentioned above, snoRNA could be divided into 5 parts: the stem region, C/D boxes, antisense element 1 (ASE1), antisense element 2 (ASE2) and the remaining sequence. The stem regions, spanning 4–5 nt at each end of a given snoRNA, form a short stem that brings the conserved C and D boxes close to one another to form the structural core motif of the snoRNA. This core motif, referred to as the stem-box structure, coordinates the specific binding of snoRNP proteins and is critical for snoRNA biogenesis and nucleolar localization. The antisense elements are complementary to rRNA or snRNA and determine the site for 2′-O-ribose methylation or for other cellular targets with other functions

Table 3. Genetic divergence of snoRNA genes and their flanking sequences in primates for both imprinted and non-imprinted snoRNAs.

	Imprinted region		Non-imprinted region	
	Kf	Ksno	Kf	Ksno
HS-PT	0.011 (0.001)	0.006 (0.002)	0.012 (0.001)	0.001 (0.001)
HS-RM	0.085 (0.002)	0.033 (0.005)	0.078 (0.003)	0.013 (0.004)

Species abbreviation: HS, humans; PT, chimpanzees; RM, rhesus. Standard deviations are shown in parentheses.

[1,25]. Figure 5 shows the distribution of the evolutionary distance, *K*, corresponding to each part of the snoRNA gene between the human and the other analyzed species. The distribution clearly shows that different parts of the genes were subjected to different levels of selection constraint. In general, the C/D boxes and the stem regions were the most conserved, with *K* values under or approximately 0.1, whereas the evolutionary rates of the other three regions varied depending on the gene family. For the HBII-52 family, the two antisense elements showed significant different evolutionary rates, with the ASE1 evolving at an average of ~4-fold faster than the ASE2 ($p<0.05$, two-tailed t-test). This pattern is consistent with the functional roles of these elements, as the ASE2, but not ASE1, region is reported to form an 18-nt long base pairing with 5-HT_{2C}R pre-mRNA to ensure its proper modification [4,19,20]. This result suggests that the ASE2 region of HBII-52 has been subjected to strong functional selective constraints. The evolution rate of the remaining sequence was between that of the ASE1 and ASE2, a result that might be partially attributed to the involvement of two putatively functional motifs, namely, the C' and D' boxes, in this region [1]. Based on our analysis, the general evolution rate of different parts of the HBII-52 snoRNA genes falls in the order of ASE1>remaining sequence>ASE2>stem region>C/D boxes. For the HBII-85 families, the ASE1 region evolved with a significant lower rate than ASE2 ($p<0.01$, two-tailed t-test) and might have been subjected to functional selection. The molecular target of HBII-85 snoRNAs has yet to be identified; and our results suggest that their potential target(s) are determined by the ASE1 region.

Positive Selection and Adaptive Signature of PWS-related Imprinted snoRNA Clusters in Euarchontoglires

Tajima's *D* test is widely used to detect positive selection on non-coding sequences [34,63]. To exam whether the imprinted snoRNA gene families in Euarchontoglires are subjected to positive selection, the Tajima's *D* test was conducted on the HBII-52 and the HBII-85 families of each species. The application is based on the consideration that the gene members within a genome are genetically closely related and originate from common ancestors. The Tajima's *D* values for the HBII-52 family in Euarchontoglires are all negative with remarkable significance (Table 4), suggesting these sequences are subjected to positive selection. Interestingly, although 70 HBII-52 gene members were found in the elephant genome, the Tajima's *D* value is non-significant (data not shown). The Tajima's *D* values for HBII-85 family of Euarchontoglires are all negative, but only significant for mice and rats, suggesting a relieved selection force on the primates genomes. For comparison, the 6 adjacent protein-coding genes in the same imprinted locus were also tested for selection. The K_a/K_s values for the six protein-coding genes were all $<<1$; half of them were suggested to subject to purifying selection (Table 5). Since these protein-coding genes are falling in the same imprinted region as the snoRNA cluster, this result suggests again that the selection is specific for the PWS-related snoRNAs rather than for the entire imprinted region.

The HBII-52 snoRNAs are thought to regulate the modification and/or splicing of 5-HT_{2C}R pre-mRNA through an 18-nt long base pairing. Since 5-HT_{2C}R is an important neural signal transduction molecule, abnormal regulation of 5-HT_{2C}R by HBII-52 was proposed to be involved in the PWS. We therefore investigated whether the regulation of 5-HT_{2C}R by HBII-52 is a universal phenomenon in eutherians. For this purpose, the binding pattern between HBII-52 and 5-HT_{2C}R mRNA was determined in a number of species (Figure 6A). The results showed that the 18-nt perfect match between HBII-52 and 5-HT_{2C}R only presented in primates and rodents. Because the D box of snoRNA is vital for snoRNPs binding, the start site of the pairing was also investigated. Similarly, in primates and rodents, most base pairing began just after the D box, while in other species the start sites were random

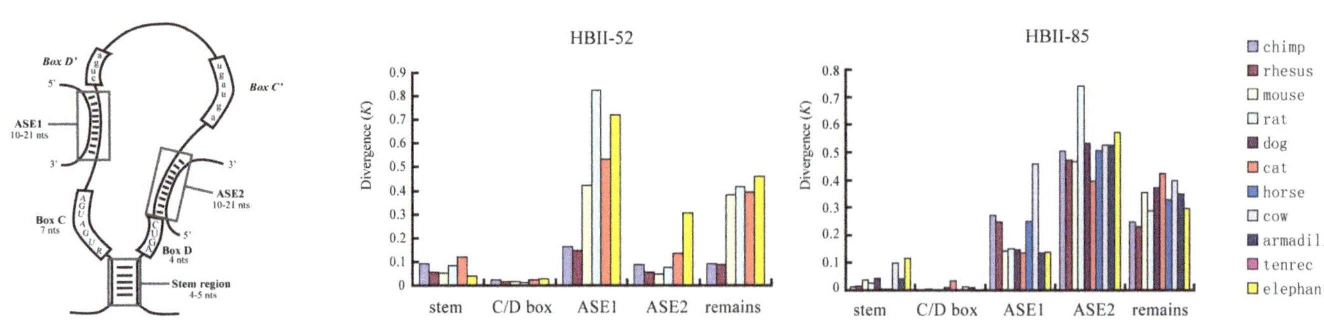

Figure 5. Evolutionary divergence rates for each part of the imprinted snoRNA genes between the human and other mammals. A diagram of box C/D snoRNA structure with different partitions was illustrated in the left [modified from 25]. *K* denotes the number of substitutions per site. Each color represents a comparison between human and another species.

Table 4. Tajima's *D* test of Euarchontoglires imprinted snoRNA gene family.

	HBII-52					HBII-85				
	HS	PT	RM	MM	RN	HS	PT	RM	MM	RN
Tajima's *D*	**−2.07**	**−1.91**	**−2.13**	**−2.48**	**−2.55**	−1.27	−1.16	−1.11	**−2.41**	**−2.48**
P value	<0.05	<0.05	<0.05	<0.01	<0.001	>0.10	>0.10	>0.10	<0.01	<0.001

Species abbreviation: HS, humans; PT, chimpanzees; RM, rhesus; MM, mice; RN, rats. The significant ($P<0.05$) test results were labeled bold.

relative to the D box. To further comparing the HBII-52:5HT_{2C}R binding patterns between primates/rodents and the other mammals, we investigated the base complementarity of the longest continuous perfect matches in humans, elephants and horses respectively (Figure 6B). The data showed that in elephants and horses there are two mismatches within the first five nucleotide positions upstream of the D box, which is important for the efficient modification of the fifth nucleotide, i.e. the C site of the A-I editing of 5HT_{2C}R mRNA [64,65]. These results therefore imply that the regulatory role of HBII-52 might first evolve in the common ancestor of primates and rodents.

Discussion

Rapid Evolution of the PWS-related Imprinted snoRNA Clusters via a Birth-and-death Process

SnoRNA genes have traditionally been regarded as conserved components of the genome, mainly due to their function of guiding the methylation and pseudouridylation of rRNA. Most snoRNA genes show great homogeneity among distant species, even between humans and yeast. In striking contrast, the snoRNAs in the imprinted PWS locus have undergone rapid evolution during the diversification of mammals. We observed substantial increases in gene number of the PWS-related imprinted snoRNAs in primates and rodents, and their gene birth rates were more than 80-fold higher than the non-imprinted snoRNAs. At the same time, their sequence evolution rate was also significantly higher than that of the non-imprinted snoRNAs. It is therefore of particular interest to study the evolutionary process underlying the rapid evolution of this imprinted snoRNA gene cluster.

Although rapid evolution of small non-coding RNAs (sncRNAs) has frequently been observed in recent years [59,60], few reports have investigated the evolutionary history of sncRNA genes in depth [59,60,66]. In this study, the remarkable evolutionary dynamics of the HBII-85 and HBII-52 gene families provide a clear example of the birth-and-death evolution. The evidence for this conclusion can be summarized as follows: (i) similar intra- and inter-genomic divergences and close phylogenetic relationships among the intra-species gene members (Table S5, Figure S1); (ii) the functional regions of the imprinted snoRNA genes are subjected to strong selective constraints; and (iii) a number of pseudogenes, representing the remnants of gene death events, were observed pervasively. Moreover, the birth-and-death process can finely explain the high variations in gene number of the PWS-related snoRNA families.

The balance between gene birth and death might have determined the gene number within the cluster in each species. In primates and rodents, the balance tends toward gene birth, while in other lineages, it leans toward gene death. The birth-and-death evolutionary model can also explain the lack of genes in the ancestral species. For such ancestral genes likely to turn to death during the diversification of mammals and were replaced by the newborn lineage-specific genes, which share few common feature among species. Thus, the ancestral gene in the common ancestor of Armadillo/Elephant and the most recent common ancestor of all placental mammals are low.

The rapid evolution of the imprinted snoRNA genes might partially due to the different way they expressed. In contrast to biallelically-expressed genes, imprinted genes are expressed from only one allele with the other allele is repressed. Because of this lack of a backup copy, similar selective force is predicted to trigger more efficient responses from imprinted genes relative to the non-imprinted counterparts. Moreover, concerning the parental interests conflict of imprinted genes, other selective constraints

Table 5. Ka/Ks tests of protein-coding genes in the PWS locus.

	Protein-coding genes					
Gene	**UBE3A**	**SNRPN**	**ATP10A**	**NDN**	**MKRN3**	**MAGEL2**
Ks	0.004	0	0.024	0.004	0.011	0.033
Ka	0	0	0.004	0	0.002	0.004
Ka/Ks	0	-	0.167	0	0.182	0.121
Purifying selection	0.151	ND	**0**	0.151	**0.045**	**0.029**

P values<0.05 for purifying selection are in bold.

might be evoked, e.g. mother-infant co-adaptation [67,68]. We therefore anticipate that other imprinted genes will exhibit rapid evolution under certain selective forces. The birth-and-death evolution of multi-gene families has played important roles during the origin and development of some complex genetic systems, such as the adaptive immune system, the homeobox super gene family and the olfactory system [69]. The PWS-related imprinted snoRNA genes have undergone a similar process at an outstanding evolutionary rate. We anticipate that they might be involved in the development of a complex system that has elaborately evolved in the mammalian lineage.

The Evolutionary Pattern of Other Imprinted Small Non-coding RNAs

Mammals have another large imprinted snoRNA cluster located in the 14q32 region of the human genome. This cluster mainly contain two multi-gene families, the 14qI and 14qII family; a rat-specific imprinted snoRNA gene family, RBII-36, is also located in this region.[4]. Our analyses revealed that the 14qI and 14qII gene family also exhibit rapid sequence substitution than non-imprinted snoRNAs (Figure 7). But significant selective constraint on the ASE was not observed (Figure 8), suggesting a lack of functional constraint on these genes. This finding is consistent with the fact that the function(s) of 14qI or 14qII families is still unknown. The phylogenetic analysis of 14qI family showed a different pattern comparing to that of the HBII-52 or HBII-85 families (Figure S6). The 14qI snoRNA genes were clustered with their orthologous, a sign of divergent evolution. Thus, our results indicate that imprinted snoRNA genes located in different genome regions are subjected to different evolution process, which might be highly related to the selective force they received. Further determining the biological functions of the imprinted snoRNA genes will help to unveil the significance of their rapid evolution.

There are studies suggest positive selection for microRNAs [70]. The microRNA cluster located in the *DLK-DIO3* imprinted locus is the largest microRNA cluster discovered thus far. A study on the evolution of this microRNA cluster revealed that most of the experimentally validated miRNAs could be detected in nearly all the sequenced eutherian genomes [71]. This finding suggests that the structure of this miRNA cluster is evolutionarily stable, which is different from the evolutionary pattern of the PWS-related snoRNA cluster in this study.

Adaptive Evolution of the PWS-related snoRNA Gene Cluster and Neural Development in Euarchontoglires

The PWS-related imprinted snoRNAs in Euarchontoglires exhibit adaptive signatures. A population data based study on the HBII-52 gene family of human suggested that positive selection and gene conversion have driven the evolution of the imprinted snoRNA gene cluster [34]. Consistent with this report, our study also detected remarkable positive selection on the HBII-52 gene family as well as the HBII-85 gene family in primates and rodents. In addition, primates and rodents have more imprinted snoRNA genes than other placental mammals (with the exception of elephants). Changes in gene number often are accompanied by the adaptation of a species to its environment. For example, an analysis of the changes in olfactory receptor gene number among mammals revealed that primates and the platypus have fewer genes than other mammals [45]. The development of the trichromatic vision system in primates greatly improved their visual sensory abilities, thereby reducing the importance of olfactory sensation. The platypus, in turn, has adapted to a semi-aquatic lifestyle and relies less heavily on olfactory sensation. Similarly, the remarkable increase in PWS-related imprinted snoRNA genes might be also needed to meet certain functional demands during mammalian evolution.

Three lines of evidence suggest that the PWS-related imprinted snoRNAs are undergoing functional evolution: (i) the imprinted snoRNAs have different expression patterns among mammals, e.g. human HBII-85 snoRNAs have a wider expression profile than MBII-85 and RBII-85 (MBII-85 and RBII-85 are the homolog of HBII-85 in mice and rats, respectively)[7], which might closely relate to their functional diversification; (ii) at the phenotypic level, the site-specific deletion of the entire murine MBII-85 gene cluster did not result in obesity, a major phenotype observed in the human PWS [24]; (iii) the interactions between the HBII-52 snoRNAs and its target, 5-HT_{2C}R mRNA, are different among mammalian lineages. Our results showed that the 18-nt perfect match between HBII-52 and 5-HT_{2C}R was only found in primates and rodents; while in other lineages, the lengths of the continuous perfect match were limited to 14 nts and with mismatches within the five nucleotide upstream box D, which region is important for the precise modification of the fifth nucleotide, i.e. the C site for the A-I editing of 5HT2CR mRNA[19]. Cavaillé and Bachellerie has indicated that two nucleotides of mismatch in the box D-proximal portion will greatly impair the modification efficiency of the target site, even in a long G/C-rich duplex [65]. While, the RNA duplex between HBII-52 and $5HT_{2C}$R mRNA is A/U rich (11/18), whose interaction is expected to be more sensitive to mismatches and a single mutation could lead to a 100-fold inhibition of the modification in a A/U rich duplex [64]. Thus, the snoRNA guided modifications is most likely invalid here. Moreover, previous studies have demonstrated that mismatches between snoRNA and exon Vb significantly influence the regulation of alternative splicing of 5-HT_{2C}R pre-mRNA [20]. Thus, we speculate that the mRNA regulatory function for HBII-52 first evolved in the common ancestor of primates and rodents. Although 70 HBII-52 gene members are found in the elephant genome, they do not have a long

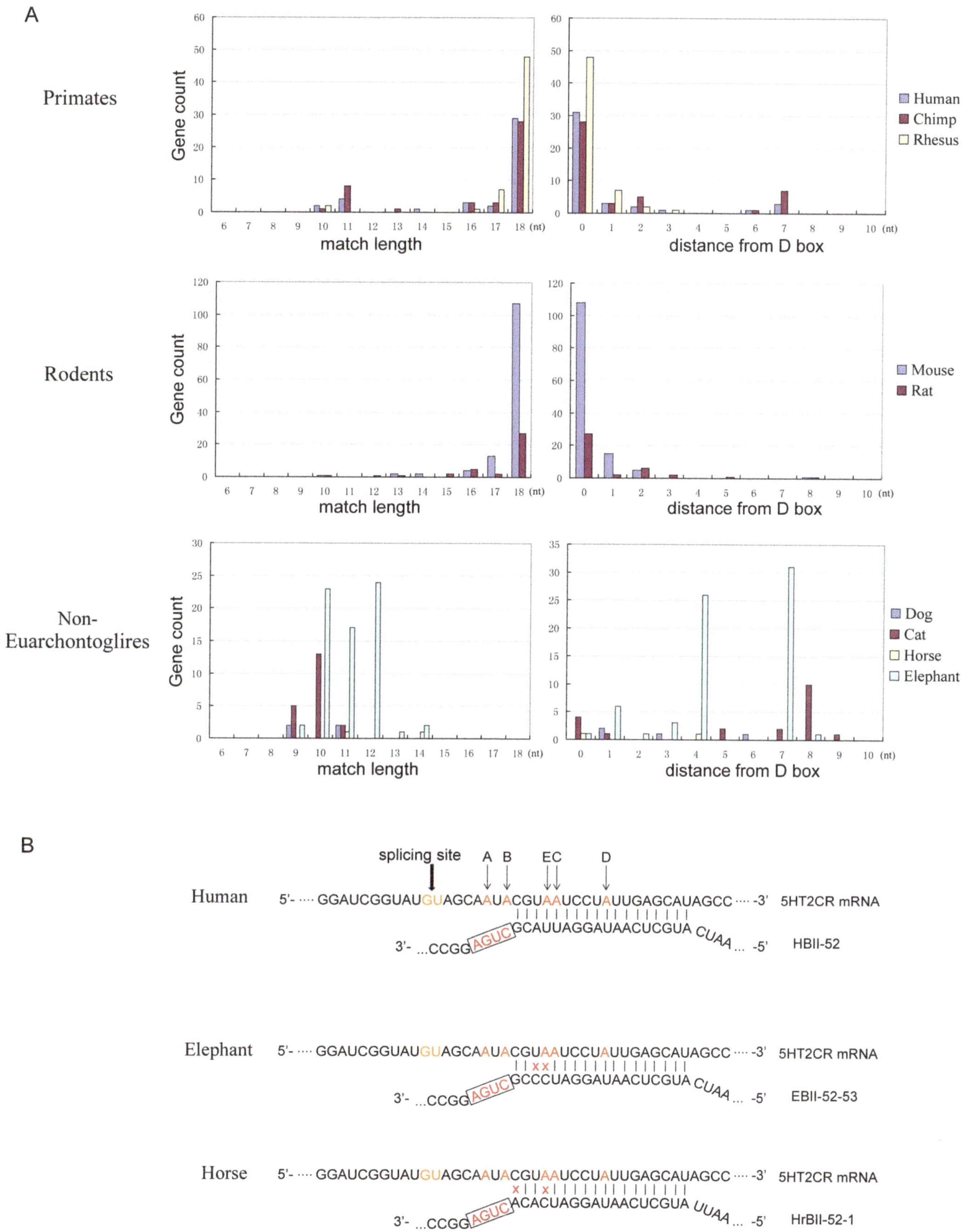

Figure 6. The binding pattern between HBII-52 and 5-HT$_{2C}$R mRNA in mammals. (A) Left panels indicate the length distribution of perfect base-pairing between HBII-52 and 5-HT$_{2C}$R mRNA. Right panels indicate the distance of base-pairing from the D box motif. Analyses of primates,

rodents and other mammals are indicated with different colors. In primates and rodents, the length of the base-pairing was centered approximately 18 nt and started just upstream of the D box. In other species, the length of the base-pairing ranged from 9–14 nt and started several nucleotides away from the D box. (B) The best base complementarity between the antisense element of the human, elephant or horse HBII-52 snoRNA and the corresponding 5-$HT_{2C}R$ receptor. Nucleotides in red indicate the A to I editing sites (A to E); nucleotides in orange indicate proximal splice site. Box, D box. Red crosses indicate mismatch within the RNA duplex.

complementary sequence to the 5-$HT_{2C}R$ pre-mRNA. It is possible that the rapid expansion of the HBII-52 gene family in the elephant genome is for other function(s).

How and to what extent the rapid evolution of PWS-related imprinted snoRNA cluster has contributed to the adaptive evolution of primates and rodents remain unclear. Many imprinted genes play important roles in brain development and functions [9]. Typically, maternally expressed imprinted genes have effects on 'higher' cognitive systems, whereas paternal genes have effects on brain systems that fill 'emotional' or autonomic functions [72]. The 15q11–13 imprinted snoRNA cluster is paternally expressed in the adult brain [73], suggesting that these genes might be involved in emotional function. Several lines of evidence suggest that the evolution of the 15q11–13 snoRNA cluster has contributed to the fine control of emotion and the related social behaviors of individuals. The HBII-52 snoRNAs could regulatethe 5-HT_2C receptor pre-mRNA the A-to-I RNA editing and alternative splicing of the 5-HT_2C receptor pre-mRNA. 5-HT (5-hydroxytryptamine, or serotonin) is a very important monoamine neurotransmitter that is well-known to contribute to feelings of well-being and happiness. The 5-HT_2C receptor is a subtype of the 5-HT receptor, which regulates mood, anxiety, feeding, and reproductive behavior [74]. HBII-52 can inhibit its RNA editing at the C-site in the nucleolus, which is crucial for decreasing the efficiency for activating the G-proteins. The functionally compromised 5-HT_2C receptor that results from this mis-editing has been detected in the depressed suicide victims [75]. HBII-52 also participates in the regulation of alternative splicing of the 5-HT_2C receptor pre-mRNA by promoting exon Vb inclusion. Without exon Vb, the mRNA does not encode a functional 5-HT_2CR receptor [20]. The evolving HBII-52 snoRNAs may therefore have participated in the fine regulation of the 5-HT neurotransmission pathway by modulating the 5-HT_2C receptor in primates and rodents.

Emotion is a vital factor affecting individual social behaviors. Two key characteristics of Angelman syndrome are an unusually sociable disposition and reduced display of negative-affect signals (stubbornness and temper tantrums) [13]. Conversely, individuals with PWS show increased negative-affect signals and are prone to mood instability and non-psychotic depression [76,77]. These findings have led to the idea that, in addition to influencing the resource allocation from the mother to the offspring through effects on, such as, suckling behavior, brain-expressed imprinted genes can also influence the social resources that the offspring receives, especially in the social animals like primates. These resources include food, protection and reproductive opportunities.

As imprinted genes play important roles in brain development and function, it has been suggested these genes might be subject to selection [9]. In this study, we revealed the rapid adaptive evolution of imprinted snoRNA genes in the PWS locus. The

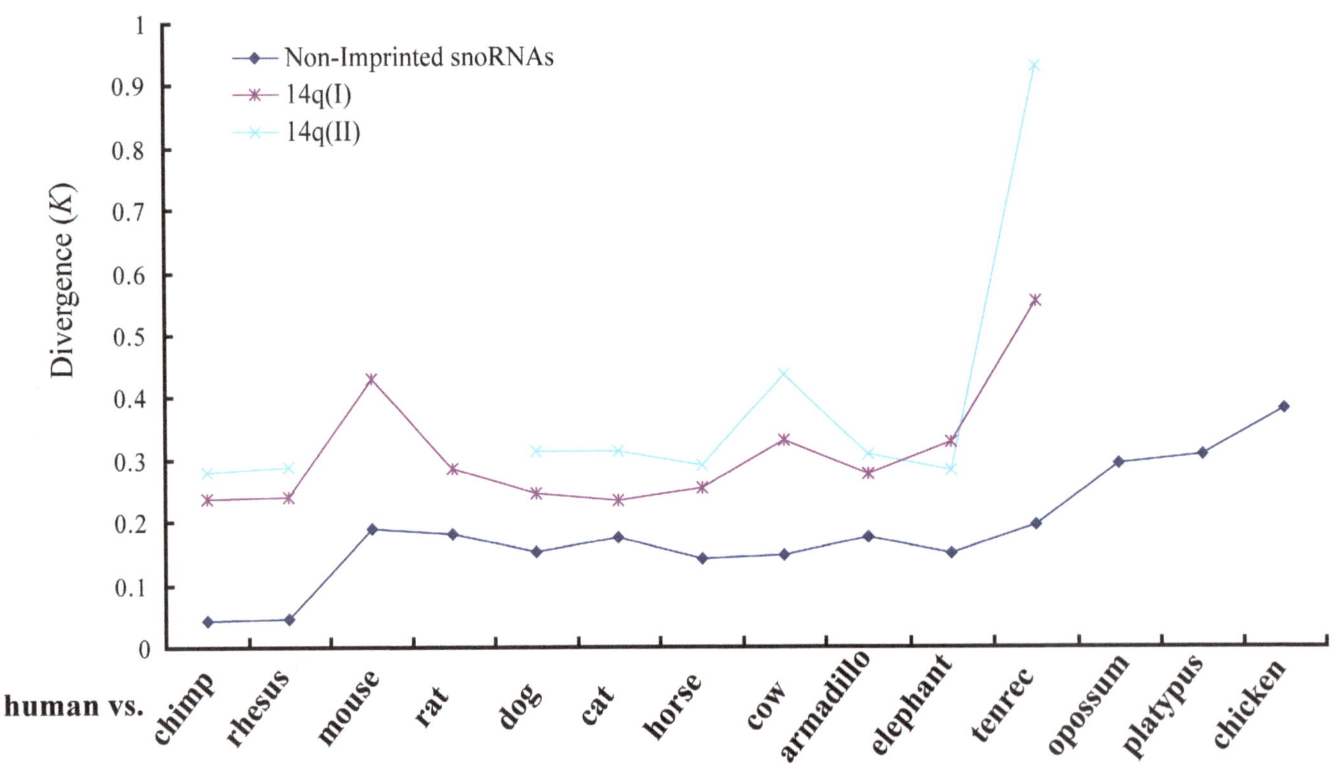

Figure 7. Evolutionary divergence of imprinted 14q(I) and 14q(II) snoRNA family vs. non-imprinted snoRNA gene families between the human and other mammals. *K* denotes the number of substitutions per site.

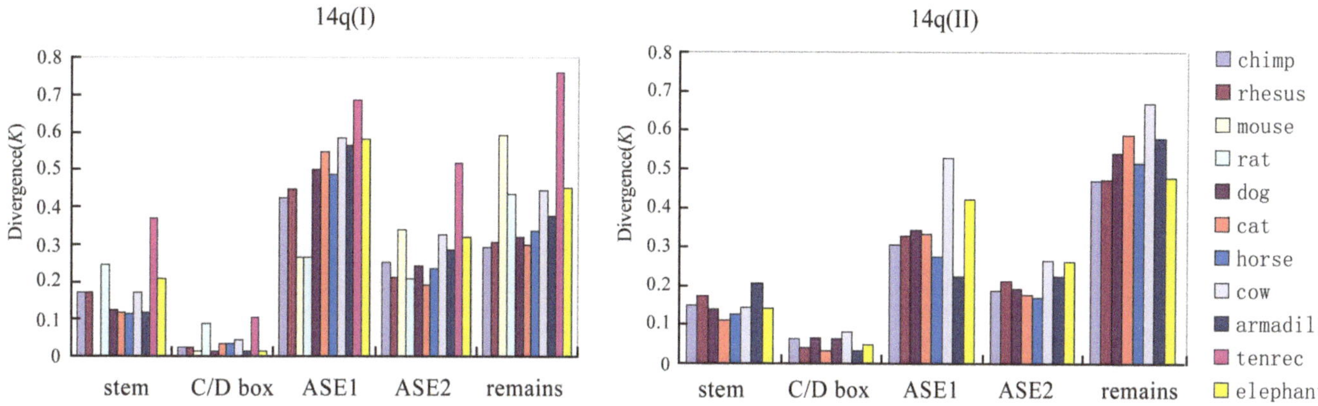

Figure 8. Evolutionary divergence rates for each part of the imprinted snoRNA genes in 14q32 region between the human and other mammals.

accelerated evolution of PWS-related imprinted snoRNA genes in Euarchontoglires might have contributed to the fine-tuning of emotion control and social behavior, which may have provided benefits such as greater allocation of social resources like reproductive opportunities. These consequences are clearly advantageous for the survival of individuals in the animal societies.

Supporting Information

Figure S1 Phylogenetic network of imprinted snoRNAs in the PWS locus. Gene families were indicted with different colors. Species abbreviations: H, human; MM, mouse; RN, rat; CF, dog; BT, cow; DN, armadillo; LA, elephant.

Figure S2 Gene numbers and changes in in the HBII-52 gene family. Numbers in black indicate the current gene number in each species, whereas red numbers indicate the ancestral gene number, blue numbers denote gained genes, and green numbers indicate lost genes.

Figure S3 Gene numbers and changes in in the HBII-85 gene family.

Figure S4 Combined physical-phylogenetic map. Imprinted snoRNA genes are depicted as vertical bar as in figure 1. The consistency between phylogenetic relationship and physical distribution suggest the genes are created by tandem duplication.

Figure S5 Human HBII-85 and HBII-52 pseudogenes. Selected pseudogenes were aligned with corresponding normal genes. Red arrow indicate loss-of-function mutations in these pseudo genes.

Figure S6 Phylogenetic analysis of 14q(I) gene family. Abbreviations: HS, human; PT, chimpanzee; RM, rhesus; MM, mouse; RN, rat; CF, dog; FC, cat; EC, horse; BT, cow; DN, armadillo; LA, elephant.

Table S1 Chromosome loci, size, and the genomic position of PWS imprinted regions in 12 eutherians.

Table S2 Pseudogenes identified in the PWS imprinted region.

Table S3 Conserved non-imprinted box C/D snoRNA gene families between human and other species.

Table S4 Nucleotide divergence between human and other species for non-imprinted and imprinted snoRNA genes.

Table S5 Intra- and inter- genomic divergence for PWS-related imprinted snoRNA genes.

Data S1 Sequences of imprinted and non-imprinted snoRNA genes.

Data S2 The sequence alignments for phylogenetic analyses.

Acknowledgments

We thank Xiao-Hong Chen, Qiao-Juan Huang and Yi-Ling Chen for technical assistance. We gratefully thank Peng Zhang for careful reading of the manuscript and helpful suggestions.

Author Contributions

Conceived and designed the experiments: YJZ LHQ. Analyzed the data: YJZ JHY QSS LLZ JL. Contributed reagents/materials/analysis tools: H.Zhou H.Zhang. Wrote the paper: YJZ LHQ.

References

1. Bachellerie JP, Cavaille J, Huttenhofer A (2002) The expanding snoRNA world. Biochimie 84: 775–790.
2. Brown JW, Echeverria M, Qu LH (2003) Plant snoRNAs: functional evolution and new modes of gene expression. Trends Plant Sci 8: 42–49.
3. Runte M, Huttenhofer A, Gross S, Kiefmann M, Horsthemke B, et al. (2001) The IC-SNURF-SNRPN transcript serves as a host for multiple small nucleolar RNA species and as an antisense RNA for UBE3A. Hum Mol Genet 10: 2687–2700.

4. Cavaille J, Seitz H, Paulsen M, Ferguson-Smith AC, Bachellerie JP (2002) Identification of tandemly-repeated C/D snoRNA genes at the imprinted human 14q32 domain reminiscent of those at the Prader-Willi/Angelman syndrome region. Hum Mol Genet 11: 1527–1538.
5. Cavaille J, Vitali P, Basyuk E, Huttenhofer A, Bachellerie JP (2001) A novel brain-specific box C/D small nucleolar RNA processed from tandemly repeated introns of a noncoding RNA gene in rats. J Biol Chem 276: 26374–26383.
6. de los Santos T, Schweizer J, Rees CA, Francke U (2000) Small evolutionarily conserved RNA, resembling C/D box small nucleolar RNA, is transcribed from PWCR1, a novel imprinted gene in the Prader-Willi deletion region, which Is highly expressed in brain. Am J Hum Genet 67: 1067–1082.
7. Cavaille J, Buiting K, Kiefmann M, Lalande M, Brannan CI, et al. (2000) Identification of brain-specific and imprinted small nucleolar RNA genes exhibiting an unusual genomic organization. Proc Natl Acad Sci U S A 97: 14311–14316.
8. Bortolin-Cavaille ML, Cavaille J (2012) The SNORD115 (H/MBII-52) and SNORD116 (H/MBII-85) gene clusters at the imprinted Prader-Willi locus generate canonical box C/D snoRNAs. Nucleic Acids Res 40: 6800–6807.
9. Wilkinson LS, Davies W, Isles AR (2007) Genomic imprinting effects on brain development and function. Nat Rev Neurosci 8: 832–843.
10. Castle JC, Armour CD, Lower M, Haynor D, Biery M, et al. (2010) Digital genome-wide ncRNA expression, including SnoRNAs, across 11 human tissues using polyA-neutral amplification. PLoS One 5: e11779.
11. Peters J (2008) Prader-Willi and snoRNAs. Nat Genet 40: 688–689.
12. Belmonte MK, Cook EH, Jr., Anderson GM, Rubenstein JL, Greenough WT, et al. (2004) Autism as a disorder of neural information processing: directions for research and targets for therapy. Mol Psychiatry 9: 646–663.
13. Bolton PF, Veltman MW, Weisblatt E, Holmes JR, Thomas NS, et al. (2004) Chromosome 15q11-13 abnormalities and other medical conditions in individuals with autism spectrum disorders. Psychiatr Genet 14: 131–137.
14. Cook EH, Jr., Scherer SW (2008) Copy-number variations associated with neuropsychiatric conditions. Nature 455: 919–923.
15. de Smith AJ, Purmann C, Walters RG, Ellis RJ, Holder SE, et al. (2009) A deletion of the HBII-85 class of small nucleolar RNAs (snoRNAs) is associated with hyperphagia, obesity and hypogonadism. Hum Mol Genet 18: 3257–3265.
16. Sahoo T, del Gaudio D, German JR, Shinawi M, Peters SU, et al. (2008) Prader-Willi phenotype caused by paternal deficiency for the HBII-85 C/D box small nucleolar RNA cluster. Nat Genet 40: 719–721.
17. Duker AL, Ballif BC, Bawle EV, Person RE, Mahadevan S, et al. (2010) Paternally inherited microdeletion at 15q11.2 confirms a significant role for the SNORD116 C/D box snoRNA cluster in Prader-Willi syndrome. Eur J Hum Genet 18: 1196–1201.
18. Runte M, Varon R, Horn D, Horsthemke B, Buiting K (2005) Exclusion of the C/D box snoRNA gene cluster HBII-52 from a major role in Prader-Willi syndrome. Hum Genet 116: 228–230.
19. Vitali P, Basyuk E, Le Meur E, Bertrand E, Muscatelli F, et al. (2005) ADAR2-mediated editing of RNA substrates in the nucleolus is inhibited by C/D small nucleolar RNAs. J Cell Biol 169: 745–753.
20. Kishore S, Stamm S (2006) The snoRNA HBII-52 regulates alternative splicing of the serotonin receptor 2C. Science 311: 230–232.
21. Kishore S, Khanna A, Zhang Z, Hui J, Balwierz PJ, et al. (2010) The snoRNA MBII-52 (SNORD 115) is processed into smaller RNAs and regulates alternative splicing. Hum Mol Genet 19: 1153–1164.
22. Doe CM, Relkovic D, Garfield AS, Dalley JW, Theobald DE, et al. (2009) Loss of the imprinted snoRNA mbii-52 leads to increased 5htr2c pre-RNA editing and altered 5HT2CR-mediated behaviour. Hum Mol Genet 18: 2140–2148.
23. Ding F, Li HH, Zhang S, Solomon NM, Camper SA, et al. (2008) SnoRNA Snord116 (Pwcr1/MBII-85) deletion causes growth deficiency and hyperphagia in mice. PLoS ONE 3: e1709.
24. Skryabin BV, Gubar LV, Seeger B, Pfeiffer J, Handel S, et al. (2007) Deletion of the MBII-85 snoRNA gene cluster in mice results in postnatal growth retardation. PLoS Genet 3: e235.
25. Kiss T (2002) Small nucleolar RNAs: an abundant group of noncoding RNAs with diverse cellular functions. Cell 109: 145–148.
26. Piekna-Przybylska D, Decatur WA, Fournier MJ (2007) New bioinformatic tools for analysis of nucleotide modifications in eukaryotic rRNA. Rna 13: 305–312.
27. Lestrade L, Weber MJ (2006) snoRNA-LBME-db, a comprehensive database of human H/ACA and C/D box snoRNAs. Nucleic Acids Res 34: D158–162.
28. Schmitz J, Zemann A, Churakov G, Kuhl H, Grutzner F, et al. (2008) Retroposed SNOfall–a mammalian-wide comparison of platypus snoRNAs. Genome Res 18: 1005–1010.
29. Edwards CA, Mungall AJ, Matthews L, Ryder E, Gray DJ, et al. (2008) The evolution of the DLK1-DIO3 imprinted domain in mammals. PLoS Biol 6: e135.
30. Rapkins RW, Hore T, Smithwick M, Ager E, Pask AJ, et al. (2006) Recent assembly of an imprinted domain from non-imprinted components. PLoS Genet 2: e182.
31. Zhang Y, Qu L (2009) Non-coding RNAs and the acquisition of genomic imprinting in mammals. Sci China C Life Sci 52: 195–204.
32. Royo H, Cavaille J (2008) Non-coding RNAs in imprinted gene clusters. Biol Cell 100: 149–166.
33. Nahkuri S, Taft RJ, Korbie DJ, Mattick JS (2008) Molecular evolution of the HBII-52 snoRNA cluster. J Mol Biol 381: 810–815.
34. Ogorelkova M, Navarro A, Vivarelli F, Ramirez-Soriano A, Estivill X (2009) Positive selection and gene conversion drive the evolution of a brain-expressed snoRNAs cluster. Mol Biol Evol 26: 2563–2571.
35. Yang JH, Zhang XC, Huang ZP, Zhou H, Huang MB, et al. (2006) snoSeeker: an advanced computational package for screening of guide and orphan snoRNA genes in the human genome. Nucleic Acids Res 34: 5112–5123.
36. Ronquist F, Huelsenbeck JP (2003) MrBayes 3: Bayesian phylogenetic inference under mixed models. Bioinformatics 19: 1572–1574.
37. Kumar S, Tamura K, Nei M (2004) MEGA3: Integrated software for Molecular Evolutionary Genetics Analysis and sequence alignment. Brief Bioinform 5: 150–163.
38. Thompson JD, Gibson TJ, Plewniak F, Jeanmougin F, Higgins DG (1997) The CLUSTAL_X windows interface: flexible strategies for multiple sequence alignment aided by quality analysis tools. Nucleic Acids Res 25: 4876–4882.
39. Huson DH, Bryant D (2006) Application of phylogenetic networks in evolutionary studies. Mol Biol Evol 23: 254–267.
40. Posada D, Crandall KA (1998) MODELTEST: testing the model of DNA substitution. Bioinformatics 14: 817–818.
41. Letunic I, Bork P (2007) Interactive Tree Of Life (iTOL): an online tool for phylogenetic tree display and annotation. Bioinformatics 23: 127–128.
42. Goodman M, Czelusniak J, Moore G, Romero-Herrera AE, Matsuda G (1979) Fitting the gene lineage into its species lineage, a parsimony strategy illustrated by cladograms constructed from globin sequences. Syst Zool 28: 132–168.
43. Page RD, Charleston MA (1997) From gene to organismal phylogeny: reconciled trees and the gene tree/species tree problem. Mol Phylogenet Evol 7: 231–240.
44. Nam J, Nei M (2005) Evolutionary change of the numbers of homeobox genes in bilateral animals. Mol Biol Evol 22: 2386–2394.
45. Niimura Y, Nei M (2003) Evolution of olfactory receptor genes in the human genome. Proc Natl Acad Sci U S A 100: 12235–12240.
46. Niimura Y, Nei M (2007) Extensive gains and losses of olfactory receptor genes in Mammalian evolution. PLoS ONE 2: e708.
47. Hedges SB, Dudley J, Kumar S (2006) TimeTree: a public knowledge-base of divergence times among organisms. Bioinformatics 22: 2971–2972.
48. Kimura M (1980) A simple method for estimating evolutionary rates of base substitutions through comparative studies of nucleotide sequences. J Mol Evol 16: 111–120.
49. Librado P, Rozas J (2009) DnaSP v5: a software for comprehensive analysis of DNA polymorphism data. Bioinformatics 25: 1451–1452.
50. Tajima F (1989) Statistical method for testing the neutral mutation hypothesis by DNA polymorphism. Genetics 123: 585–595.
51. Koonin EV, Wolf YI (2010) Constraints and plasticity in genome and molecular-phenome evolution. Nat Rev Genet 11: 487–498.
52. Tamura K, Dudley J, Nei M, Kumar S (2007) MEGA4: Molecular Evolutionary Genetics Analysis (MEGA) software version 4.0. Mol Biol Evol 24: 1596–1599.
53. Royo H, Basyuk E, Marty V, Marques M, Bertrand E, et al. (2007) Bsr, a nuclear-retained RNA with monoallelic expression. Mol Biol Cell 18: 2817–2827.
54. Sandovici I, Kassovska-Bratinova S, Vaughan JE, Stewart R, Leppert M, et al. (2006) Human imprinted chromosomal regions are historical hot-spots of recombination. PLoS Genet 2: e101.
55. Murphy WJ, Eizirik E, O'Brien SJ, Madsen O, Scally M, et al. (2001) Resolution of the early placental mammal radiation using Bayesian phylogenetics. Science 294: 2348–2351.
56. Leister D (2004) Tandem and segmental gene duplication and recombination in the evolution of plant disease resistance gene. Trends Genet 20: 116–122.
57. Richly E, Kurth J, Leister D (2002) Mode of amplification and reorganization of resistance genes during recent Arabidopsis thaliana evolution. Mol Biol Evol 19: 76–84.
58. Meyers BC, Kozik A, Griego A, Kuang H, Michelmore RW (2003) Genome-wide analysis of NBS-LRR-encoding genes in Arabidopsis. Plant Cell 15: 809–834.
59. Meunier J, Lemoine F, Soumillon M, Liechti A, Weier M, et al. (2013) Birth and expression evolution of mammalian microRNA genes. Genome Res 23: 34–45.
60. Zheng GX, Ravi A, Gould GM, Burge CB, Sharp PA (2011) Genome-wide impact of a recently expanded microRNA cluster in mouse. Proc Natl Acad Sci U S A 108: 15804–15809.
61. Lu J, Shen Y, Wu Q, Kumar S, He B, et al. (2008) The birth and death of microRNA genes in Drosophila. Nat Genet 40: 351–355.
62. Lu J, Fu Y, Kumar S, Shen Y, Zeng K, et al. (2008) Adaptive evolution of newly emerged micro-RNA genes in Drosophila. Mol Biol Evol 25: 929–938.
63. Wang Y, Shen D, Bo S, Chen H, Zheng J, et al. (2010) Sequence variation and selection of small RNAs in domesticated rice. BMC Evol Biol 10: 119.
64. Cavaille J, Nicoloso M, Bachellerie JP (1996) Targeted ribose methylation of RNA in vivo directed by tailored antisense RNA guides. Nature 383: 732–735.
65. Cavaille J, Bachellerie JP (1998) SnoRNA-guided ribose methylation of rRNA: structural features of the guide RNA duplex influencing the extent of the reaction. Nucleic Acids Res 26: 1576–1587.
66. Zhang R, Wang YQ, Su B (2008) Molecular evolution of a primate-specific microRNA family. Mol Biol Evol 25: 1493–1502.
67. Plagge A, Gordon E, Dean W, Boiani R, Cinti S, et al. (2004) The imprinted signaling protein XL alpha s is required for postnatal adaptation to feeding. Nat Genet 36: 818–826.

68. Curley JP, Barton S, Surani A, Keverne EB (2004) Coadaptation in mother and infant regulated by a paternally expressed imprinted gene. Proc Biol Sci 271: 1303–1309.
69. Nei M, Rooney AP (2005) Concerted and birth-and-death evolution of multigene families. Annu Rev Genet 39: 121–152.
70. Li J, Liu Y, Xin X, Kim TS, Cabeza EA, et al. (2012) Evidence for positive selection on a number of MicroRNA regulatory interactions during recent human evolution. PLoS Genet 8: e1002578.
71. Glazov EA, McWilliam S, Barris WC, Dalrymple BP (2008) Origin, evolution, and biological role of miRNA cluster in DLK-DIO3 genomic region in placental mammals. Mol Biol Evol 25: 939–948.
72. Keverne EB, Martel FL, Nevison CM (1996) Primate brain evolution: genetic and functional considerations. Proc Biol Sci 263: 689–696.
73. Runte M, Kroisel PM, Gillessen-Kaesbach G, Varon R, Horn D, et al. (2004) SNURF-SNRPN and UBE3A transcript levels in patients with Angelman syndrome. Hum Genet 114: 553–561.
74. Heisler LK, Zhou L, Bajwa P, Hsu J, Tecott LH (2007) Serotonin 5-HT(2C) receptors regulate anxiety-like behavior. Genes Brain Behav 6: 491–496.
75. Gurevich I, Englander MT, Adlersberg M, Siegal NB, Schmauss C (2002) Modulation of serotonin 2C receptor editing by sustained changes in serotonergic neurotransmission. J Neurosci 22: 10529–10532.
76. Soni S, Whittington J, Holland AJ, Webb T, Maina E, et al. (2007) The course and outcome of psychiatric illness in people with Prader-Willi syndrome: implications for management and treatment. J Intellect Disabil Res 51: 32–42.
77. Cassidy SB (1997) Prader-Willi syndrome. J Med Genet 34: 917–923.

Substrate-Dependent Evolution of Cytochrome P450: Rapid Turnover of the Detoxification-Type and Conservation of the Biosynthesis-Type

Ayaka Kawashima, Yoko Satta*

Department of Evolutionary Studies of Biosystems, The Graduate University for Advanced Studies (Sokendai), Shonan Village, Hayama, Kanagawa, Japan

Abstract

Members of the cytochrome P450 family are important metabolic enzymes that are present in all metazoans. Genes encoding cytochrome P450s form a multi-gene family, and the number of genes varies widely among species. The enzymes are classified as either biosynthesis- or detoxification-type, depending on their substrates, but their origin and evolution have not been fully understood. In order to elucidate the birth and death process of cytochrome *P450* genes, we performed a phylogenetic analysis of 710 sequences from 14 vertebrate genomes and 543 sequences from 6 invertebrate genomes. Our results showed that vertebrate detoxification-type genes have independently emerged three times from biosynthesis-type genes and that invertebrate detoxification-type genes differ from vertebrates in their origins. Biosynthetic-type genes exhibit more conserved evolutionary processes than do detoxification-type genes, with regard to the rate of gene duplication, pseudogenization, and amino acid substitution. The differences in the evolutionary mode between biosynthesis- and detoxification-type genes may reflect differences in their respective substrates. The phylogenetic tree also revealed 11 clans comprising an upper category to families in the cytochrome P450 nomenclature. Here, we report novel clan-specific amino acids that may be used for the qualitative definition of clans.

Editor: Vladimir N. Uversky, University of South Florida College of Medicine, United States of America

Funding: This work was supported by Japan Society for the Promotion of Science (JSPS) Grants-in-Aid for Scientific Research (B) Grant Number 21370106 and for JSPS Fellows. This work was also supported in part by the Center for the Promotion of Integrated Sciences (CPIS) of Sokendai. The funders had no role in study design, data collection and analysis, decision to publish, or preparation of the manuscript.

Competing Interests: The authors have declared that no competing interests exist.

* Email: satta@soken.ac.jp

Introduction

Enzymes in the cytochrome P450 (CYP) family are heme-binding proteins. The first CYP protein was discovered in rat liver microsomes [1], and it was later functionally characterized as a monooxygenase [2]. Monooxygenases incorporate one of the two atoms of molecular oxygen into the substrate, which results in hydroxylation in the most cases. *CYP* genes form a multi-gene family and encode proteins with amino-acid sequence identities higher than 40%. Each family comprises subfamilies with amino-acid sequence identities higher than 55%. In the classification of CYPs, a clan is defined as a higher-order category of CYP families [3]. Although clans can be useful for defining the relationships among *CYP* genes in different phyla within each kingdom [4], the definition of "clan" is rather arbitrary compared with the definitions of "family" and "subfamily."

CYP genes are present in vertebrates, invertebrates, plants, fungi, and even some prokaryotes [5]. The number of known *CYP* genes in metazoan, plant, and fungus genomes is moderately large. For example, there are 115 *CYP* genes in the human genome, 97 in the sea squirt (*Ciona intestinalis*) (Cytochrome P450 Homepage: http://drnelson.uthsc.edu/cytochromep450.html [6]), 120 in the sea urchin (*Strongylocentrotus purpuratus*) [7], 457 in rice (*Oryza sativa*) [6], 272 in *Arabidopsis thaliana* [8], and 159 in *Aspergillus oryzae* [9]. In contrast, there are relatively a small number of *CYP* genes in eubacteria or Archaea, ranging from none in *Escherichia coli* to 33 in *Streptomyces avermitilis* [9]. Among metazoan *CYP* genes, *CYP51* is particularly conserved and participates in the synthesis of cholesterol, which is an essential component of the eukaryotic cell membrane. A possible prokaryotic homolog (*CYP51B1*) to the metazoan *CYP51* is reported in the genome of *Mycobacterium tuberculosis* [10]. It is therefore thought that *CYP51* is the most ancient *CYP* gene. Although the functional role of CYPs in prokaryotes is not well defined [11,12,13], the presence of eukaryotic *CYP* genes in prokaryotes indicates that the emergence of *CYP*s preceded the origin of eukaryotes [14]. However, it has also been suggested that bacterial *CYP51* arose through lateral transfer from plants [15]. Indeed, the absence of *CYP* genes in some bacteria, such as *E. coli*, suggests that they are not essential in prokaryotes.

CYPs are classified into two types, the detoxification type (D-type) and the biosynthesis type (B-type), on the basis of their substrates [4,16,17]. In humans, the D-type detoxifies xenobiotics such as plant alkaloids, aromatic compounds, fatty acids, and especially drugs. On the other hand, the B-type is involved in the biosynthesis of physiologically active chemicals such as steroids, cholesterols, vitamin D3, and bile acids. In general, however, only a few CYPs have well-defined substrates, and even in humans, the substrates of some CYPs remain unidentified. For this reason, the nomenclature of vertebrate CYPs (family or subfamily) is largely determined by significant phylogenetic clustering with known functional sequences in humans. We also adopt this phylogenetic method to classify each vertebrate gene as either B- or D-type.

Fission yeasts have two B-type *CYP* genes (*CYP51F1* and *CYP61A3*) but no D-type genes. D-type genes are therefore thought to have emerged from a B-type gene in eukaryotes [4,14,16]. Nelson *et al.* [9] performed a phylogenetic analysis of 1,572 metazoan *CYPs* and reported the presence of 11 clans. They further investigated the origin and evolutionary processes of these clans. However, the length of intermodal branches connecting the different clans were too short to unambiguously discern the phylogenetic relationships between clans, making it difficult to fully characterize early *CYP* evolution.

In vertebrates, the number of *CYP* genes per genome varies greatly between species; for example, humans have 115, mice have 185, and zebrafish have 81 (Cytochrome P450 Homepage). Among the 115 human *CYPs*, 57 are functional and 58 are pseudogenes, suggesting rapid gene turnover. The 115 genes constitute 18 families, and the number of subfamilies within each family ranges from 1 to 13. These *CYP* genes are distributed on all chromosomes except chromosomes 5, 16, and 17. Five clusters of closely related genes are located on chromosomes 1, 7, and 10 (one cluster each) and chromosome 19 (two clusters) [18,19]. One such cluster on chromosome 19 has been studied in primates and rodents as well from an evolutionary viewpoint, indicating that an initial tandem duplication occurred in an early mammalian ancestor and that gene duplications and/or rearrangements frequently occurred in a lineage-specific manner [20].

In this paper, we aim to elucidate the birth and death processes of vertebrate *CYP* genes. In particular, we compare and contrast the origin and evolution of B- and D-types, and present an evolutionary model of vertebrate *CYP* genes.

Materials and Methods

Sequence datasets and identification of B- and D-type genes in vertebrates and invertebrates

The nucleotide sequences of 115 *CYP* genes in the human genome were obtained from the Cytochrome P450 Homepage. Using these sequences as queries, we performed a basic local alignment search tool (BLAST) search by using BLASTn and downloaded coding sequences (CDS) of homologous nucleotide sequences from 14 vertebrate species (*Pan troglodytes*: CHIMP2.1.4, *Macaca mulatta*: MMUL_1, *Callithrix jacchus*: C_jacchus3.2.1, *Bos taurus*: UMD 3.1, *Canis lupus familiaris*: CanFam3.1, *Mus musculus*: GRCm38.p2, *Rattus norvegicus*: Rnor_5.0, *Monodelphis domestica*: BROADO5, *Gallus gallus*: Galgal4, *Taeniopygia guttata*: 3.2.4, *Anolis carolinensis*: AnoCar2.0, *Xenopus tropicalis*: JGI 4.2, *Oryzias latipes*: MEDAKA1.70, and *Danio rerio*: Zv9) from NCBI (http://www.ncbi.nlm.nih.gov/) or ENSEMBL databases (http://www.ensembl.org/index.html). In the BLAST search, the top two hits and the top five hits were retrieved when B- and D-type genes were used as queries, respectively. The nucleotide sequences of ref-seq from NCBI were obtained, and sequences from ENSEMBL were filtered by length (>1000 bp) and their identity with human genes. The extent of sequence identity was dependent on the divergence time between each vertebrate species and humans. For example, in fish, we filtered out sequences with identity >60%. Orthology was confirmed by the presence of a syntenic region and the presence of adjacent loci, if any.

The following invertebrate species were included in the analysis: amphioxus (*Branchiostoma floridae*), sea squirt (*C. intestinalis*), sea urchin (*S. purpuratus*), sea anemone (*Nematostella vectensis*), water flea (*Daphnia pulex*), and fruit fly (*Drosophila melanogaster*). Protein sequences obtained from the Cytochrome P450 Homepage were used for the analysis of invertebrate CYPs. Only protein sequences >350 amino acids in length were included in the phylogenetic analysis. Because of the too extensive sequence divergence between vertebrate and invertebrate *CYP* genes, BLAST searches of the NCBI and ENSEMBL databases were not performed.

Molecular evolutionary analysis

Vertebrate nucleotide sequences and invertebrate amino acid sequences in *CYP* coding regions were aligned separately using ClustalW [21] implemented in MEGA5 [22], and each alignment was further edited by hand. In the alignment of the vertebrate nucleotide sequences, we first translated them into the amino acid sequences and after checked by eye, reconverted them to the nucleotide sequences. We excluded sites at which >20% of the operating taxonomic units (OTUs) showed gaps. As a result, 28.7% of the aligned sites showed >60% identity, 48.5% showed >50% identity, and 71.9% showed >30% identity (data not shown). We then constructed Neighbor-joining (NJ) trees [23] using either nucleotide differences per site (p-distance) [24] or amino acid distances (JTT distance) [25]. We performed missing data treatment under both the pairwise deletion and complete deletion options. The maximum likelihood (ML) [26] method was used to test the tree topology. All methods for tree construction were implemented in MEGA5 [22].

Pseudogenization or deletion of genes

The nucleotide sequences of the *CYP* pseudogenes in the human genome were obtained from the Cytochrome P450 Homepage. We selected genes containing >1000 bp out of the 1500 bp CDS. We retrieved orthologous genes from other vertebrate genomes by performing BLAST searches, using the human sequences as queries. The orthologous sequences were aligned with their human counterparts by ClustalW. Based on this alignment, we searched other vertebrates for nonsense or frame-shift mutations found in humans. To estimate the time of pseudogenization, we calculated the ratio of non-synonymous substitutions to synonymous substitutions, always per site, for pairs comprising a pseudogene and an orthologous functional gene. Using this ratio, we estimated the pseudogenization time for all *CYP* pseudogenes from the formula in Sawai *et al.* [27]. We used the TimeTree (http://www.timetree.org/index.php [28]) as a reference for calibrating species divergence time. When an orthologous gene is absent (deletion) in any non-human vertebrate, we searched for the syntenic region in the genome in order to confirm the deletion.

Estimation of functional constraint

In order to compare the amino-acid substitution rate for each functional *CYP* gene in primates, we normalized the rate with the synonymous substitution rate. This normalization for each gene measured the degree of "functional constraint". To be complete, we assumed that the gene tree is the same as the species tree for four primates (humans, chimpanzees, rhesus macaques, and marmosets) and placed the numbers of synonymous and non-synonymous substitutions on each branch by the least squares method [29]. The degree of functional constraint $1 - f$ is obtained from the ratio (f) of the sum of non-synonymous substitutions to that of synonymous substitutions of all branches in each tree. Finally, we compared the degree of functional constraint or directly the f value between B-type and D-type genes by using the Mann–Whitney U test [30].

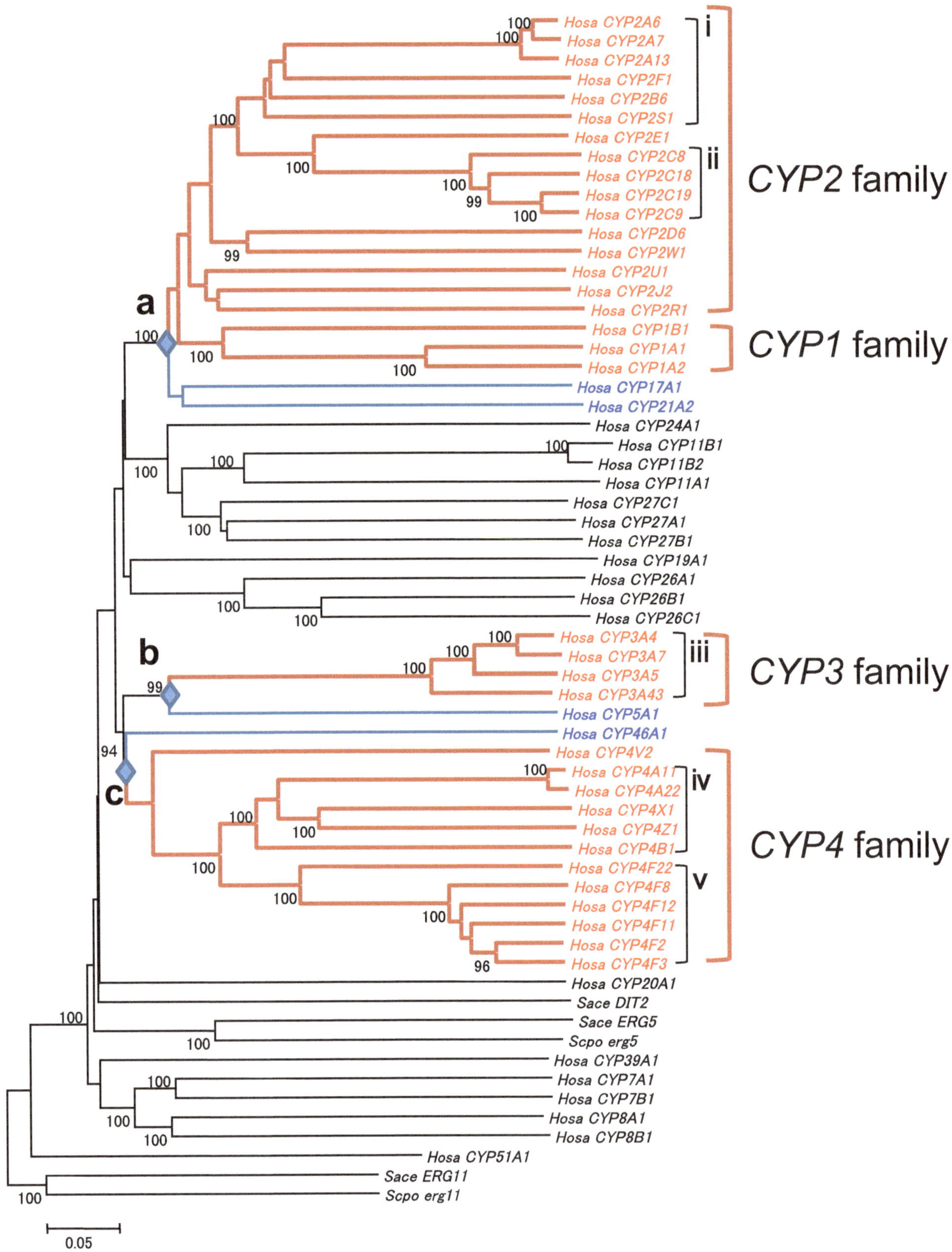

Figure 1. Phylogenetic tree of *cytochrome P450* genes in humans. The tree includes all functional *CYP* genes in humans (*Hosa*) and all yeasts (*Sace, Saccharomyces cerevisiae*; *Scpo, Schizosaccharomyces pombe*). The tree was constructed using the NJ method for nucleotide differences between the CDS and rooted with yeast *CYP51* gene sequences (*Sace ERG11* and *Scpo erg11*). Red text indicates D-type *CYP* genes, and black and blue text indicate B-type *CYP* genes. The *CYP1–4* families are indicated by a red bracket on the right side of the tree. Three diamonds (*a*, *b*, and *c*) indicate duplications of B- and D-type genes. The B-type genes that were the ancestors of D-type genes are indicated with a blue line and character. Black

brackets and roman numerals (i–v) at the tips of the tree show five clusters of D-type genes: i, the *CYP2* family on chromosome 19q; ii, the *CYP2C* subfamily on chromosome 10q; iii, the *CYP3A* subfamily on chromosome 7q; iv, the *CYP4* family on chromosome 1p; v, the *CYP4F* subfamily on chromosome 19p. The number near each node indicates the bootstrap value (>94%) supporting the node.

Results

Vertebrate D-type genes emerged independently three times from B-type genes

Among the 57 functional *CYP* genes in the human genome, 35 are D-type genes and 22 are B-type genes. This classification is based on the description of the enzyme substrate [31], if any, and subfamily or family classification [6]. D-type genes constitute four *CYP* families: *CYP1* (3 genes), *CYP2* (16 genes), *CYP3* (4 genes), and *CYP4* (12 genes). B-type genes are grouped into 14 families: *CYP5* (1 gene), *CYP7* (2 genes), *CYP8* (2 genes), *CYP11* (3 genes), *CYP17* (1 gene), *CYP19* (1 gene), *CYP20* (1 gene), *CYP21* (1 gene), *CYP24*

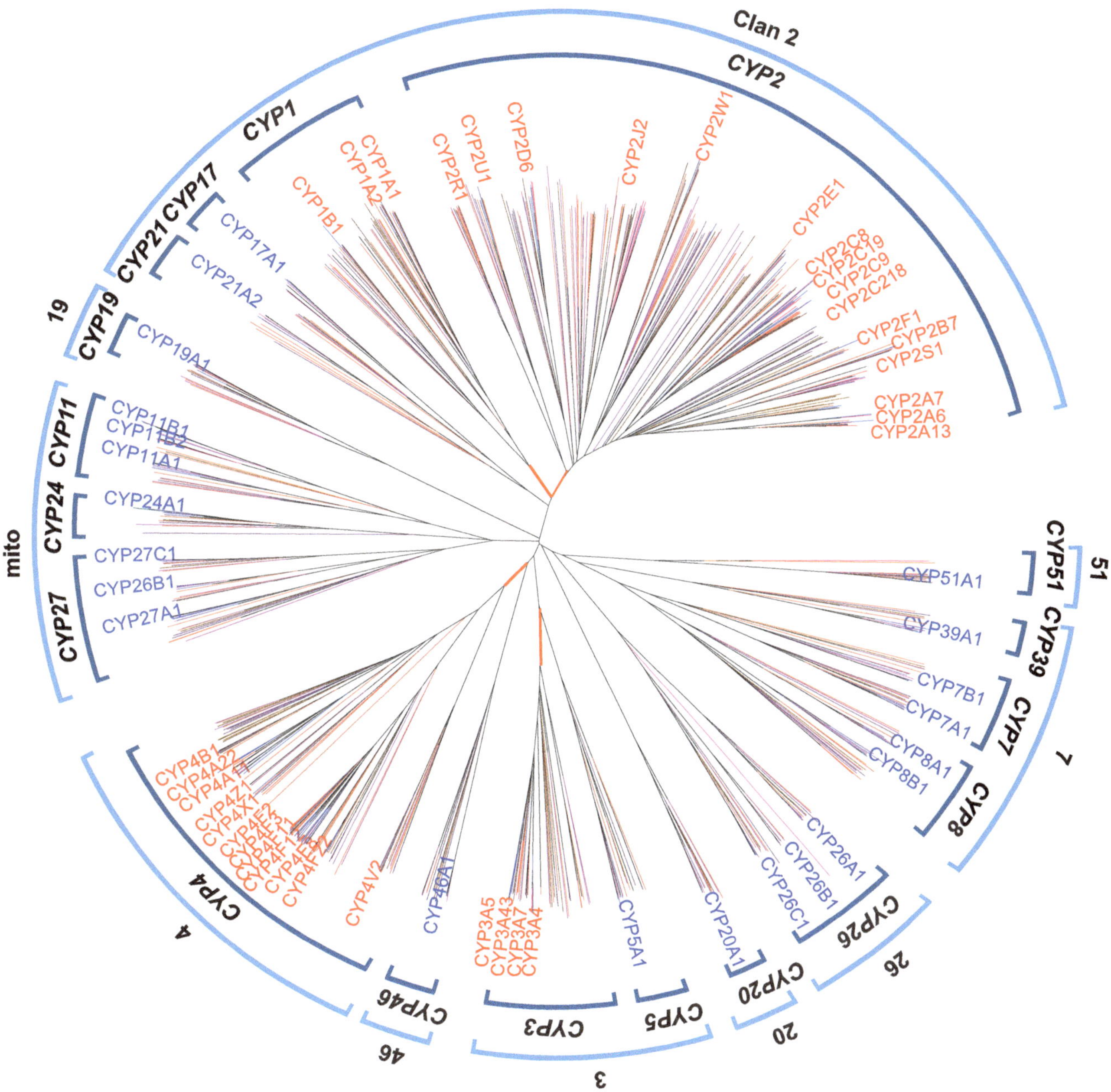

Figure 2. The phylogenetic tree of B- and D-type *CYP* genes in vertebrates. An internal bracket at the tips of the tree indicates the *CYP* family in vertebrates, and an external bracket indicates clusters for a clan. D-type *CYP* genes in humans are shown in red, and B-type *CYP* genes are shown in blue. Red-shaded branches indicate the divergence of D-type from B-type. Text colors indicate the following: red for D- and blue for B-type in humans; dark brown for *Bos taurus*; light blue for *Canis lupus familiaris*; pink for *Mus musculus*; aqua for *Rattus norvegicus*; dark red for *Monodelphis domestica*; dark orange for *Gallus gallus*; purple for *Taeniopygia guttata*; brown for *Anolis carolinensis*; blue-purple for *Xenopus tropicalis*; red-purple for *Oryzias latipes*; orange for *Danio rerio*.

Table 1. Presence or absence of vertebrate orthologs to human *CYP* genes.

CYP family		*Species name*													
	genes	*Patr*	*Mamu*	*Caja*	*Bota*	*Cafa*	*Mumu*	*Rano*	*Modo*	*Anca*	*Gaga*	*Tagu*	*Xetr*	*Orla*	*Dare*
1	A1	1	1	1	1	1	1	1							
	A2	1	1	1	1	1	1	1							
	A1 or A2								2	3	1	1	1	1	1
	B1	1	1	1	1	1	1	1	1	1	0	1	1	1	1
	Others	0	0	0	1	0	0	0	1	2	1	0	3	2	3
2	A6	0	1	0	0	0									
	A7	1	2	0	0	1									
	A13	1	2	1	1	1									
	Other A*						4	2	1	0	0	0	0	0	0
	B6	1	1	1	1	1	4	4	2	0	0	0	0	0	0
	C8	1	1	1											
	C9	1	1	1											
	C18	1	1	1											
	C19	1	1	1											
	Other C*				7	2	9	6	8	0	0	0	0	0	0
	D6	1	1	1	2	0	5	5	1	1	1	1	5	0	0
	E1	1	1	1	1	1	1	1	1	0	0	0	0	0	0
2	F1	1	1	1	1	3	1	1	1	1	0	0	0	0	0
	J2	1	1	1	5	1	6	3	6	1	4	2	1	0	0
	R1	1	1	1	1	1	1	0	1	1	1	1	1	1	1
	S1	1	1	1	1	1	2	1	1	0	0	0	0	0	0
	U1	1	1	1	1	1	1	1	1	0	0	1	1	1	1
	W1	1	1	1	1	1	1	1	1	2	1	1	0	0	0
	Others	1	1	1	1	2	1	2	1	15	7	7	18	10	20
3	A4	1	1	1											
	A5	1	1	1											
	A7	1	1	1											
	A43	1	1	1											
	Other A*				3	4	8	4	4	3	2	1	5	1	1
	Others	0	0	0	0	0	0	0	0	0	0	0	0	2	5
4	A11	1	1	1	2	1									
	A22	1	1	1	2	1									
	A11 or 22						5	1	0	0	0	0	0	0	0
	OtherA*	0	0	0	0	3									
	B1	1	1	1	1	1	2	1	1	6	2	2	4	2	3
	F2	1	1	1											
	F3	1	1	1											
	F8	1	1	1											
	F11	1	1	1											
	F12	1	1	1											
	F22	1	1	1											
	Other F*				6	3	9	8	6	1	0	1	3	1	1
	V2	1	1	1	1	1	1	1	0	1	1	0	2	1	2
	X1	1	1	1	1	2	1	1	2	0	0	0	0	0	0
	Z1	1	1	1	0	0	0	0	0	0	0	0	0	0	0
	5A1	1	1	1	1	1	1	1	1	1	1	1	1	1	1
7	A1	1	1	1	1	1	1	1	1	1	1	1	1	1	1
	B1	1	1	1	1	1	1	1	1	1	1	1	0	0	1

Table 1. Cont.

CYP family		*Species name*													
	genes	*Patr*	*Mamu*	*Caja*	*Bota*	*Cafa*	*Mumu*	*Rano*	*Modo*	*Anca*	*Gaga*	*Tagu*	*Xetr*	*Orla*	*Dare*
8	A1	1	1	1	1	1	1	1	1	0	0	0	1	1	1
	B1	1	1	1	1	1	1	1	2	1	1	1	2	1	3
11	A1	1	1	1	1	1	1	1	1	0	1	1	1	1	1
	B1	1	1	1	1	0	1	1	1	1	0	0	0	0	1*
	B2	1	1	1	0	1	1*	1*	0	0	0	0	0	0	0
	17A1	1	1	1	2	1	1	1	0	1	1	1	1	1	2
	19A1	1	1	1	1	1	1	1	1	1	1	1	1	2	2
	20A1	1	1	1	1	1	1	1	1	1	1	1	1	1	1
	21A2	1	1	1	1	1	1	1	1	1	1	1	1	2	2
	24A1	1	1	1	1	1	1	1	1	1	0	2	2	1	1
26	A1	1	1	1	1	1	1	1	1	1	1	1	1	1	1
	B1	1	1	1	1	1	1	1	0	0	1	1	1	1	1
	C1	1	0	1	1	1	1	1	1	1	1	1	1	1	1
27	A1	1	1	1	1	1	1	1	1	1	1	0	2	1	2
	B1	1	1	1	1	1	1	1	1	1	0	0	1	1	1
	C1	1	1	1	1	1	0	0	1	1	1	1	1	1	1
	39A1	1	1	1	1	1	1	1	1	1	1	1	1	0	1
	46A1	1	1	1	1	1	1	1	0	1	1	1	2	2	2
	51A1	1	1	1	1	1	1	1	1	1	1	1	1	1	1

Others: *CYP* genes are not orthologs to the human *CYP* genes listed here but are subfamily members belonging to each family.
*: Genes are included in the subfamily, but the subfamily number differs from that in humans.

(1 gene), *CYP26* (3 genes), *CYP27* (3 genes), *CYP39* (1 gene), *CYP46* (1 gene), and *CYP51* (1 gene) (Table S1). Using the definition proposed by Nelson [32], the 57 CYPs can be classified into 10 clans: clans 2, 3, 4, mito, 7, 19, 20, 26, 46, and 51. Clan "mito" contains genes encoding enzymes that operate in mitochondria. Of the 10 clans, 6 (2, 3, 4, mito, 7, and 26) contain more than two families, whereas 4 (19, 20, 46, and 51) contain only one single family. The amino acid alignment of the 57 functional *CYP* genes showed that four amino aid sites are conserved. Two of these (310F and 316C) sare located near the heme-binding region (Figure S1). The latter site (316C) is known to be structurally close to the iron ion in the heme-binding region and to operate as an active center of the enzyme [33]. This conserved cysteine is said as the proximal Cys [33]. The other two sites (242E and 245R) are located about 80 amino acids upstream from the proximal Cys. Although it is unknown whether these amino acids are involved in any specific function, their conservation suggests some evolutionary or functional importance. Furthermore, several clan-specific amino acids were found in the 57 functional human *CYP*s (Figure S2). Some of them were conserved not only in vertebrates but also in metazoans, although the number of conserved sites correlates with the number of genes in each clan (data not shown).

To characterize the phylogenetic relationships among the 57 functional *CYP* genes in the human genome, an NJ tree was constructed based on the total nucleotide differences (p-distances) between the CDSs (Figure 1). In the resulting tree, members of each family formed monophyletic groups with respect to other families, and each monophyletic group was supported by a relatively high bootstrap value. The phylogeny showed that D-type genes emerged independently from B-type genes at least three times: first, an ancestral gene of *CYP17A1* and *CYP21A1* was duplicated, generating the ancestor of the *CYP1* and *CYP2* families (node *a* in the tree, Figure 1). Second, the *CYP3A* subfamily arose from the common ancestor of *CYP3* and *CYP5* (node *b* in the tree). Third, an ancestor of *CYP46A1* was duplicated, generating the ancestor of the *CYP4* family (node *c* in the tree). All nodes (*a*, *b*, and *c*) were supported by high bootstrap values (94~100% in Figure 1). In addition to these bootstrap values, amino acids that could distinguish B- from D-type genes were also identified (Figure S3). For example, an amino acid site in the middle of the sequence supported node *a*. In the D-type genes, F was shared by all members of the *CYP1* family whereas V was shared by all members of the *CYP2* family. In contrast, the B-type *CYP*s, *CYP17A1*, and *CYP21A1*, shared T at that site. Similarly, several other amino acid changes that support nodes *a*, *b*, and *c* were observed (Figure S3).

To investigate the duplicaion times of three major D-types from their ancestral B-types, orthologs and paralogs of human B-type and D-type *CYP* genes were retrieved from 14 vertebrate genomes. This resulted in a total of 710 *CYP* nucleotide sequences so that we examined twice as many vertebrate sequences as in the previous study (388) [9]. The presence or absence of vertebrate orthologs to the 57 functional human genes is summarized in Table 1, showing that almost all 14 genomes contain orthologs of B-type genes. We used the pairwise deletion option and constructed a phylogenetic tree (Figure 2); its topology readily confirmed the orthologous relationship between human and other vertebrate B-type genes. However, it was difficult to identify orthologous relationships between D-type genes from humans and other vertebrates, especially in the *2A*, *2C*, *3A*, and *4F* subfamilies, owing to frequent species-specific duplications. Nevertheless, monophyletic relationships within each D-type family (*CYP1–4*) were observed with

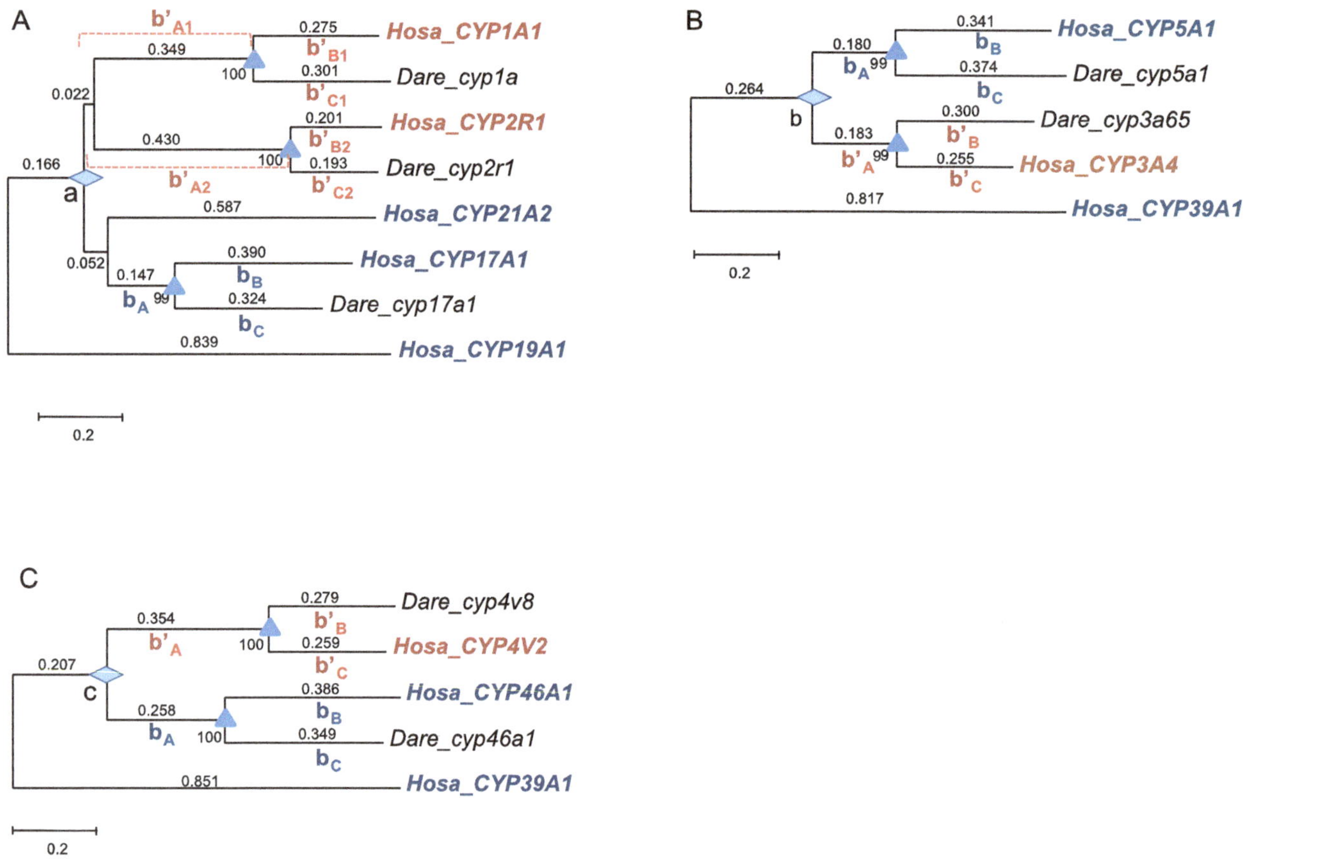

Figure 3. Duplication time of B-type and D-type genes. A) The divergence between *CYP1/2* and *17A1/21A2*. B) The divergence between *CYP3A* and *5A1*. C) The divergence between *CYP4* and *46A1*. The divergence between humans and zebrafish was used as a calibration time (=400 mya), and is shown as a triangle in each tree. The duplication event is shown as a diamond. b_A, b_B, and b_C represent the branch length between the duplication event and species divergence. The branches after species divergence in B-type genes and the branches b'_A, b'_B, and b'_C represent the length of D-type genes. The number near each branch shows the branch length.

relatively high bootstrap values (>80%), so that vertebrate genes in each monophyletic group are classified as the D-type. The phylogenetic analysis revealed that human D- and B-type genes had already emerged when vertebrates diverged, and that three duplication events occurred in the B-type genes from which the D-type genes were originated.

Assuming a molecular clock and that zebrafish and humans diverged 400 million years ago (mya) (TimeTree; http://www.timetree.org/), we calculated the total branch lengths leading to both B- and D-type genes (branch b_A, b_B, and b_C to B-type and b'_A, b'_B, and b'_C to D-type in Figure 3A–C) to estimate the timing of the emergence of the *CYP1–4* families (nodes *a*, *b*, and *c* in Figure 1). Since b_B, b_C, b'_B, and b'_C correspond to 400 million years (myr), each ratio of (b_A+b_B) to b_B, (b_A+b_C) to b_C, (b'_A+b'_B) to b'_B, and (b'_A+b'_C) to b'_C yielded an estimate of the duplication time. The estimates varied from 623–1316 mya for *a*, 601–664 mya for *b*, and 681–926 mya for *c*. To be conservative, we used the youngest estimate for each node: 623±35 mya for *a*, 601±34 mya for *b*, and 681±37 mya for *c*. As anticipated, these estimates preceded the emergence of vertebrates (608 mya, TimeTree) but occurred after the divergence of vertebrates and chordates (774 mya, TimeTree). This finding suggests that invertebrates do not possess orthologs to vertebrate D-type genes, despite the presence of D-type *CYPs* in insects, which function in insecticide resistance and detoxification of plant alkaloids [34].

Evolutionary relationship between invertebrate and vertebrate *CYPs*

To further examine the relationships between vertebrate and invertebrate D-type *CYP* genes, we searched for homologs of human D-type *CYPs* in six invertebrate species (amphioxus: *B. floridae*, sea squirt: *C. intestinalis*, sea urchin: *S. purpuratus*, sea anemone: *N. vectensis*, water flea: *D. pulex*, and fruit fly: *D. melanogaster*). A total of 543 *CYP* amino acid sequences were retrieved from the Cytochrome P450 Homepage. A preliminary search to determine the phylogenetic position of vertebrate *CYPs* in the tree that included both vertebrate and invertebrate *CYPs* revealed that each vertebrate *CYP* family formed a monophyletic group. To simplify the phylogenetic analysis, amino acid sequences from these invertebrates were aligned only with sequences from humans, as a representative vertebrate, and the tree was constructed on the basis of amino acid distances (Figure 4).

The amino-acid distance tree shows that 10 clans (clans 2, 3, 4, mito, 7, 19, 20, 26, 46, and 51) are common to vertebrates; the tree also reveals one *Drosophila*-specific clan. A previous study of 1,572 *CYP* sequences also identified 11 clans in metazoans, but with inclusion of clan 74, which was present only in lancelets, sea anemones, and Trichoplax, but absent in vertebrates [9]. In the present analysis, despite the inclusion of both lancelet and sea anemone, clan 74 was not detected. However, a further phylogenetic analysis that included only yeasts, humans, lancelets,

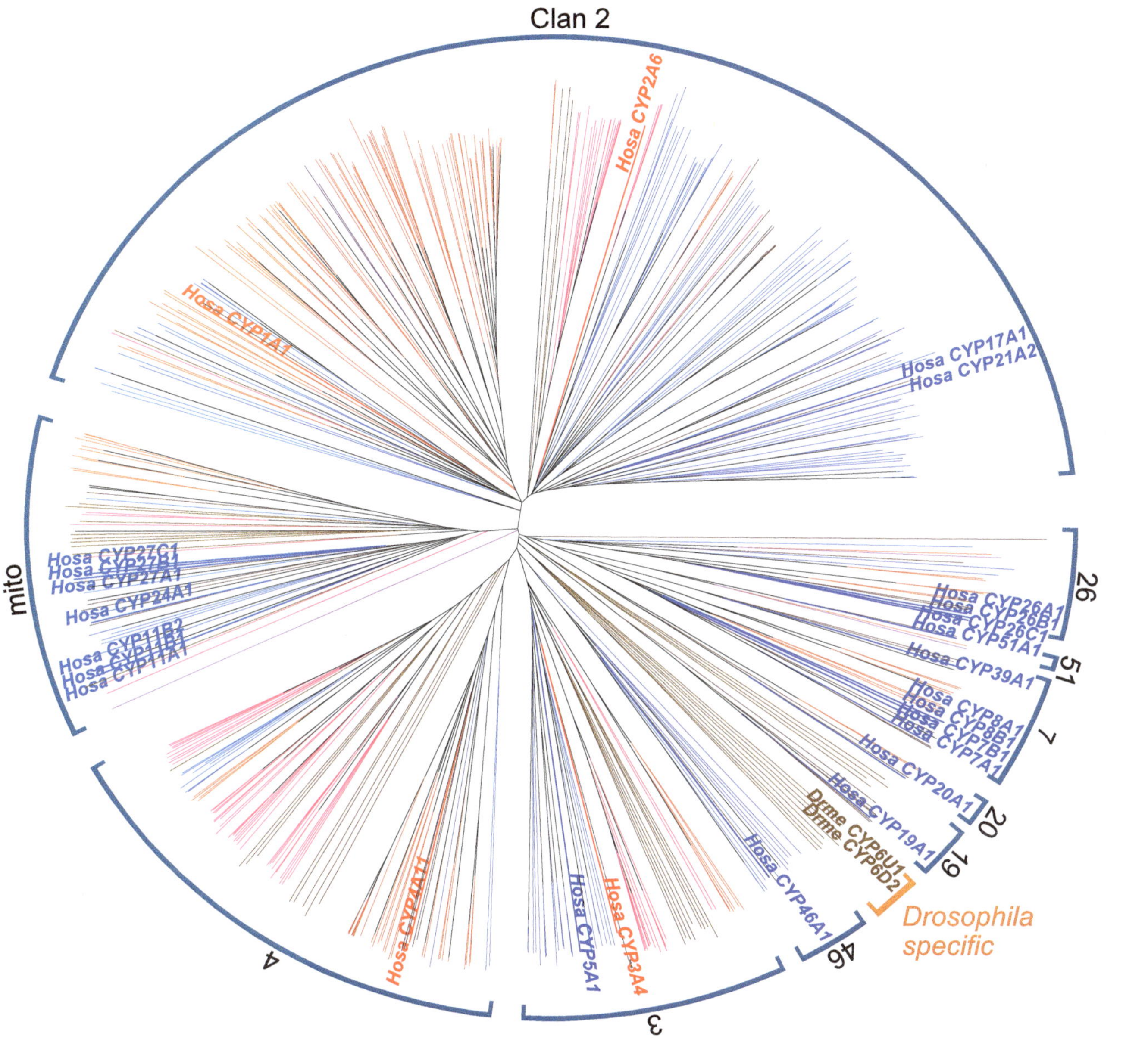

Figure 4. NJ tree of all invertebrate *CYP* genes. D- and B-type *CYP* genes in humans are shown in red and blue text, respectively. The numbers near the brackets indicate clans. Orange character indicates the *Drosophila*-specific clan. Abbreviations and their color (in parentheses) are defined as follows: *Hosa, Homo sapiens* (red for D- and blue for B-type); *Brfl, Branchiostoma floridae* (dark brown); *Neve, Nematostella vectensis* (light blue); *Dapu, Daphnia pulex* (pink); *Stpu, Strongylocentrotus purpuratus* (aqua); *Ciin, Ciona intestinalis* (dark red); *Drme, Drosophila melanogaster* (brown); *Sace, Saccharomyces cerevisiae*; *Scpo, Schizosaccharomyces pombe* (purple). Gene names of *CYP6D2* and *6U1* in *D. melanogaster* are shown in clan 19. Bootstrap values supporting nodes of clusters mentioned in the text are shown.

and sea anemones identified clan 74, although it was supported by a relatively low bootstrap value (55%). In addition, the genes that comprised the *Drosophila*-specific clan (*CYP6D2*, *6U1*, *28A5*, *28C1*, *28D1*, *308A1*, *309A1*, *350A1*, and *317A1*) were all included in clan 3 [9]. This holds true when we draw trees with different methods (maximum likelihood), although the bootstrap value for this clan is too low (<20%) to confirm this inclusion. We also observed some other differences from the previous study [9]: clan 51 did not include any sea-urchin gene, and clan 20 included neither sea urchin nor sea-anemone gene (Figure 4). The absence of a sea-urchin *CYP51* ortholog can be explained by the incompleteness of the database used here. In fact, a blast search of the NCBI database using human *CYP51* as a query identified a *CYP51* gene (Accession number: NM_001001906) in the recently published sea urchin genome. However, clan 20-like genes were absent from the sea urchin and the sea-anemone genomes in the database. In addition, clan 19 in the present tree appeared to include the *Drosophila* genes (*313A1*, *313B1*, *316A1*, and *318A1*) that were included in clan 4 in the previous study. In fact, the *Drosophila*-specific genes in clan 19 shared 16 of 433 amino acids with human *CYP19* (Figure S4), and these 16 amino acids were conserved among vertebrate CYPs. However, an ML tree supported the presence of the *Drosophila* sequences in clan 4, with very low bootstrap support (6%).

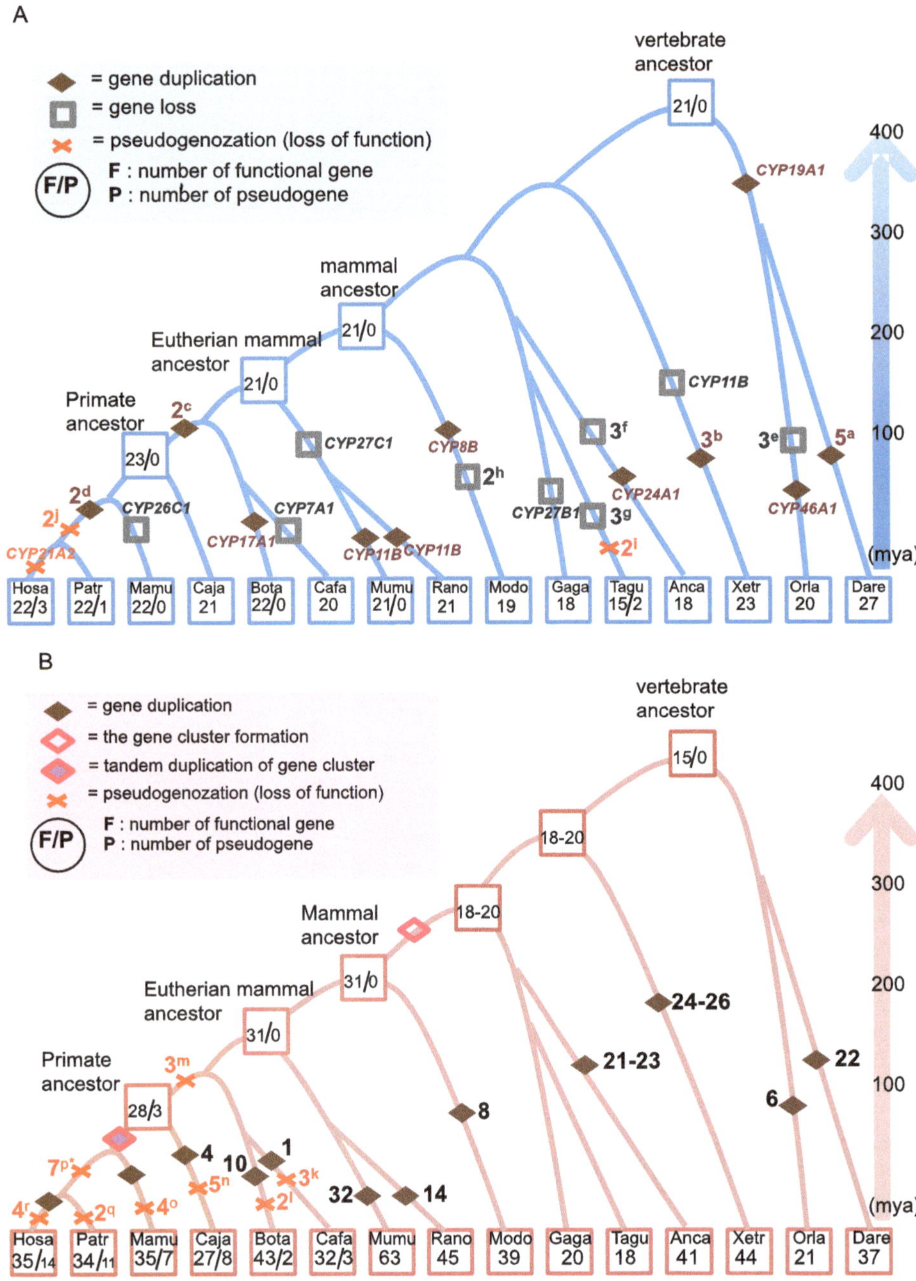

Figure 5. The birth and death processes of *CYP* genes in vertebrates. A) B-type *CYP* genes and B) D-type *CYP* genes. In both figures, numbers inside squares represent the number of functional genes and pseudogenes in each species and its ancestors. Diamonds, crosses, and rectangles indicate gene duplication, pseudogenization, and deletion events, respectively. The number adjacent to each symbol represents the number of events. The letter adjacent to the number indicates the list of *CYP* genes, as follows. a: *CYP8B2*, *CYP8B3*, *CYP17A1*, *CYP27A1*, and *CYP46A1*, b: *CYP8B*, *CYP27A1*, and *CYP46A1*, c: *CYP11B* and *CYP21A2*, d: *CYP51* (two genes), e: *CYP7B*, *CYP11B*, and *CYP39A1*, f: *CYP11A*, *CYP21A2*, and *CYP26*, g: *CYP11B*, *CYP21A2*, and *CYP27* (A or B), h: *CYP17A1* and *CYP26B1*, i: *CYP24A1* and *CYP27A1*, j: *CYP21A1P* and *CYP51* (two genes), k: *CYP4F9P*, *CYP4F23P*, and *CYP4F24P*, l: *CYP4A11* and *CYP2F1P*, m: *CYP2T2P*, *3P*, and *CYP2G1P*, n: *CYP4A11*, *CYP4B1*, *CYP4F22*-like (two genes), and *CYP4F23P*, o: *CYP2A7P1*, *CYP2A13*, *CYP2B6P*, and *CYP4F11*, p: *CYP2B7P1*, *CYP2D8P1*, *CYP2F1P*, *CYP4F9P*, *CYP4F23P*, and *CYP4F24P*, **CYP1D1P* were found in *Pan paniscus*, but were absent from *Pan troglodytes*, q: *CYP2B6* and *CYP2C18*, r: *CYP2A7P1*, *CYP2G2P*, and *CYP4Z2P*.

Clans including invertebrate *CYP* genes were supported by low bootstrap values, and clan definitions were dependent on the methods used for tree construction. Thus, the notion of clan becomes ambiguous and ill-defined for distantly related metazoan *CYP* genes.

The origin of D-type genes in vertebrates and invertebrates

Clan 2 included human *CYP17A1* and *CYP21A2* (B-type) as well as members of the *CYP1* and *CYP2* families (D-type). Similarly, clan 3 included both types of *CYP* genes: *CYP5A1* (B-type) and *CYP3A* subfamilies (D-type). These two cases indicate that the emergence of D-type from B-type genes occurred after the

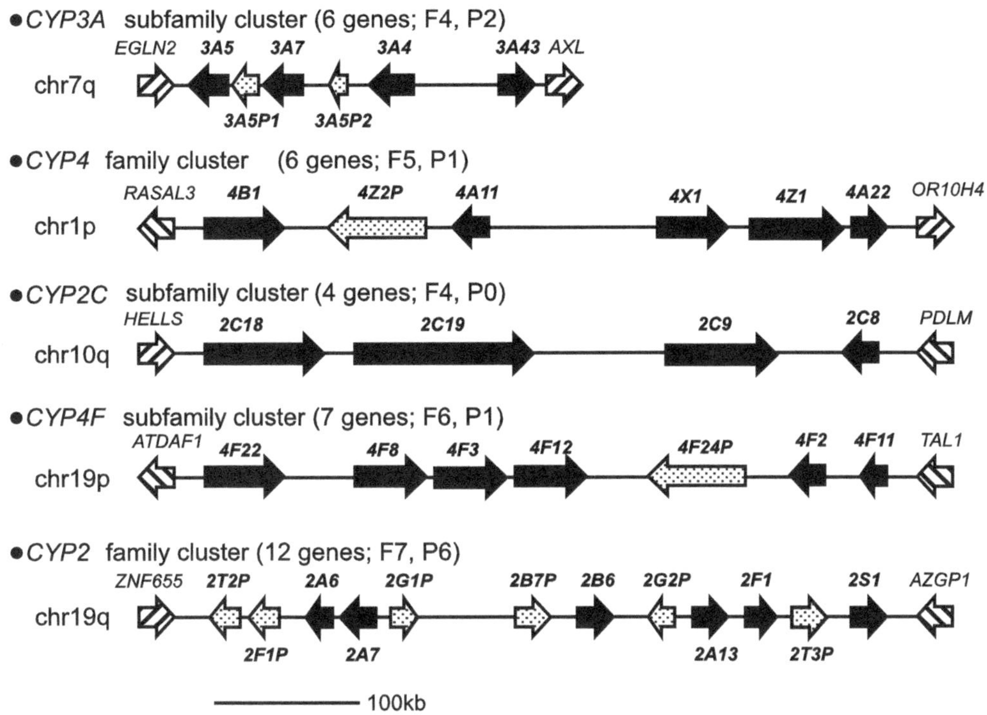

Figure 6. *CYP* gene clusters in the human genome. A striped arrow represents an anchor gene in a syntenic region of each cluster. Black arrows and dotted arrows represent functional *CYP* genes and pseudogenes, respectively. The length of each gene cluster is approximately 500 kb, except the *CYP3A* cluster (250 kb). The number of total genes, functional genes, and pseudogenes in each cluster are shown after the cluster name.

emergence of the clan. However, clan 4 included only the *CYP4* family from humans but not *CYP46A1*, an ancestor of the *CYP4* family. This is the only case where the emergence of the D-type predates clan emergence. In addition, clan 4 included both vertebrate and invertebrate genes. Vertebrate *CYP4* likely acquired its detoxification function in the stem lineage of vertebrates when invertebrate sequences were B-type; alternatively, the ancestor of clan 4 may have already possessed D-type functions when invertebrate genes in clan 4 encoded D-type enzymes.

Fruit flies are known to possess two D-type *CYP* genes, *CYP6D2* and *CYP6U1*, which function in insecticide metabolism. In the tree generated in this study, these *CYP* genes were distantly related to human D-type genes, suggesting that D-type genes in fruit flies emerged independently from those in vertebrates.

Gene duplications and losses in the B- and D-type lineages during vertebrate evolution

Nearly all of the 14 vertebrate genomes examined here contained 21 orthologs to the 22 functional human B-type genes. On the basis of the presence or absence of *CYP* genes in each vertebrate genome, we parsimoniously estimated the number of genes in each ancestor of amniotes, mammals, eutherian mammals, primates, catarrhini, and hominoids, as well as the number of gains of genes in each taxonomic lineage. The number of ancestral genes remained stable throughout the evolution of vertebrates: the number of genes in each vertebrate ancestor did not change over the course of evolution until the emergence of a primate ancestor. A gene-duplication event occurred in the primate ancestor, generating *CYP11B2*. In the ancestor of hominoids, emergence of new genes occurred twice, generating the ancestors of the *CYP51P1* and *CYP51P2* genes (Figure 5A).

In contrast to the rather stable mode of evolution observed in the stem, lineage-specific gains and losses of genes occurred relatively frequently. For instance, a shared duplication of *CYP19A1* occurred in the lineage leading to the common ancestor of zebrafish and medaka. In addition, lineage-specific gene duplications occurred in the zebrafish (*CYP8B2*, *CYP8B3*, *CYP17A1*, *CYP27A1*, and *CYP46A1*), medaka (*CYP46A1*), frog (*CYP8B1*, *CYP27A1*, and *CYP46A1*), green anole (*CYP24A1*), and opossum (*CYP8B1*) lineages. Interestingly, gene duplications of *CYP8B*, *CYP46A1*, and *CYP27A1* occurred independently several times in a species-specific manner. Similarly, lineage-specific gene losses (deletions) were observed; for instance, deletions occurred in a lineage leading to the medaka (*CYP7B1*, *CYP11B1*, and *CYP39A1*), frog (*CYP11B1*), green anole (*CYP11A1*, *CYP21A2*, *CYP26A1*), chicken (*CYP27B1*), zebra finch (*CYP11B1*, *CYP21A2* and *CYP27*), and opossum (*CYP17A1* and *CYP26B1*). Several deletions affecting the same genes (*CYP11B1* and *CYP21A2*) occurred independently in medaka, frog, green anole, and zebrafinch.

Although only a limited number of genomic sequences are available, we identified 19 gene gains and 16 losses among the 15 available genomes, including the human genome. Assuming that the total branch length in the vertebrate tree is 2,685 myr (for individual species divergence times, see Figure S5), we estimated the rate of gene gains and losses to be 0.7 and 0.6 per 100 myr, respectively.

Using a similar analysis, we also examined the 14 vertebrate genomes for the presence of paralogs and orthologs of 35 human D-type genes. This analysis revealed that the number of genes varies from 15 to 31 in ancestral species, and from 18 to 63 in extant species (Figure 5B). In contrast to the relatively stable evolutionary mode of B-type genes, D-type genes underwent more frequent gene duplications and pseudogenization (Table 1).

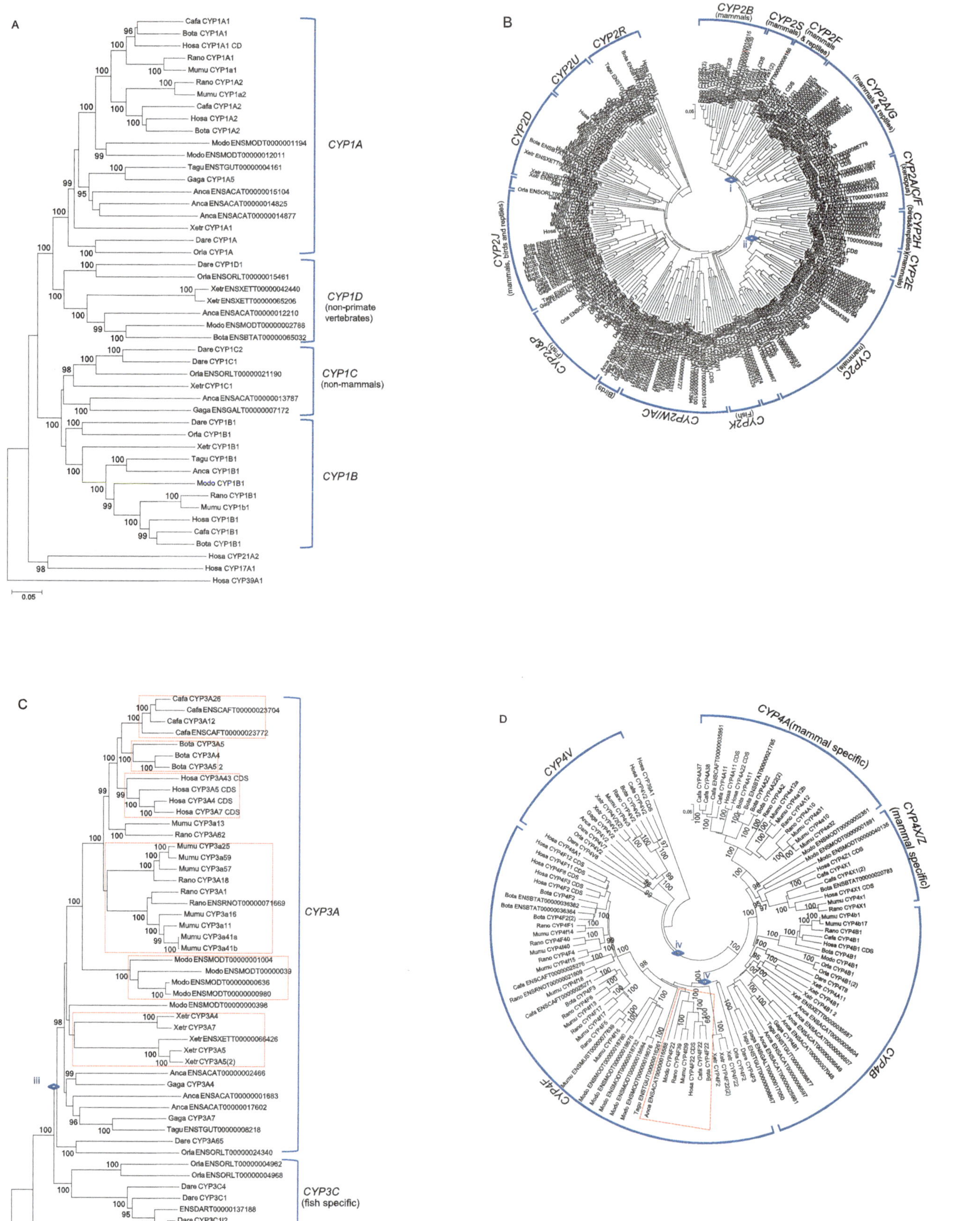
A
Cafa CYP1A1
Bota CYP1A1
Hosa CYP1A1 CD
Rano CYP1A1
Mumu CYP1a1
Rano CYP1A2
Mumu CYP1a2
Cafa CYP1A2
Hosa CYP1A2
Bota CYP1A2
Modo ENSMODT0000001194
Modo ENSMODT00000012011
Tagu ENSTGUT00000004161
Gaga CYP1A5
Anca ENSACAT00000015104
Anca ENSACAT00000014825
Anca ENSACAT00000014877
Xetr CYP1A1
Dare CYP1A
Orla CYP1A
CYP1A
Dare CYP1D1
Orla ENSORLT00000015461
Xetr ENSXETT00000042440
Xetr ENSXETT00000065206
Anca ENSACAT00000012210
Modo ENSMODT00000002788
Bota ENSBTAT00000065032
CYP1D
(non-primate vertebrates)
Dare CYP1C2
Dare CYP1C1
Orla ENSORLT00000021190
Xetr CYP1C1
Anca ENSACAT00000013787
Gaga ENSGALT00000007172
CYP1C
(non-mammals)
Dare CYP1B1
Orla CYP1B1
Xetr CYP1B1
Tagu CYP1B1
Anca CYP1B1
Modo CYP1B1
Rano CYP1B1
Mumu CYP1b1
Hosa CYP1B1
Cafa CYP1B1
Bota CYP1B1
CYP1B
Hosa CYP21A2
Hosa CYP17A1
Hosa CYP39A1
0.05
B
CYP2B
(mammals)
CYP2S
(mammals)
CYP2F
(mammals & reptiles)
CYP2A/G
(mammals & reptiles)
CYP2A/C/F
(Xenopus)
CYP2H
(birds&reptiles)
CYP2E
(mammals)
CYP2C
(mammals)
CYP2K
(Fish)
CYP2W/AC
(Birds)
CYP2J&P
(Fish)
CYP2J
(mammals, birds and reptiles)
CYP2D
CYP2U
CYP2R
i
ii
C
Cafa CYP3A26
Cafa ENSCAFT00000023704
Cafa CYP3A12
Cafa ENSCAFT00000023772
Bota CYP3A5
Bota CYP3A4
Bota CYP3A5 2
Hosa CYP3A43 CDS
Hosa CYP3A5 CDS
Hosa CYP3A4 CDS
Hosa CYP3A7 CDS
Mumu CYP3a13
Rano CYP3A62
Mumu CYP3a25
Mumu CYP3a59
Mumu CYP3a57
Rano CYP3A18
Rano CYP3A1
Rano ENSRNOT00000071669
Mumu CYP3a16
Mumu CYP3a11
Mumu CYP3a41a
Mumu CYP3a41b
Modo ENSMODT00000001004
Modo ENSMODT00000039
Modo ENSMODT00000000636
Modo ENSMODT00000000980
Modo ENSMODT00000000398
Xetr CYP3A4
Xetr CYP3A7
Xetr ENSXETT00000066426
Xetr CYP3A5
Xetr CYP3A5(2)
Anca ENSACAT00000002466
Gaga CYP3A4
Anca ENSACAT00000001683
Anca ENSACAT00000017602
Gaga CYP3A7
Tagu ENSTGUT00000008218
Dare CYP3A65
Orla ENSORLT00000024340
CYP3A
iii
Orla ENSORLT00000004962
Orla ENSORLT00000004968
Dare CYP3C4
Dare CYP3C1
ENSDART00000137188
Dare CYP3C1l2
Dare CYP3C1l2 2
CYP3C
(fish specific)
Hosa CYP5A1
Hosa CYP39A1
0.05
D
CYP4A(mammal specific)
CYP4X/Z
(mammal specific)
CYP4B
CYP4F
CYP4V
iv
v

Figure 7. Phylogenetic tree of the D-type family. A) *CYP1* family, B) *CYP2* family, C) *CYP3* family, and D) *CYP4* family. Each NJ tree was based on the total nucleotide substitutions among members. The origin of each of the five clusters (corresponding to i–v in Figure 1) is indicated with a diamond in Figure B–D. Each subfamily is indicated by a bracket. In Figure B, the *CYP2T* subfamily is not shown because no functional gene belonging to this subfamily is present in the human genome. In Figures C and D, the red dashed rectangle outlines a specific clade.

One important difference between D- and B-type genes is that D-type genes cluster on chromosomes, and these clusters are composed of closely related genes. This difference is reflected in the phylogeny, which shows that the genes in each cluster are monophyletic (Figure 1). In the human genome, five clusters have been identified: the *CYP2* family clusters on chromosome 19q [18,19], the *CYP2C* subfamily clusters on chromosome 10q, the *CYP3A* subfamily clusters on chromosome 7q, the *CYP4* family clusters on chromosome 1p, and the *CYP4F* subfamily clusters on chromosome 19p. Each cluster region occupies approximately 500 kb, with the exception of *CYP3A*, which occupies 250 kb. Each cluster included the following number of genes: 12 for *CYP2*, four for *CYP2C*, six for *CYP3A* and *CYP4*; and seven for *CYP4F* (Figure 6).

Using the phylogenic analyses of each *CYP1–4* family in vertebrates, we identified several species-specific gene duplications. The topology of the tree for the *CYP1* family revealed four subfamilies (*1A*, *1B*, *1C*, and *1D*), and showed that these subfamilies diverged in the ancestor of vertebrates. The *CYP1A* and *1B* subfamilies were conserved from fish to humans, whereas primates lacked *CYP1D*, and mammals lacked *CYP1C* (Figure 7A). The *CYP2* family was shown to be composed of 16 subfamilies (*CYP2A*, *2B*, *2C*, *2D*, *2E*, *2F*, *2G*, *2H*, *2J*, *2K*, *2R*, *2S*, *2T*, *2U*, *2W*, and *2AC*), three of which (*CYP2B*, *2E* and *2S*) were specific to mammals, while the *2A/G* and *F* subfamilies were present only in mammals and reptiles. These five subfamilies (except the *CYP2E* subfamily) diverged successively to form the *CYP2* cluster in an ancestor of mammals (Figure 7B). However, *CYP2U* and *2R* were shown to be common to all vertebrates. The *CYP3* family tree contained only two subfamilies, *CYP3A* and the fish-specific *3C* family (Figure 7C). *CYP3A* comprised amphibian-, bird-, and mammal-specific clades. In each taxonomic group, members of the *CYP3A* subfamily appear to have been duplicated independently. The tree constructed for the *CYP4* family included six subfamilies (*4A*, *4B*, *4F*, *4V*, *4X*, and *4Z*) (Figure 7D). *CYP4A* and *4X/Z* were specific to mammals, whereas the other three subfamilies (*4B*, *F*, and *V*) were common to all vertebrates. In particular, the members of the *4F* subfamily formed several species-specific clusters, except *CYP4F22*. It is unclear, however, whether these species-specific clusters resulted from gene conversion or from recent duplication of the subfamily in each species. The evolution of D-type genes has involved frequent species-specific gene duplications, compared to B-type genes (Figure 5B). In D-type genes, it is unclear how many gene duplications occurred before eutherian divergence. We estimated the rate of duplication subsequent to the eutherian radiation, which revealed 53 duplications in 432 myr, or the rate of 12.7 duplications per 100 myr. No gene loss was observed. These results are in contrast to the results for B-type genes.

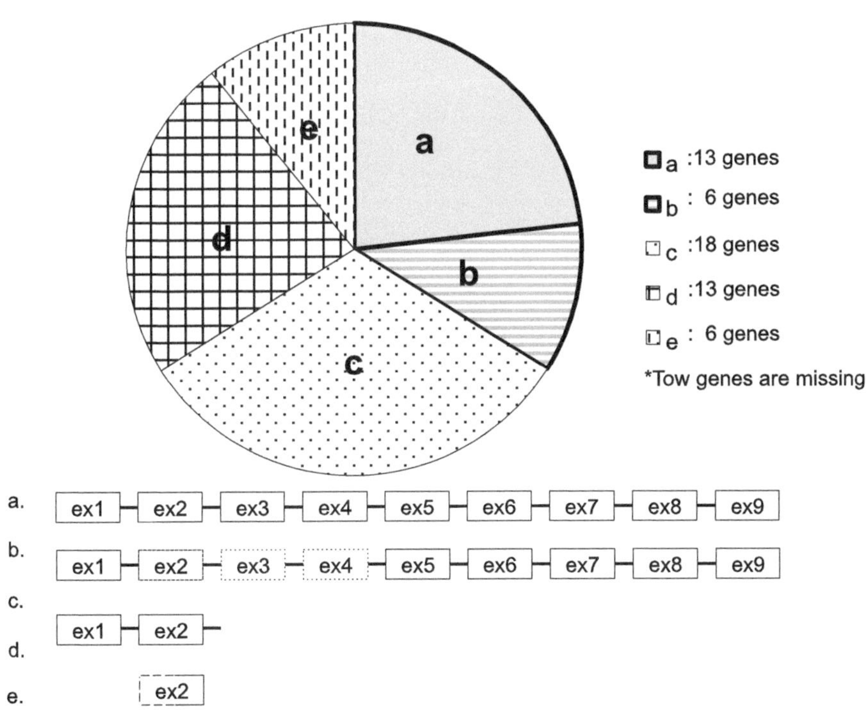

Figure 8. Categorization of the 58 human *CYP* pseudogenes. Among the 58 pseudogenes, paralogs were detected by a BLAST search. a. The number of exons and introns is the same as in the paralogous genes (13 genes), b. They contain greater than half the number of exons and introns of their paralogs (6 genes), c. One or two exons or introns remained (18 genes), d. A portion of an exon remained (13 genes), e. The BLAST search returned no hits (6 genes). *Two pseudogenes were absent from the human genome databases. Approximately one-third (a and b) of the human *CYP* pseudogenes were used for phylogenetic analysis.

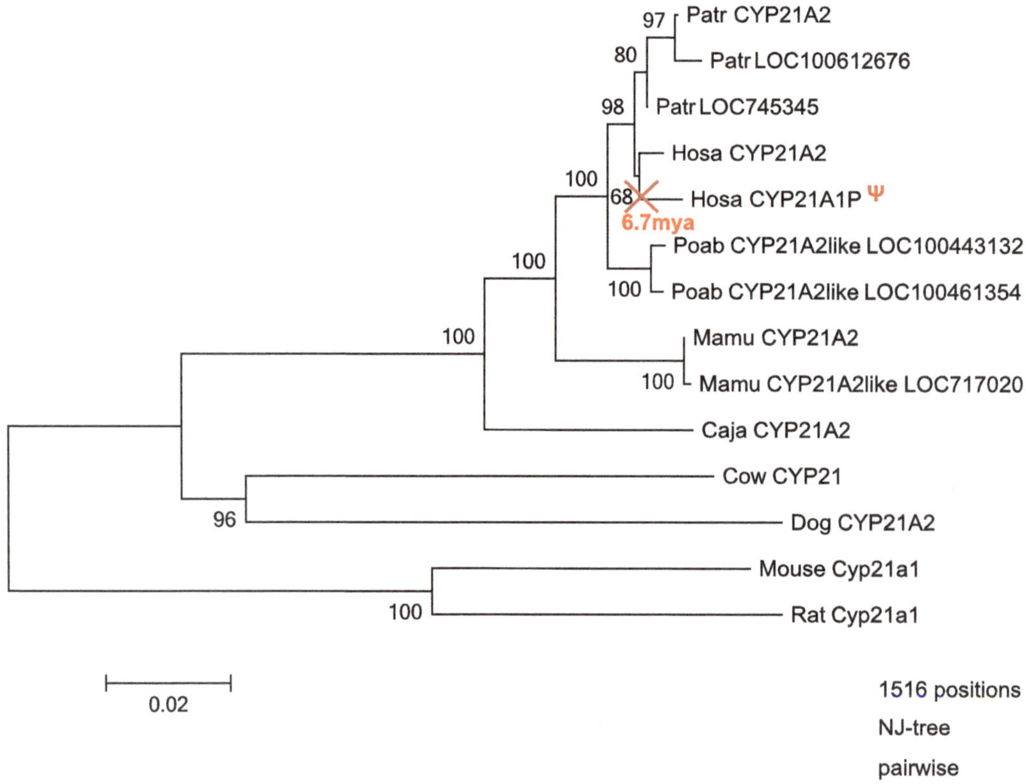

Figure 9. Time of pseudogenization of *CYP21A1P* in humans. The phylogenetic tree was obtained using the NJ method, using the CDS. The pseudogene is represented by Ψ. The cross shows the time at which function was lost.

B- and D-type *CYP* pseudogenes

The evolutionary modes of D- and B-type *CYP* genes differed also in pseudogenization, which is defined as a loss of gene function. Among the 58 pseudogenes present in the human genome, more than half (41 of 58) are fragmented, with few exons and introns remaining. The total length of such pseudogenes is less than one-tenth of that of a functional *CYP* gene, which prevented identification of several of the original genes (Figure 8). We identified the original functional genes for 17 pseudogenes, among which 3 were B-type (*CYP21A1P*, *CYP51P1*, and *CYP51P2*) and 14 were D-type (*CYP1D1P*, *CYP2A7P1*, *CYP2B7P1*, *CYP2D7P1*, *CYP2D8P1*, *CYP2F1P*, *CYP2G1P*, *CYP2G2P*, *CYP2T2P*, *CYP2T3P*, *CYP4F9P*, *CYP4F23P*, *CYP4F24P*, and *CYP4Z2P*) (Table S1). Of the 3 B-type pseudogenes, *CYP51P1* and *CYP51P2* are processed pseudogenes, and the biological causes of their pseudogenization are not related to a relaxation of functional constraints. In this sense, *CYP21A1P* is only a pseudogene due to relaxation of functional constrains. Rhesus macaques, orangutans, and humans have two copies of *CYP21A*, and chimpanzees have three (Figure 9). However, a pseudogene for *CYP21A* is present only in humans, and the time of pseudogenization was estimated to be 6.7 mya, around the divergence of humans from chimpanzees. The presence of this pseudogene is clinically significant in humans: partial gene conversion from a pseudogene to a functional gene causes 21-hydroxylase deficiency; furthermore, copy number variation has been observed in the region containing *CYP21A* and the neighboring *C4A* in the *HLA* region of human chromosome 6 [35].

In contrast, among the 14 D-type pseudogenes, four (*CYP2G1P*, *2G2P*, *2T2P*, and *2T3P*) have been reported to be human-specific, on the basis of a comparison between humans and mice [36]. We searched for orthologs to the human pseudogenes in other primate genomes and found that all but *CYP2G2P* are pseudogenized in other primates as well, but are functional in non-primate vertebrates (Figure S6). Our findings showed that *CYP2G1P*, *2T2P*, and *2T3P* are primate-specific pseudogenes, whereas *CYP2G2P* is a human-specific pseudogene. Using an accelerated non-synonymous substitution rate in pseudogenes [27], we calculated that *CYP2G2P* emerged 2.6 mya. In addition to *CYP2G2P*, further analysis revealed a single human-specific pseudogene, *4Z2P*, with a pseudogenization time of 6.4 mya. On the basis of the results of our analysis, *CYP2D7P1* also appeared to be a human-specific pseudogene. Interestingly, however, pseudogenization of this ortholog has also been found in orangutans, but the cause is different from that for humans [37]. It appears that this gene lost its function in humans and orangutans independently. In D-type genes, in addition to the 14 pseudogenes present in the human genome, 7 pseudogenes were identified in chimpanzees, macaques, marmosets, dogs, and cows. Among the seven, six were specie-specific, one (*2C18*) to chimpanzees, two (*2A13* and *4F11*) to macaques, and three (*4B1* and two *4F22*-like genes) to marmosets. The remaining *CYP2B6P* was pseudogenized independently in chimpanzees and macaques. Among the 11 pseudogenes, with the exception of the three human-specific pseudogenes, *CYP2A7P1* was pseudogenized in macaques and humans independently, at 28.4 mya and 5.9 mya, respectively. Pseudogenization of the remaining 10 genes occurred in the primate or hominoid stem lineage.

It is unclear how many times pseudogenization occurred in D-type genes before eutherian divergence. We estimated the rate subsequent to the eutherian radiation at 30 pseudogenizations over 432 myr, yielding a rate of 6.9 per 100 myr. In contrast, the

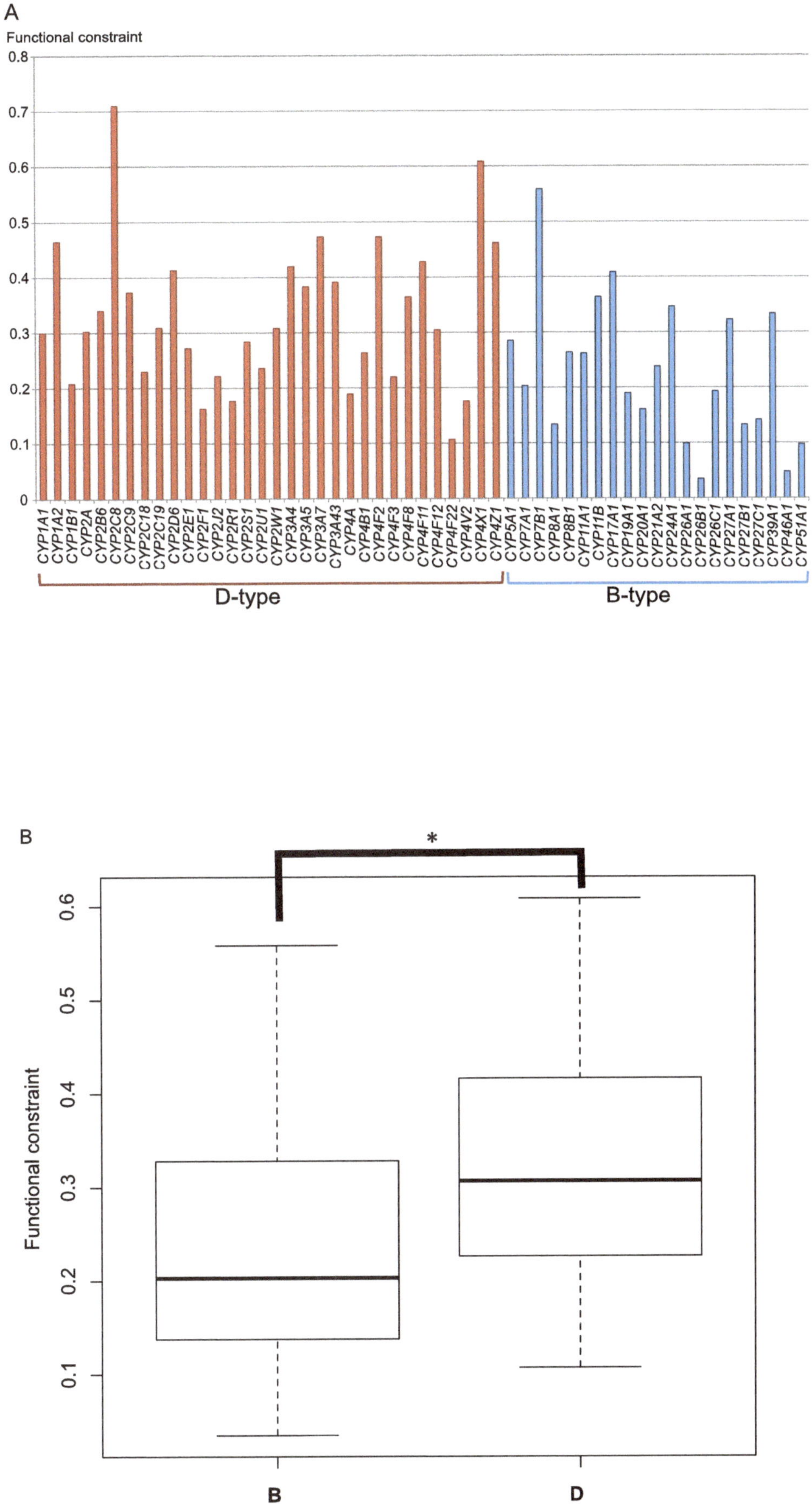
A
Functional constraint
0.8
0.7
0.6
0.5
0.4
0.3
0.2
0.1
0
CYP1A1
CYP1A2
CYP1B1
CYP2A
CYP2B6
CYP2C8
CYP2C9
CYP2C18
CYP2C19
CYP2D6
CYP2E1
CYP2F1
CYP2J2
CYP2R1
CYP2S1
CYP2U1
CYP2W1
CYP3A4
CYP3A5
CYP3A7
CYP3A43
CYP4A
CYP4B1
CYP4F2
CYP4F3
CYP4F8
CYP4F11
CYP4F12
CYP4F22
CYP4V2
CYP4X1
CYP4Z1
CYP5A1
CYP7A1
CYP7B1
CYP8A1
CYP8B1
CYP11A1
CYP11B
CYP17A1
CYP19A1
CYP20A1
CYP21A2
CYP24A1
CYP26A1
CYP26B1
CYP26C1
CYP27A1
CYP27B1
CYP27C1
CYP39A1
CYP46A1
CYP51A1
D-type
B-type
B
*
Functional constraint
0.6
0.5
0.4
0.3
0.2
0.1
B
D

Figure 10. Functional constraint of *CYP* genes. A) Functional constraint was estimated for each *CYP* gene. The *y*-axis shows the functional constraint obtained via the ratio of per-site non-synonymous substitutions to synonymous substitutions (D_N/D_S). Red bars indicate D-type genes, and blue bars indicate B-type genes. B) Comparison of median values for functional constraints between primate B- and D-type genes. "B" indicates functional constraint of B-type genes, and "D" indicates functional constraint of D-type genes. The *P*-value was 0.01282 (significance was defined as $P<0.05$). The *P*-value was calculated using the Mann–Whitney *U* test.

number of pseudogenization events in B-type genes was estimated to be only five over 2,685 myr, yielding a rate of 0.19 per 100 myr.

Evolutionary rate of B- and D-type genes

Our results revealed that births and deaths of genes were more frequent in D-type genes than in B-type genes. As such, it was important to compare the evolutionary rate of B- and D-type genes. For this comparison, the non-synonymous substitution rate was normalized to the synonymous rate, and the ratio (*f*) for each B- and D-type gene was calculated in primates (see Materials and Method, Figure 10A). The average *f*-values for B- and D-type genes were calculated to be 0.24±0.14 and 0.33±0.13, respectively, and the median values for B- and D-type genes were determined to be 0.23 and 0.31, respectively (Figure 10B). The average and median values for D-type genes were significantly greater than those for B-type genes (Wilcoxon's test, *P*-value = 0.0173), suggesting that the degree of functional constraint (1-*f*) is stronger in B-type than in D-type genes. These results are consistent with the rapid birth and death process of D-type genes.

Discussion

The origin of D-type *CYP* genes

The origin of B-type genes is assumed to be single and ancient, because fission yeast possesses B-type genes and because a possible ortholog to the B-type gene *CYP51* is present even in prokaryotic genomes. However, D-type genes are thought to have different origins. The present phylogenetic analysis demonstrated that four D-type families are conserved among all vertebrates, and that the D-type families were derived from three gene-duplication events of B-type genes in the stem lineage of vertebrates. Based on the molecular clock hypothesis, B- to D-type gene duplications occurred approximately 600–700 mya, consistent with the phylogenetic analysis. However, the D-type *CYPs* impart resistance to insecticides in invertebrates; in fruit flies, two such enzymes are CYP6U1 and CYP6D2. The phylogenetic analysis of both human and fruit fly *CYP* genes indicated an independent emergence of D-type genes. Moreover, other invertebrate genomes contain human D-type-like genes, but orthology has not been confirmed. It appears that D-type genes in vertebrates and insects evolved independently from different origins, which is consistent with the idea of a rapid turnover of D-type genes.

In this paper, we focused on an early stage of *CYP* gene diversification in vertebrates and showed the emergence of D-type from B-type genes. However, some exceptions should be noted. For example, *CYP2R1* is categorized as a B-type gene on the basis of its function, but the nucleotide sequence showed that it is closely related to other D-type *CYP2* genes. From this observation, it appears that *CYP2R1* has been converted from a D-type to a B-type *CYP* gene. This is supported by the observation that the amino acid sequence of CYP2R1 is highly conserved in all vertebrates, reflecting the high degree of functional constraint against the gene.

The evolution of *CYP* genes is driven by substrate specificity

The birth and death (pseudogenization) rates of B- and D-type genes differed in magnitude: the rates in B-type genes were 0.7 and 0.2 per 100 myr, respectively, whereas those in D-type genes were 12.7 and 6.9 per 100 myr, respectively. Compared with D-type genes, the evolution of B-type genes was highly conserved with regard to their mode of birth and death processes as well as amino acid substitutions. The substrates of B-type enzymes are chemicals that play important roles in metabolism of vitamin D, steroids, and cholesterol. In contrast, the substrates of D-type enzymes are xenobiotics such as plant alkaloids. In light of this substrate specificity, we hypothesize that the conserved evolutionary pattern observed in B-type enzymes reflects the importance and conservation of their substrates, whereas the rapid evolution of D-type enzymes indicates that their substrates are flexible and highly dependent on environmental factors. Future studies of the evolution of the substrate-recognition sites will be required to confirm this hypothesis.

Supporting Information

Figure S1 Conserved amino acids at a position. The *x*-axis indicates amino-acid positions in an alignment of 710 vertebrate *CYP* genes (after excluding gaps), and the y-axis indicates the ratio of conserved amino acids at each position. Red bars indicate highly (>95%) conserved positions. The chart below the bar graph reports the approximate position of six substrate recognition sites (SRS): SRS1–6. The bracket represents the heme-binding region (~10 amino acids) of the *CYP* gene.

Figure S2 Conserved amino acids within each clan of vertebrate CYP genes. Conserved amino acids within each clan are shown in colored columns: the red column is specific for clan 2; aqua, mito; orange, 4; blue, 3; yellow, 26, and brown, 7. The four sites in purple (E, R, G, and C) are highly conserved (>95%) sites among all vertebrate species. This figure shows only human *CYP* genes: positions are 275–285 and 350–360.

Figure S3 Amino acids distinguishing B-type from D-type genes within each clan. The yellow column indicates the amino acids in B-type genes that differ from D-type genes. The other colors are the same as in Figure S2. A, Clan 2; B, Clan 3; C, Clan 4 and 46.

Figure S4 Conserved amino acids between human and *Drosophila* genes in a *CYP19* clan. Human (*CYP19A1*) and Drosophila (*CYP313A1*, *313B1*, *316A1*, and *318A1*) genes share the same amino acids at 16 of 423 aligned sites. The shared sites are shown in red.

Figure S5 Phylogenetic tree and species-divergence time. The phylogenetic tree was constructed on the basis of the divergence time of 15 species (time tree). The species names at the tip of the tree are abbreviated as in Table 1. *Hosa*, *Homo sapiens*. The number at each node represents species divergence time in mya. The scale under the tree indicates time in myr.

Figure S6 The cause of functional loss in human-specific *CYP* pseudogenes. There are four human-specific *CYP* psuedogenes (*CYP2G1P, 2P, CYP2T2P*, and *3P*). Possible causal mutations, premature stop codons (red) and frame-shift mutations (blue), were identified in human and other primate CYP nucleotide and amino acid alignments. The row labeled "exon" for *CYP2G1P* and *2P* shows the number of exons in which mutations were found, and the row labeled "bp" indicates the nucleotide position of the CDS in functional genes from rat and mouse.

Table S1 The number of *CYP* gene in Human. After exclusion of truncated pseudogeens each category includes genes as below. **a**: *CYP1A1, CYP1A2, CYP1B1, CYP2A6, CYP2A7, CYP2B6, CYP2C8, CYP2C9, CYP2C18, CYP2C19, CYP2D6, CYP2E1, CYP2F1, CYP2J2, CYP2R1, CYP2S1, CYP2U1, CYP2W1, CYP3A4, CYP3A5, CYP3A7, CYP3A43, CYP4A11, CYP4A20, CYP4A22, CYP4B1, CYP4F2, CYP4F3, CYP4F8, CYP4F11, CYP4F12, CYP4F22, CYP4V2, CYP4X1*, **b**: *CYP1D1P, CYP2A7P1, CYP2B7P1, CYP2D7P1, CYP2D8P1, CYP2F1P, CYP2G1P, CYP2G2P, CYP2T2P, CYP2T3P, CYP4F9P, CYP4F23P, CYP4F24P, CYP4Z2P*, **c**: *CYP5A1, CYP7A1, CYP7B1, CYP8A1, CYP8B1, CYP11A1, CYP11B1, CYP11B2, CYP17A1, CYP19A1, CYP20A1, CYP21A2, CYP24A1, CYP26A1, CYP26B1, CYP26C1, CYP27A1, CYP27B1, CYP27C1, CYP39A1, CYP46A1, CYP51A1*, **d**: *CYP21A1P, CYP51P1, CYP51P2.*

Acknowledgments

We thank Dr. Takahata for his critical discussion of this manuscript.

Author Contributions

Conceived and designed the experiments: AK YS. Performed the experiments: AK YS. Analyzed the data: AK YS. Contributed reagents/materials/analysis tools: AK YS. Wrote the paper: AK YS.

References

1. Klingenberg M (1958) Pigments of rat liver microsomes. Arch Biochem Biophys 75: 376–386.
2. Omura T, Sato R (1962) A new cytochrome in liver microsomes. J Biol Chem 237: PC1375–PC1376.
3. Nelson DR (1998) Metazoan cytochrome P450 evolution. Comp Biochemi and Phys Part C: Pharm Toxic Endocr 121: 15–22.
4. Gotoh O (2012) Evolution of cytochrome p450 genes from the viewpoint of genome informatics. Biol Pharm Bull 35: 812–817.
5. Munro AW, Lindsay G (1996) Bacterial cytochromes P-450. Mol Microbiol 20: 1115–1125.
6. Nelson DR (2009) The cytochrome p450 homepage. Human Genomics 1:59–65.
7. Sea Urchin Genome Sequencing Consortium (2006) The genome of the sea urchin *Strongylocentrotus purpuratus*. Science 314: 941–952.
8. Mao G, Seebeck T, Schrenker D, Yu O (2013) CYP709B3, a cytochrome P450 monooxygenase gene involved in salt tolerance in *Arabidopsis thaliana*. BMC Plant Biol 13: 169.
9. Nelson DR, Goldstone JV, Stegeman JJ (2013) The cytochrome P450 genesis locus: the origin and evolution of animal cytochrome P450s. Phil Trans R Soc B 368: 20120474.
10. Quaderer R, Omura S, Ikeda H, Cane DE (2006) Pentalenolactone biosynthesis. Molecular cloning and assignment of biochemical function to PtlI, a cytochrome P450 of Streptomyces avermitilis. J Am Chem Soc 128: 13036–13037.
11. Aoyama Y, Horiuchi T, Gotoh O, Noshiro M, Yoshida Y (1998) CYP51-like gene of *Mycobacterium tuberculosis* actually encodes a P450 similar to eukaryotic CYP51. J Biochem 124: 694–696.
12. Yoshida Y, Aoyama Y, Noshiro M, Gotoh O (2000) Sterol 14-demethylase P450 (CYP51) provides a breakthrough for the discussion on the evolution of cytochrome P450 gene superfamily. Biochem Biophys Res Comm 273: 799–804.
13. Debeljak N, Fink M, Rozman D (2003) Many facets of mammalian lanosterol 14α-demethylase from the evolutionarily conserved cytochrome P450 family CYP51. Arch Biochem Biophys 409: 159–171.
14. Qi X, Bakht S, Qin B, Leggett M, Hemmings A, et al. (2006) A different function for a member of an ancient and highly conserved cytochrome P450 family: From essential sterols to plant defense. Proc Natl Acd Sci USA 103: 18848–18853.
15. Nelson DR (1999) Cytochrome P450 and the individuality of species. Arch Biochem Biophys 369: 1–10.
16. Rezen T, Debeljak N, Kordis D, Rozman D (2004) New aspects of lanosterol 14a-demethlase and cytochrome P450 evolution: Lanosterol/cycloartenol diversification and lateral transfer. J Mol Evol 59: 51–58.
17. Nebert DW, Dalton TP (2006) The role of cytochrome P450 enzymes in endogenous signaling pathways and environmental carcinogenesis. Nature Rev Cancer 6: 947–960.
18. Hoffman SMG, Hu S (2006) Dynamic evolution of the CYP2ABFGST gene cluster in primates. Mutation Res 616: 133–138.
19. Hu S, Wang H, Knisely AA, Raddy S, Kovacevic D, et al. (2008) Evolution of the CYP2ABFGST gene cluster in rat, and a fine-scale comparison among rodent and primate species. Genetica 133: 215–226.
20. Thomas JH (2007) Rapid birth-death evolution specific to xenobiotic cytochrome P450 genes in vertebrates. PLoS Genetics 3: e67
21. Larkin MA, Blackshields G, Brown NP, Chenna R, McGettigan PA, et al. (2007) Clustal W and Clustal X version 2.0. Bioinformatics 23: 2947–2948.
22. Tamura K, Peterson D, Peterson N, Stecher G, Nei M, et al. (2011) MEGA5: Molecular evolutionary genetics analysis using maximum likelihood, evolutionary distance, and maximum parsimony methods. Mol Biol Evol 28: 2731–2739.
23. Saitou N, Nei M (1987) The neighbor-joining method: a new method for reconstructing phylogenetic trees. Mol Biol Evol 4: 406–425.
24. Nei M and Kumar S (2000). Molecular Evolution and Phylogenetics. Oxford University Press, New York.
25. Jones D, Taylor WR, Thronton JM (1992) The rapid generation of mutation data matrices from protein sequences. Comput Appl Biosci 8: 275–282.
26. Felsenstein J (1981) Evolutionary trees from DNA sequences: a maximum likelihood approach. J Mol Evol 17: 368–76.
27. Sawai H, Go Y, Satta Y (2008) Biological implication for loss of function at major histocompatibility complex loci. Immunogenetics 60: 295–302.
28. Hedges SB, Dudley, Kumar S (2006) Time Tree: a public knowledge-base of divergence times among organisms. Bioinformatics 23: 2971–2.
29. Rzhetsky A, Nei M (1992) Statistical properties of the ordinary least-squares, generalized least-squares, and minimum-evolution methods of phylogenetic inference. J Mol Evol 35: 367–75.
30. Mann HB, Whitney DR (1947) On a test of whether one of two random variables is stochastically larger than the other. Ann Math Statist 18: 50–60.
31. Ohmura T, Ishimura Y, Fujii Y (2009) Molecular biology of P450. 2nd ed.
32. Nelson DR, Koymans L, Kamataki T, Stegeman JJ, Feyereisen R, et al. (1996) P450 superfamily: update on new sequences, gene mapping, accession numbers and nomenclature. Pharmacogenetics, 6(1): 1–42
33. Meunier B, de Visser SP, Shaik S (2004) Mechanism of oxidation reactions catalyzed by cytochrome P450 enzymes. Chem Rev 104: 3947–3980.
34. Feyereisen R (2011) Arthropod CYPomes illustrate the tempo and mode in P450 evolution. Bioch et Biophys Acta 1814: 19–28.
35. Urabe K, Kimura A, Harada F, Iwanaga T, Sasazuki T (1990) Gene conversion in steroid 21-hydroxylase genes. Am J Hum Genet 46: 1178–1186.
36. Nelson DR, Zeldin DC, Hoffman SM, Maltais LJ, Wain HM, et al. (2004) Comparison of cytochrome P450 (CYP) genes from the mouse and human genomes, including nomenclature recommendations for genes, pseudogenes and alternative-splice variants. Pharmacogenetics 14(1): 1–18.
37. Yoshiki Yasukochi, Yoko Satta (2011) Evolution of the CYP2D gene cluster in humans and four non-human primates. Genes Genet Syst 86: 109–116.

PERMISSIONS

All chapters in this book were first published in PLOS ONE, by The Public Library of Science; hereby published with permission under the Creative Commons Attribution License or equivalent. Every chapter published in this book has been scrutinized by our experts. Their significance has been extensively debated. The topics covered herein carry significant findings which will fuel the growth of the discipline. They may even be implemented as practical applications or may be referred to as a beginning point for another development.

The contributors of this book come from diverse backgrounds, making this book a truly international effort. This book will bring forth new frontiers with its revolutionizing research information and detailed analysis of the nascent developments around the world.

We would like to thank all the contributing authors for lending their expertise to make the book truly unique. They have played a crucial role in the development of this book. Without their invaluable contributions this book wouldn't have been possible. They have made vital efforts to compile up to date information on the varied aspects of this subject to make this book a valuable addition to the collection of many professionals and students.

This book was conceptualized with the vision of imparting up-to-date information and advanced data in this field. To ensure the same, a matchless editorial board was set up. Every individual on the board went through rigorous rounds of assessment to prove their worth. After which they invested a large part of their time researching and compiling the most relevant data for our readers.

The editorial board has been involved in producing this book since its inception. They have spent rigorous hours researching and exploring the diverse topics which have resulted in the successful publishing of this book. They have passed on their knowledge of decades through this book. To expedite this challenging task, the publisher supported the team at every step. A small team of assistant editors was also appointed to further simplify the editing procedure and attain best results for the readers.

Apart from the editorial board, the designing team has also invested a significant amount of their time in understanding the subject and creating the most relevant covers. They scrutinized every image to scout for the most suitable representation of the subject and create an appropriate cover for the book.

The publishing team has been an ardent support to the editorial, designing and production team. Their endless efforts to recruit the best for this project, has resulted in the accomplishment of this book. They are a veteran in the field of academics and their pool of knowledge is as vast as their experience in printing. Their expertise and guidance has proved useful at every step. Their uncompromising quality standards have made this book an exceptional effort. Their encouragement from time to time has been an inspiration for everyone.

The publisher and the editorial board hope that this book will prove to be a valuable piece of knowledge for researchers, students, practitioners and scholars across the globe.

LIST OF CONTRIBUTORS

Martin J. Thompson
Life Sciences Department, Natural History Museum London, London, United Kingdom
Department of Zoology, University of Cambridge, Cambridge, United Kingdom

Martijn J. T. N. Timmermans
Life Sciences Department, Natural History Museum London, London, United Kingdom
Division of Biology, Imperial College London, London, United Kingdom

Wei Liu
Department of Agronomy, Zhejiang University, Hangzhou, Zhejiang, China

Wei Li
Department of Agronomy, Zhejiang University, Hangzhou, Zhejiang, China

Qiuling He
Department of Agronomy, Zhejiang University, Hangzhou, Zhejiang, China

Muhammad Khan Daud
Department of Biotechnology and Genetic Engineering, Kohat University of Science and Technology, Kohat, Pakistan

Jinhong Chen
Department of Agronomy, Zhejiang University, Hangzhou, Zhejiang, China,

Shuijin Zhu
Department of Agronomy, Zhejiang University, Hangzhou, Zhejiang, China

Ana Tavares
Laboratory of Molecular Genetics, Instituto de Tecnologia Química e Bioló gica (ITQB), Oeiras, Portugal

Jesper B. Nielsen
Dept. of Clinical Microbiology 445, Copenhagen University
Hospital, Hvidovre, Denmark,

Kit Boye
Dept. of Clinical Microbiology 445, Copenhagen University
Hospital, Hvidovre, Denmark

Susanne Rohd
Dept. of Clinical Microbiology 445, Copenhagen University
Hospital, Hvidovre, Denmark

Ana C. Paulo
Molecular Microbiology of Human Pathogens, ITQB, Oeiras, Portugal

Henrik Westh
Dept. of Clinical Microbiology 445, Copenhagen University
Hospital, Hvidovre, Denmark
Faculty of Health and Medical Sciences, University of Copenhagen,
Copenhagen, Denmark

Kristian Schønning
Dept. of Clinical Microbiology 445, Copenhagen University
Hospital, Hvidovre, Denmark
Faculty of Health and Medical Sciences, University of Copenhagen,
Copenhagen, Denmark

Hermı´nia de Lencastre
Laboratory of Molecular Genetics, Instituto de Tecnologia Química e Bioló gica (ITQB), Oeiras, Portugal
Laboratory of Microbiology and Infectious Diseases, The Rockefeller University, New York, New York, United States of America

Maria Miragaia
Laboratory of Molecular Genetics, Instituto de Tecnologia Química e Bioló gica (ITQB), Oeiras, Portugal,
Laboratory of Bacterial Evolution and Molecular Epidemiology, ITQB, Oeiras, Portugal

Wen-Yan Li, Xiang Wang, Ri Li, Wen-Qiang Li and Kun-Ming Chen
State Key Laboratory of Crop Stress Biology in Arid Areas, College of Life Sciences, Northwest A&F University, Yangling, Shaanxi, China

Francisco X. Nascimento
Universidade Federal de Santa Catarina, Departamento de Microbiologia, Laborato´ rio de Microbiologia do Solo, Floriano´ polis SC, Brazil,

Márcio J. Rossi
Universidade Federal de Santa Catarina, Departamento de Microbiologia, Laborato´ rio de Microbiologia do Solo, Floriano´ polis SC, Brazil

Cláudio R. F. S. Soares
Universidade Federal de Santa Catarina, Departamento de Microbiologia, Laborato´ rio de Microbiologia do Solo, Floriano´ polis SC, Brazil

Brendan J. McConkey
University of Waterloo, Department of Biology, Waterloo, Ontario, Canada

Bernard R. Glick
University of Waterloo, Department of Biology, Waterloo, Ontario, Canada

Kazufumi Hosoda
Institute for Academic Initiatives, Osaka University, Suita, Osaka, Japan

Masumi Habuchi
Department of Frontier Biosciences, Graduate School of Frontier Biosciences, Osaka University, Suita, Osaka, Japan

Shingo Suzuki
Quantitative Biology Center, RIKEN, Suita, Osaka, Japan

Mikako Miyazaki
Department of Biotechnology, Graduate School of Engineering, Osaka University, Suita, Osaka, Japan

Go Takikawa
Department of Bioinformatic Engineering, Graduate School of Information Science and Technology, Osaka University, Suita, Osaka, Japan

Takahiro Sakurai
Department of Bioinformatic Engineering, Graduate School of Information Science and Technology, Osaka University, Suita, Osaka, Japan

Akiko Kashiwagi
Faculty of Agriculture and Life Science, Hirosaki University, Hirosaki, Aomori, Japan

Makoto Sueyoshi
Department of Bioinformatic Engineering, Graduate School of Information Science and Technology, Osaka University, Suita, Osaka, Japan

Yusuke Matsumoto
Department of Bioinformatic Engineering, Graduate School of Information Science and Technology, Osaka University, Suita, Osaka, Japan

Ayako Kiuchi
Department of Biotechnology, Graduate School of Engineering, Osaka University, Suita, Osaka, Japan

Kotaro Mori
Department of Frontier Biosciences, Graduate School of Frontier Biosciences, Osaka University,
Suita, Osaka, Japan

Tetsuya Yomo
Department of Frontier Biosciences, Graduate School of Frontier Biosciences, Osaka University, Suita, Osaka, Japan
Department of Bioinformatic Engineering, Graduate School of Information Science and Technology, Osaka University, Suita, Osaka, Japan
Exploratory Research for Advanced Technology (ERATO), Japan Science and Technology Agency, Suita, Osaka, Japan

María Arribas
Centro de Astrobiología (INTA-CSIC), 28850 Torrejón de Ardoz, Madrid, Spain

Kirina Kubota
Facultad de Ciencias, Universidad Autónoma de Madrid, 28049 Cantoblanco, Madrid, Spain

Laura Cabanillas
Centro de Astrobiología (INTA-CSIC), 28850 Torrejón de Ardoz, Madrid, Spain

Ester Lázaro
Centro de Astrobiología (INTA-CSIC), 28850 Torrejón de Ardoz, Madrid, Spain

Tomohiro Osugi
Laboratory of Integrative Brain Sciences, Department of Biology, Waseda University, and Center for Medical Life Science of Waseda University, Tokyo, Japan

Tomoki Okamura
Laboratory of Integrative Brain Sciences, Department of Biology, Waseda University, and Center for Medical Life Science of Waseda University, Tokyo, Japan

You Lee Son
Laboratory of Integrative Brain Sciences, Department of Biology, Waseda University, and Center for Medical Life Science of Waseda University, Tokyo, Japan

Makoto Ohkubo
Laboratory of Integrative Brain Sciences, Department of Biology, Waseda University, and Center for Medical Life Science of Waseda University, Tokyo, Japan

Takayoshi Ubuka
Laboratory of Integrative Brain Sciences, Department of Biology, Waseda University, and Center for Medical Life Science of Waseda University, Tokyo, Japan

Yasuhisa Henmi
Aitsu Marine Station, Center for Marine Environmental Studies, Kumamoto University, Kumamoto, Japan

Kazuyoshi Tsutsui
Laboratory of Integrative Brain Sciences, Department of Biology, Waseda University, and Center for Medical Life Science of Waseda University, Tokyo, Japan

Ryan M. Ames
Computational and Evolutionary Biology, Faculty of Life Sciences, The University of Manchester, Manchester, United Kingdom

Daniel Money
Department of Ecology and Evolutionary Biology, University of Kansas, Lawrence, Kansas, United States of America

Simon C. Lovell
Computational and Evolutionary Biology, Faculty of Life Sciences, The University of Manchester, Manchester, United Kingdom

Ronny Kellner
MPRG Fungal Biodiversity, Max Planck Institute for Terrestrial Microbiology, Marburg, Germany
Department of Geobotany, Ruhr-Universität Bochum, Bochum, Germany

Christian Hanschke
Department of Geobotany, Ruhr-Universität Bochum, Bochum, Germany

Dominik Begerow
Department of Geobotany, Ruhr-Universität Bochum, Bochum, Germany

Yanguang Chu, Qinjun Huang, Bingyu Zhang and Changjun Ding, Xiaohua Su
State Key Laboratory of Tree Genetics and Breeding, Research Institute of Forestry, Chinese Academy of Forestry, Key Laboratory of Tree Breeding and Cultivation, State Forestry Administration, Beijing, China

Xiaozhuo Han
School of Applied Mathematics, Guangdong University of Technology, Tianhe District, Guangzhou, China

Cang Hui
Centre for Invasion Biology, Department of Mathematical
Sciences, University of Stellenbosch, Matieland, South Africa
Mathematical and Physical Biosciences, African Institute for Mathematical Sciences, Muizenberg, South Africa

Jennifer R. Mandel
Department of Plant Biology, University of Georgia, Athens, Georgia, United States of America

Edward V. McAssey
Department of Plant Biology, University of Georgia, Athens, Georgia, United States of America

Savithri Nambeesan
Department of Plant Biology, University of Georgia, Athens, Georgia, United States of America

Elena Garcia-Navarro
Instituto de Agricultura Sostenible, Consejo Superior de Investigaciones Científicas, Córdoba, Spain

John M. Burke
Department of Plant Biology, University of Georgia, Athens, Georgia, United States of America

Brenna A. McLeod
Biology Department, Saint Mary's University, Halifax, Nova Scotia, Canada
Nova Scotia Museum of Natural History, Halifax, Nova Scotia, Canada

Timothy R. Frasier
Biology Department, Saint Mary's University, Halifax, Nova Scotia, Canada

Zoe Lucas
Nova Scotia Museum of Natural History, Halifax, Nova Scotia, Canada
The Friends of the Green Horse Society, Halifax, Nova Scotia, Canada

Khidir W. Hilu , Chelsea M. Black.a and Dipan Oza.b
Department of Biological Sciences, Virginia Tech, Blacksburg, Virginia, United States of America

Shane D. Lavery
School of Biological Sciences, University of Auckland, Auckland, New Zealand,
Institute of Marine Science, University of Auckland, Auckland, New Zealand

Ahmad Farhadi
Dept. of Fisheries, University of Tehran, Karaj, Iran

Hamid Farahmand
Dept. of Fisheries, University of Tehran, Karaj, Iran

Tin-Yam Chan
Institute of Marine Biology and Center of Excellence for the Oceans, National Taiwan Ocean University, Keelung 202, Taiwan

Ashkan Azhdehakoshpour
IFRO- Chabahar Branch, Chabahr, Iran

Vibhavari Thakur
School of Biological Sciences, University of Auckland, Auckland, New Zealand

Andrew G. Jeffs
Institute of Marine Science, University of Auckland, Auckland, New Zealand

Xinliang Yu
College of Chemistry and Chemical Engineering, Hunan Institute of Engineering, Xiangtan, Hunan, China
State Key Laboratory of Chemo/Biosensing and Chemometrics, College of Chemistry and Chemical Engineering, Hunan University, Changsha, Hunan, China

Yixiong Yu
G1302, Lushan International Experimental School, Changsha, Hunan, China

Qun Zeng
Department of Neurological Surgery, Xiangtan Central Hospital, Xiangtan, Hunan, China

Daniel R. O'Donnell
Program in Ecology, Evolutionary Biology, and Behavior, Michigan State University, East Lansing, Michigan, United States of America
Department of Zoology, Michigan
State University, East Lansing, Michigan, United States of America

Abhijna Parigi
Program in Ecology, Evolutionary Biology, and Behavior, Michigan State University, East Lansing, Michigan, United States of America
Department of Zoology, Michigan
State University, East Lansing, Michigan, United States of America

Jordan A. Fish
Department of Zoology, Michigan
State University, East Lansing, Michigan, United States of America
Center for Microbial Ecology, Michigan State University, East Lansing, Michigan, United States of America

Ian Dworkin
Program in Ecology, Evolutionary Biology, and Behavior, Michigan State University, East Lansing, Michigan, United States of America
Department of Zoology, Michigan State University, East Lansing, Michigan, United States of America
BEACON Center for the Study of Evolution in Action, Michigan State University, East Lansing, Michigan, United States of America

Aaron P. Wagner
BEACON Center for the Study of Evolution in Action, Michigan State University, East Lansing, Michigan, United States of America

Yi-Jun Zhang
Key Laboratory of Gene Engineering of the Ministry of Education, State Key Laboratory of Biocontrol, and School of Life Sciences, Sun Yat-sen University, Guangzhou, P. R. China

Jian-Hua Yang
Key Laboratory of Gene Engineering of the Ministry of Education, State Key Laboratory of Biocontrol, and School of Life Sciences, Sun Yat-sen University, Guangzhou, P. R. China

Qiao-Su Shi
Laboratory of Liver Disease Hospital, The Third Affiliated Hospital, Sun Yat-sen University, Guangzhou, P. R. China

Ling-Ling Zheng
Key Laboratory of Gene Engineering of the Ministry of Education, State Key Laboratory of Biocontrol, and School of Life Sciences, Sun Yat-sen University, Guangzhou, P. R. China

Jun Liu
Institute of Human Virology, Zhongshan School of Medicine, Sun Yat-sen University, Guangzhou, P. R. China

Hui Zhou
Key Laboratory of Gene Engineering of the Ministry of Education, State Key Laboratory of Biocontrol, and School of Life Sciences, Sun Yat-sen University, Guangzhou, P. R. China

Hui Zhang
Institute of Human Virology, Zhongshan School of Medicine, Sun Yat-sen University, Guangzhou, P. R. China

Liang- Hu Qu
Key Laboratory of Gene Engineering of the Ministry of Education, State Key Laboratory of Biocontrol, and School of Life Sciences, Sun Yat-sen University, Guangzhou, P. R. China

Ayaka Kawashima and Yoko Satta
Department of Evolutionary Studies of Biosystems, The Graduate University for Advanced Studies (Sokendai), Shonan Village, Hayama, Kanagawa, Japan

Index

A

Acc Deaminase Gene, 49-50, 63
Adaptation Of A Cyanobacterium, 66
Adaptation To Fluctuating Temperatures, 72
Adaptative Evolution Of Virulence-related Genes, 105
Alpha-hemolysin (hla), 21
Anti-predator Strategies, 191, 199
Autotrophic Ability Of The Evolved Strain, 68

B

Biochemically Rich Environment, 66-67, 69-70
Bioinformatics Analyses, 86
Black Poplar (populus Nigra), 115

C

Candidate Genes, 2, 117, 127, 137-140
Chloroplast-like State, 66
Chromosomal Locations, 11, 36, 46
Cloning Of Cdnas, 116
Complete Yeast Genomes, 94
Conservation Of The
Biosynthesis-type, 218
Crop-related Traits, 137, 144

D

Demographic History, 105, 109-110, 127
Discontinuous Polymorphism, 1
Dna Polymorphism, 20, 112-113, 127, 216

E

Ecological Significance, 49
Endosymbiosis, 66
Environmental Gradients, 128-131, 135-136
Environmental Stresses, 17, 20, 35, 115
Evolution Of Genome Content, 94
Evolution Of Mimicry, 1
Evolutionary Analyses, 109, 113, 137, 203-204
Evolutionary Divergence Of Geographic Subspecies, 173
Evolutionary Time, 36, 94, 107
Expression Analysis, 10, 17-20, 116, 126
Extensive Museum Collection, 1

F

Formation Of Fitness Valley, 128
Fungal Isolates, 110
Fungal Plant Parasites, 105
Fungus-plant Interactions, 105

G

Gene Duplications, 11, 16, 103, 205, 219, 227, 229, 232
Gene Family Histories, 94-95
Gene Identification, 100, 111, 138
Gene Identification And Primer Design, 138
Gene Retrieval And Annotation, 11
Gene Structure Prediction, 11
Genome-wide Survey, 10, 48
Gnih Gene, 83, 85, 90-91
Gonadotropin-inhibitory Hormone (gnih), 83, 92
Gossypium Raimondii, 10-11

H

High Evolutionary Capability, 105
Hospital And Community Origin, 21
House-keeping Genes, 105-109
Hypothalamic Neuropeptide, 83, 92
Hypothalamic-pituitary-gonadal (hpg), 83

I

Immunoaffinity Purification, 86-87, 91
Impact Of Gene Molecular Evolution, 161
Insights Into Alpha-hemolysin (hla) Evolution, 21

L

Lineage Specific Expansions, 94-95, 97, 100
Lineage-specific Gene Family, 94
Linkage Disequilibrium (ld), 108, 115, 121, 124, 144
Long Distance Dispersal, 105
Low Temperature Stress Treatment, 11

M

Mimetic Butterflies, 1, 7
Molecular Cloning, 47, 86, 92, 233
Molecular Evolution, 32, 113, 115-116, 125-126, 137, 139, 145, 152, 160-161, 163, 171, 216, 233

N

Natural History Museum (nhm), 3
Neural Development, 202-203, 212
Neuropeptide Ff (npff), 83
Niche Construction, 128-131, 133, 135-136
Non-secreted Proteins, 106-107, 112
Nucleotide Diversity, 23, 26, 106-108, 111-113, 115, 117,

119, 121-127, 138-140, 143, 146, 152, 155, 177

Nucleotide Sequence Of Hla, 23, 30

P

Papilio Dardanus Wing Patterns, 1

Patterns Of Variation, 105

Peptide Extraction, 87, 92

Phenotypic Diversity, 1-2

Photosynthetic Function, 66-68, 70

Phylogenetic Analysis, 10-11, 14-15, 21, 23, 35, 41-42, 44, 47-49, 59, 63, 83, 85, 87, 89-90, 92, 104, 113, 117, 154, 170-171, 177, 207, 212, 215, 218-219, 224, 229, 232

Phylogenetic Reconstruction, 161-162, 169-170

Plant Growth And Yield, 115

Plant-biotrophic Fungi, 105

R

Rapid Birth-and-death Evolution, 202

Rfamide Peptide Family, 83, 92

Rna Virus, 72-73, 81-82

S

Sequence Alignment, 37, 44, 46, 48, 62, 64, 104, 111, 117, 152, 160, 162, 177, 184, 204, 216

Single Nucleotide Polymorphisms, 106-107, 112, 116-117

Standing Genetic Variation, 191, 193, 199-200

Staphylococcus Aureus Clones, 21, 33

Stratified Genotypic Distributions, 128

Stringent Selective Pressure, 72, 75, 79

Substrate-dependent Evolution, 218

Support Vector Machine Classification, 186-187

T

Tests Of Neutrality, 117, 119, 127

Tryptic Soy Agar Plates (tsa), 22

Tryptic Soy Broth (tsb), 22

U

Ustilago Maydis Virulence Clusters, 105

W

Whole-genome Resequencing, 66, 68

Wing Pattern Determination, 1